R 35 35 3

Y0-FCH-202

WITHDRAWN

Laser Interaction

and Related Plasma Phenomena

Volume 4B

A Continuation Order Plan is available for this series. A continuation order will bring delivery of each new volume immediately upon publication. Volumes are billed only upon actual shipment. For further information please contact the publisher.

Laser Interaction

and Related Plasma Phenomena

Volume 4B

Edited by

Helmut J. Schwarz

Professor of Physics
Rensselaer Polytechnic Institute
Troy, New York

and

Heinrich Hora

Professor of Theoretical Physics
The University of New South Wales
Kensington, Sydney, Australia

PLENUM PRESS • NEW YORK AND LONDON

Library of Congress Cataloging in Publication Data

Main entry under title:

Laser interaction and related plasma phenomena.

Proceedings of the 1st-4th workshops held at Rensselaer Polytechnic Institute.
Includes bibliographies.
1. High temperature plasmas—Congresses. 2. Controlled fusion—Congresses. 3. Lasers—Congresses. I. Schwarz, Helmut J., 1915- ed. II. Hora, Heinrich, ed. III. Rensselaer Polytechnic Institute, Troy, N. Y.
QC718.5.H5L37 530.4'4 79-135851
ISBN 0-306-37154-5 (v. 4b)

Second half of the Proceedings of the Fourth Workshop on Laser Interaction and Related Plasma Phenomena held at Rensselaer Polytechnic Institute, Troy, New York, November 8-12, 1976.*

ADVISORY BOARD

N. G. BASOV
P. N. Lebedev Institute, USSR

J. L. BOBIN
Commisariát à l'Energie Atomique, France

K. BOYER
Los Alamos Laboratories, New Mexico

A. J. DeMARIA
United Technologies Research Center

J. EMMETT
Lawrence Livermore Laboratories

A. H. GUENTHER
Kirtland Air Force Base

P. HARTECK
Rensselaer Polytechnic Institute

R. HOFSTADTER
Stanford University

R. E. KIDDER
Lawrence Livermore Laboratories

M. LUBIN
University of Rochester

D. PFIRSCH
Max-Planck Institut für Plasma Physik, Germany

A. M. PROKHOROV
P. N. Lebedev Institute, USSR

CH. YAMANAKA
Osaka University, Japan

*Partially supported by the US Energy Research and Development Administration

QC
718.5
H5
L37
1977
V.4B

© 1977 Plenum Press, New York
A Division of Plenum Publishing Corporation
227 West 17th Street, New York, N.Y. 10011

All rights reserved

No part of this book may be reproduced, stored in a retrieval system, or transmitted, in any form or by any means, electronic, mechanical, photocopying, microfilming, recording, or otherwise, without written permission from the Publisher

Printed in the United States of America

PREFACE

Since the third Workshop on "Laser Interaction and Related Plasma Phenomena" in 1973, one area within the scope of this conference received increased attention: laser fusion. This possibility was emphasized in February 1977 in a Seminar on US energy policies at The Hartford Graduate Center by John F. O'Leary, Head of the Federal Energy Administration, who said that "by the year 2100, ... laser fusion will be coming along, giving us a new age of choice". Efforts in research and development were stepped up to investigate new concepts of laser ignition of controlled nuclear reactions. Here, one expects no radioactive waste from fuel. The deuterium-tritium reaction - the only one which may be possible with magnetic field confinement in tokamaks - has a highly radioactive tritium cycle, while, in principle, laser reactions are possible with pure deuterium, hydrogen-boron or others.

The worldwide progress in laser compression was not only stimulated by the energy crisis, but also by its advancements. In our first Workshop in 1969 F. Floux of the French Limeil Laboratories described his experiments, which led, only one month later, to the production of fusion neutrons in such large numbers as had not been achieved up to then (see appendix of Vol. I these Proceedings). At the Workshop in 1971, Sklizkov, co-worker of Nobel Laureate Basov, Lebedev Institute Moscow, reported on their nine-beam laser system which exceeded the former results by producing up to 10^7 neutrons per pulse. At the same conference the compression of plasma by lasers was discussed and a better version of an earlier compression scheme presented. This might have stimulated the detailed numerical calculations on laser compression of plasmas as worked out by Nuckolls. At the third Workshop in 1973 details of the newly performed two-dimensional compression calculations and their stability were presented by Nuckolls of the Lawrence Livermore Laboratories; also K. Brueckner (at that time with KMS - Fusion) discussed similar results. A steady progress in experimentation was reported concerning neutron production and various diagnostic techniques.

The progress that has taken place since the third Workshop in 1973 is reflected in the Proceedings of the fourth Workshop in 1976 presented in these volumes; one example is the neodymium glass laser. In 1973 the power of the 3 to 5 ns pulses of the laser development of the Lebedev group (Basov, Sklizkov) and of KMS (Brueckner et al) was about 300 GW, whereas now Livermore reaches 2.5 terawatt in one beam. Livermore's Shiva glass system will be described here by Trenholme and Boyle. In these volumes we also find the first detailed description of the 216-beam glass system of Basov, Sklizkov et al under construction. The Garching group of Germany (Hohla, Witte et al) arrived at 1 terawatt from iodine lasers; the Los Alamos group (Singer et al) described the design of their CO_2 laser system aimed at 200 Terawatt by 1981, presently 150 J with 10^{15} W/cm^2. The achievements of neutron yields of 10^9 from 100 kJ incident laser energy on microballoons and of the increase of these yields by 100 at condensation of the DT mixture (Roy Johnson, KMS) are milestones of the present advances.

As the title of the Workshop implies, our objective is discussion of the fundamental interaction processes, of which laser fusion is only one, but a most important aspect. Our first item on the agenda dealt with the most advanced laser systems for interaction studies. Apart from the already mentioned systems like Shiva, 216-beam and iodine lasers, the program covered damage and basic design problems (R. Carman, Los Alamos), CO_2 lasers of 200 J - 1 ns (Los Alamos), 30 ps CO_2-pulses (Yablonovitch, Harvard University), nuclear pumped CO_2 and other gas lasers (Miley, University of Illinois), x-ray lasers of 117 Angstrom (Jaeglé, University of Paris), the possibilities of gamma-ray lasers and Q-switching methods.

For programming presentations in the field of interaction at moderate intensities, it was necessary to make the appropriate selection from the numerous worldwide efforts. One example of sophisticated spectroscopic diagnostics is the contribution by Tondello et al (University of Padova, Italy), while on gas breakdown the measurement of the polarization dependent emission of electrons (Yablonovitch) is another example.

Although there exists some overlap, the experiments at high laser intensities were divided into the fusion-oriented results and those related to the generation of fast ions. Extensive experiments were reported by research centers such as Livermore, Los Alamos, KMS, Lebedev-Moscow, Limeil-Paris, Osaka University, Rochester University, Naval Research, and Garching (using iodine lasers): neutron production, backreflection, emission of x-ray spectra, spatial and temporal resolution of x-rays (see for example, original methods of Sklizkov, Moscow and those of Salzmann, SOREQ, Israel), ion collection, etc. It became evident that the nonlinear effects are responsible for the appearance of the fast ions and are dominant at high laser intensities. The 60 keV ions, as observed at KMS two years ago (see Slater et al), were probably not due to self-focusing, while, however, self-focusing could not be excluded in the case of Yamanaka's 50 keV ions. This has given a new aspect to the concept of laser compression. Therefore, the generation of these fast ions was discussed with special attention; the 6 MeV Au^{38+} ions of Luther-Davies et al (Australian National University, Canberra) and the 1.5 MeV protons of Livermore and Los Alamos from laser irradiated targets were of special interest. The Los Alamos results represented many years of data collecting showing that the nonlinear effects occur at the same intensities, regardless whether 20-ps pulses from a neodymium glass laser or 1-ns pulses from a CO_2-laser were applied.

Discussions of the theory of instabilities concentrated on a contribution by Bobin (Limeil) with methods for differentiating in experimental signals. Generation of magnetic fields was covered experimentally and theoretically. Discussions of relativistic effects included radiation forces, known also as nonlinear, or ponderomotive forces or electrostriction (Brueckner). Self-focusing due to these forces (including thermal effects as reviewed by Sodha et al) and self-focusing due to relativistic effects were discussed; numerical examples for the nonlinear forces were presented (Lawrence et al). Several contributions were devoted to the electron motion (relativistic) in very high laser fields, where - as this was done also at the last Workshop in 1973 - hypothetical ideas were discussed beyond nuclear fusion, namely towards pair production and generation of gravitational waves.

The theory of nuclear reactions by laser compression was treated in several contributions. Kidder, Livermore, treated the fusion-fission reactor; Nishihara, Osaka, reported fusion yield calculations; similar aspects were described by Niu and Yabe, Tokyo. Ray et al (University of New South Wales, Sydney) reported new results on the penetration of reaction products and on the reheat in fusion calculations.

In concluding this preface we like to emphasize that all major laboratories working towards laser fusion are in the process of building lasers reaching the 100 terawatt range where the wavelength does not seem to make much difference contrary to what one thought even during the last Workshop. These efforts appear to enable us to decide soon whether laser fusion is feasible; the next Workshop should, therefore, bring us closer to the answer to this crucial question.

Our thanks are first due to all speakers and participants who contributed to the success of this Workshop with their excellent presentations and enthusiastic discussions. We are also indebted to the US Energy Research and Development Administration for their partial financial support.

We are grateful for the counsel of the following advisors: N. G. Basov (P. N. Lebedev Physical Institute of the Academy of Sciences, Moscow, USSR), J. L. Bobin (Commissariat à l'Énergie Atomique, France), K. Boyer (Los Alamos Scientific Laboratories, Los Alamos, New Mexico), A. J. DeMaria (United Technologies Research Center, East Hartford, Connecticut), J. Emmett (Lawrence Livermore Laboratory, University of California), A. H. Guenther (Kirtland Air Force Base, New Mexico), P. Harteck (Rensselaer Polytechnic Institute, Troy, New York), R. Hofstadter (Stanford University, California), R. E. Kidder (Lawrence Livermore Laboratory, University of California), M. Lubin (University of Rochester, New York), D. Pfirsch (Max-Planck-Institut für Plasmaphysik, Garching, Germany), A. M. Prokhorov (P. N. Lebedev Physical Institute of the Academy of Sciences, Moscow, USSR), and Ch. Yamanaka (Osaka University, Japan).

We would also like to thank Dr. Robert G. Loewy, Vice President for Academic Affairs and Provost of Rensselaer Polytechnic Institute, for taking time off to welcome the participants in his opening address at the first session which again, as in our previous Workshops, was chaired with stimulating vigor by Professor Paul Harteck.

We thank Rensselaer Polytechnic Institute's Office of Continuing Studies, its Director, Richard J. Teich, as well as its Director of Special Programs, G. Marvin Clark, and their staff for their invaluable efforts in the preparation and administration of this Workshop-Conference. Last but not least we are grateful to our secretaries, Mrs. Ellen Riley, Mrs. Judy Rohan, and Mrs. Judy Stanic, who worked with great dedication in our Hartford office and the latter at the University of New South Wales assisting in the organization of the Workshop as well as in preparing the Proceedings.

March 1977

Helmut Schwarz
Heinrich Hora

CONTENTS OF VOLUME 4B

CONTENTS OF VOLUME 4A

I. LASERS FOR FUSION

TWO NEW CHARGED-PARTICLE DIAGNOSTICS FOR LASER-FUSION EXPERIMENTS*

D.C. Slater and F.J. Mayer

KMS Fusion, Inc.

Ann Arbor, Michigan 48106

INTRODUCTION

Illuminating a gas-filled spherical shell pellet by a high power laser is now a conventional laser-fusion experiment. During laser energy deposition, the target material can be conveniently divided into three regions - the outer corona, consisting of a hot plasma; the tamper, which is the relatively cool remainder of the shell; and the fuel gas at the interior. The fusion burn reaction is most efficient if the fuel can be compressed to a very high density before appreciable heating has occurred. A number of experiments have demonstrated that compression and heating do occur, and that neutrons are generated by the fusion reaction[1], but many questions remain about the exact nature of the energy transfer process from the laser to the fuel.

The ions emitted from such target experiments might be associated with their region of origin. The hot corona generates the fast ions - consisting of ions from the shell material with energies greater than 5 keV per nucleon. These are followed by the bulk of the ionized shell and fuel material, characterized by somewhat lower temperatures. In addition, a number of light ion reaction products are created in the target interior with energies in the few MeV region. They carry information on the fuel and tamper conditions at the time of peak reaction rate. Precise measurements of these three groups of ions can be a useful tool in understanding the dynamics of the absorption and implosion processes.

*Presented at the Fourth Workshop on "Laser Interaction and Related Plasma Phenomena" held at RPI, Troy, New York, November 8-12, 1976.

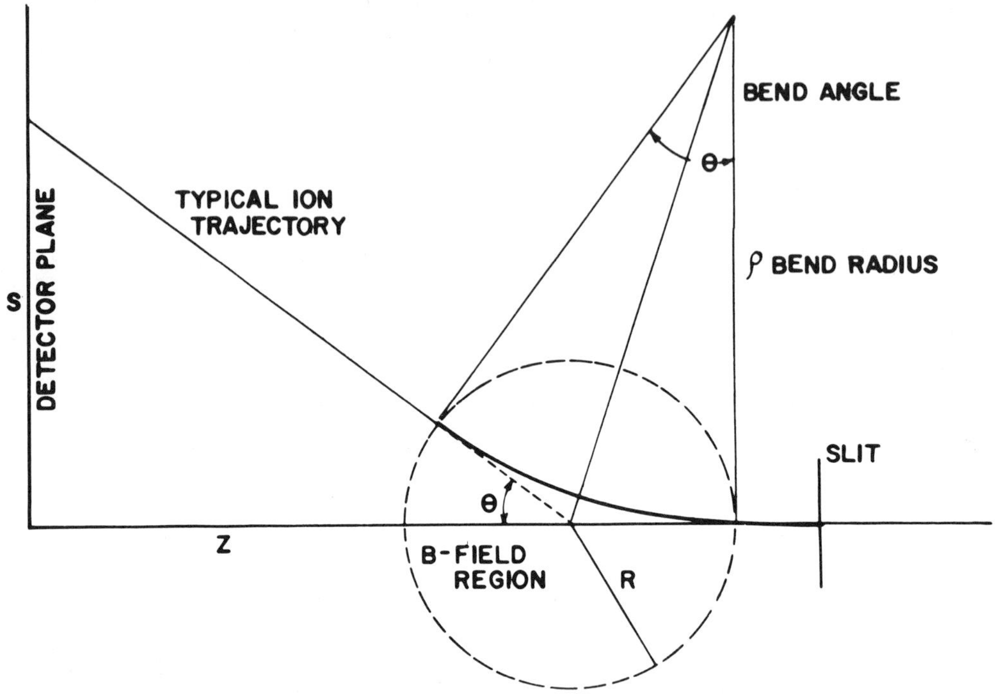

Figure 1 Schematic diagram of the ion spectrograph.

Here we describe two magnetic spectrographs with passive detection media which allow individual ion counting. One system is used to record fast corona ions, the other is designed for measuring energy spectra of protons, tritons, and alpha particles from DD and DT fusion reactions.

FAST ION SPECTROGRAPH

The development of solid materials to record nuclear tracks has proceeded rapidly in recent years[2]. Etching of these materials occurs preferentially along the damage trail left by the passage of heavily ionizing radiation. They are insensitive to light, X-rays, and electrons, which makes them uniquely suited for fast-ion detection in a plasma environment. Figure 1 shows a schematic diagram of a simple magnetic spectrograph designed to use cellulose-nitrate-coated foil detectors. Ions travel 115 cm from the target to a slit aperture 80 μm x 1 cm. They then enter a uniform magnetic field region of circular cross section, traveling along constant radius trajectories to the exit boundary of the field, and then along straight paths to the detector. There is no focusing by the field. The very narrow entrance slit provides good resolution over most of the detector surface. The very small

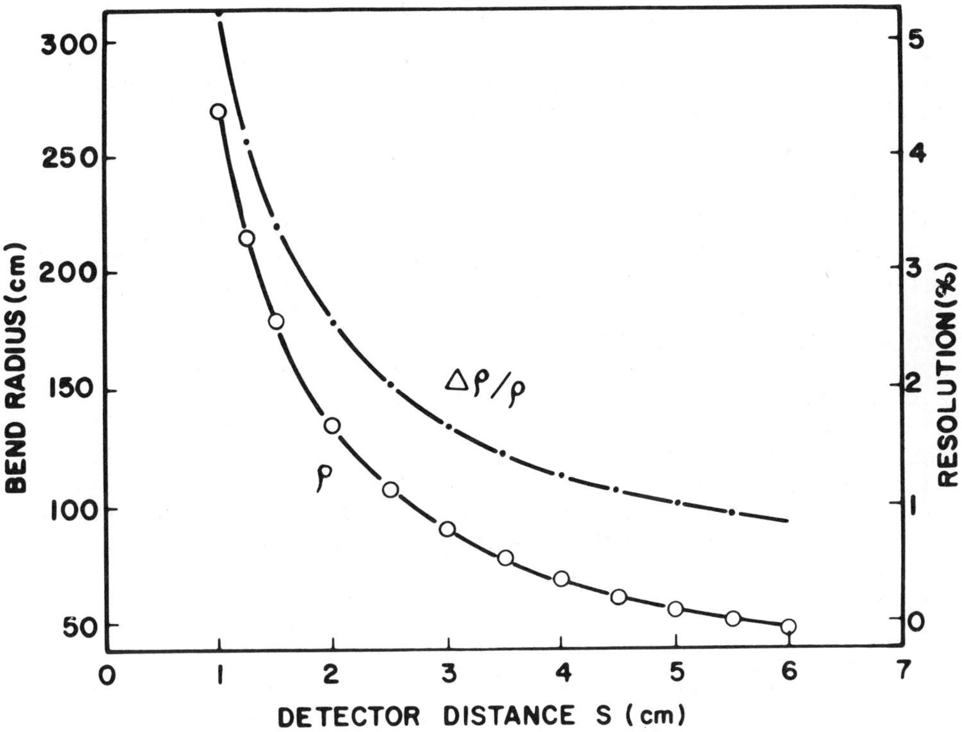

Figure 2 Bending radius and resolution versus distance along the detector plane for the ion spectrograph.

solid angle (6×10^{-7} sr) is necessary to keep the track density at the detector below $10^5/cm^2$.

The detector plane position S is related geometrically to the bending radius ρ by:

$$\rho = R/\tan\left[(1/2)\tan^{-1}(S/Z)\right] \quad (1)$$

where R is the radius of the magnet pole piece and Z is the distance from the magnet to the detector (see Figure 1). In our device, R = 5 cm and Z = 27 cm. The dispersion curve ρ versus S and the resolution curve $\Delta\rho/\rho$ versus S are shown in Figure 2. Here the resolution is computed for $\Delta S = 0.05$ cm, which corresponds to one field of view in the microscope used for scanning the film. The resolution limit imposed by the entrance slit is somewhat better.

The field-radius product is a measure of the ion momentum per unit charge. For fully stripped ions of A/Z = 2,

$$B\rho(\text{kG-cm}) = 20.9 \times 10^{-8} V \text{ (cm/sec)} \quad (2)$$

where V is the ion velocity. The energy per nucleon can also be determined from the velocity:

$$E/ \text{ (keV)} = 5.17 \times 10^{-16} V^2 \quad . \quad (3)$$

The detector foils are 2 x 8 cm strips of Kodak LR115 type II film. They consist of about 12 μm of red-dyed cellulose nitrate on a polyester backing. After exposure, particle tracks are revealed by etching in a 10% NaOH solution at 55° C for 60 minutes. Foils are scanned under a 250-power microscope using dark field illumination. Using this technique, heavy ions of velocity greater than 2×10^8 cm/sec can be seen. This value corresponds to a range of about 1 μm for the ion in the C-N. For protons, track registration occurs only around the peak value of the energy loss rate dE/dx. In these experiments, protons in the energy range 100 to 200 keV are routinely observed. The diameter of the proton track is much less than the tracks of the heavy ions, providing easy discrimination. No attempt has been made to identify heavy ions by species, although some techniques for doing so have been reported[3].

The targets in these experiments were spherical glass shells of 50- to 55-μm diameter and 0.6- to 0.8-μm wall thickness filled with 10 atm of deuterium-tritium gas. The lens and ellipsoidal mirror illumination system provided near-normal spherical illumination. On-target laser powers were about 0.4 terawatt. Four typical spectra are shown in Figure 3. Here the number of ions per unit velocity is plotted as a function of ion velocity. In the velocity range 2 to 7×10^8 cm/sec these spectra fall more than two orders of magnitude. The data are well represented by the form $dN/dV = A \exp(-V/V_o)$, with V_o around 0.8×10^8 cm/sec and A around 4×10^{15} ions per 10^8 cm/sec. The total energy of these fast ions, extrapolated into 4π steradians, is approximately 1 to 2 joules. These results are typical of some 20 analyzed spectra from similar experiments.

The observed inverse exponential velocity distribution can be associated with the spectrum derived from a simple similarity model of an isothermally expanding corona[4]. In this model, the ion velocity distribution at the end of the laser pulse has the exponential form with the scale velocity given by $C = [(Z\theta_e + \theta_i)/M]^{\frac{1}{2}}$ where Z and M are the ion atomic number and mass, and θ_e and θ_i are the electron and ion temperatures. At this time, 3/8 of the total energy is thermal and will be converted into additional kinetic energy when the ions have reached the detector. Hence, the observed V_o is greater by a factor of $\sqrt{8/5}$ than the value at the end of the laser pulse. Setting $\theta_i = 0$, the range of electron temperatures

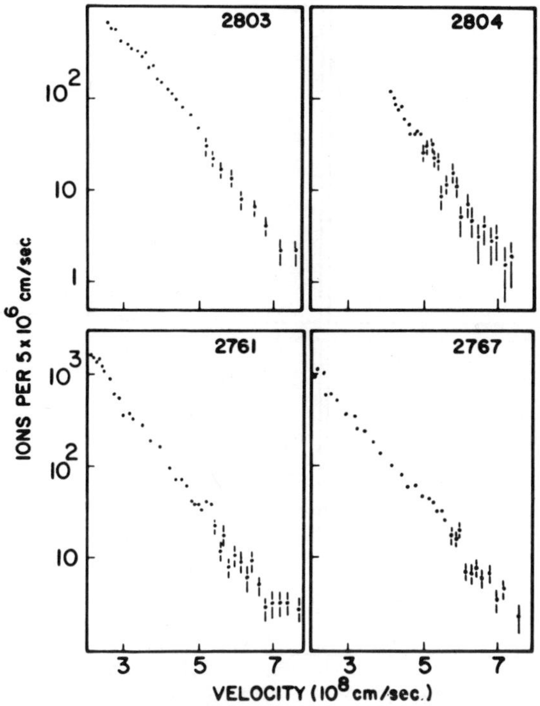

Figure 3 Fast ion velocity spectra. All targets 50-55 μm diameter glass shells, 0.6 - 0.8 μm thick. Incident laser power approximately 0.4 TW.

deduced from the measured scale velocities is 7 to 11 keV, which is substantially greater than the 1 to 2 keV predicted by simple inverse bremsstrahlung absorption. These corona temperatures are in agreement with those deduced from measurements of the hard X-ray spectra in similar experiments.

The total energy in these fast ions is approximately half the total energy absorbed by the target, yet the number of fast ions represents only about 6% of the target mass. This phenomenon therefore represents an important component in the energy balance. If fast ion generation requires an anomalously hot corona, then some mechanism must be invoked which limits thermal conduction into the pellet and simultaneously enhances the absorption in the critical density region relative to inverse bremsstrahlung. Simulations[4] using the one-dimensional hydrodynamic code TRHYD can reproduce the ion spectrum if half the absorption is made to occur in a thin outer layer where the thermal conductivity is reduced 1000 times from its classical value. The inability of the fast

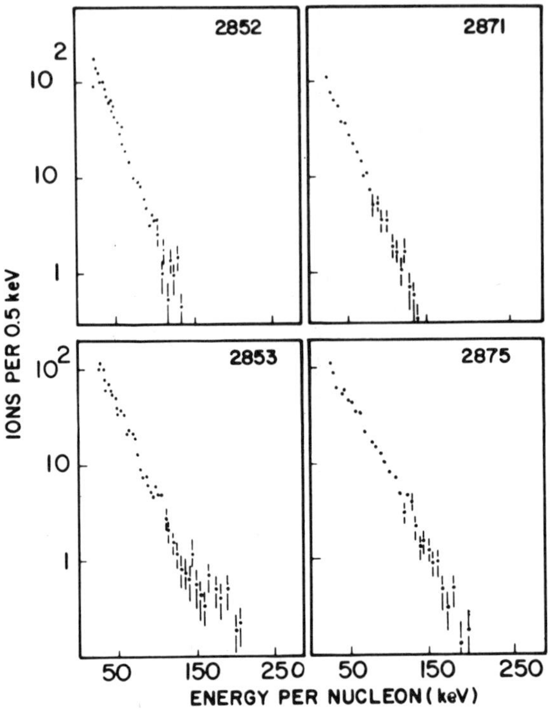

Figure 4 Fast ion energy spectra. Shots 2852, 2853, and 2871 were PVA shell targets of 59-90 μm diameter and 1.0 to 2.7 μm thickness. Shot 2875 was a solid sphere of 49-μm diameter.

ions to provide much recoil momentum means that these simulations show weaker implosions and lower neutron yields than are predicted by purely classical calculations in which the same total energy is absorbed.

While the isothermal-rarefaction model with a hot corona provides a natural explanation for the velocity distributions, a few of the measured spectra can be better represented as distributions which fall off exponentially with the energy per nucleon variable. This was true of the three shots made with polyvinyl alcohol shell targets. These spectra are shown in Figure 4. Here the energy scale factor is about 20 keV/nucleon. While we have no simple model for this alternative parametrization, it does appear satisfactory for a subset of the data and cannot be ruled out.

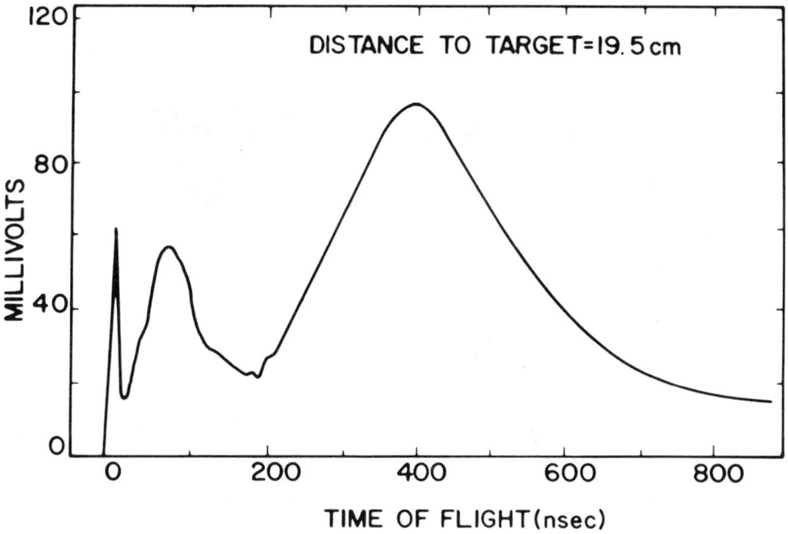

Figure 5 Oscilloscope voltage trace from a biased charge collector.

We have made some comparisons between fast ion data obtained with the spectrograph and with our conventional charge collectors. These detectors record the ion current as a function of time, and must be corrected for secondary electron emission. An example is given in Figure 5. The initial peak is due to electrons photo-emitted from the collector at the time of the laser pulse. The second peak is due to fast ions, and the large broad feature represents the arrival of the bulk of the ionized target material.

This voltage trace was converted into an ion velocity spectrum and compared with the spectrograph data. This comparison is shown in Figure 6, with the spectrograph data normalized by the solid angle ratio. The absolute agreement is good at a velocity of 2×10^8 cm/sec, which is the lower limit of the C-N foils, but the two measurements diverge quickly from this point. There are at least three probable explanations for this difference. First, the C-N foil measurements include no contributions from protons. In the narrow velocity band in which protons are observed, they account for about half the discrepancy. Second, the secondary emission correction of the charge collector data is extrapolated from measurements made at velocities below 2×10^8 cm/sec.[5] In these extrapolations, the secondary emission coefficients are essentially

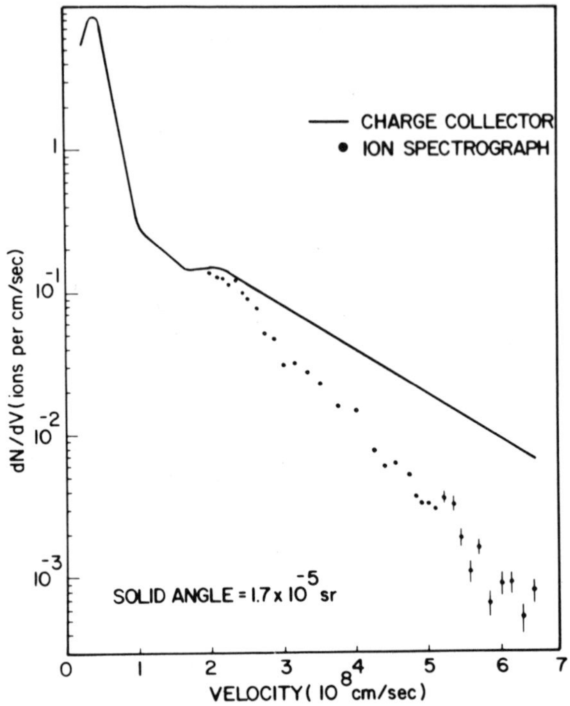

Figure 6 Ion velocity spectrum from a charge collector compared to the ion spectrograph. Spectrograph points have been normalized by the solid angle ratio.

constant. Other data, however, indicate they may in fact be increasing[6]. Finally, no subtraction of the tail from the zero-time spike has been made in the charge collector trace. This omission contributes substantial uncertainty in the high velocity data.

The ion energy per unit velocity is plotted in Figure 7. We have assumed the mass distribution to be that of the glass, i.e., silicon and oxygen in 1:2 ratio. The distinction between fast and slow components is evident, with the division occurring at 1×10^8 cm/sec. The integrated energies in this particular case show 2/3 of the total ion energy in the fast ion group. This result is typical in experiments with similar targets and laser powers.

We compute the absorbed laser energy as the sum of the ion kinetic energy, the X-radiation energy, and the ionization energy of the target. The fast ion spectrograph results indicate that about half of this total is in the fast ion group. This division

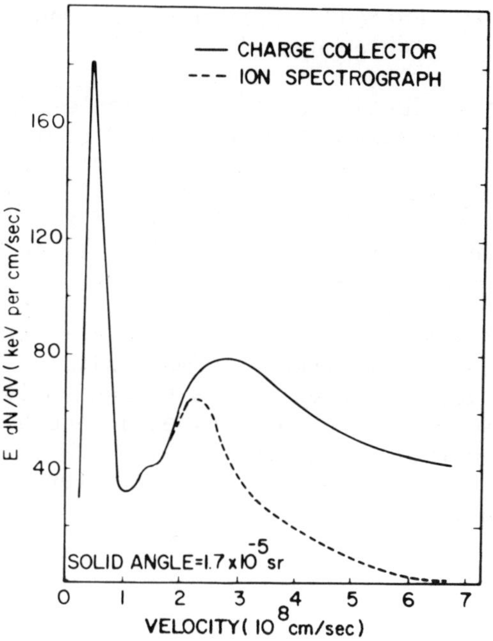

Figure 7 Ion energy distribution in particle velocity as determined by the charge collector and the ion spectrograph.

of energy places strong constraints on models of the laser-plasma interaction physics. Continuing studies of fast ion spectra should provide additional insight into the dependence of these absorption and transport mechanisms on intensity, wavelength, polarization, and atomic number.

REACTION PRODUCT SPECTROMETER

Charged particles from nuclear reactions provide another interesting diagnostic tool for laser-fusion studies. These particles are created in the pellet interior as a result of the compression and heating of the fuel gas. Once created, they may be sufficiently energetic to reach a distant detector, providing a means of probing the surrounding material from the inside out. The most convenient reaction products are the 3.5 MeV alpha particles from DT fusion, and the 3.0 MeV protons and 1.0 MeV tritons from DD fusion. Alpha particles have been observed by researchers from the Lawrence Livermore Laboratories[7], the Los Alamos Scientific

Laboratory[8], and KMS Fusion[9]. Proton measurements have been reported by the two latter groups.

The energy distribution of the reaction products has been calculated for ions in a Maxwellian velocity distribution[10]. The spectrum is approximately gaussian, with a slight upward shift in central energy compared to the nominal reaction value. The energy spread reflects the velocity distribution of the fuel ions, and is proportional to the square root of the ion temperature. As the newly created particles leave the target core, they lose energy by collisions with the surrounding fuel, tamper, and corona material. Calculations indicate the energy loss to electrons in the dense tamper material dominates other loss mechanisms. Because the ratio of the DT to DD reaction rates varies strongly with temperature, the alpha-to-proton yield ratio provides a second independent determination of the fuel ion temperature. Once this temperature is known, the density of the fuel gas can be inferred from the total yield, so that an estimate of the fuel compression can be made.

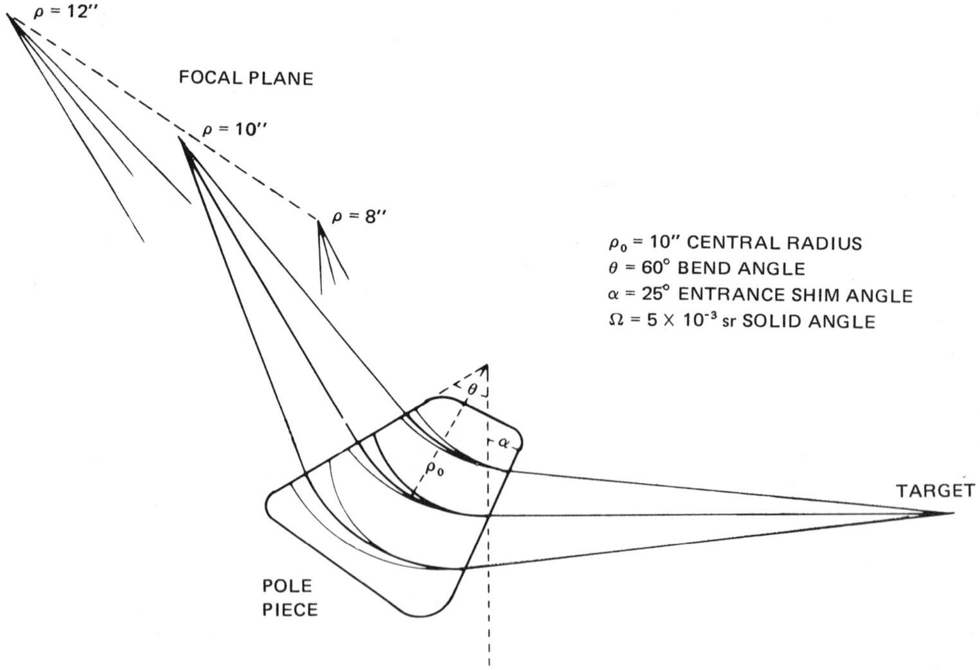

Figure 8 Magnetic field configuration of the magnetic spectrograph. Distance from the target to the field boundary is 76 cm.

Table I summarizes the quantities which can be deduced from charged reaction product energy spectra. These straightforward deductions ignore the temporal and spatial dependencies of the entire implosion process. The KMSF hydrodynamic simulation code TRHYD is presently being modified to compute the reaction product spectra so that detailed comparisons can be made.

Previously published measurements have all used scintillator-photomultiplier (SPM) detectors to measure the time-of-flight of the reaction products. The long flight paths needed to obtain good energy resolution are in general inconsistent with the large solid angles desired for good statistical accuracy. We have recently replaced our SPM device with a magnetic spectrograph, which allows substantial improvement in all of the measured parameters. Table II shows a comparison of the spectrograph and the SPM.

The spectrograph was designed and built to our specifications by Industrial Coils, Inc. of Middleton, Massachusetts. The basic design parameters of the magnetic field region are shown in Figure 8. All particles of a given energy within the acceptance angle are focused to a particular line. These lines collectively determine the focal surface where the detector is placed. A slight

TABLE I

REACTIONS

$D + T \rightarrow \alpha$ (3.5 MeV) + N (14.1 MeV)

$D + D \rightarrow P$ (3.0 MeV) + T (1.0 MeV)

MEASURED QUANTITIES

Yields

Energy Losses

Energy Spreads

INFERRED QUANTITIES

Fuel Ion Temperature

Fuel Compression

Tamper Density Integral

Tamper Electron Temperature

TABLE II

Comparison of Scintillator-Photomultiplier[a] and Magnetic Spectrograph for DT Alpha Particle Detection

	SPM	SPECTROGRAPH
Acceptance	1/150,000	1/4,000
Energy Resolution	5%	0.5%
Absolute Energy	4%	0.1%
Energy Loss[b]	40%	1 %
Yield - Instrumental	15%	5 %
- Statistical[c]	12%	2 %

(a) Reference 9

(b) Assuming mean energy of 3.2 MeV

(c) Assuming total fusion yield of 10^7 neutrons

curvature in the entrance field boundary provides second order corrections to the focusing of the central momentum ray. The large solid angle (5 msr) is obtained by weak focusing in the transverse direction. Figure 9 shows a sketch of the spectrograph, which weighs 4000 pounds and can provide a maximum field of 15 kG at 150 amps.

Particles are detected with either nuclear emulsion plates or cellulose-nitrate etchable foils. To date, only the C-N foils have been used. They are quite convenient for recording alpha particles but are insensitive to 3 MeV protons. After exposure, the foils are etched in NaOH and scanned under a microscope where particle tracks are counted as a function of position along the focal plane. The spectrograph alignment and calibration are being done with a ^{148}Gd radioactive source which emits a single-energy alpha particle at 3.2 MeV. While a full characterization of the focal plane is not yet complete, the best energy resolution obtained so far is 20 keV, consistent with the design goals of the device.

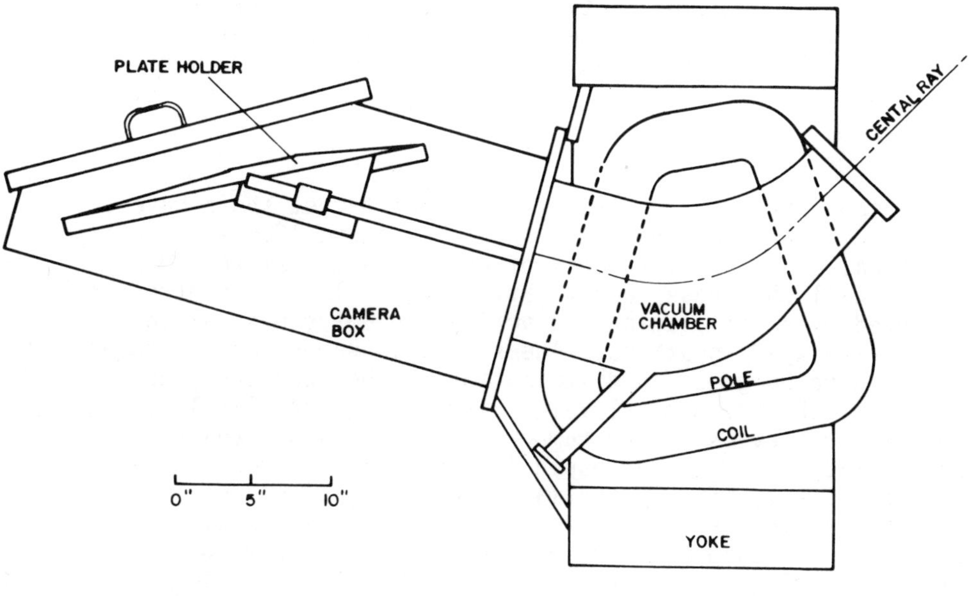

Figure 9 Schematic drawing of the magnetic spectrograph. The target is 76 cm from the entrance pole edge.

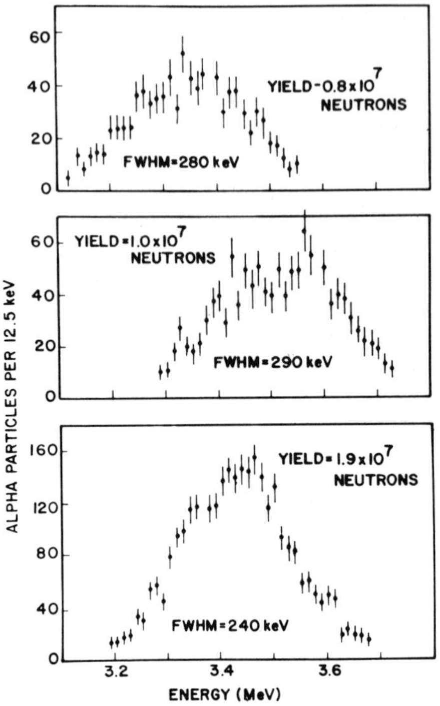

Figure 10 Three alpha particle energy spectra. All targets were 50 to 55 μm diameter glass spherical shells filled with 10 atm of DT gas. On-target laser powers were approximately 0.4 TW.

Alpha particle spectra from several target shots have been analyzed. Three of these are shown in Figure 10. All are from spherical glass shell targets of about 52-μm diameter, 0.6 - to 0.7-μm wall thickness, and 10 atm of DT fill gas. Neutron yields measured by scintillation and activation detectors varied from 0.8 to 1.9 x 10^7 neutrons. The spectral shapes can be adequately represented by gaussian distributions. The mean energies are very near the 3.52 MeV nominal value, indicating very little energy loss while traversing the tamper. The energy spreads vary from 240 to 290 keV FWHM, corresponding to fuel ion temperatures of 1.8 to 2.5 keV, if only thermal broadening is considered. Other line broadening mechanisms, such as nonuniform energy loss and straggling, should be smaller than the mean energy loss and are probably negligible in these experiments. Using a simple similarity model[11] we can estimate the fuel compression necessary to obtain these yields at the measured temperatures. For the lower spectrum,

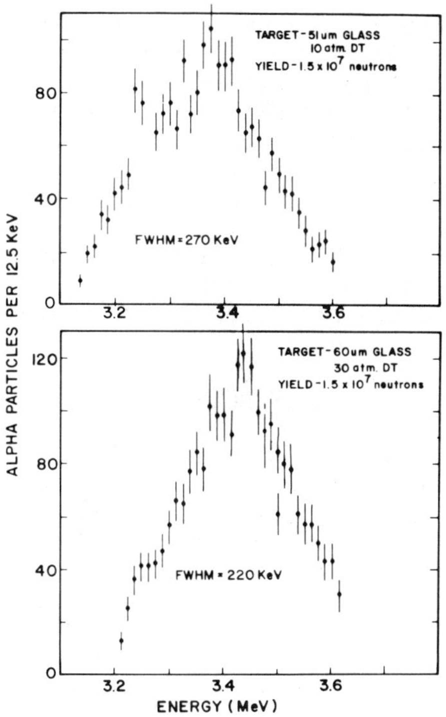

Figure 11 Two additional alpha particle energy spectra. Note that for similar target sizes and neutron yields, the target with 30 atm of fill gas gave a narrower peak than the 10 atm target, indicating a lower fuel temperature.

a volume compression of several thousand is indicated. For the upper spectrum, the compression would be about 200.

These few data points suggest that the lower yield shots actually achieve slightly higher temperatures than higher yield shot. Since the reaction rate is a very strongly increasing function of temperature, this result is surprising. More data are needed to fully define the relationship between yield and temperature for these experiments.

Two additional spectra shown in Figure 11 for similar size targets producing similar neutron yields illustrate the effect of higher fill pressure. The narrower spectrum from the target with the higher pressure is what would be expected if equal amounts of thermal energy are transferred to the fuel in each case.

To summarize, accurate measurements of charged reaction products can be an important diagnostic tool for studying the interior of imploding pellets. Experiments to date indicate that moderate heating and substantial compressions of the fuel have been achieved. With expected improvements in energy calibration and the use of proton and triton sensitive nuclear emulsion detectors, the magnetic spectrograph should be capable of making more accurate determinations of the fuel and tamper conditions in the near future. Also, detailed comparisons between these measurements and hydro code simulations should help to elucidate some of the underlying physical processes in the implosion dynamics.

REFERENCES

1. G. Charatis, J. Downward, R. Goforth, B. Guscott, T. Henderson, S. Hildum, R. Johnson, K. Moncur, T. Leonard, F. Mayer, S. Segall, L. Siebert, D. Solomon, C. Thomas, Plasma Physics and Controlled Nuclear Fusion Research (Proc. Int. Conf., Tokyo, 1974) IAEA, Vienna (1975) Vol. II, 317.
   J. Holtzrichter, H. Ahlstrom, E. Storm, J. Swain, IEEE Second International Conference on Plasma Science, Ann Arbor, Michigan (1975) 85.
2. R.L. Fleischer, P.B. Price, R.M. Walker, Nuclear Tracks in Solids, Univ. of Calif. Press (1975).
3. F.H. Ruddy, H.B. Knowles, G.E. Tripard, Phys. Rev. Lett. 37 (1976) 826. Also, Ref. 2, Ch. 3.
4. P.M. Campbell, P. Hammerling, R.R. Johnson, J.J. Kubis, F.J. Mayer and D.C. Slater, Proceedings of the 6th International Conference on Plasma Physics and Controlled Nuclear Fusion (IAEA) Berchtesgaden, Germany (1976).
5. R.R. Goforth, Rev. Sci. Instrum. 47 (1976) 451.
6. Karl Heinz Krebs, Fortschritte der Physik 16 (1968) 451.
7. V.W. Slivinsky, H.G. Ahlstrom, K.S. Tirsell, J. Larsen, S. Glass, G. Zimmerman, H. Shay, Phys. Rev. Lett. 16 (1975) 1803.
8. G.H. McCall, Tai Ho Tan, A.H. Williams, Bull. Am. Phys. Soc. 20 (1975) 1318.
9. R.R. Goforth, F.J. Mayer, H. Brysk and R.A. Cover, J. Appl. Phys. 47 (1976) 4850.
10. H. Brysk, Plasma Phys. 15 (1973) 611.
11. F.J. Mayer, R.R. Johnson, D.J. Tanner, KMSF Report U486 (to be published in Physics of Fluids).

This research was supported by the United States Energy Research and Development Administration under Contract EY-76-C-02-2709.

MULTIWAVE PICOSECOND LASER INTERFEROMETRY OF CO_2-LASER-PRODUCED PLASMAS*

R. L. Carman, A. G. Engelhardt, and N. Clabo

University of California
Los Alamos Scientific Laboratory
Los Alamos, New Mexico 87545

Recently there has been much discussion centered around the production of fast ions and electrons in laser-produced plasmas. Malone et al.,[1] at Los Alamos, found that the observed particle distributions could be explained by anomalously low thermal conductivity which would lead to decoupling of the corona from the dense core, overheating of the corona, and thus producing very non-Maxwellian particle distributions. Kidder[2] actually speculated on this same type of decoupling for different reasons much earlier. The crucial assumption being made was that the density changed slowly in space from the underdense portion of the plasma to well above critical. The role of photon pressure or ponderomotive forces was probably underestimated in this earlier work. More recent theoretical work[3] indicates that many of the same effects can be obtained if one speculates that severe density profile modifications occur due to the photon pressure exceeding the plasma pressure. Models including these new forces predict that the density can rise from less than 1/10 of critical to more than 10 times critical in a spatial extent of about 1 μm The University of Rochester[4] time-resolved data on harmonic generation in laser produced plasmas as well as data presented at this meeting seems to support this conjecture of large density gradients. The 2ω light is expected to arise from the critical surface while the 3/2 ω light is believed to come from the 1/4 critical surface. The convergence of the two surfaces for intermediate times in the driving laser pulse suggests that 1/4 n_c and n_c coexist within the experimental resolution.

What is required to unequivocally determine what is happening is a more direct determination of the hydrodynamic motion during the laser illumination. A number of Laboratories have apparently come to the same conclusion as exemplified by the holographic interferometry work done at LLL by Attwood and Coleman,[5] the Schlieren work done at Garching by Ariga and Sigel,[6] the interferometry work at Limeil of Martineau et al.,[7] and the more recent work at the National Research Council by Fedojevs et al.[8] In the previous work, a number of limitations can be identified. First the resolution of the interferogram or Schlieren photograph is usually limited by the illumination source, namely its coherence and/or duration. The other limitations relate to the plasma physics or data acquisition. The interferogram is usually converted to plasma density profiles by assuming cylindrical symmetry and performing an Abel inversion.[9] Also, the interferogram is taken at one time during the plasma evolution and then correlated with interferograms taken on other laser shots; thus great demands are placed on reproducibility from both the plasma and laser viewpoint. As a final comment, there are usually many steps in the process of taking and subsequently reducing the data that leads to losses in ultimate resolution. This comment is even more relevant to holographic interferometry where reconstruction must also be included among the many steps.

Let us now explore each of these points one by one, in order to find out if other viable solutions can be found to the problem density fronts.

PLASMA ILLUMINATION SOURCE AND SYCHRONIZATION WITH THE FUSION LASER

Typical density front velocities of $\gtrsim 10^7$ to 10^8 cm/s imply that pulse durations of < 1 to 10 ps/μm resolution are required. Since we expect the shock fronts to steepen to the point where the full spatial extent of the important density profiles is $\sim$ 1 μm, we wish to have a spatial resolution of $\lesssim$ 1 μm. Certainly this resolution should be dictated by the optics rather than the illumination source. Finally, spatial rippling of the critical surface and the production of flares in numerical simulations appear to occur within a few picoseconds and indicate that even shorter illumination pulse widths are useful. This ideal situation is to be contrasted with the current experimental conditions where $\sim$ 10 μm spatial resolution is being obtained using probe laser pulse durations of between 5 ps (dye lasers)[5,6] to 30 ps (NRC glass laser).[8] While analysis of fringe shapes and their change across the field of view can provide information in these long pulse experiments, we would prefer to obtain the hydrodynamic information more directly.

About 8 years ago,[10] we developed a glass laser for studying transient stimulated scattering and self-focusing phenomena which had many of the desirable laser characteristics required here. In Table I we summarize the typical pulse characteristics for this laser. Under optimum conditions, we have obtained pulses as short as 0.2 ps of 0.1 TW peak power with a brightness of $\sim 2 \times 10^{18}$ W/cm^2-ster. Careful studies[11] of this laser have lead to some understanding of how it works. In Figure 1, we present an illustration of the oscillator layout. This system's operation is based on the premise

TABLE I

NORMAL OUTPUT			SINGLE PULSE (COMPRESSED)		
λ	=	1.06 μm	λ	=	1.06, .53, or .35 μm
Δk	=	200 cm^{-1}	Δk	=	200 cm^{-1}
Θ_{DIV}	=	200 μrad	Θ_{div}	=	200 μrad
E_{PULSE}	$\simeq$	10 mj/pulse	E_{pulse}	$\sim$	1-5 mj/pulse
E_{TRAIN}	$\sim$	1 J	t_{pulse}		.5-5 ps
T_{TRAIN}	=	1-3 μs			
$T_{INTERPULSE}$	$\simeq$	10 ns			
t_{PULSE}	$\simeq$	5 ps			

BEAM SIZE - 5mm x 10mm Elliptical cross section
Spatial Distribution - Flat topped, with hard zeroes at the edge

that self-focusing occurs in the laser glass and therefore, a non-linear feedback mechanism compensating for the spatial nonlinearity is established. While the intracavity lenses are adjusted to cause the mirrors to appear to be plane parallel to low recirculating intensities, for high intensities the resonator becomes unstable. The rate of change in the cavity instability always exceeds the self-focusing instability causing the beam to spatially expand. Meanwhile, the time dependence of the refractive index causes the pulse bandwidth to expand up to and somewhat beyond that of the gain bandwidth as shown in Fig. 2. When the pulse round-trip-gain is just equal to the pulse losses due to output coupling, diffraction, spatial beam expansion, and pulse frequency bandwidth expansion outside the fluorescence linewidth, a stable equilibrium is

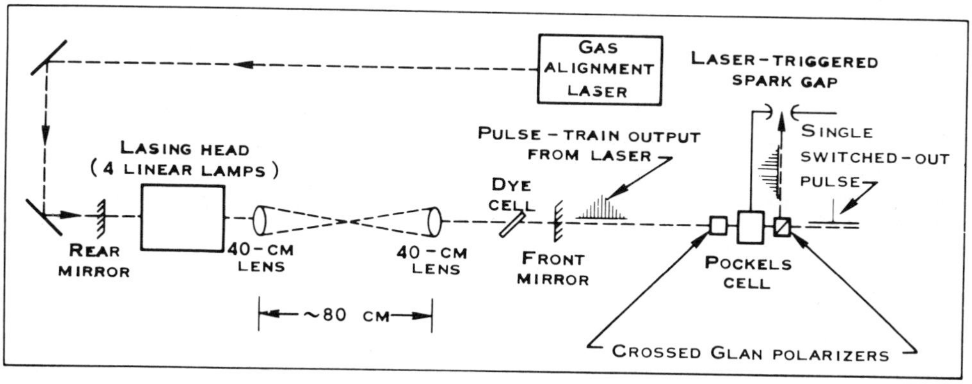

Fig. 1

Schematic of the picosecond oscillator cavity and the single pulse selector.

established. At this time the pulse recirculates virtually unchanged inside the cavity until depletion of the upper level population causes the output pulse energy to fall (see Fig. 3). While this description provides some insight into how the oscillator works, many other subtleties occur which have not been mentioned.[11] Due to the highly linear frequency chirp which occurs in the majority of the output pulses (center of train), grating pairs can be very successfully employed to compress these pulses in time[12] down to ≲ .5 ps, as was done in our earlier experiments.[10] If the oscillator is operated in the linear optics (low intensity) limit, 1 ps pulses of 15 cm^{-1} bandwidth are directly produced, as shown in Fig. 4.

Utilizing this laser source as an illuminator for an interferometer introduces many complications which until recently had not been completely explored. The problem lies in the lack of temporal coherence. The 200 cm^{-1} pulse bandwidth implies that we are essentially dealing with white light. Thus an achromatic interferometer is required. Fortunately, "opticians" have been busily developing such interferometers.[13] In Fig. 5, we show an example of such an interferometer. The use of gratings as beamsplitters makes it possible to easily and reproducibly operate an equal path length interferometer where the equality of paths is established simply by crossing the two beams accurately at Grating 2. Lens 1 renders the zeroth and first orders parallel while focusing each beam to a small spot. Lens 2 crosses both orders at Grating 2 while imaging the plasma with magnification onto the film. Utilizing 75 mm f1.9 lenses, we have been able to obtain 300 line pairs/mm resolution or 1.6 μ/resolution point when operating with an image magnification of ≈ 35.

In order for the illumination laser pulse duration not to appreciably change the ultimate spatial resolution, the plasma must not move more than ∿ 0.5 μm during the pulse, which implies we require a pulse duration of ≤ 5 psec even at 10^7 cm/sec velocities. While the previously quoted resolution was obtained using a HeNe laser source (see Fig. 6) and we do not expect any surprises when the Nd:glass source is substituted, we have not as yet taken any interferograms or made any resolution tests utilizing the glass laser.

A closely related issue to the above deals with the synchronization of independent laser systems without precluding attainment of optimum performance for each device. In this particular case the desired Nd:glass system performance, high energy in a very short duration pulse, exceeds that normally attainable, and implies that a minimum of interference in this system can be tolerated. On the other hand, heavily saturated amplifiers and pulse width requirements of 100 ps to 1 ns suggest that many pulse generation schemes for the CO_2 radiation are feasible. While pulses from traditional accousto-optical mode locking of CO_2 oscillators satisfy system requirements, the synchronization problem is difficult, especially when the total CO_2 laser timing requirements, which occur over a period of ∿ 7 μs, are fully included.

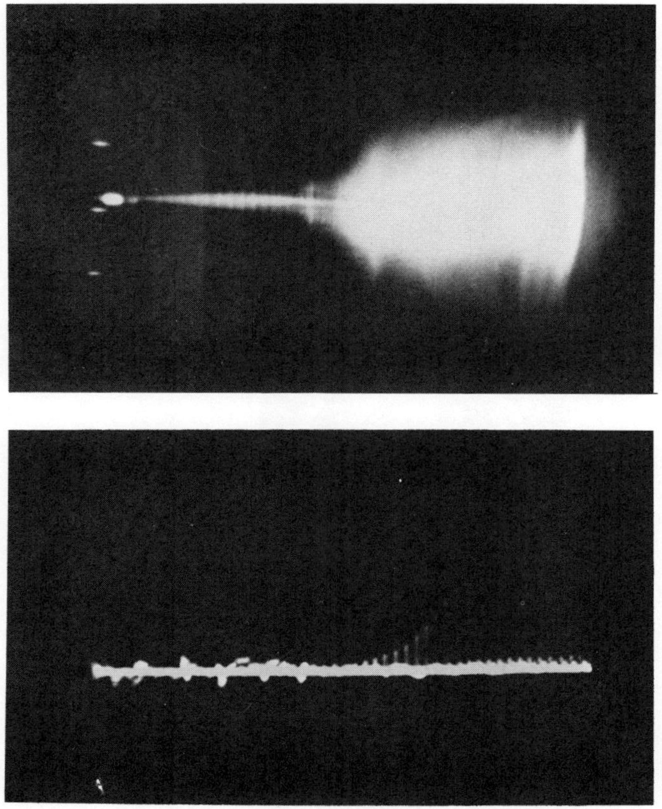

Fig. 2.
Sudden onset of self phase modulation as seen from streaked output pulse spectrum (top) with 100 Å fiducials (side) and photodiode output (bottom) on same time scale.

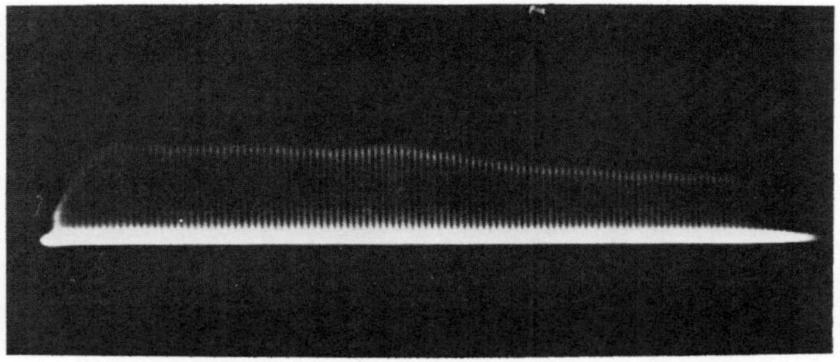

Fig. 3.

Typical pulse train with 10 ns interpulse time illustrating pulse energy saturation.

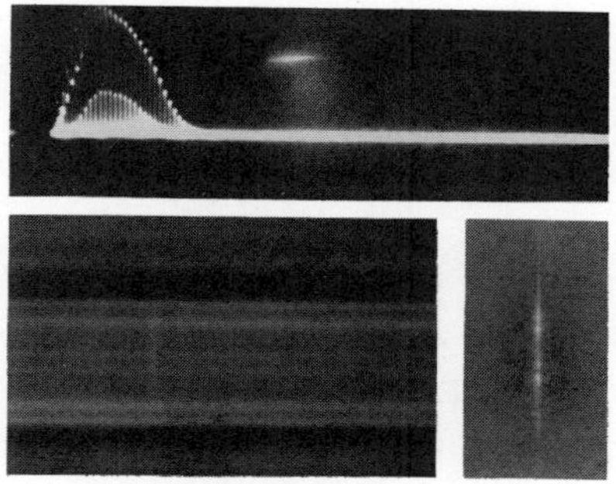

Fig. 4.
Pulse train (top), 1 ps pulse measured by TPF bandwidth (bottom right) generated directly by oscillator.

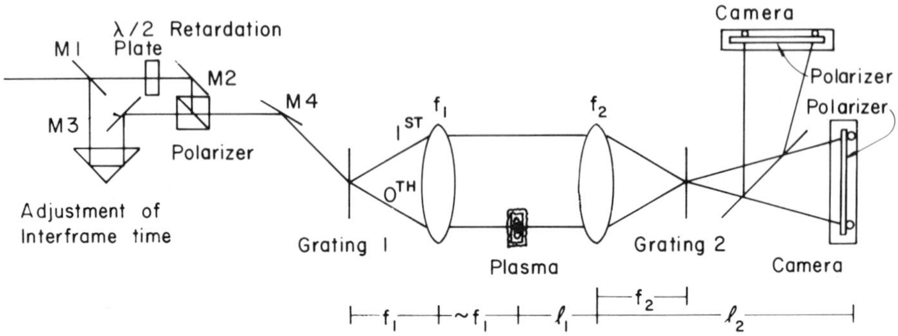

Fig. 5.

Schematic of double-grating interferometer allowing two interferograms to be taken at different times during a single CO_2 laser pulse.

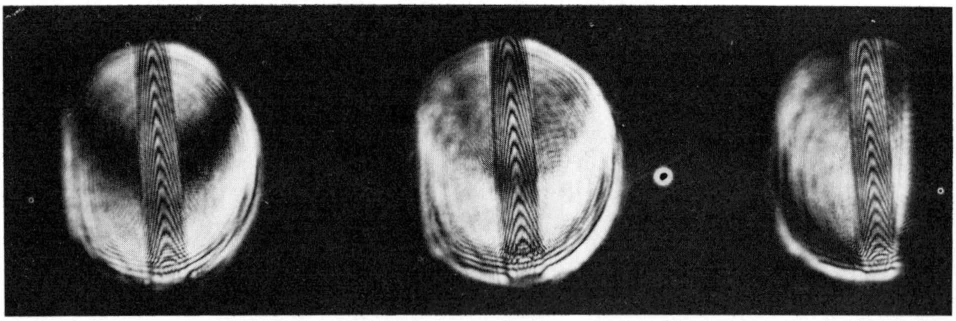

Fig. 6.

Interferogram taken with HeNe laser and grating interferometer of a 100 μ diameter glass stalk.

We would like to digress to describe our pulse selection, clipping, and temporal shaping schemes which provide a viable alternative.[14] Figure 7 illustrates the basic idea. If full-wave retardation is initially established on a Pockels cell, the process of switching the voltage to zero leads to a time varying transmission window, where

$$T(t) = \sin^2 \pi \frac{V(t)}{V_\lambda} .$$

The initial leakage of light can be reduced to arbitrarily low values by cascading several Pockels cells in series. Also, the shape of the transmission window can be programmed by programming the electrical voltage waveforms. Further, these Pockels cells can be dispersed throughout the amplifier chain to satisfy isolation requirements. Figure 8 shows that the pulse width can be controlled simply by changing one resistor value, where the Pockels cell is simply a capacitor in the RC circuit, while Fig. 9 shows that somewhat more complex networks allow the pulse shape to be easily modified. In Fig. 10, an example of such a pulse clipping process is shown, while in Fig. 11, we show a demonstration of consecutive double-pulse-clipping to generate a .75 ns pulse from an initial 300 ns pulse. The intermediate pulse duration is 12 ns.

By using many Pockels cells in series, and programming their respective voltage waveforms independently, it is possible to generate any optical pulse shape in time, currently down to durations of ∿ 100 ps. Further, this process is so general, that an analog computer based on these techniques could be constructed. While analog computers are well beyond the scope of this paper, we do wish to show how the electrical programming has been achieved. In Fig. 12, we show schematically a pulse generator capable of driving up to seven

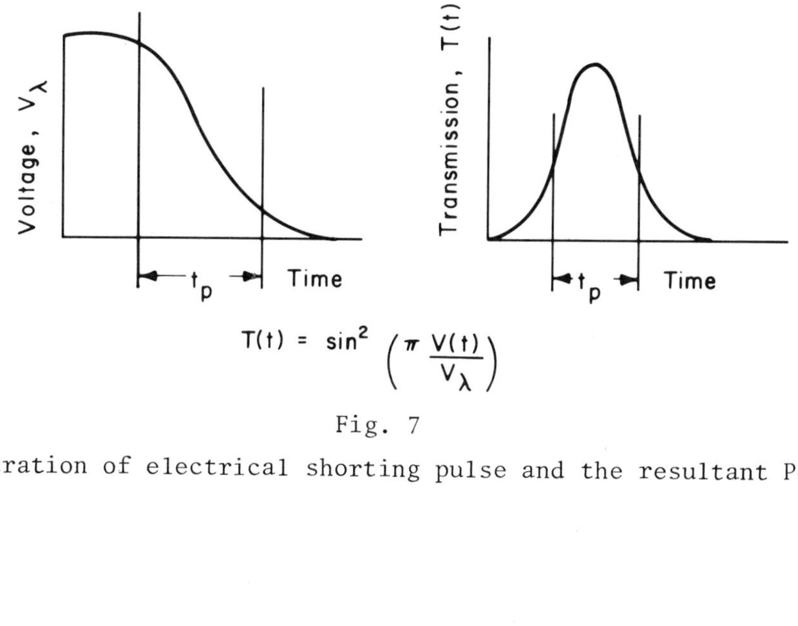

Fig. 7

Illustration of electrical shorting pulse and the resultant Pockels cells.

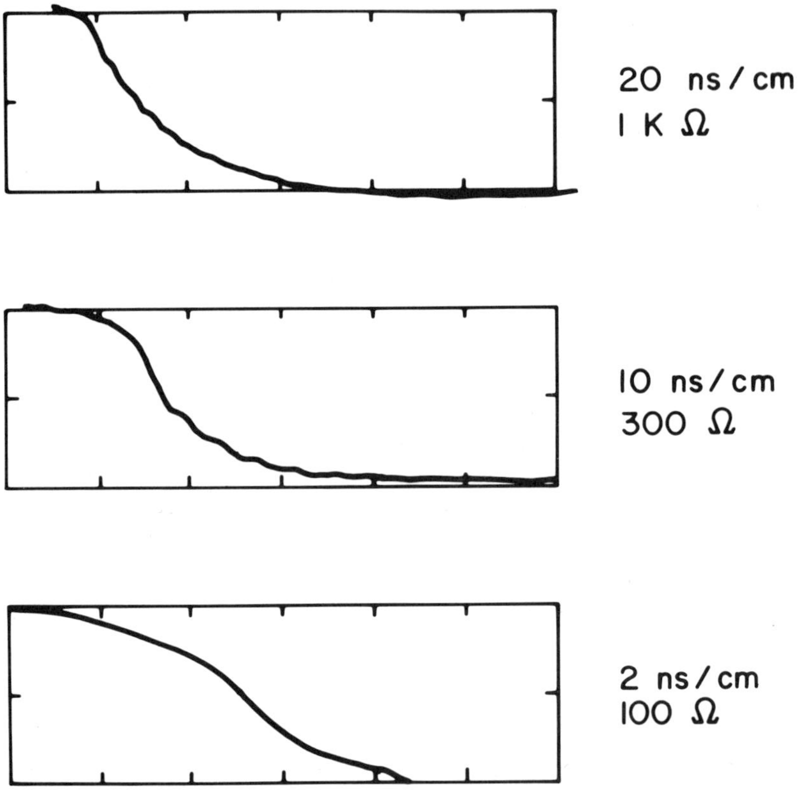

Fig. 8.

Voltage shorting pulses obtained across a Pockels cell by changing only the series resistor.

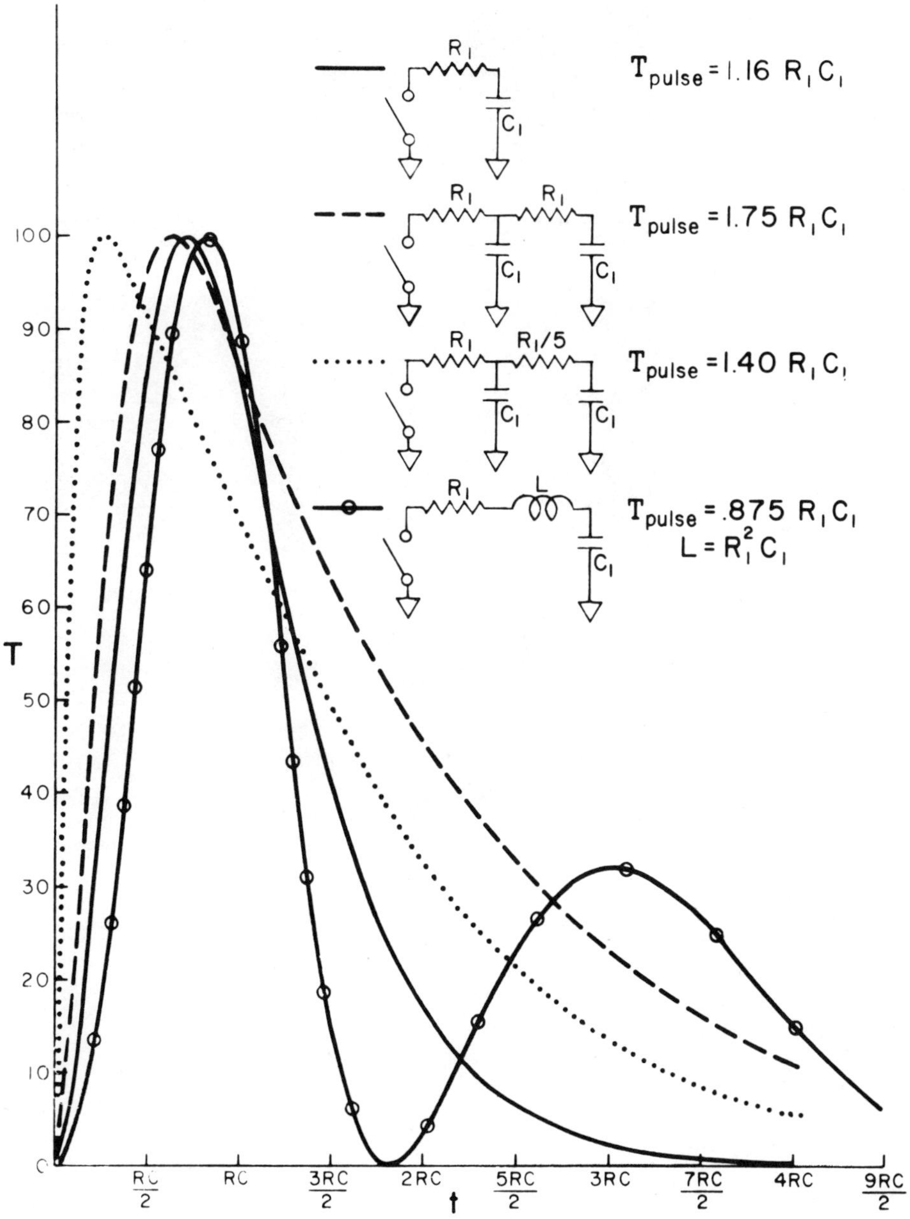

Fig. 9.

Various achievable Pockels cell transmission functions utilizing the LRC shorting circuits shown at the top.

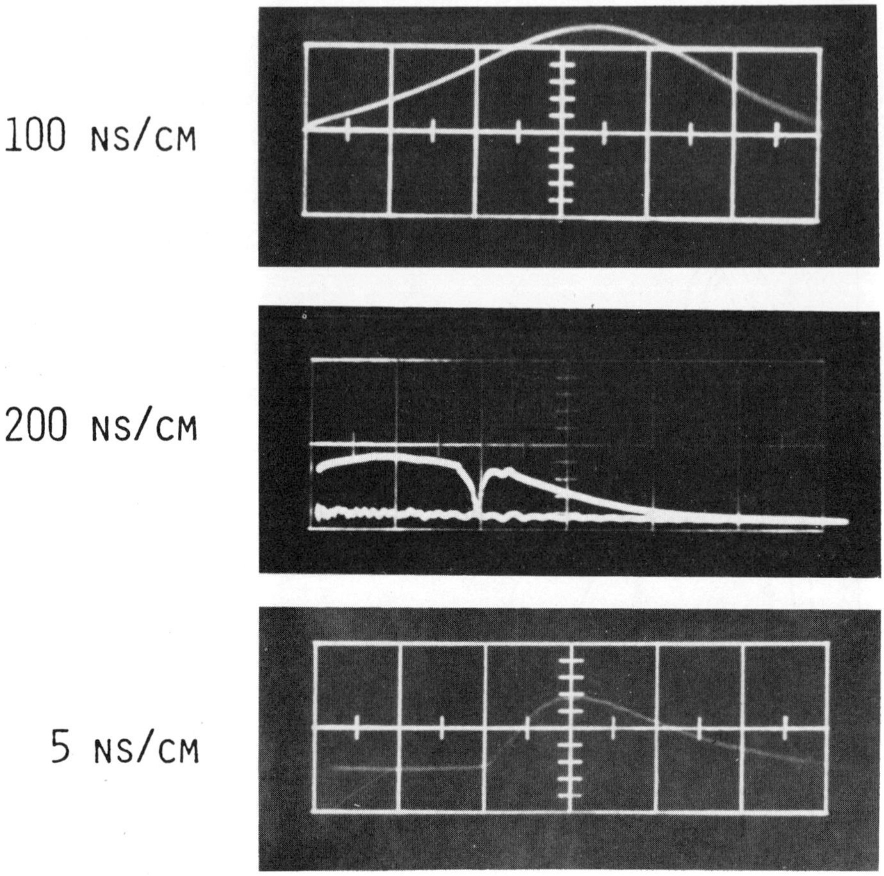

Fig. 10.

A 300 ns FWHM pulse of < 10^{-4} cm^{-1} bandwidth (Top), generated in the oscillator, is Pockels cell clipped to 12 ns FWHM (Bottom). The rejected optical pulse is also shown (Center).

Pockels cells optically in series, where independent but time synchronized voltage waveforms are produced. A spark gap is used to launch an electrical pulse simultaneously into the seven cables, where the duration of this pulse is determined by the R-C time constant at the launching side. By then employing nonlinear electrical pulse propagation in various lengths of delay lines inserted into the cables, it is possible to sharpen the electrical pulse risetime. This is done by causing the higher current portions of the

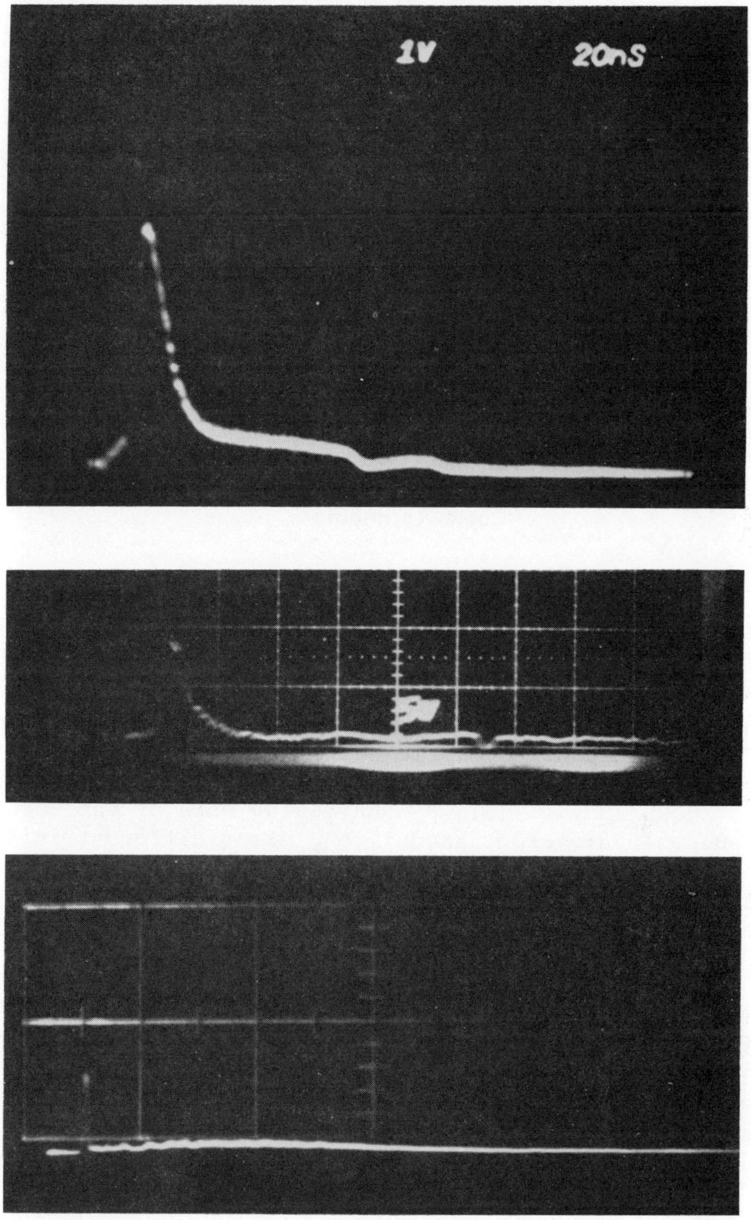

Fig. 11.

The previously pulse clipped pulse of 12 ns NWHM (Top) is again pulse clipped to .75 ns FWHM (Bottom). The rejected pulse is shown in the center.

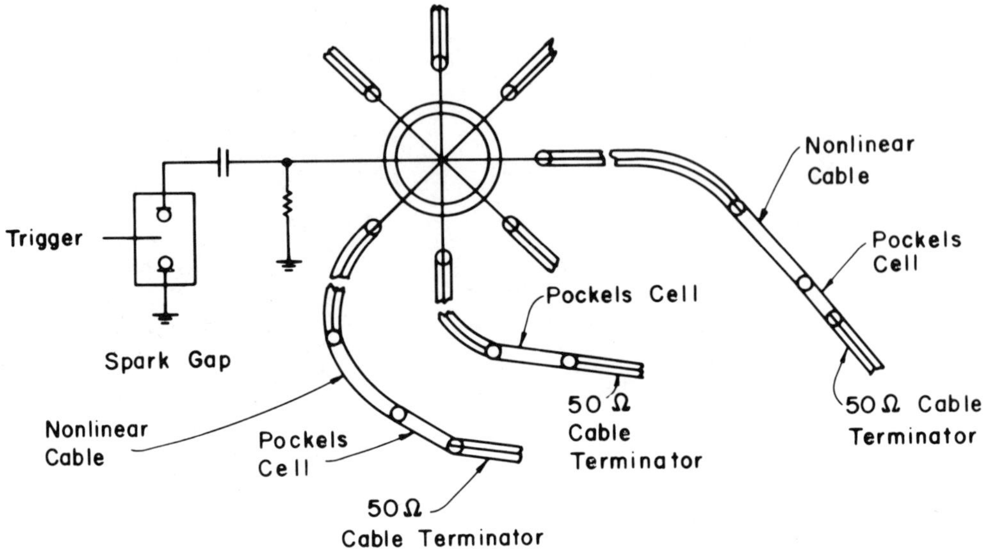

Fig. 12.

Schematic of the electrical pulse generator used to program and synchronize the voltages applied to as many as 7 Pockels cells.

pulse to travel faster than the lower current portions. In our case, we achieved this characteristic using ferrite cores with a high but saturable permeability. A schematic of the construction of such a cable is shown in Fig. 13. By sharpening each of the Pockels cell driver pulses a different amount, and using different initial DC biases on the cells, it is possible to generate optical pulses with the desired temporal distributions. In Fig. 4, we show examples of electrical pulse steepening obtained, where we initially launched a 60 ns fall-time pulse and sharpen it in this case to < 2 ns fall-time. These techniques are neither new nor unexplored[15] as electrical pulse risetimes down to ∿ 100 ps have been demonstrated. In Fig. 15, using these techniques we show an example of an optical pulse generation which has a 30 ns base width and a .75 ns spike duration produced using three Pockels cells. This particular pulse was emitted by the "LASL" ruby laser. The pulse generation system of this ∿ .1 TW device is shown in Fig. 16, and is capable of producing pulses from ∿ 1 ps duration up to 50 ps with the mode locked system and 100 ps up to 700 ns utilizing the long pulse oscillator, pulse selection and pulse clipping system. To answer questions dealing with reliability, this system has been in operation for in excess of two years, running essentially every day. Pulse reproducibility is better than can be determined by an oscilloscope while re-alignment is typically required about once every six months.

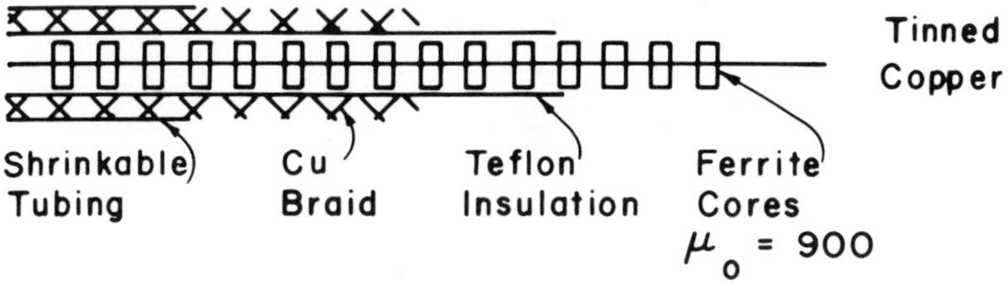

Fig. 13.

Schematic of the construction of the electrical pulse sharpening cables.

Let us come back now from this digression to the main subject. A CO_2 oscillator employing a smoothing tube is capable of high reproducibility in generating pulses in the range of 30 ns to 150 ns duration; several are currently in operation at LASL. While accurate temporal pulse shaping at the oscillator is less necessary in multikilojoule CO_2 systems because of the gain saturation effects in the amplifier chain, it is necessary to produce pulses with high contrast ratio of about 100 ps to 1 ns width. E. McLellan at LASL, has been working on Pockels cells for CO_2 lasers and has successfully developed one system which is capable of 250 ps pulse clipping while systems 3 to 4 times faster are currently being designed. Moreover, pulse clipped CO_2 oscillators are currently being employed as pulse generators for a few of our smaller CO_2 systems.

Another piece of technology which is required is a very low-jitter high-voltage electrical switch. Laser triggered spark gaps of the conventional design have jitters from one to several nanoseconds, and therefore are of no use. A new type of laser triggered spark gap has been developed at LASL[16] which employs a solid dielectric.[16] By replacing high pressure gas with one mil or less of mylar between the electrodes, trigger jitters of $\sim$ 30 ps have been obtained. Electrical pulse risetimes are as short as 100 ps. The triggering is accomplished with 1.06 μm light of duration between 5 ps and 50 ps. Typical delay times are also correspondingly shortened in the solid dielectric system. In Fig. 17, we show a schematic of such a spark gap.

The final item required to accomplish the full sychronization is the employment of a high gain photomultiplier plus narrow band filter whose transmission band is centered on the Nd:glass fluorescense band. By placing the photomultiplier at the rear mirror of the Nd:glass oscillator, the pulse buildup from noise can be detected as a break in the normal fluorescence. Combining the detection

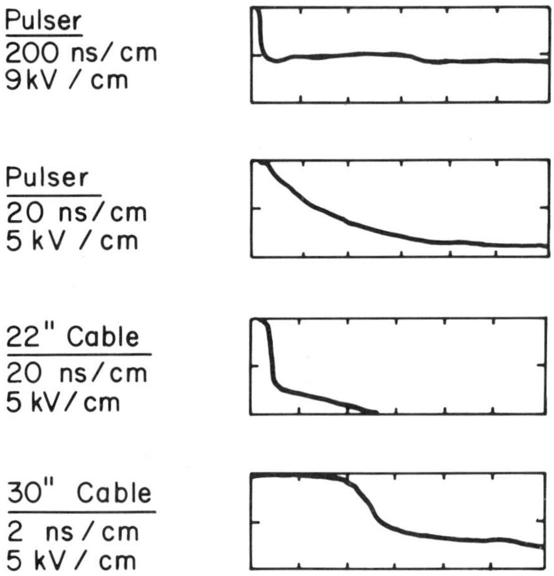

Fig. 14.

Illustration of the electrical pulse sharpening achieved by using various lengths of nonlinear cables.

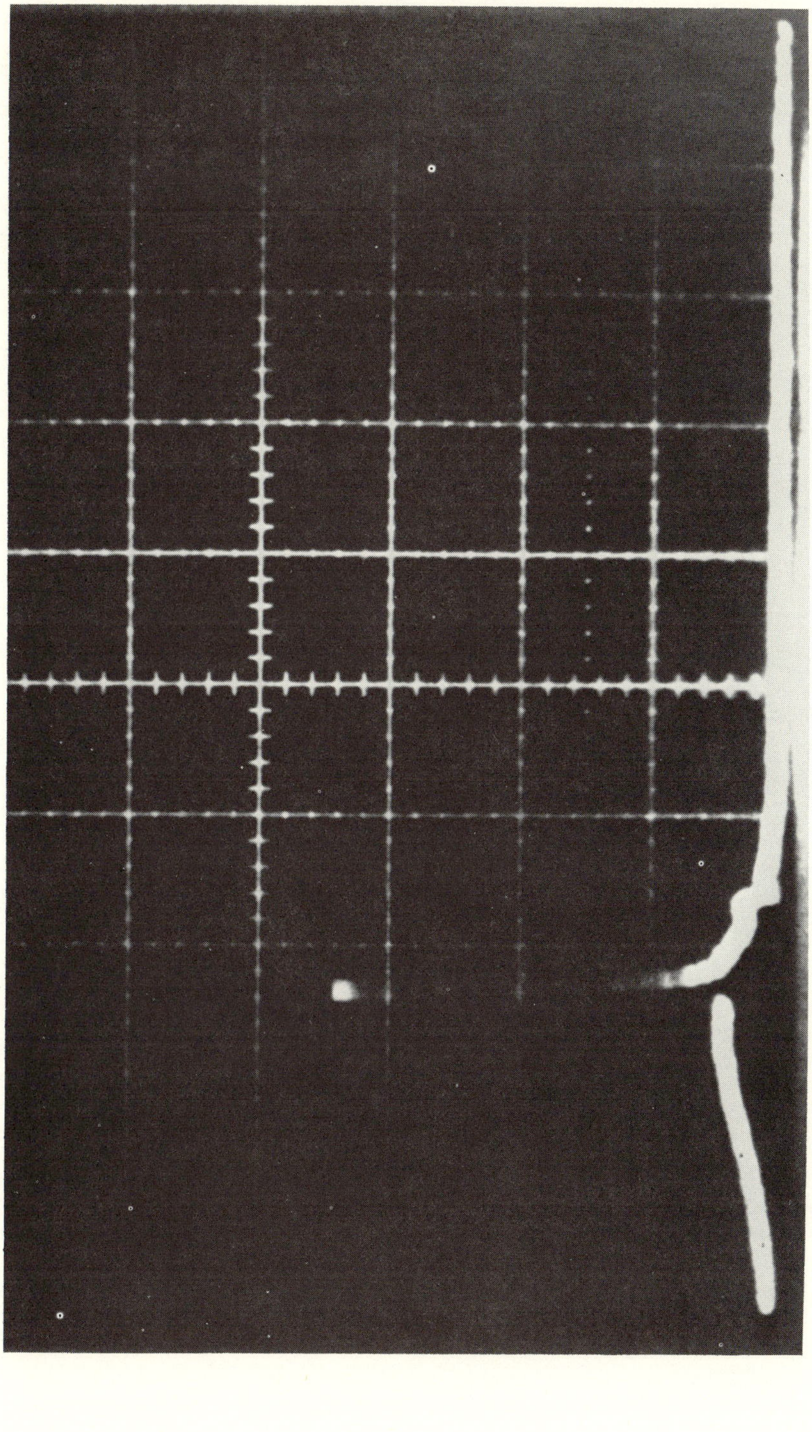

Fig. 15.
Example of an optical pulse generated by three series Pockels cells and tab electrical pulse generator illustrated in Fig. 12. Base 30 ns risetime, spike .75 ns FWHM.

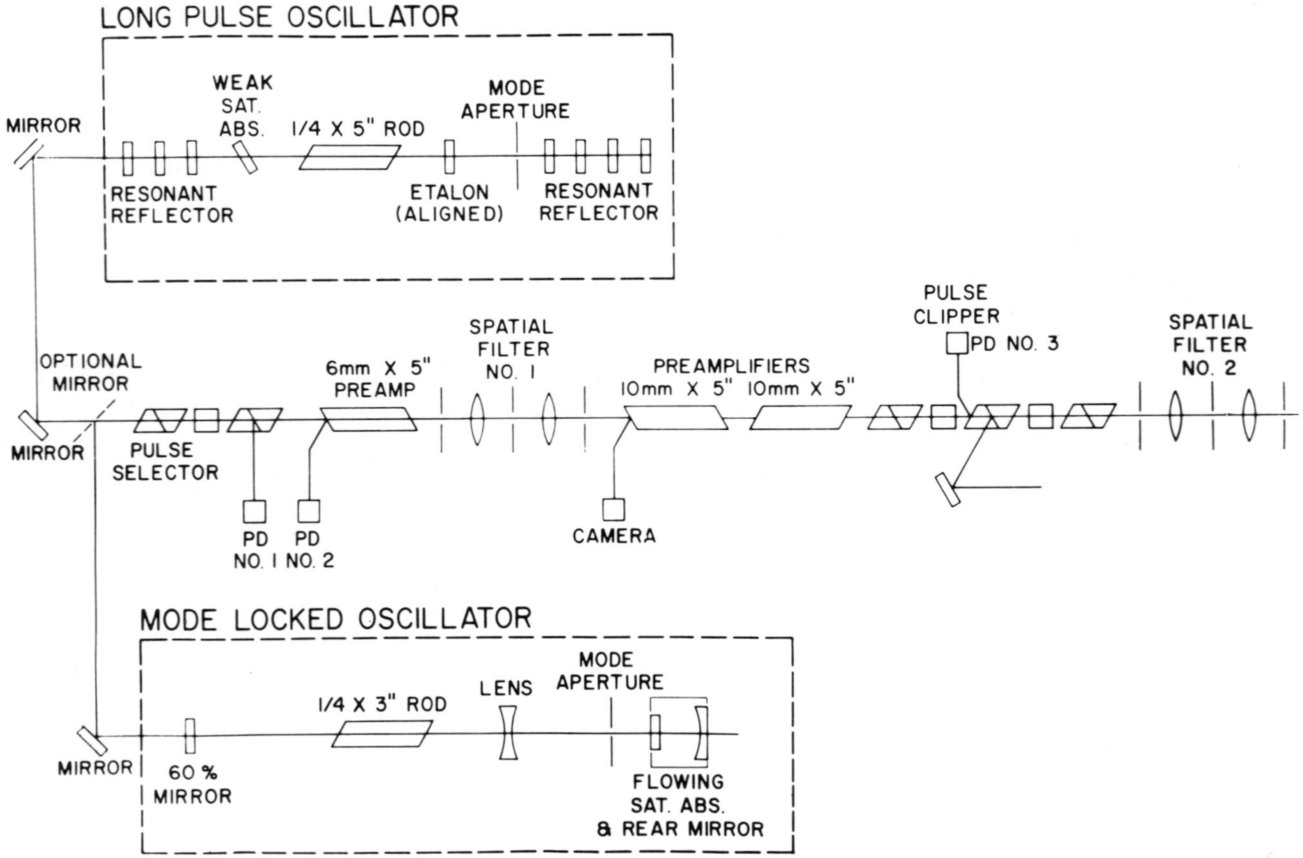

Fig. 16.
Pulse generation system employing two oscillators, two spatial filters, and three Pockels cells used to drive the "LASL" high power ruby laser. Pulses from 1 ps to 700 ns are produced with full temporal pulse shaping over most of this range.

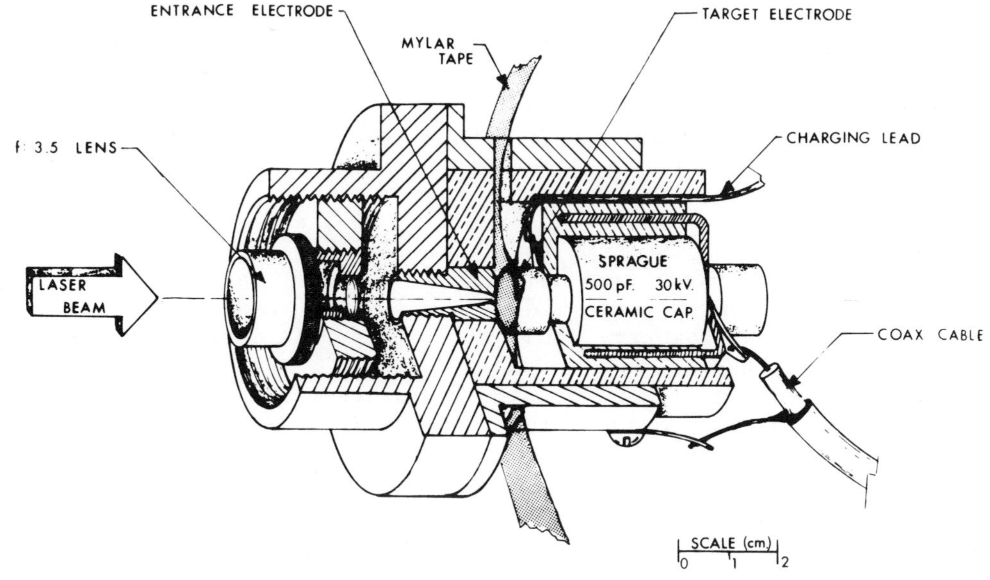

Fig. 17.

Schematic of low-jitter laser-triggered solid dielectric spark gap. (Ref. 16).

system with a high frequency electrical bandpass filter, we have successfully detected the oscillator pulse output approximately e^{15} below the final output intensity, corresponding to approximately 3.5 μs of anticipation with a jitter of less than 200 ns. Since the Nd:glass pulse trains are typically 1-2 μs long, there are no additional difficulties introduced by this 200 ns jitter.

Putting together all of this technology, with previously published work,[17] we believe that low-jitter, nearly optimum synchronization of these two systems is possible. At this time we are in the initial check-out phase. We have not yet had everything working simultaneously for the case of CO_2 and Nd:glass, although in another experiment involving a ruby laser and an iodine laser, the complete synchronization has been successfully demonstrated.

PLASMA PHYSICS AND DATA ACQUISITION

At this point, we wish to leave the subjects dealing with laser technology, and return to plasma physics and the three remaining questions raised at the beginning of this paper. The process of converting plasma interferograms into electron density contours involves calculating a function whose path integral is known. In general this is not a well posed mathematical problem. If, however, cylindrical symmetry about an axis normal to the interferometer axis can be assumed, then the integral equation has a unique solution, namely the Abel inversion.[9] Therefore, the existence of cylindrical symmetry is typically assumed. In this work, we plan to employ two independent interferometers which will view the plasma in two orthogonal directions at the same time. We can thereby appraise the extent of the departures from cylindrical symmetry, and thus some error analysis can be performed. We also plan to search for an axis where cylindrical symmetry does exist, to permit inversion of the second interferogram.

In connection with numerical simulations, we hope to use the interoferograms to provide a check on the full hydrodynamics. For part of this work, inversions of the data to produce density profiles is unnecessary as numerically generated interferograms of the numerically preducted particle distributions can be directly compared with our experimental data. However, the significance of this type of work depends on three situations, Namely: What signal to noise ratio can be obtained for the digitized interoferograms and can it be improved with image processing? How many interoferograms can be taken at different times during the CO_2 pulse? And Finally, how readily can the data be symmetrized to allow correlations with one dimensional hydrodynamic calculations?

For purposes of comparing with instability calculations, we need answers to such questions as: What is the magnitude of the density gradients? Is there any rippling of the critical surface, and if so, what is the scale? And finally, how asymmetric is the plasma evolution, and when do the asymmetries occur?

We plan to address each of these questions, but clearly one interferogram per interferometer per CO_2 laser shot is totally inadequate to answer any of these modeling questions. Also, since the times at which very steep density gradients occur will present problems in the resolution and analysis of the interferograms, we must rely on the computer predictions at these times having fit both the earlier and later time interferometric data. As a result we will spend an ever increasing amount of time exploring methods of increasing the number of interferograms which can be taken at different times during the one CO_2 pulse. Our initial installation will allow two different times during the CO_2 pulse to be examined by taking advantage of the freedom to chose the polarization of the interferometer illumination source. By first employing vertically and then later horizontally polarized light, each interferometer can yield two interferograms during one CO_2 pulse. We are investigating many other options which will allow us to increase the number of interferograms from two to as many as 10 or 20.

Another degree of freedom introduced by using two interferometers is the possibility of doing two wavelength interferometry.

In this way it is possible to separate the electron distributions from the heavy particle distributions. These experiments require that we know how the plasma varies with observation direction. Hence preliminary work will be done only at one wavelength, but because eventually we hope to be able to do two wavelength interferometry, we have a severe requirement on high signal to noise ratios for our data.

This leads to the question of how to best couple the direct interferogram to a digitizer with the minimum introduction of noise. Initially we plan to take all our data on film and then digitize the film data. For the case of two interferometers each producing two interferograms per CO_2 laser shot, a 1.5 μm resolution, and a 500 μm field of view, we require the digitizing with adequate bookkeeping of about 450,000 data points. As we expand the number of interferograms by about an order of magnitude, we quickly reach the point where speed in data handling also becomes imperative. Therefore, we are already beginning work on more efficient and noise-free methods of handling the data. Also, it is just this problem which has led us to direct interferometry, as opposed to holographic interferometry. A second reason for this decision was the lack of ideas on how to maintain high diffraction efficiency on the reconstruction light source over the whole hologram in the face of the large variations in the light intensities with refraction angle at the plasma. A related issue is the small dynamic range of film which provides only a small limitation on direct interferometry but severe limitations on holographic interferometry.

Acknowledgments

The authors wish to express their appreciation to D. Kohler[18] of LASL for some very illuminating discussions on interferometry. This work was performed under the auspices of the U. S. Energy Research and Development Administration.

* Presented at the Fourth Workshop on "Laser Interaction and Related Plasma Phenomena," held at RPI, Troy, New York, November 8-12, 1976.

1. R. C. Malone, R. L. McCrory, and R. L. Morse, Phys. Rev. Lett. 34, 721 (1975).

2. R. E. Kidder and J. W. Zink, Nucl. Fusion 12, 325 (1972).

3. D. Henderson, private communication (October 1976).

4. S. Jackel, J. Albritton, and E. Goldman, Phys. Rev. Lett. 35, 514 (1975).

5. D. T. Attwood and L. W. Coleman, Appl. Phys. Lett. 24, 408 (1974); see also "Ultraviolet Probing of Laser Produced Plasmas with Picosecond Pulses." D. T. Attwood, Lawrence Livermore Laboratory Report No. UCRL-11144 (July 14, 1976).

6. S. Ariga and R. Siegel, "Picosecond Microphotography as a Diagnostic Method for Laser Produced Plasma, "Max-Planck-Gasellshaft, 8046, Garching, FRG.

7. J. Martineau, this meeting.

8. R. Fedosejvs, I. V. Tomov, and M. C. Richardson, Bull. Amer. Phys. Soc. 21, 772 (1976).

9. W. Lochte-Holtgreven, Plasma Diagnostics, W. Lochete-Holtgreven, ed. (North Holland Publishing Co., Amsterdam 1968) ch. 3, p 184.

10. J. F. Reintjes, R. L. Carman, and F. Shimizu, Phys. Rev. A8, 1486 (1973).

11. R. L. Carman, LLL Laser Fusion Semi-Annual Report, UCRL-500021-81, p 4 (1971).

12. E. B. Treacy, IEEE J. Quantum Electron QE-5, 454 (1969); also Phys. Lett. A28, 34 (1968).

13. See, for example, B. J. Chang, Optics Comm. 9, 357 (1973); B. J. Chang, R. Alferness, and E. N. Leith, Appl. Optics 14, 1592 (1975).

14. R. L. Carman, Bull. Amer. Phys. Soc. 20, (1975); R. L. Carman, Conference on Lasers and Nonlinear Optics, Tbilisi, USSR, May 24-28, 1976; R. L. Carman, Los Alamos Reports LA-UR-76-1013 and LA-UR-75-1945.

15. I. G. Kateyev, Electromagnetic Shock Waves, (London Iliffe Books Ltd., London (1966)), Translated by Scripta Technica Ltd.

16. H. D. Sutphin, R. F. Benjamin, A. J. Lieber, and C. B. Webb, "Low Jitter Laser Triggered Spark Gap Using a Thin Solid Dielectric," Los Alamos Report No. LA-UR-75-1783.

17. R. L. Carman, J. Reintjes, and H. Furumoto, IEEE J. Quantum Elec. QE-6, 751 (1970).

18. Earlier work at LASL on this subject is described in David Kohler, D. Giovanielli, R. P. Godwin, G. H. McCall, and M. M. Miller, Bull. Amer. Phys. Soc. 19, 854 (1974).

GENERATION AND SUPPRESSION OF FAST-IONS FROM A LASER-PLASMA AND RELATED SELF-FOCUSSING EFFECTS *

B.Luther-Davies,M.R.Siegrist,**V.del Pizzo,J.L.Hughes

Research School of Physical Sciences, The Australian National University, Canberra, A.C.T. 2601, Australia

ABSTRACT

We report studies of fast-ion emission from high and low-Z targets irradiated by 25 or 100 psec pulses from a Neodymium laser. We have studied the functional dependence of the ion kinetic energy on laser intensity, and shown that fast-ions are suppressed if a mode-locked pulse train irradiates the target at the same maximum intensities used with a single pulse. When fast ions are suppressed anomalous groups of high energy ions are emitted parallel to the target surface. An explanation of this effect is closely related to ponderomotive force self-focussing in laser-produced plasmas and recent results on this phenomenon will be presented.

*Presented at the Fourth International Workshop Conference on "Laser Interaction and Related Plasma Phenomena" held at R.P.I., Troy, New York,Nov 8-12 1976.

**Present address : CRPP-EPPL
Av. de Bains
1007 Lausanne, Switzerland

INTRODUCTION

Laser-plasma studies at the Australian National University are concentrated on establishing the conditions whereby we expect to generate electron-positron pairs within a plasma via the 'trident effect'. The threshold for this process in laser-produced plasmas has been considered by Bunkin and Kazakov (1) and was also the subject of a paper presented at the last R.P.I. meeting by Shearer *et al* (2). We recall some of Shearer's results here. Firstly the cross-section (σ_T) for pair production via the trident effect from electrons with energies $\geq 2\, m_o c^2$ was given by :

$$\sigma_T = 9.6 \times 10^{-4} \, (\alpha r_o Z)^2 \left(\frac{E_v}{m_o c^2} - 2 \right)^{3.6} \qquad 1.$$

where m_o the rest mass of the electron,
c the velocity of light in vacuum,
α the fine structure constant,
r_o the classical radius of the electron,
Z the nuclear charge,
E_v the oscillation energy of the electron.

Secondly, for electrons oscillating relativistically in the electric field of a circularly polarised laser beam, the oscillation energy has an intensity dependence given by :

$$\frac{E_v}{m_o c^2} = \left(\frac{\lambda^2 I}{9 \times 10^{18}} \right)^{0.657} \qquad 2.$$

where λ is the vacuum wavelength in microns,
I is the laser intensity in W/cm^2.

Using 1 and 2 the rate of pair production is :

$$\frac{d N_p}{dt} = \frac{0.15\, Z^2\, l^3}{\lambda Z_i} \left\{ \frac{N_e}{N_c} \right\}^2 (\gamma - 3)^{3.6} \frac{(\gamma^2 - 1)^{0.5}}{\gamma} \qquad 3.$$

$$\gamma = 1 + \frac{E_v}{m_o c^2}$$

$\frac{N_e}{N_c}$ is the ratio of the electron density to cut-off density,

Z_i is the ion charge $= N_e/N_i$ with N_e, N_i the electron and ion number densities respectively,

l is the scale length of the interaction volume.

The approximate pair generation threshold was deduced by Shearer to be given by $I\lambda^2 = 3 \times 10^{19}$ W/cm^2 μm^2. Inspection of equation 3 shows that for a fixed plasma density and volume the pair generation rate is proportional to the nuclear charge Z.

This threshold is well in excess of the values of laser flux density which can be achieved by focussing the output from even the most powerful lasers using a simple lens. Even supposing that the threshold was reduced because of electron heating by plasma instabilities as mentioned by Shearer special techniques would still be required to obtain the necessary flux densities. Our approach to this problem has been to attempt to use whole-beam self-focussing within the plasma to increase the in-plasma laser flux density by the required amount above the vacuum value. For example, it is envisaged that a beam of some 30-40 μm diameter and flux density $\simeq 3 \times 10^{16}$ W/cm^2 could be self-focussed to a filament about 1 μm in diameter with a resulting thousand-fold increase in intensity.

In this paper we describe initial experiments towards pair production. Our work, to date, has concentrated on the study of "fast-ion" emission from laser-produced plasmas as a technique for monitoring the hot electron temperature (3) in high-Z plasmas. Also because of the sensitivity of the fast-ion energy to in-plasma laser flux density (4,5,6), we expect that the onset of self-focussing will be accompanied by an anomalous increase in the energy of fast-ions.

In the latter part of this paper we present some of our latest results on computer simulations of self-focussing in plasmas and relate these to anomalies that we have already observed in our study of fast-ions.

APPARATUS AND EXPERIMENTS

We have constructed a medium-power Neodymium laser system for this work. In order to simulate as best we can the conditions used to predict self-focussing in our computer codes, the laser has been constructed as a versatile unit capable of producing pulse durations of 5, 25 or 100 psec and able to deliver single pulses or multiple pulse trains onto a target. We expect to add a second oscillator to the system which will be synchronised with the existing mode-locked oscillator to provide a source of nanosecond prepulses for target experiments (7).

The present laser is based around a Neodymium : Yag mode-locked oscillator which provides trains of either 25 or 100 psec pulses. The oscillator output is fed to a pockels cell which selects a single pulse from the train with a contrast ratio of typically 1000.

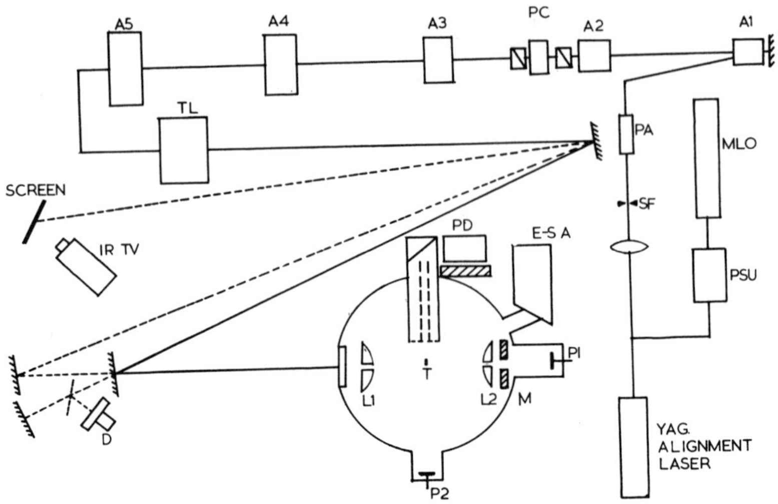

Figure 1. Apparatus:MLO - mode-locked oscillator; PSU - pulse switch-out unit; SF - spatial filter; PA - pre-amplifier; A1-A5 - Nd : Glass amplifiers; TL - telescope; PC - pockels cell; IRTV - infra-red TV camera; D - vacuum photodiode; PD - positron detector; E-SA - electro-static analyser; P1, P2 - ion probes; L1, L2 - 75 mm F = 1 lenses; M - turning mirror.

The beam then undergoes spatial filtering before being amplified by a single Nd : Yag pre-amplifier and a five-stage Nd : Glass chain utilizing 12.5 cm long ED-2 rods. A second pockels cell in the chain serves both to increase the contrast ratio of the pulse to greater than 10,000 and for back-reflection isolation. Typical output powers are 50 GW in a 25 psec pulse at which power the total 'B-integral' value (8) for the system is 1.5.

For operation at 5 psec the Nd : Yag rods are replaced by Hoya LHG-5 which can be directly substituted without need for other changes in the system. The operation of LHG-5 has found to be reliable with the oscillator mode-locked with Kodak 9740 dye. As yet, no target experiments have been performed at 5 psec.

The laser output is fed to our single-beam target chamber which contains two F = 1, 75 mm focussing lenses and a turning mirror which allow the target to be illuminated from the rear with respect to the direction of the incoming laser beam. The whole layout of the system is shown in figure 1.

The 1/e electric field diameter of the focussed spot on-target was determined to be 35 ± 5 μm and laser flux densities up to about

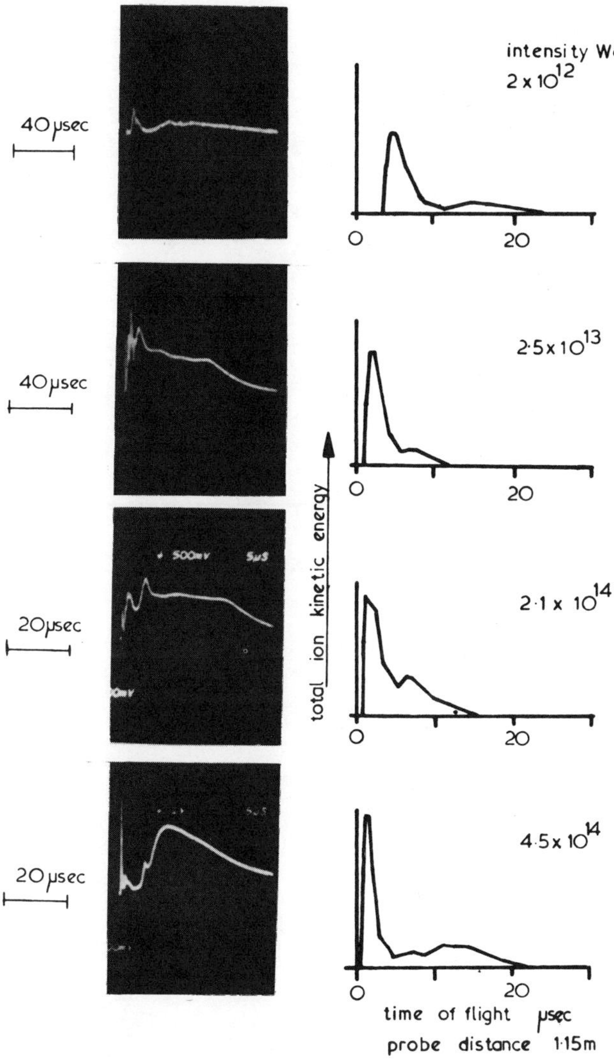

Figure 2. (a) Left-hand side : ion probe signals from P1 for different laser intensities. (b) Right-hand side : corresponding total ion kinetic energy as a function of time.

3×10^{15} W/cm^2 have been used so far.

Our study of fast-ion emission has used two ion probes, one of which views the target through apertures in the focussing lens and turning mirror in a direction along the axis of the incoming laser

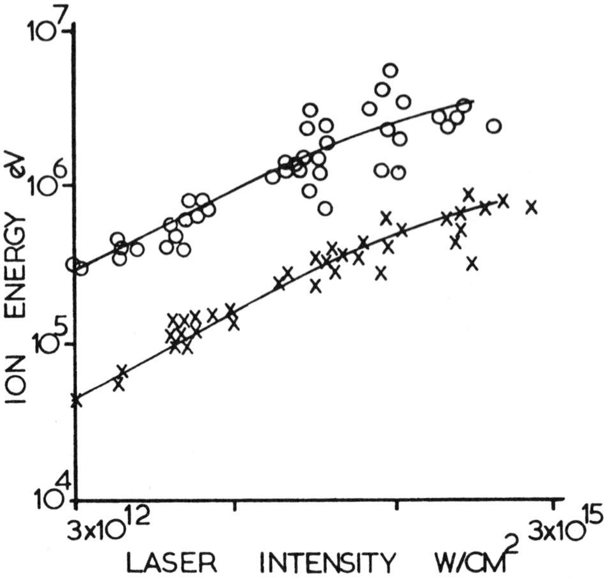

Figure 3. The variation of fast-ion energies from gold as a function of laser intensity. The upper trace corresponds to the fastest ion energies observed; the lower ion energies at the peak of the fast-ion signal.

beam. The second probe views the target at ninety degrees to that direction. Both probes were either plane biassed collectors or deep Faraday-cup probes.

As already mentioned above we have aimed to study fast-ion emission from high-Z targets as a means of monitoring the plasma hot-electron temperature. It has been shown by Fabre *et al* (9) in an experimental study of fast-ion emission from polyethylene irradiated by a CO_2 laser, that the fast-ion energy and hot electron temperatures are directly related. Since evaluation of electron temperatures by measurement of X-ray Bremmstrahlung emission from high-Z plasmas cannot be done using simple K-edge filter X-ray spectrometers, fast-ion emission offers an easy method for monitoring for the presence of electrons with sufficient energy for pair production within the plasma.

From the non-linear force theory of Hora (10) and the flux limit model in the presence of reduced electron thermal conductivity of Morse and Nielson (3) the energy of the fast ions is related to the hot electron energy by the ionic charge Z_i. The flux limit condition also predicts that the ion energies should increase in proportion to the two thirds power of the laser flux density. This dependence of fast-ion energy has been observed for plasmas created using nano-second pulses from CO_2 lasers (5,9). Experiments using 40 psec neodymium laser pulses to irradiate thin polyethylene foils have produced a somewhat different result (11).

Our experiments have been performed with several different target materials including Gold, Tungsten, Cadmium, Copper, Aluminium and Polyethylene, and either 25 or 100 psec duration laser pulses. Although with high-Z targets the relation between fast-ion energy and laser intensity approaches the 2/3 power law predicted by the flux-limit model, the results for low-Z materials are different with a much weaker variation. Our experimental results are shown in figures 2 - 7.

Figure 2 shows a set of typical ion current traces from the in-line ion probe for different laser intensities. The pulse duration was 100 psec, and the target a 250 μm diameter gold wire. Even at the lowest laser intensities ($\simeq 10^{12}$ W/cm^2) fast-ions are generated as evidenced by the double-peaked ion current trace. At low intensities ($I < 3 \times 10^{14}$ W/cm^2) the signals were very repeatable whilst at higher values both the amplitude of the fast-ion signals and the velocity of the ions varies unpredictably from shot to shot. We note that the amplitude of the fast-ion peaks reduces with respect to the slower ion components at high intensities, this is weakly reflected in the variation of the integrated ion kinetic energy as a function of time which is also shown for each trace in figure 2.

In figure 3 we plot the variation of fast-ion kinetic energy from gold as a function of laser intensity for a 25 psec laser pulse. Two curves are shown, the upper corresponds to the maximum ion energy and the lower to the ion energy which corresponds to the first peak of the fast-ion signal. Most of the kinetic energy of expansion of the fast-ions is carried by ions with energies between these two limits. Several points are worthy of note. Firstly, the slope of the curves shown is ≃0.55 a value significantly below that predicted by the flux limit model. Secondly an anomalous region of the curves exists around 3×10^{14} W/cm^2 where the ion energy varies erratically from pulse to pulse, and peak energies up to 6 MeV were observed. At higher intensities the fast-ion signal develops strong modulations, which are only weakly apparent at lower intensities (fig. 4) and which we attribute to the presence of different ionization states within the plasma. We will return to this anomaly in the discussion.

With knowledge of the plasma ionization state, an estimate of the maximum value of the hot electron temperature can be made from these results. Measurements using an electro-static analyser to determine plasma charge states (q) shows the presence of ions with $q > +15$ at

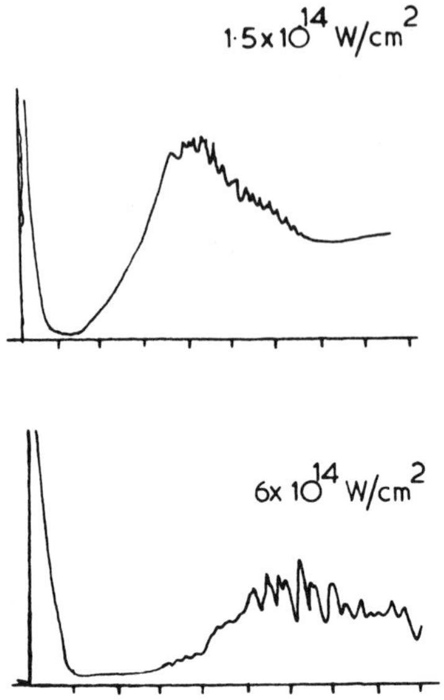

Figure 4. Fast-ion signals within the anomalous region shown in figure 3, showing reduction of amplitude and increased modulation at higher intensities.

10^{14} W/cm^2. The maximum value of the hot electron temperature in the anomalous region would be, therefore, 400 keV.

Figure 5 shows data for a range of target materials with both high and low atomic numbers. The data corresponds to the low energy curve of figure 3 and, thus, maximum fast-ion energies are about a factor of ten higher than those shown. The slope of these curves reduces to about 0.3 for Carbon with intermediate slopes for the other materials. Fast ion energies reduce with the atomic number of the target, and this is shown more clearly in figure 6. According to Hora (10) and Morse and Nielson (3) for a fixed plasma hot electron temperature the ion kinetic energy should be linearly dependent on Z_i , the ion charge. At high intensities our results support this result if we assume $Z_i = Z$. For lower laser intensities we can interpret the deviation from the linear dependence of ion energy on atomic number as due to weak ionization of the plasma. By extrapolation of the curves in Figure 6 to Z = 1 we can estimate the value and intensity dependence of the hot-electron temperature. We find peak temperatures ≈100 keV varying in proportion to $I^{0.3}$.

Several other features of fast-ion emission which are consistent with observations from other laboratories were noted. For example, the ion energies measured with both in-line and side probes were identical. The fast-ion signals were strongly peaked in the direction of the target normal and most of the energy of expansion of the plasma was carried by the fast-ion component.

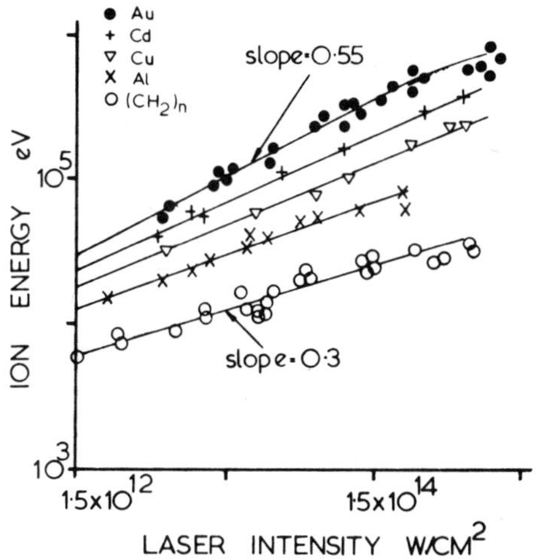

Figure 5. Data as in figure 3 but for a range of target materials. Ion energies at the peak of the current traces are shown.

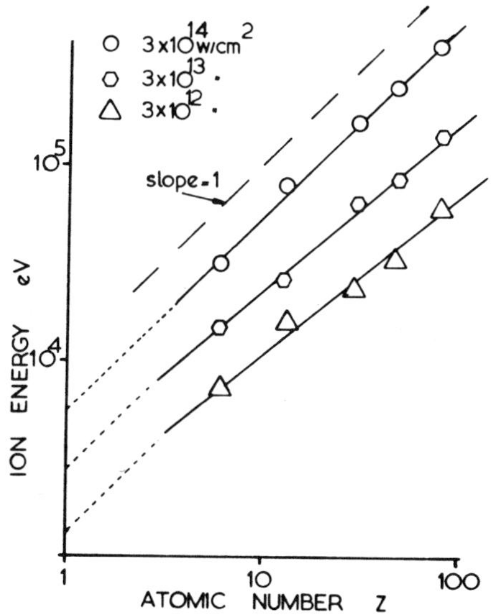

Figure 6. The variation of ion kinetic energy as a function of atomic number Z for different intensities.

We compared ion energies generated from a gold target by both 25 and 100 psec duration pulses and found no observable difference between them for a particular laser intensity.

Our experiments have also demonstrated that in certain conditions of irradiation, fast-ion emission can be totally suppressed. We irradiated a target with the full pulse train from our oscillator at the same maximum intensities as used with a single pulse and observed no fast-ions. This effect is shown in figure 7. Here the upper of the two sets of traces correspond to irradiation with a single laser pulse, whilst the lower pulse train with the same maximum intensity. The upper and lower signals in each case correspond to the in-line and side probe current traces. Note that whilst fast-ions are evident with the single pulse they are completely suppressed with the pulse train. In this latter case, however, the ion emission is anisotropic with a group of ions with energies between 1 and 10 keV being emitted parallel to the target surface. We discuss this effect below.

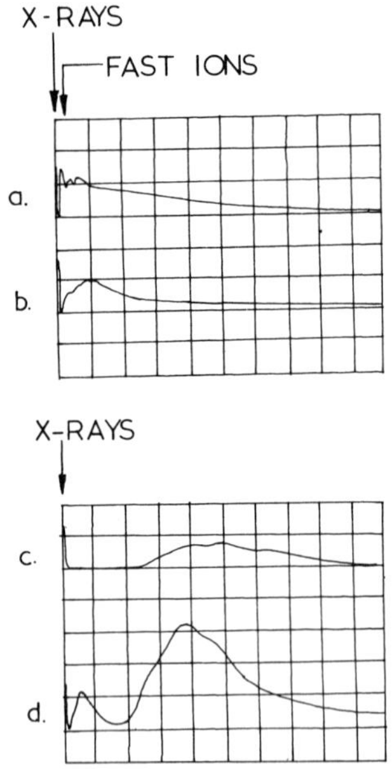

Figure 7. Ion emission from gold irradiated at 5 x 10^{13} W/cm^2 peak flux density. Upper traces single pulse; lower traces mode-locked pulse train. Each pair of traces represents the current from in-line (upper) and side (lower) ion probes. Target to probe distances are 1.15m the in-line probe and 0.22m for the side probe. CRO sweep rates are 20 μsec/div and 5 μsec/div for the two probe traces.

DISCUSSION

We concentrate our discussion on two aspects of our results which are significant for our work. Firstly the occurrence of the large shot-to-shot variation in fast-ion energy for laser intensities $\simeq 3 \times 10^{14}$ W/cm^2. The maximum ion energies observed in this region would correspond to an effective increase in laser intensity of about one order of magnitude over the measured value if the dependence of ion kinetic energy on laser intensity observed at lower intensities holds in this region. We examine the possibility that this anomaly is due to the onset of self-focussing. Secondly we consider the suppression of fast ions when a pulse train irradiates the target and the occurrence of the intermediate energy ions ejected from the plasma parallel to the target surface. Several explanations are possible for the suppression of fast-ions. The different plasma density profile near the critical region which exists due to the pre-pulsing effect of the pulse train could cause light from subsequent pulses to be absorbed in the cold plasma corona thus leaving little laser energy for fast-ion generation. Alternatively, fast ions could still be generated near the critical region but due to the existence of a thick (>200 μm) high density (10^{19} - 10^{21} cm^{-3}) ablation halo the fast-ions are subsequently thermalised by collisions with other ions. This is reasonable since the mean free path would be less than 2 μm in this situation. Our experiments do not allow judgement in this matter. Of particular interest, however, is the group of sideways-directed ions, which due to their higher ion energies contain most of the expansion energy of the plasma in that direction. A possible explanation is that they are generated by the action of ponderomotive forces in the plasma corona which occur due to the non-uniform field distribution of the beam in the plasma. Such forces are very important in self-focussing of beams in plasmas and we treat this subject in more detail in the following section.

SELF-FOCUSSING OF LASER BEAMS IN PLASMAS

Self-focussing of pulses of nanosecond duration or shorter in plasmas can occur due to the operation of two different mechanisms, both of which form a refractive-index gradient in the plasma which is favourable to self-focussing.

The first mechanism, known as ponderomotive force self-focussing (12, 13, 14) occurs due to the electrodynamic part of the Landau Lifshitz equation which produces a force on the electrons proportional to $\nabla \langle E^2 \rangle$ ($\langle\ \rangle$ denotes averaging over one optical cycle). Due to the approximately Gaussian shape of a focussed laser beam propagating in an initially uniform plasma, this causes a radial force which acts to eject electrons (and ions because of their strong electrostatic attraction to the electrons), from the beam centre. The plasma

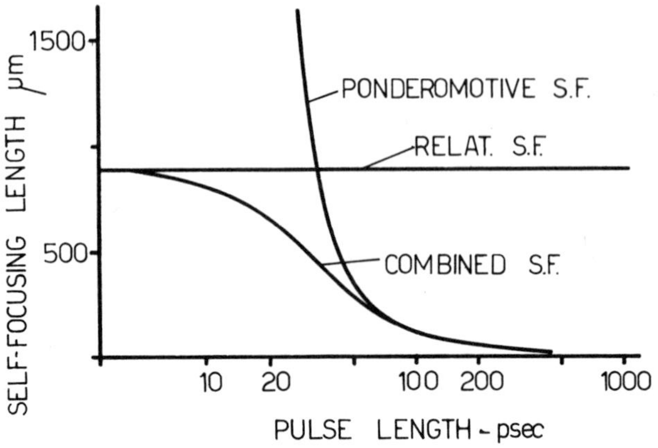

Figure 8. Variation of self-focussing distance versus pulse duration for ponderomotive relativistic and combined self-focussing for pulses of 10^{15} W/cm^2, initial diameter 35 μm, with $N_e = 0.5\ N_c$, $T_e = 50$ eV and absorption neglected.

responds to this force in a time given approximately by the ratio of the beam diameter to the velocity of sound in the plasma - typically 1 nsec. The resulting perturbation of the plasma density causes a variation in real part of the plasma refractive index (n_r) with $n_r \simeq 1$ on axis and $n_r < 1$ off axis thus forming a positive lens-like medium.

The second mechanism has been discussed extensively by Hora (15) and Max *et al* (16) and arises from a relativistic modification to the plasma cut-off density which occurs due to the coherent oscillation of electrons in the field of an electromagnetic wave. Due to the non-uniformity of the beam intensity this can also result in lensing of the plasma to cause self-focussing. In comparison, however, with the ponderomotive force mechanism its action is essentially instantaneous.

The two mechanisms, therefore, have basically two time scales of operation. For nanosecond pulses ponderomotive force self-focussing dominates whilst for picosecond times relativistic self-focussing occurs. We show, however, in figure 8(17) the results of threshold calculations for both mechanisms as a function of pulse duration for the plasma parameters shown. It can be seen that even for pulses as short as 20 psec ponderomotive force self-focussing cannot be neglected and in all the following results both mechanisms are included.

The data for figure 8 was generated using our comprehensive

a

b

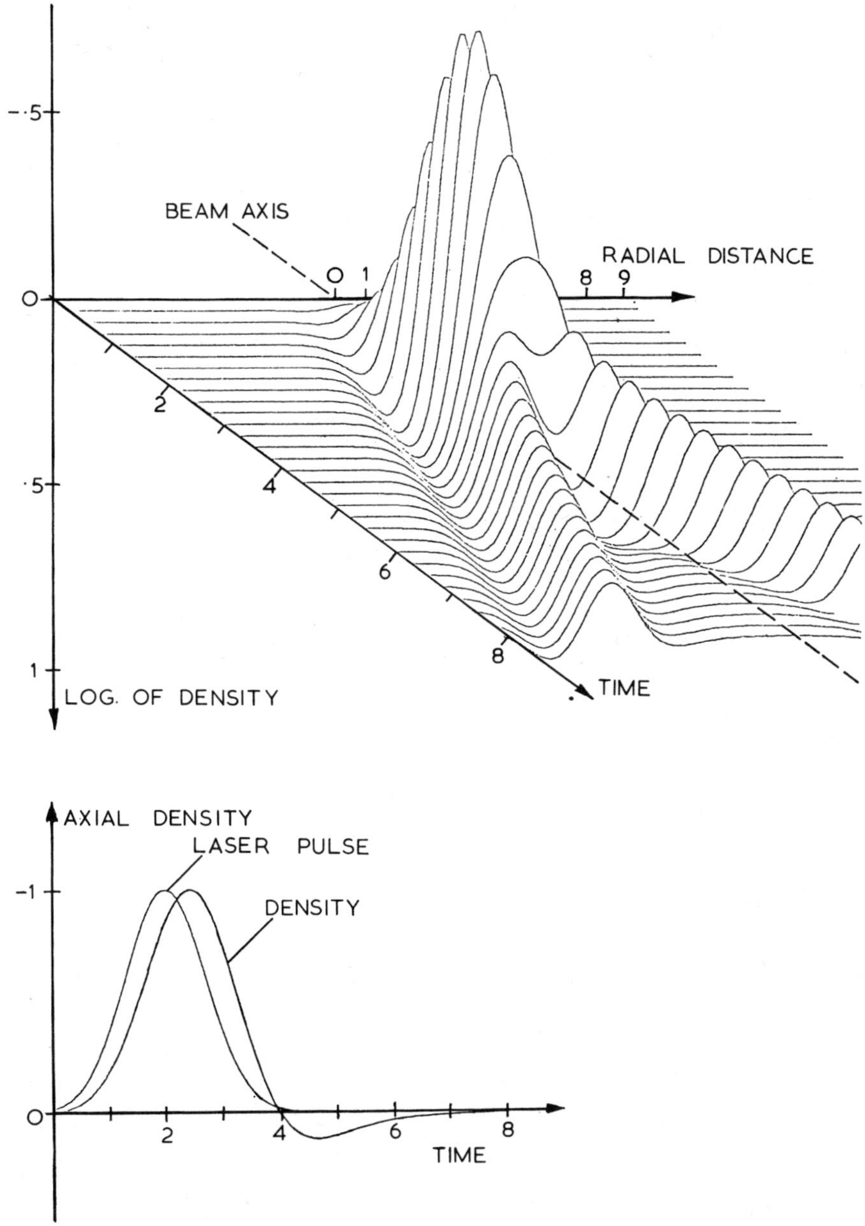

Figure 9. The radial plasma density profile in the cross-section of a Gaussian laser pulse developing in time. Time and space units correspond to the laser pulse half-width in time and space.

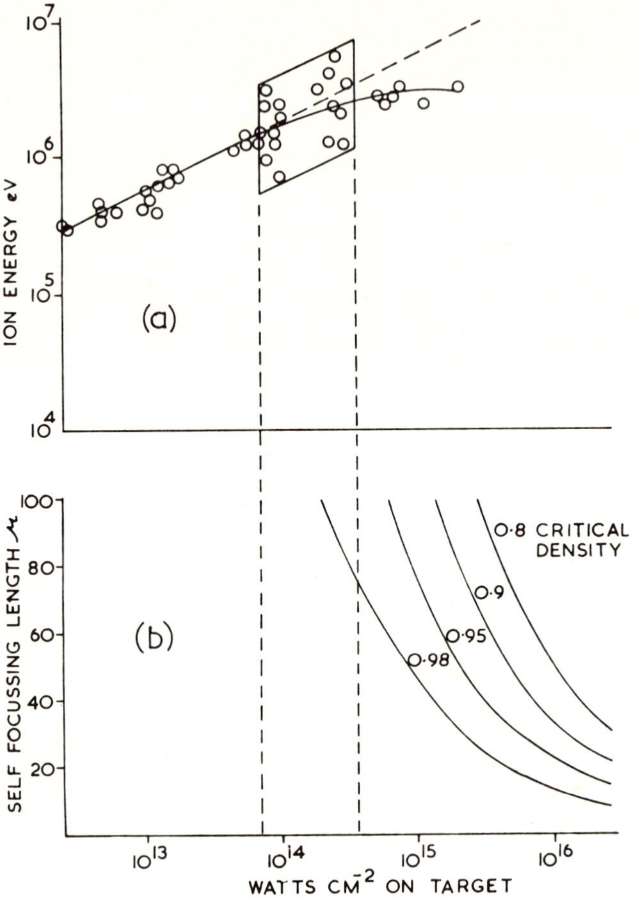

Figure 10. A comparison of the threshold of fast-ion anomalies and self-focussing lengths in non-absorbing plasmas.

computer model of self-focussing in plasmas. As well as treating both self-focussing mechanisms the effect of diffraction, absorption, and the dynamic response of the plasma to the ponderomotive force are included. The details of this programme have been published in the literature (16,17).

In figure 9a we show a calculation illustrating the action of the ponderomotive force in a plasma (18). Here a pulse of approximately 2 nsec duration and 2 x 10^{13} W/cm^2/eV propagates through an initially uniform plasma. The action of the ponderomtive force causes a time-dependent density depression to occur on the axis of the beam as shown in figure 9b, as the plasma is ejected. Calculations relevant to short pulses propagating in gold plasmas show that the plasma is ejected from the beam centre with a velocity of 3-4 x 10^6 cm sec^{-1} which is comparable with the velocity of the sideways directed ion group in our experiments. We, therefore, suggest that this ion group is evidence of the action of the ponderomotive force in the plasma halo.

We have used our codes to predict self-focussing lengths in high density plasmas for laser intensities within the range 10^{14}-10^{16}W/cm^2.

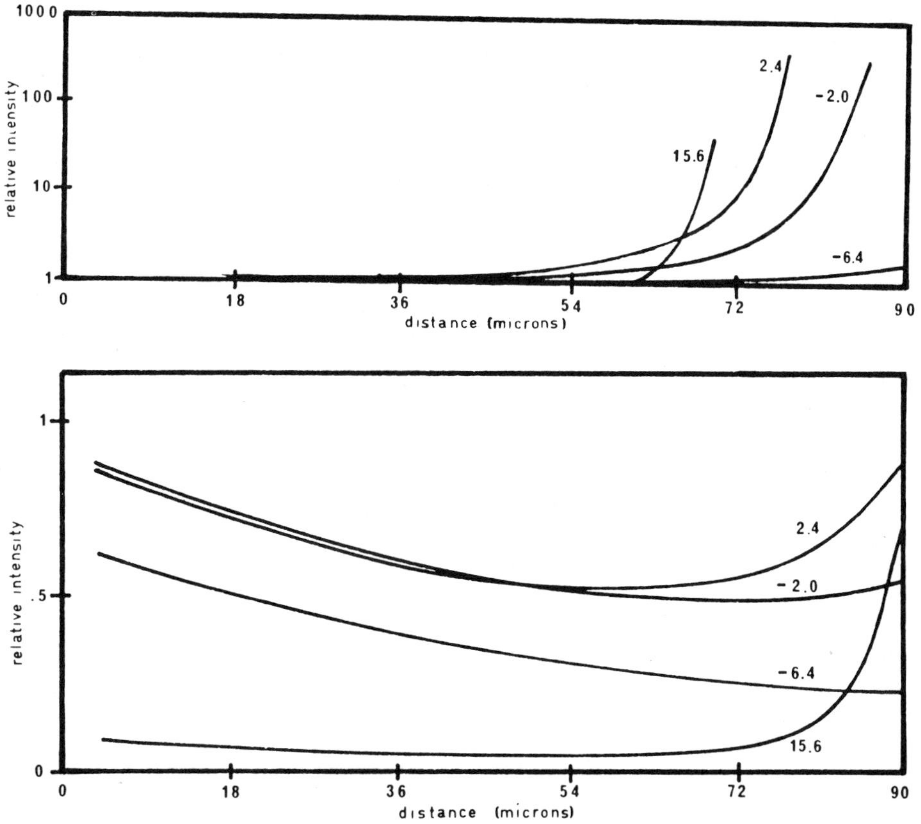

Figure 11. Time and space history of the intensity of a beam which undergoes self-focussing in a plasma. The intensity relative to a peak input value normalised to unity is shown as a function of distance of propagation (X-axis) for times wrt the pulse maximum (t=0). The full pulse width (e^{-1} intensity) was 20psec and T_e=500 eV, I_{max}=4 x 10^{15}W/cm^2. In the upper curves absorption is neglected - note the difference in y-scales.

In figure 10 we compare the results with the intensity region where we observe the anomalous increase in fast-ion energy. The self-focussing lengths reduce to less than about 80 μm for plasma densities close to critical and intensities in the range of interest. This behaviour is consistent with self-focussing being responsible for this anomaly particularly since the experimental value of the average laser intensity is only known within a factor of two at best, and hot spots in the focussed beam may substantially affect the focussing behaviour.

We note, however, that absorption was neglected in these calculations. This approximation has been found to be valid for electron temperatures as low as 200 eV, since self-focussing lengths are primarily determined by the refractive index profile near the input to the plasma. Once the wave fronts have curved towards a focus absorption has little effect on the actual self-focussing length. In contrast, however, absorption strongly affects the intensity achieved in the self-focussed filament. This is illustrated in figure 11a and 11b where calculations of beam intensity as a function of propagation distance and time in a gold plasma of 0.9 N_c are shown. The peak laser intensity in this case is 4×10^{15} W/cm^2 and the pulse duration is 20 psec. The two sets of data represent the calculations when absorption is neglected (11a) and included (11b) with an electron temperature of 500 eV. In the latter case no intensity enhancement occurs because of the influence of absorption. In figure 12 we show the effect of plasma absorption on the self-focussing threshold for a gold plasma with density 0.9 N_c and thickness 180 μm. Two pulse durations 25 and 100psec have been used. In this case we define self-focussing to have occurred when the instantaneous output intensity exceeds the input intensity to the plasma by a factor of ten. The time at which this occurs during the pulse is plotted as a function of laser intensity. The time scale runs from -tp to +tp where tp is the 1/e intensity duration of the pulse. A minimum ten-fold increase in the peak input intensity is achieved in the self-focussed filament when t = 0 (the peak of the pulse). Figure 13 shows the variation of laser intensity in a 20 psec pulse required for a ten-fold increase in peak input intensity as a function of electron temperature. Figures 12 and 13 show the strong sensitivity of the in-filament intensity to small variations in either input laser intensity or electron temperature. This sensitivity could adequately explain the very poor shot to shot repeatability observed in the fast-ion signals for laser intensities within the anomalous region. The time dependence of the ponderomotive force self-focussing mechanism is illustrated in figure 12 by the marked reduction in self-focussing thresholds at increased pulse durations.

CONCLUSIONS

In this paper we have outlined our work which is aimed at electron-positron pair generation in laser-produced plasmas. We

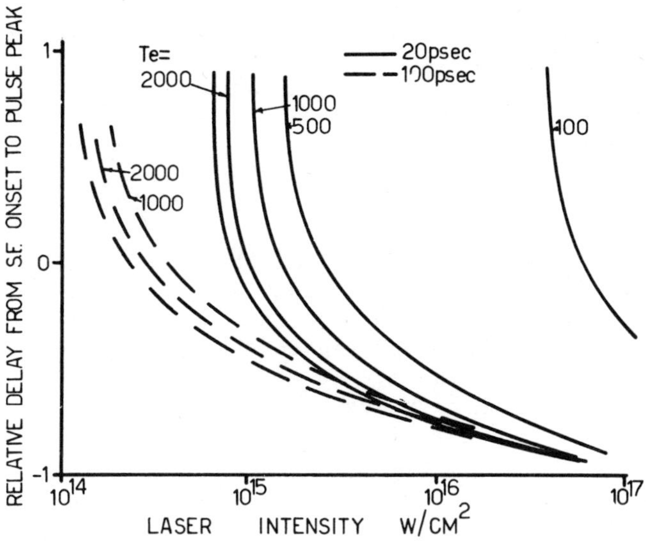

Figure 12. The effect of temperature-dependent absorption on self-focussing. The time relative to the pulse maximum for a ten-fold enhancement in the exit intensity of a beam propagating through a 180 μm thick plasma is plotted as a function of laser intensity, pulse duration and plasma temperature.

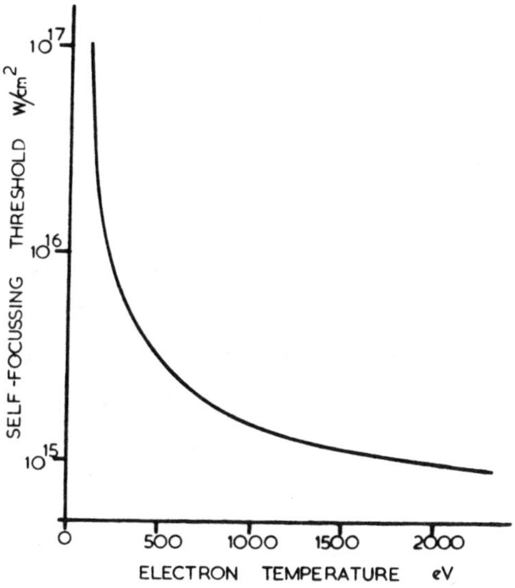

Figure 13. The laser intensity required for self-focussing at the peak of a 20 psec laser pulse in a 180 μm thick plasma in the presence of temperature-dependent absorption.

believe that the study of fast-ion emission provides us with a sensitive means of detecting the onset of self-focussing in plasmas - a phenomenon essential for pair generation with current laser technology.

Anomalies in emission of fast high-Z ions have been observed which are consistent with, but do not categorically confirm, the onset of self-focussing in a laser-plasma. Further studies involving measurements of hard X-ray emission, the spatial characteristics of X-ray emission and electron emission from plasmas should help determine if self-focussing has occurred. If this is achieved, the emission of X-rays in the MeV region should be accompanied by pair generation within the plasma. We are presently turning our attention to the detection of those pairs.

ACKNOWLEDGEMENTS

The authors wish to thank Pat Donohue, Jimmy Edwards, Colin Steele, Peter Smith and their colleagues for the valued technical assistance to this project. Mrs Helen Hawes deserves special thanks for her patience in typing this manuscript.

REFERENCES

1. Bunkin F.V., Kazakov A.E., Soviet Phys. Dokl. 15 758 (1971).
2. Shearer J.W., Garrison J., Wong J., Swain J.E. In Laser Interaction and Related Plasma Phenomena Vol. 3B. 803 1974 Plenum Press.
3. Morse R.L., Nielson C.W., Phys. Fluids 16 909 (1973).
4. Yamaka C., Yamanaka T., Sasaki T., Yoshida K., Waki K., Kang H.B., Phys. Rev. A6 2335 (1972).
5. Ehler A.W., Jrnl. Appl. Phys. 46 2464 (1975).
6. Luther-Davies B., Hughes J.L., Optics Comms. 18 351 (1976).
7. Reintjes J., Eckardt R.C., De Rosa J.L., IEDN Technical Digest 326 (1974).
8. Holzrichter J.F., Speck D.R., Jrnl. Appl. Phys. 47 2459 (1976).
9. Fabre E., Garban C., Popovics C., Poquerusse A., Stenz C., paper T8, IXth International Quantum Electronics Conference, Amsterdam June 1976.
10. Hora H. in Laser Plasmas and Nuclear Energy, p.58, Plenum Press (1975).
11. Pearlman J.S., Anthes J.P., Appl. Phys. Letts. 27 581 (1975).
12. Hora H.Z., Physik 226 156 (1969).
13. Kaw P., Schmidt G., Wilcox T., Phys. Fluids 16 1522 (1973).
14. Siegrist M.R.,Optics Comms. 16 402 (1976).
15. Hora H. J.,Opt. Soc. Am. 65 882 (1975).
16. Max C.E., Arons J., Langdon A.B., Phys. Rev. Letts. 33 209 (1974).
17. Siegrist M.R., Luther-Davies B., Hughes J.L., Optics Comms. 18 603 (1976).
18. Siegrist M.R., to be published in Jrnl. Appl. Phys.

FAST ION AND NEUTRON EMISSION IN CO_2 LASER IRRADIATION OF SOLID DEUTERIUM TARGETS *

J.L. Bócher, J.P. Elie, J. Martineau, M. Rabeau
and C. Patou

C.E.A. - Centre d'Etudes de Limeil
B.P. n° 27 - 94190 - Villeneuve-Saint-Georges, France

ABSTRACT

Experimental results obtained from 10.6 μm irradiation of plane targets have shown a fast ion distribution whose peak occurs at an energy value of a few keW and whose maximum energy is higher than 60 keV. Of the 25 to 30 % of the incident laser energy absorbed by the plasma only 6 to 8 % was directed toward plasma heating. This fact indicates that 60 % of the absorbed energy was carried out by ions with energies greater than 10 keV. For the yield of 10^4 neutrons with 2.45 MeV energy, the fastest of these ions are responsible. Interferometric measurements have shown a significant density profile modification. These results lead to the conclusion that resonant absorption becomes predominant over bremsstrahlung absorption for intensities greater than 10^{13} W/cm^2. The associated ponderomotive force could be considered as a possible mechanism for fast ion generation.

* Presented at the Fourth Workshop on "Laser Interaction and Related Plasma Phenomena" held at Rensselaer Polytechnic Institute, Troy, New York, November 12-18, 1976

I. INTRODUCTION

Large experiments based on the concept of compression have been carried out using neodymium glass lasers ($\lambda = 1.06\ \mu m$) as well as carbon dioxyde lasers ($\lambda = 10.6\ \mu m$) [1-9]. Preliminary experiments indicated that it was clearly impossible to explore every phenomenon that might occur in laser-imploded plasma without dealing with single beam experiments, more accessible for the experimentalists. Under these conditions interaction experiments can bring new and relevant information [10-14] on laser light absorption [15-19], plasma heating [20-21] and thermal conduction[22-24].

Among the parameters which have a real influence on the experimental results, one can mention the nature and geometry of the target, the laser flux and the laser wavelength. If the use of a long wavelength laser seems more appropriate to heat underdense magnetically confined plasmas [25-29] the $10.6\ \mu m$ wavelength cannot be excluded in the laser interaction studies with solid targets. In fact most instability thresholds [30] are reduced by a factor 10^2 - 10^3 when using CO_2 lasers instead of neodymium glass lasers. Consequently with incident CO_2 laser fluxes above 10^{12} W/cm^2 observations of many flux dependent phenomena are possible. When increasing laser flux above 10^{12} - 10^{13} W/cm^2, the electron distribution function is strongly perturbed by the large amplitude of the pump wave. Consequently, most of the invoked classical mechanism involve maxwellian distribution functions no longer apply. Furthermore, these conditions of irradiation can induce a high electrostatic field. The resulting ponderomotive force [31-34] modifies the density profile [35-39] and consequently enhances resonant absorption. The highly non thermal electron distribution plays a major role in the laser plasma coupling efficiency near the critical density [40-41] as well as in heat transport through the underdense and into the overdense plasma.

Our experimental results show the interest of the $10.6\ \mu m$ wavelength to study the consequence of the above considerations such as fast ion generation [42-44] hard X-ray and neutron emission [18] and perturbed density profiles [39] easily attainable at this wavelength by interferometry. The investigations start by an accurate measurement of the energy balance. This information comes from both reflection and charge collection measurements. The observed neutron emission may be correlated with the presence of a significant production of fast ions. The interferometry gives account of the perturbations which appear in the density profile near the maximum of the laser flux. In a second part, we try to interpret these experimental results, showing that at 10^{13} W/cm^2 resonant absorption becomes predominant. Then the associated ponderomotive force can be taken into account in the interpretation of fast ion generation [42-43]. Other possible mechanisms are also mentionned. Numerical calculations describe the plasma hydrodynamic behavior in the presence of fast ion propagation into the dense region [44].

II. EXPERIMENTAL SET-UP AND DIAGNOSTICS

The laser

The nanosecond 10 J, CO_2 TEA laser used for these experiments was developed and set up by C.G.E. The laser consisted of an oscillator delivering 200-300 mJ in 40 ns. This pulse was preamplified up to 600-800 mJ. Then a 1.5 ns short pulse was chopped from the 40 ns pulse. For this purpose we used a double Pockels cell composed of two AsGa crystals placed between crossed polarizers. The laser output energy was about 10 J contained in a 1.7 ns pulse after amplification through double discharge C.G.E. modules [45]. The amplification was performed with three one meter long amplifiers of 4.5 x 4.5 cm^2 aperture followed by three other amplifiers of 7.0 x 7.0 cm^2 aperture. SF_6 cells and p-doped germanium, used as saturable absorbers, contributed to a contrast ratio of 10^5 at the output of the laser. The fragile parts of the laser were protected from the backscattered light by two different systems : an afocal with real focus and Fresnel prisms combined with Brewster angle polarizers. In such a configuration the laser beam was circularily polarized. The beam was focused using a twenty centimeter focal length NaCl lens with an aperture of f/3. The measured focal spot in vacuum did not exceed 120 μm in diameter. This result was in good agreement with numerical calculations including aberrations of the lens. When taking into account optical losses due to beam splitters, window and lens of the interaction chamber, the available energy on the target was about 6 to 7 Joules. Thus the maximum laser flux in vacuum corresponded to 3×10^{13} W/cm^2. We controlled the energy contained in the laser prepulse in the range of less than 1 to 50 mJ by a suppressing SF_6 cell or p-doped Germanium (see fig. 1).

The target and the diagnostics

The experimental set-up is presented in figure 2. The target was a solid deuterium prism with a one millimeter square section. This rod was produced by a classical cryogenic device. The interaction chamber was evacuated to 10^{-7} torr. The target position with respect to the laser beam axis was controlled by an optical system, which looked in two perpendicularly crossed planes. Incident and reflected energies and pulses shapes were recorded around the target at 0°, 45° and 90°. Charge collection measurements were carried out using Faraday cups biased to - 80 V. The particle emission was monitored in differend directions at 25°, 45° and 67° on both sides with respect to the laser beam axis. At 45°, on one side, an electrostatic charged particle analyzer was associated with the collector.

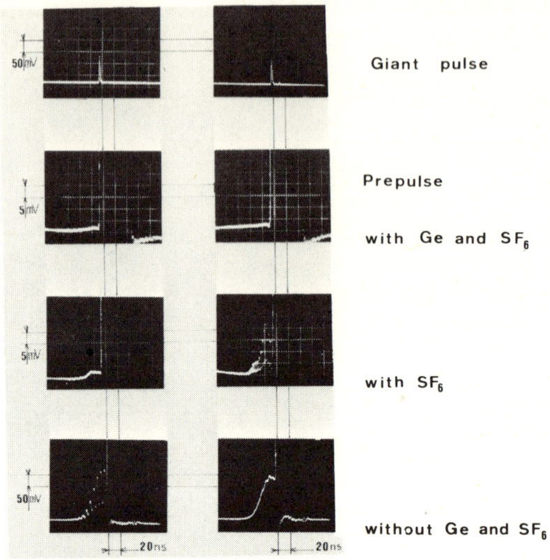

Fig. 1 The laser prepulse measurements

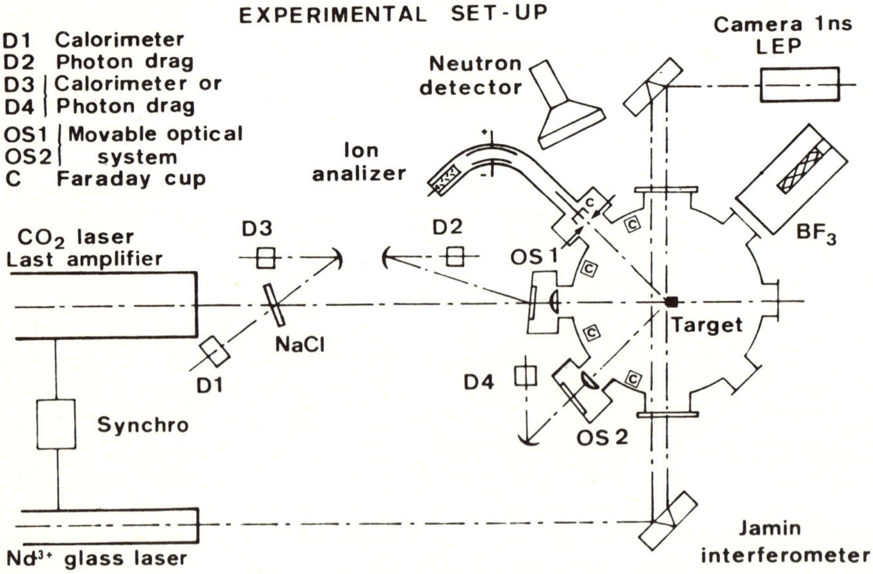

Fig. 2 Experimental set up

Using the second harmonic ($\lambda = 0.53$ μm) of a 30 ns neodymium glass laser connected to a Jamin interferometer, it was possible to obtain the spatial distribution of the plasma density. The interferograms were taken with a camera of a one nanosecond aperture. The spatial resolution of the interferometer was better than 20 μm. The synchronization of the CO_2 and Neodymium glass lasers suffered from a jitter of 60 ns. The camera was triggered by the CO_2 laser pulse. Neutron detection was obtained by one large plastic phosphor detector connected through light pipes with an XP 1040 Radiotechnique photomultiplier. The detector was located at 70 cm from the target and completely shielded by one to three lead sheets with a five millimeter thickness. The signals were recorded on Tektronix type 7904 and 485 oscilloscopes. Taking into account the photomultiplier time delay ($\sim$ 40 ns), the chronology of the X-ray and neutron signals was measured with an accuracy of about 5 ns. The incident laser pulse signal taken at the entrance of the interaction chamber was chosen as the time origin of the events. A second detector was also utilized in order to make neutron time-of-flight measurements. Assuming isotropy for neutron emission, a BF_3 detector was also used to make an order of magnitude estimate of their total number emitted in 4 π sr. The overall efficiency of this detector was determined using a known neutron flux from a Ra-D-Be source. One pulse corresponded to 3000 neutrons at 30 cm in the whole space.

III. EXPERIMENTAL RESULTS

The major part of the experimental work has focused on the study and the understanding of the laser-plasma energy transport mechanisms at 10.6 μm. The energy balance of the interaction was determined by two methods : one consisted of evaluating the total reflection in the whole space. The other way which used Faraday cups to measure number, energy and distribution of the ions, was complementary. Interferometric measurements performed at 0.53 μm gave a more precise idea of the hydrodynamic behavior of the plasma. Hard X-ray and neutron emission measurements brought interesting information for the identification of the processes involved in this interaction.

Reflection

The reflectance through the solid angle of observation was defined as the ratio between the reflected energy and the incident energy. Observations were made using identical optical lenses with the same aperture of f/3. The pulse shape of the reflected light was not very different from the incident laser pulse shape. The percentage of backscattered and/or reflected energy through the focusing lens was about 6 %. This value was reduced to 1 % at 45°. At 90° the reflectance was always lower than 0.5 %. Spatial distribution of the reflection is shown in figure 3. On integrating over the solid angle in the relevant half space, we obtain an order of magnitude estimate for the total reflectance

$$R_t = 2\pi \int_0^{\pi/2} \rho(\theta) \sin\theta \, d\theta$$

where $\rho(\theta)$ is the reflectance per unit of solid angle in the θ direction. Figures 3 and 4 indicate that with increasing incident laser energy, the total reflectance varied from 50 % up to 75 %. The total reflected energy was not modified significantly by the presence of a prepulse (about 50 mJ in 30 ns). However the spatial distribution of the reflectance was more peaked in the direction of irradiation.

Ion time-of-flight measurements

Charges particle collectors associated with an electrostatic charged particle analyzer showed without ambiguity the presence of fast ions with a kinetic energy up to 60 keV, followed by thermal ions corresponding to a plasma temperature of few hundred eV. In the range of 10^{12} to 10^{13} W/cm^2, fast ion emission was always observed. The average value of the total number of collected ions was 5.7 x 10^{15}. The results presented quite good symmetry for ion emission. A total of 1.6 x 10^4 ions had a kinetic energy greater than 10 keV i.e. : about 4 % of the total number of ions. However

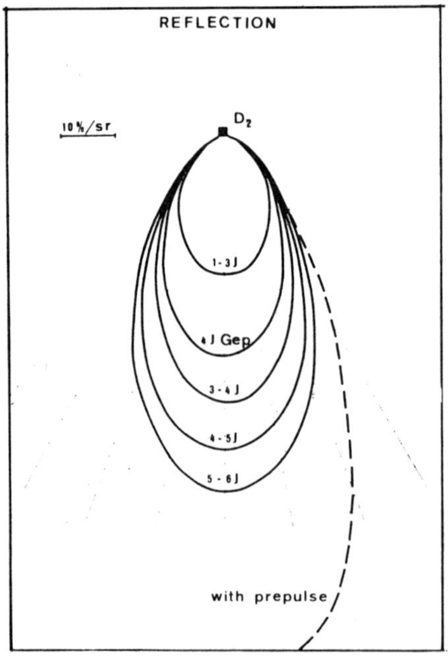

Fig. 3 Spatial distribution of the reflection

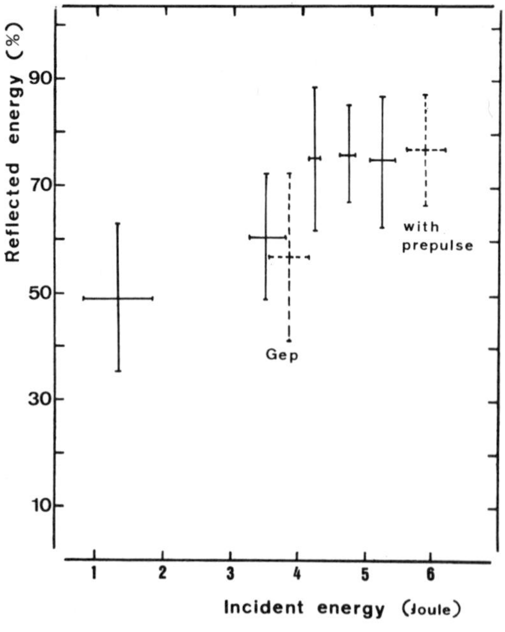

Fig. 4 Reflected energy versus laser energy

these ions carried out 15.4 % of the incident laser energy i.e. : 60 % of the total absorbed laser energy (see fig. 5). In the presence of a 50 mJ - 30 ns laser prepulse we do not observe such fast ions. Figures 6 and 7 show the occurence and the variation of the maximum kinetic energy of the fast ions as a function of the incident laser energy. This occurence cannot be taken as a physical threshold but rather as a minimum limit of experimental detection. The maximum kinetic energy of the fast ions was confirmed by measurements using the electrostatic charged particle analyzer.

Neutron emission

In 80 % of all laser shots, signals from neutron detectors were recorded with deuterium targets. The diagnostics used gave a measurement of the neutron time-of-flight, the relative amplitude of the different neutron signals and the estimate of the total number of neutrons emitted in the whole space. The histogram plotted in figure 8 indicates that most of neutrons were produced during the laser pulse and came directly from the target to the detectors (neutron time-of-flight $\sim$ 35 ns). The other signals recorded after approximately 100 ns were attributed to scattering processes and/or neutron emission due to collisions of fast ions

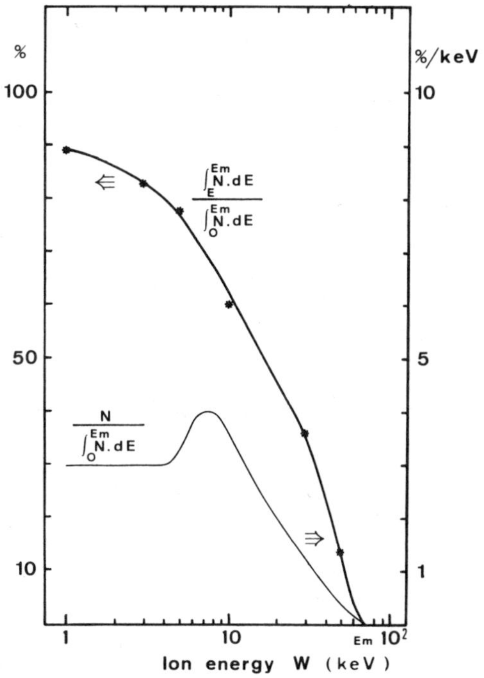

Fig. 5 The percentage of absorbed laser energy carried out by fast ions of a given energy

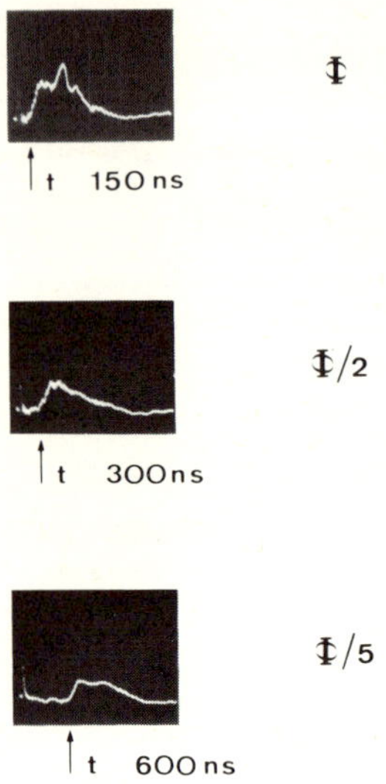

Fig. 6 Occurence of fast ions vs incident laser energy

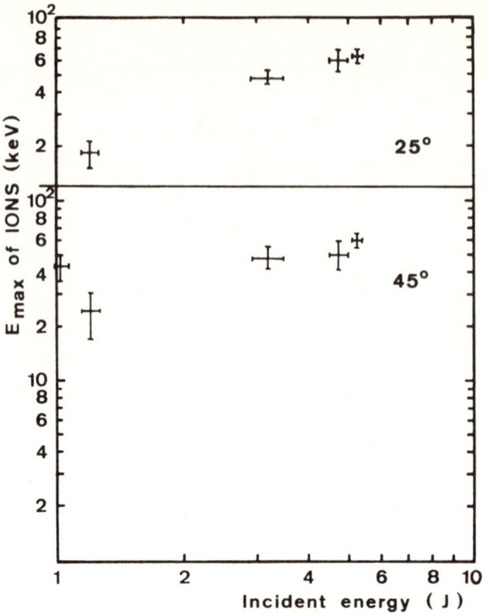

Fig. 7 Maximum kinetic energy vs incident laser energy

with a thin deuterium layer on the chamber wall (for 60 keV ions ; time-of-flight $\sim$ 100 ns from the target to the wall). The dashed rectangle corresponded to hard X-ray emission ($\sim$ 100 keV). Neutron time-of-flight measurements performed by moving one of the plastic scintillator detectors, were consistent with 2.45 MeV neutrons produced by D-D reaction. The total number of neutrons measured by the BF_3 detector did not exceed 10^4 neutrons when assuming isotropy of neutron emission. The experimental set-up allowed the use of deuterium or hydrogen solid targets alternatively. No signal occured when replacing deuterium by hydrogen.

The experimental values for the total D-D cross sections were found to be best fitted by Gamow's formula in the range from 10 to 100 keV [46]

$$\sigma_{D\cdot D}(w) = \frac{288}{w}\exp\left(-\frac{45.8}{w^{1/2}}\right)$$

where $\sigma_{D\cdot D}(w)$ is in barns (10^{-24} cm^2) and the deuteron energy is in keV. From this formula and the ion energy spectrum deduced from the experimental data, we have estimated the number of neutrons produced by collisions on a 100 μm thin deuterium target with a 10^{19} cm^{-3} density, using the relation :

$$\frac{dN}{dt} = N_i(w)\, n_t\, \sigma(w)\, v$$

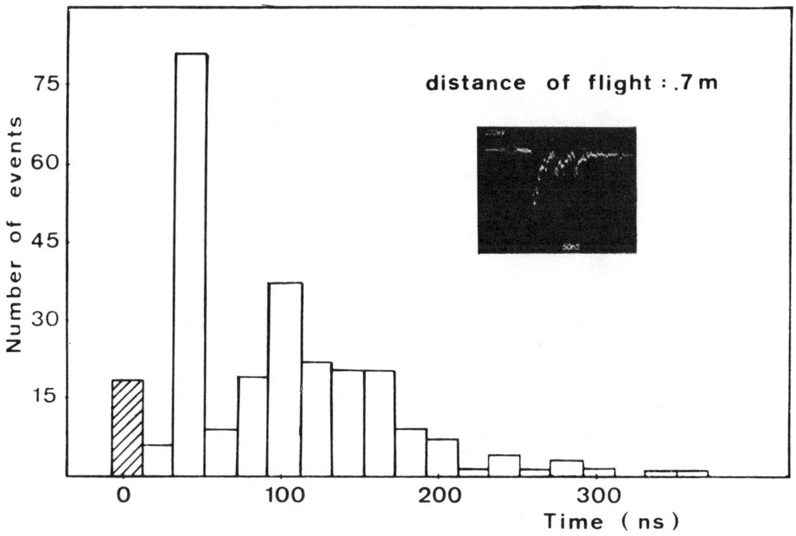

Fig. 8 Histogram of neutrons

where N_i (w) is the number of ions at the energy w , n_t is the target density and **v** the ion velocity. Taking into account the mean free path for a D-D reaction ($\sim 10^5$ m), the fastest ions are responsible for the neutron yield. On integrating the number of neutrons between 10 to 60 keV (which is upper limit in energy, for the collected ions), we have obtained an evaluation of the total number of neutrons emitted in 4 π sr compatible with the experimental result($\sim 10^4$ neutrons).

Interferometry

One of the major problems of this diagnostic was the synchronization of the CO_2 and neodymium glass lasers. If the jitter was about 60 ns, the accuracy of the measurement was within 1 ns. It has been possible to obtain sequences of interferograms built up from single events for the plasma produced from deuterium targets. Three interferograms are presented in figure 9. These interferograms were taken at different times : t = - 3 ns ; t = 0 and t = + 15 ns. The time origin was chosen at the maximum of the laser pulse. Inside the focal spot ($\sim$ 5 mm in the interferograms), the hydrodynamic expansion seems to be planar. The first interferogram correspons to plasma creation by the incoming laser prepulse. The critical density is located 150 μm in front of the target surface. During the laser pulse, near the peak power, we observe an important modification of the density gradient. From the vacuum towards the target, the density increases above the critical density increases above the critical density, now located at 450 μm,

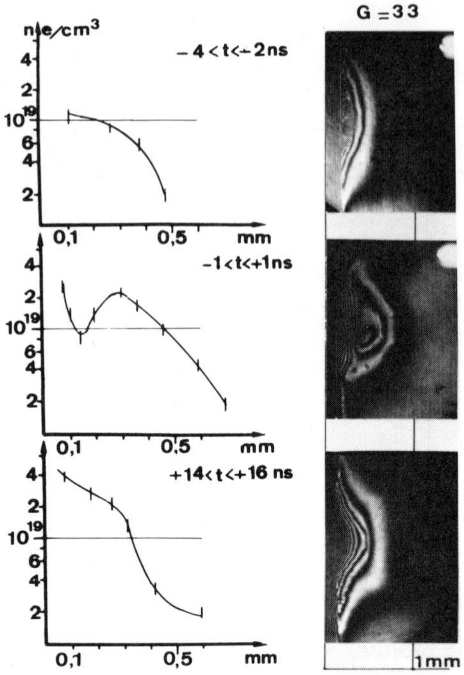

Fig. 9 Interferograms

reaches a maximum value of 2×10^{19} cm^{-3} before decreasing to a minimum and decreasing again. A complete spatial Abel inversion confirmed the presence of a hole of density inside the plasma. A similar phenomenon has been reported by T.P. Donaldson et alii in an experiment using a long CO_2 laser pulse interacting with a carbon target. On the last interferogram, the hole vanished. The density gradient near the critical value becomes steeper.

IV. NUMERICAL CALCULATIONS AND INTERPRETATION

Absorption of the laser light

The incoming laser light which penetrates into the density gradient must be absorbed or reflected at/or below the critical density n_c. This well-defined underdense region between $n_c/4$ and n_c is the so called "absorption zone" and gives the scale length of the density profile. For our experimental conditions, numerical calculations using a one dimensionnal hydrodynamic code as well as interferometric measurements indicate that the density gradient scale length varies constantly. The inverse bremsstrahlung absorption is plotted as a function of time in figure 10. The total

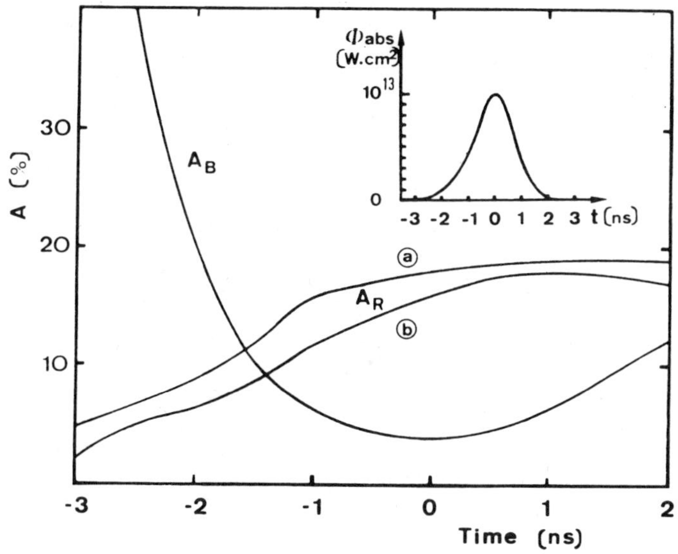

Fig. 10 Resonant and bremsstrahlung absorption as a function of time

time integrated absorption by this mechanism does not exceed 5 % of the incident laser energy. This result is not in agreement with the experimental values which give 25 % of absorbed laser energy. Consequently, inverse bremsstrahlung is not the only mechanism contributing to the laser light absorption.

Resonant absorption

Due to the focusing system, even with a circularly polarized laser beam, there is always a longitudinal component of the electric field. The transverse component is reflected at $n_c = n_c \cos^2\theta$ where θ is the angle of incidence with respect to the density gradient [31]. The electric vector E in the plane of incidence is partially converted into a plasma wave. The wave frequency is precisely equal to the frequency of plasma oscillations at the critical density. The plasma waves carry with them some part of the energy of the standing electromagnetic wave. Ultimately the energy of the plasma waves goes to heat the plasma by dissipation [47]. The total percentage of absorption by this mechanism is obtained by integrating over the solid angle of irradiation

$$A(t) = \frac{1}{1-\cos\theta_{max}} \int_0^{\theta_{max}} A(\theta, t) \sin\theta \, d\theta$$

Where $A(\theta, t)$ is the resonant absorption coefficient, assuming a linear density profile near the critical density. A (t) is represented in figure 10.for two different values of the characteristic scale length. In the early stage of the interaction, the resonant absorption is relatively weak before increasing up to a maximum of 20 % of the incident laser energy. The total time integrated resonant absorption is about 16 to 18 %. Taking into account both bremsstrahlung and resonant absorption, a maximum of 22 % of the incident laser energy is transfered to the plasma. This value is in good agreement with the experimental result.

Fast ion generation

We can admit that most of the absorbed energy is transfered to the electrons. "Hot" electrons corresponding to a very non maxwellian distribution may give their energy to some ions of the plasma. Various mechanisms can lead to hot electron production. In particular :

- Electron temperature can be locally very high because of self focusing of the laser beam, due either to a non uniformity of the beam itself or to electron density fluctuations in the plasma[48-49]
- At high laser fluxes, electrostatic instabilities may enhance locally the transfer to the electrons. This results in a strongly perturbed and non thermal electron distribution
- Density fluctuations as well as self generated magnetic fields may reduce the thermal conduction below the classical value of the flux limiter. Then a certain amount of energy is available near the critical density and may be used to accelerate ions by means of the fast electrons [50-52]
- Resonant absorption plays an important role in the interpretation of the energy transfer in our experiment. In fact the ponderomotive force associated with the resonant field can also be the origin of fast electrons and then fast ions. The mean kinetic energy of fast electrons due to this process is given by

$$< \Sigma > = \left(\frac{m_e \, \omega_0 \, \Phi_a \, L}{n_c} \right)^{1/2}$$

  where Φ_a is the absorbed laser flux and L the characteristic scale length [47]. Di Bona et al. [43] have calculated the fast ion distribution function, taking into account the temporal and the spatial evolution of the incident laser pulse. The resonant field is always located at the critical density. When the fast ions leave the resonant zone, the density varies locally and thus the critical density moves. The mechanism which is invoked here is actually non stationary. However different density profiles have been used in the calculations. Results showed that there was no significant discrepancy in the range of 50 to 300 μm inhomogeneity scale length. Then the theoretical treatment does not include a temporal evolution of the density profile.

Upon traversing the force field, the ions gain the energy Σ defined as

$$n_i(\Sigma - \Sigma_S) = \int_{z_c}^{z_0} \left(-\frac{\partial}{\partial z} \frac{E^2(z)}{16\pi} \right) dz$$

z_0 is the point where $E(z_0)$ is equal to the amplitude of the electric field in vacuum. The model has been compared with our experimental results. Within the framework of these comparisons, the theoretical results obtained corroborate that ion acceleration by the ponderomotive force is able to play a leading part in fast ion generation.

Energy transport

If the electron-electron collision frequency is sufficiently high to thermalize their own distribution, the energy transport can be derived from kinetic equations. The thermal flux is written as

$$\Phi_c = - K_c \nabla T_e$$

where $K_c \propto T_e^{5/2}$ is the classical conductivity. These calculations assume a weakly-perturbed maxwellian electron distribution function. Then the thermal electrons contribute through classical thermal conduction to the transfer of energy to the target. Under these classical conditions, the thermal flux cannot exceed an upper limit

$$\Phi_{LIM} = n_e k T_e \langle v_e \rangle$$

In fact, when intense laser light is absorbed near the critical density, "hot" electrons belonging to a strongly perturbed distribution interact with the plasma. The notion of temperature is no longer valid and the flux limited formula breaks. The energy transport should be insured by a non classical mechanism. To approach this problem, different ways are possible. One consists of matching transport coefficients using a phenomenological dimensionless parameter to give account of the experimental results [53]. This first approach leads to reduce the upper limit of the thermal conduction. This interpretation must be taken delicately. A more realistic and complicated theoretical approach would be a microscopic treatment dealing with a known maxwellian distribution function. This method implies a complete description of the energy transport by "hot" particles which could seriously modify, the laser plasma coupling parameters and which could influence the wavelength scaling laws [54].

In one dimensional numerical hydrodynamic code we have considered the influence of the fast ions on the hydrodynamic behavior. Thermal electrons contribute to electron thermal conduction. We assume that the "hot" electrons transfer their energy to ions via electrostatic field. The rest of "hot" electrons are run away. Experiments on thin targets have shown the possibility for fast electrons to pass through the dense plasma. A similar situation can occur when an overdense bump appears in the density gradient. Then we can consider this bump as a thin layer and allow on purpose to fast ions to propagate on both sides with respect to the critical density plane away and towards the dense region (see figure 9). In the code, we do not presume of the physical mechanism which can generate fast ions. We only assume that they are produced continuously at the critical density with a kinetic energy directly related to the absorbed laser power. Numerical results corresponding to a total absorbed laser energy of 30 % are presented in figure 11. In this particular case, only 15 % of the incident laser energy contribute to direct plasma heating near the critical density. The fast ions which leave the absorption zone carry away 12 % of the incident laser energy while 3 % is available for overdense plasma heating. Effectively, the fast ions propagating towards the solid target are slowed down by cold electrons and ions near the ablation front. Figure 11 represents the density profile at different times. A strong modification of the density profile appears during the rise time of the laser pulse (t = - 1 ns). At the maximum power (t = 0), the density gradient near the critical density becomes steeper. The significant change in density correlated with the mass, momentum and energy transfer of the fast ions, has to be compared to the experimental observations. Numerical calculations taking into account the effective slowing down of the fast ions indicate that they can propagate up to density as high as 5×10^{21} cm^{-3}, where they loose their kinetic energy [55-56]. This mechanism which contributes to wide the electron temperature plateau and to increase the ion temperature is equivalent to by-pass the critical density and to supply the thermal conduction. This phenomenon increases the pressure near the solid density (see fig. 12).

V. CONCLUSION

In this paper, we have evidenced experimentally fast ion and neutron emissions as well as strong perturbed density profile. We have shown that at 10^{13} W/cm^2 interpretations using classical mechanisms are no longer valid. Resonant absorption takes a leading part in the absorbed energy processes. It has been demonstrated that most of the absorbed energy is carried out by fast ions. The neutron yield is essentially due to collisions of the fastest ions with the underdense plasma. The modifications of the density gradient is explained in terms of fast ion generation or radiation pressure.

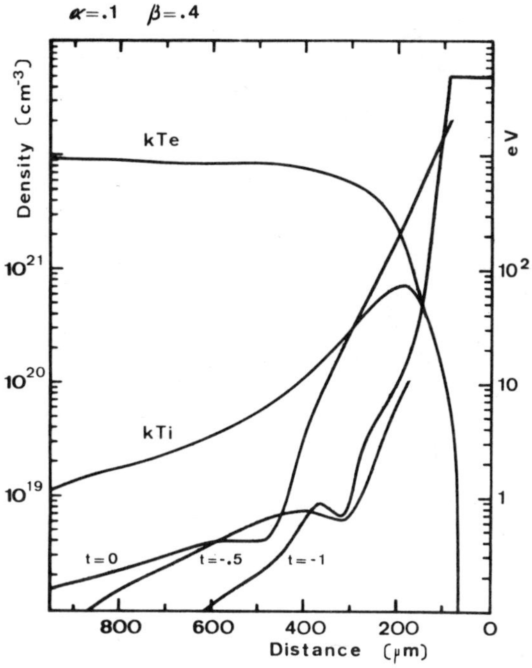

Fig. 11 Density profiles from numerical calculations

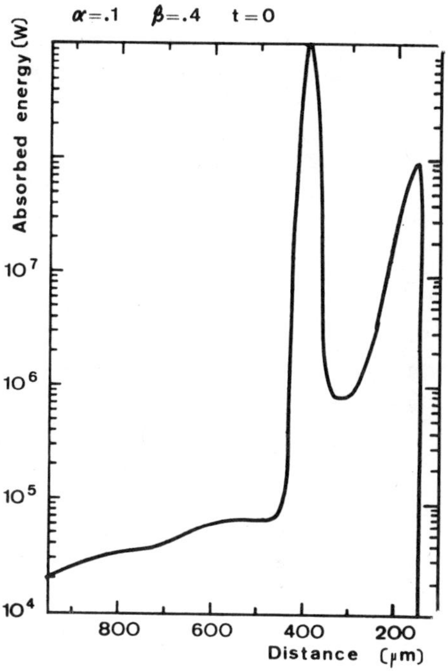

Fig. 12 Spatial distribution of the absorbed energy

ACKNOWLEDGEMENTS

The authors are indebted to Drs. P. Guillaneux, G. Di Bona, D. Babonneau and P. Chelle for helpful discussions. They wish to thank E. Goetschy, G. Niérat and M. Rostaing for their technical assistance in the experiment and C. Jourdain for his contribution to laser working. They thank A.G. Engelhardt and C. Carman for constructive remarks.

REFERENCES

1. J. Nuckolls, L. Wood, A. Thiessen, G. Zimmerman, Nature 239, 139 (1972)

2. G. Charatis, J. Downard, R. Goforth, B. Guscat, T. Henderson, S. Hildum, R. Johnson, K. Moncur, T. Leonard, F. Mayer, S. Segall, L. Siebert, D. Solomon and C. Thomas.- Proceedings 5th IAEA - Plasma Fusion Conference, Tokyo, Japan (1974)

3. E.K. Storm, H.G. Ahlstrom, M.J. Boyle, D.W. Phillion, V.C. Rupert, K.G. Tirsell, D.R. Speck, F. Rainer, L.W. Coleman, R.A. Lerche, H.N. Kornblum, LLL UCRL 78581 (1976)

4. N.G. Basov, E.G. Gamaly, O.N. Krokhin, Yu. A. Mikhailov, G.V. Sklizkov, S.I. Fedotov, Third Workshop on "Laser Interaction and Related Plasma Phenomena", Ed. by H. Schwarz and H. Hora, Plenum Press, Vol. 3B, 553 (1974)

5. T.F. Stratton, G.F. Erickson, C.A. Fenstermaker and E.O. Swickard, I.E.E.E. J. Quantum Electron. QE-9, 157 (1973)

6. J. Soures, L.M. Goldman and M. Lubin, Third Workshop on "Laser Interaction and Related Plasma Phenomena", Ed. by H. Schwarz and H. Hora, Plenum Press, Vol. 3B, 607 (1974)

7. D. Billon, D. Cognard, P.A. Holstein, J. Launspach, C. Patou, J.M. Reisse, D. Schirmann, Fourth Workshop on "Laser Interaction and Related Plasma Phenomena" at RPI, Troy, New-York, November 8-12 (1976) These Proceedings, pp. 503-534

8. C. Yamanaka, M. Yokoyama, S. Nakai, T. Yamanaka, Y. Izawa, Y. Kato, T. Sasaki, T. Mochizuki, Y. Kitagawa, M. Matoba, K. Yoshida, J. Mizui and N. Yamaguchi, IAEA, Berchtesgaden, W. Germany, October 6-13 (1976)

9. S. Kaliski, Third Workshop on "Laser Interaction and Related Plasma Phenomena", Ed. by H. Schwarz and H. Hora, Plenum Press, Vol. 3B, 495 (1974)

10. F. Floux, D. Cognard, L.G. Denoeud, G. Piar, D. Parisot, J.L. Bobin, F. Delobeau and C. Fauquignon, Phys. Rev. A1, 821 (1970)

11. J.L. Bobin, M. Decroisette, B. Meyer and Y. Vitel, Phys. Rev. Lett. 30, 594 (1973)

12. A. Salères, Thesis to be published Orsay (1977)

13. J.A. Stamper, K. Papadopoulos, R.N. Sudan, S.O. Dean, E.A. Mc Lean and J.M. Dawson, Phys. Rev. Lett. 26, 1012 (1971)

14. M.G. Drouet and R. Bolton, Phys. Rev. Lett. 36, 591 (1976)

15. A.W. Ehler, J. Appl. Phys. 46, 2464 (1975)

16. J. Martineau, P. Paranthoën, M. Rabeau, C. Patou, Opt. Comm. 15, 404 (1975)

17. H.A. Baldis, H. Pépin and T.W. Johnston, Opt. Comm. 15, 311 (1975)

18. J. Martineau, M. Rabeau, J.L. Bocher, J.P. Elie, C. Patou, Opt. Comm. 18, 347 (1976)

19. K. Eidmann and R. Sigel, Third Workshop on "Laser Interaction and Related Plasma Phenomena", Ed. H. Schwarz and H. Hora, Plenum Press, Vol. 3B, 667 (1974)

20. P.E. Dyer, D.J. James, G.J. Pert, S.A. Ramsden and M.A. Skipper Third Workshop on "Laser Interaction and Related Plasma Phenomena" Ed. H. Schwarz and H. Hora, Plenum Press, Vol.3A,191(1974)

21. E. Fabre, C. Stenz, Phys. Rev. Lett. 32, 823 (1974)

22. J.S. Pearlman and J.P. Anthes, Appl. Phys. Lett. 27, 581 (1975)

23. D.V. Giovanielli, J.F. Kephart and A.H. Williams, J. Appl. Phys. 47, 2907 (1976)

24. R.A. Haas, W.C. Mead, W.L. Kruer, D.W. Phillion, H.N. Kornblum, J.D. Lindl, D. Mc. Quigg, V.C. Rupert and K.G. Tirsell, LLL UCRL 77930 (1976), to be published in Phys. of Fluids

25. J.M. Dawson, R.E. Kidder, A. Hertzberg, G.C. Vlases, H.C. Ahlstrom and L.C. Steinhauer, IAEA, Vienna 1, 673 (1971)

26. J. Martineau, Phys. Fluids 17, 131 (1974)

27. P.A. Pincosy, P. Bournot, D. Dufresne, J.P. Caressa, M. Autric, Applied Physics Lett. 28, 583 (1976)

28. S. Denus, Z. Dzwigalski, J. Farni, S. Kaliski, M. Kielesinski, J. Kostecki, J. Kudicki, S. Nagrava, J. Wolowsli, E. Woryna, Preprint, Institute of Plasma Physics, Varsaw, Poland (1976)

29. P.A. Holstein, CEA Report 4721 (1976)

30. M. Porkolab, Physica 826, 86 (1976) and references there in

31. V.L. Ginzburg, Pergamon Press, 2nd Edition, 260 (1970)

32. A.Y. Wong, R.L. Stenzel, Phys. Rev. Lett. 34, 727 (1975)

33. H. Hora, Phys. Fluids, 12, 182 (1969)

34. H. Hora, 2nd Workshop on "Laser Interaction and Related Plasma Phenomena" Ed. by J. Schwarz and H. Hora, Plenum Press (1972)

35. K. Brueckner, Preprint to be published

36. T.P. Donaldson and I.P. Spalding, Phys. Rev. Lett. 36, 467 (1976)

37. O. Krokhin, Yu. A. Mikhailov, A. Rupasov, G. Skliskov, A. Shikanov, Yu. A. Zakharenko, N. Zorev, ICPIG, 349 (1975)

38. M.C. Richardson, N.H. Burnett, H.A. Baldis, G.D. Enright, R. Fedosejevs, N.R. Isenor and I.V. Tomov, Fourth Workshop on "Laser Interaction and Related Plasma Phenomena " at RPI, Troy, New-York, November 8-12 (1976) These Proceedings, pp. 161-180

39. J. Martineau, J.L. Bocher, J.P. Elie, M. Rabeau, European Conference on Laser Interaction, Palaiseau, France (1976)

40. R.L. Morse and C.W. Nielson, Phys. Fluids 16, 909 (1973)

41. K. Brueckner and S. Jorna, Rev. of Mod. Phys. 46, 325 (1974)

42. V.P. Silin, Sov. Phys. JETP Lett. 21, 152 (1975)

43. D. Babonneau, G. di Bona, P. Chelle, M. Decroisette and J. Martineau, Phys. Lett. 57A, 247 (1976)

44. J.P. Elie, D. Babonneau, J.L. Bocher, J. Martineau and M. Rabeau, Internal Report, Unpublished (1976)

45. R. Dumanchin, M. Michon, J.C. Farcy, G. Boudinet and J. Rocca Serra, IEEE J. Quant. Electron. QE-8, 163 (1972)

46. S. Glasstone, R.H. Lovberg, Controlled Thermonuclear Reactions 20, D. Van Nostrand Company Inc. (1960)

47. J.P. Friedberg, R.W. Mitchell, R.L. Morse and L.I. Rudsinski, Phys. Rev. Lett. 28, 795 (1972)

48. R.A. Haas, M.J. Boyle, K.R. Manes, E. Swain, Journal of Applied Physics 47, 1318 (1976)

49. J.W. Shearer and J.L. Eddleman, Phys. Fluids 16, n° 10, 1753 (1973)

50. J.A. Stamper and D.A. Tidman, Phys. Fluids 16, 2024 (1973)

51. J.J. Thomson, C.E. Max and K. Estabrook, Phys. Rev. Lett. 35, 663 (1975)

52. X. Fortin, Internal Report, Unpublished (1976)

53. R.C. Malone, R.L. Mc Crory and R.L. Morse, Phys. Rev. Lett. 34, 721 (1975)

54. D.V. Giovanielli, J.F. Kephart, G.H. Mc Call, IEEE on Plasma Science, Austin, U.S.A. (1976)

55. I.P. Shkarosfky, T.W. Johnston, M.P. Bachinsky, "The particle kinetics of Plasmas", Ed. Addison-Wesley (1966)

56. D.V. Sivukhin, Reviews of Plasma Physics, Edited by M.A. Leontovich, Consultants Bureau, New-York, London (1970)

MEASUREMENT OF HIGH-ENERGY CHARGED PARTICLES FROM LASER-PRODUCED PLASMAS*

Tai Ho Tan, G. H. McCall, A. Williams,
and D. Giovanielli

University of California
Los Alamos Scientific Laboratory
Los Alamos, New Mexico 87545

High-energy charged particles are copiously produced from the interactions of high power laser beams and matter. These particles may carry with them the bulk of the energy involved as well as the relevant information concerning the mechanisms and the history of the interaction processes.

Two types of high-energy charged particles of very different origin will be discussed here. First, we will discuss briefly some measurement of alpha particles and protons in the MeV region which are products of nuclear fusion reactions. The present level of fusion yield is around 10^7 to 10^8 from the DT-filled microballoon targets. Measurements of these particles could yield valuable information on the degree of compression, fuel temperature, and other relevant data. Second, we will present some recent measurement of very fast ions whose velocity is larger than 5×10^8 cm/s and electrons above 100 keV in energy. The emphasis of this paper will be on the many surprising features of the ultra-fast ion measurement which have until now escaped detection. These measurements are done with glass microballoon targets, flat foils of various thickness, and double foils of different separation. These experiments are carried out using both Nd:glass laser and CO_2 laser. In the case of single and double flat foil targets, both the front and back emission are measured. Angular distribution information is also obtained by tilting the flat targets at different angles with respect to the laser beam direction. These measurements yield much needed information on the general features of laser-target interactions, laser wavelength effects, laser energy absorption, thermal conduction, energy transport, target preheating, etc., which ultimately lead to a clearer understanding of the mechanisms involved in laser-induced fusion.

NUCLEAR FUSION PRODUCTS

Let us begin by summarizing some measurement of protons and alpha particles produced from d-d and d-t reactions, respectively. The detection of these particles was done by using the time-of-flight thin film detector illustrated in Fig. 1. The detail of this detector has been published.[1] These experiments were carried out at KMSF by the LASL group, under an ERDA contract.[2] A "clam shell" focusing system was employed.[3] The laser pulse was composed of a train of one to six pulses with a nominal duration of 30 ps per pulse. The maximum intensity on target was estimated at approximately 3×10^{15} W/cm^2. The targets used were DT-filled glass microballoons approximately 50 μm in diameter with a wall thickness of 0.7 μm. The solid angle subtended by the detector was 9.4×10^{-3} steradian and, typically, 1000 - 5000 alpha particles per shot were collected. A typical alpha-particle spectrum showing the background below 2 MeV is given in Fig. 2(a) and a linear plot of the distribution function above 2 MeV is shown in Fig. 2(b) indicating a maximum at around 2.5 MeV. A maximum energy shift of 1.5 MeV for the alpha particles gives a ρR value of 1.4×10^{-3} g/cm^2. The measured energy shift for the proton was about 160 ± 40 keV and ρR of $(2 \pm 1) \times 10^{-3}$ g/cm^2 was obtained. Fuel densities of 0.6 to 2.0 g/cm^3 are implied which indicates a compression of 1000 times. The ratio of d-t to d-d reactions was also determined to be 80 ± 20 which implies the fuel ion temperature to be 2 to 6 keV. By comparison with the calculation of Fraley and Mason,[4] the experiments indicate that optimum conditions are not achieved and effects from radiative and suprathermal electron preheating may be fairly serious. The experimental ρR value is smaller and the fuel temperature is much higher than those predicted.

It is interesting to note that a similar measurement was carried out by the KMSF group.[5] The results indicate a smaller shift than ours. However, their solid angle of detection is much smaller and an energy cut-off in the spectrum was introduced due to dispersion in the magnetic field. Furthermore, some differences may also be expected since their laser intensities were several times higher. Without the precise knowledge of the electron temperature and the energy loss in plasmas, the estimated values of ρR must at best be considered preliminary. Hence, comparison of data under various conditions of laser-matter interactions must be handled with care. In any case it is obvious that the present fusion yield of 10^7 - 10^8 is very far from the break-even requirement to achieve our energy goal. The necessity of employing megajoules from higher power lasers may compel us to use only those relatively high efficiency laser systems such as CO_2 or HF lasers. It is therefore imperative that the wavelength effects in laser fusion be thoroughly understood. Just as important, the problems of electron preheating the target core must be overcome in order to achieve the desired degree of compression for efficient nuclear fusion. Finally, it is extremely relevant to understand the mechanisms that lead to

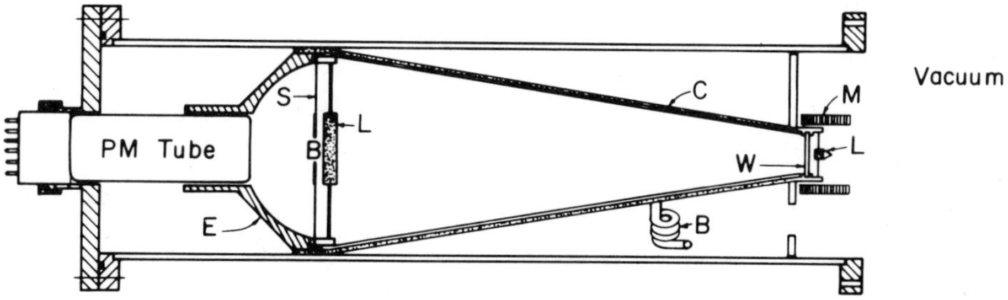

Fig. 1.
Thin-film detector assembly employing very fast plastic scintillator and photo multiplier tube and highly efficient light reflector.

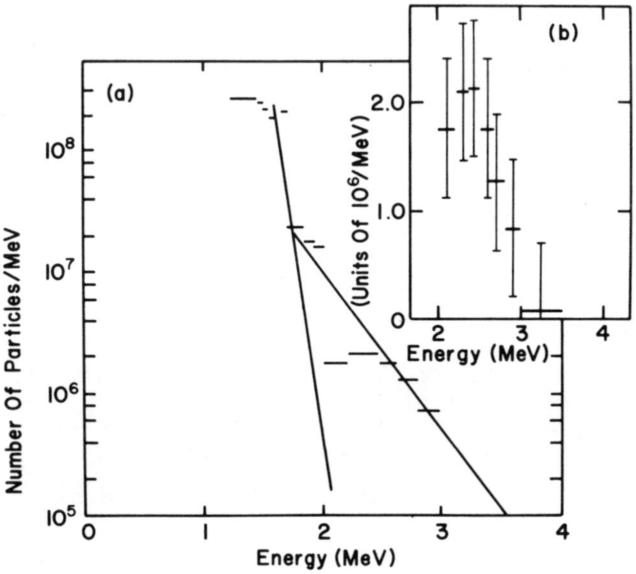

Fig. 2.
(a) Typical alpha-particle spectrum showing the background below 2 MeV. (b) Alpha-particle energy distribution above 2 MeV.

those observed fusion reactions even at the present laser intensity level of $\leq$ 1 TW. Toward these ends, we turn next to the discussions of fast ion and electron measurement.

FAST IONS AND ELECTRONS

The detectors are similar to the TOF detector shown in Fig. 1 with the exception that the light window is a 4000-Å-thick titanium layer coated directly on the 8-μm-thick scintillator. Only one laser beam is used for these experiments.

Figure 3 shows four typical ion spectra emitted from the laser-induced plasmas as observed from the front and back sides of the target. [(a) and (b) are front and back emission from a Nd:glass laser shot, 70 ps and $\sim$ 1 J in a 60 μm diameter; (c) and (d) are front and back emission from a CO_2 laser shot, 1.2 ns, 150 J in $\sim$ 150 μm diameter.] The most striking feature is the steep rise in the leading edge of the ion spectrum. Several important correlations were observed for the front emission alone (front is the laser side of the target).

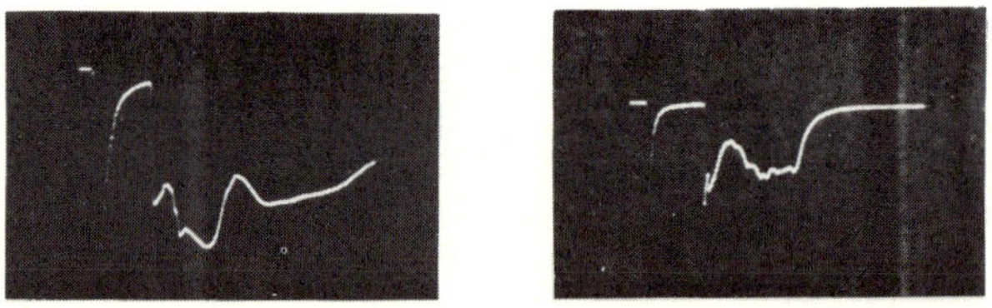

(a) FRONT (b) REAR

ND: GLASS LASER, 70 ps, 1.5 J, 60 μm DIAMETER SPOT, 5000 Å THICK CELLULOSE ACETATE TARGET.

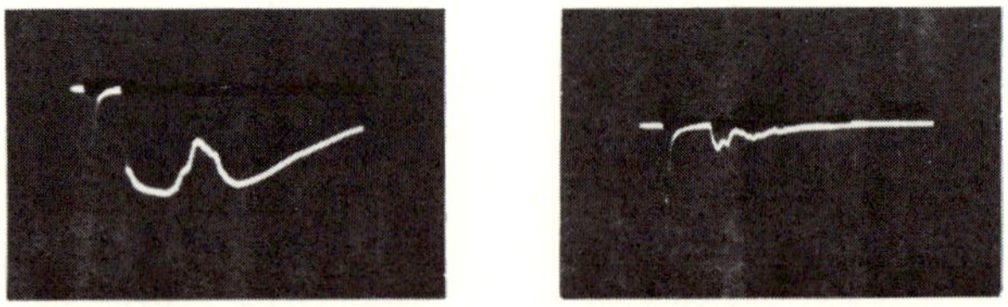

(c) FRONT (d) REAR

CO_2 LASER, 1.2 ns, 150 J, 150 μm DIAMETER, 50 μm THICK MYLAR TARGET.

Fig. 3.

Oscilloscope traces of ion velocity spectrum from a Nd:glass laser shot and a CO_2 laser shot. The first pulse represents a time fiducial marker from x-rays.

1. Its presence implies a steep rising edge in the laser pulse. Adjustment of the quantity of saturable absorber in the laser system severely alters this feature. The presence of any prepulse will wipe out this feature.
2. It is directly dependent on the focused intensity on the target.
3. With the ball-and-disc target, it correlates with implosion observed by x-ray pinhole cameras. A similar co-relation was observed with the number of produced neutrons.

The angular distribution for both the front and the back emission appears to be within a 25° half angle cone, which is similar to the previously reported ion angular distribution for lower energy ions. With the intensity in the neighborhood of $\lesssim 10^{15}$ W/cm^2 for Nd:glass and CO_2 lasers, the observed fastest velocity is typically in the range of 0.8 to 1.0 x 10^9 and 0.8 to 1.5 x 10^9 cm/s, respectively, for the front emission.

In a given laser energy range (within a factor of 3 or 4), the observed maximum velocity appears to be only weakly dependent on the amount of absorbed energy. However, the number of the fastest ions seems to increase proportionally with laser energy. It is particularly interesting to note that the fast ion components behave very similar to the energy absorption pattern recorded by the calorimeters when targets were illuminated under different angular orientation with respect to the polarization direction of the laser beam. Obviously, these fast ions carried away a great bulk of the total absorbed energy for a given unit of laser energy on focus. The number of the highest energy ions appears to be somewhat larger for the CO_2 shots. However, plasma evolution due to a much larger exposure time to the laser beam is expected to play a role.

We have also observed that the value of the maximum velocity as well as the spectrum shape in the front emission from the single flat foil target is independent of the foil thickness. This feature enables us to make meaningful comparison of front and back emissions as a function of foil thickness while keeping the laser energy fairly constant. The ratio of the ion velocity from the fastest component between the rear and the front emission is plotted in Fig. 4. We see the curve dropping rapidly in the vicinity of a foil thickness of 7 or 8 mg/cm^2. This foil thickness is much too large for the observed ions to propagate through (ion average energy > 1.5 MeV would be necessary). Unless some extremely large electric fields are set up inside the target to accelerate and decelerate the same particles as it propagates from the region where laser energy is absorbed to the rear surface, it will be more reasonable to assume that ions are accelerated off both faces of the target. Propagation of hot electrons through the target could set up such conditions. Another supporting evidence for this can be derived as follows. We notice that the fastest emitted ions both in the front and the rear are protons. This was done by comparing the observed dE/dx of the ions passing through foils of different thickness with the known table. Shots from plastic foil target (hydrogen intensive) and platinum foil target (in principle no hydrogen but may acquire some from the pump oil on both surfaces) of the same area density give similar results. In the case of platinum target, the only way the protons at the rear surface can acquire energy is from electrons that stream through the target.

It is obviously of interest to understand the abrupt appearance of the fastest ion component from the plasmas. A model based on an isothermal expansion of a collisionless plasma[6] appears to provide a reasonable description of such behavior. Furthermore, an electron temperature in the neighborhood of 10 keV can be inferred from our observed maximum velocity of the front emitted protons.

It is again worth pointing out that the rising edge of the fast ions generally appears steeper in the rear side component then those in the front emission. Also, in Fig. 4, it is interesting to note that the velocity of the fastest rear emitted ions is greater than those from the front when the targets are between 0.02 to 0.1 mg/cm^2 in thickness. When the target becomes thinner, however, the rear side ion velocity seems to slow down. This effect can be explained by considering the momentum given to the entire target by the laser radiation pressure.

The abrupt falling off in the back emission in the double foil data illustrates the insulation property of the vacuum separating the foils. It will be discussed later.

Figure 5 shows the plot of the ratio of the number of fastest ions emitted from the rear surface to those emitted from the front surface vs the target thickness. If we assume that the energy carried by the fastest ions is proportional to the amount of energy remaining in the hot electron distribution after it traverses from the front to the rear of the target then we can compare the electron deposition in the target to the number of ions as a function of that electron temperature. In a simple model, assuming the target material behind the absorption region to be field free and cold, we can use the stopping power of electrons in matter to determine the fraction of a given Maxwellian electron distribution that would not be absorbed in a single pass through the target foil. Curves for energy remaining in the electrons at the rear of the foil for initial temperatures of 5, 10, and 20 keV are plotted on the figure. It is clear that an electron temperature in the neighborhood of 10 keV appears to agree well with the experimental data. It is worth pointing out the agreement with the implied temperature resulted from the isothermal expansion model, as mentioned a while ago. It is also worth noting that previous measurement of high-energy x-ray spectra[7] from both CO_2 and Nd:glass lasers suggested hot electron temperatures of the order of 15 keV. If not by accident this may shed some light on the mechanisms involved during the energy transport.

So far, we have noted great similarity between the data from both the CO_2 and Nd:glass laser shots. Manifestation of laser wavelength dependence, is weak.[8]

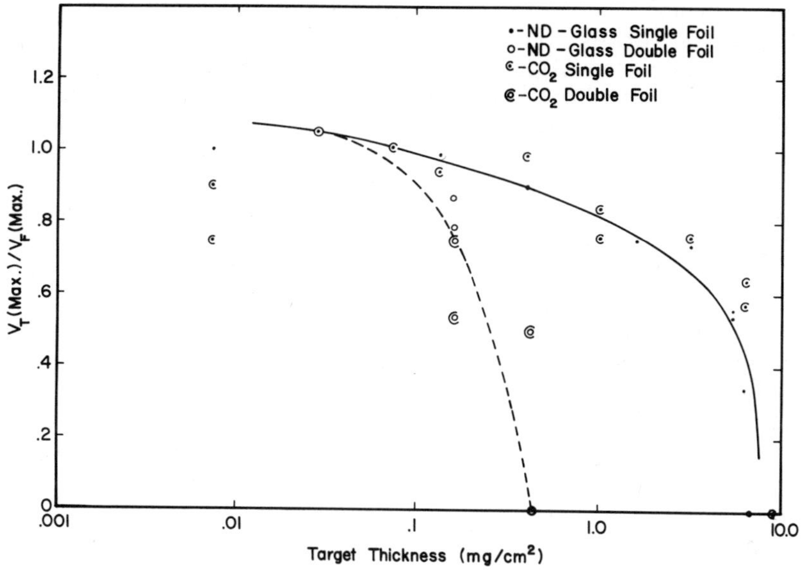

Fig. 4.

Ratio of the fastest rear emitted ion velocity to the fastest front emitted ion velocity plotted against flat target area density.

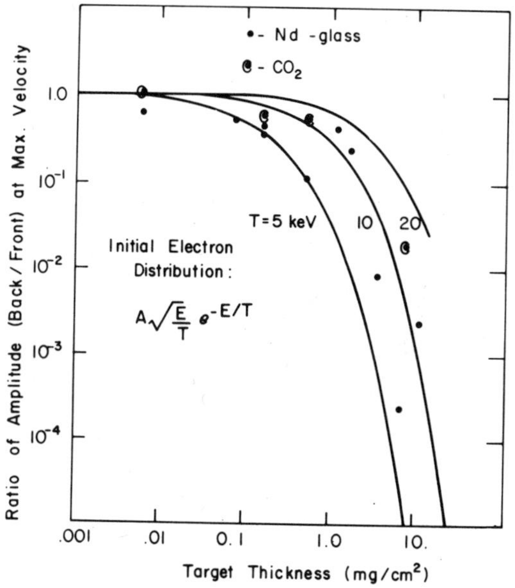

Fig. 5.

Ratio of the number of the fastest rear emitted ions to the number of the fastest front emitted ions plotted against target area density.

Hot electrons have been observed to leave the target freely.[9] Figure 6 shows a series of electron spectra with energies greater than 100 keV. For the highest energy components, the electron "temperature" appears to be about 100 keV, the same for both CO_2 and Nd:glass. Understanding of these energetic electrons may be complicated by the absence of any knowledge on the large time-dependent potential. In any case only a small fraction (< 1%) of the absorbed incident laser energy is re-emitted in the form of very energetic electrons. If one could limit the electrons which preheat the target core only to those which escape the absorbing plasma, its effects on subsequent target compression may not be too serious. How these energetic electrons are related to the observed fast ions and the roles they play in plasma evolution can not be ascertained at the present time.

Because of their large number and their shorter range, electrons in the tens of keV regime are expected to play a significant role both in thermal conduction and energy transport, as well as pre-heating the target interior. In the next set of data, we shall address ourselves to the crucially important preheat problems.

If we can prevent the energetic electrons from penetrating the target core by erecting an effective barrier, then we might indeed be able to reduce the amount of target core preheating.[10] Double plastic flat foils separated by a vacuum were used as target as shown in Fig. 7(a). Here F_1 is a foil thin enough to insure the same front and back ion emission while making sure that no laser

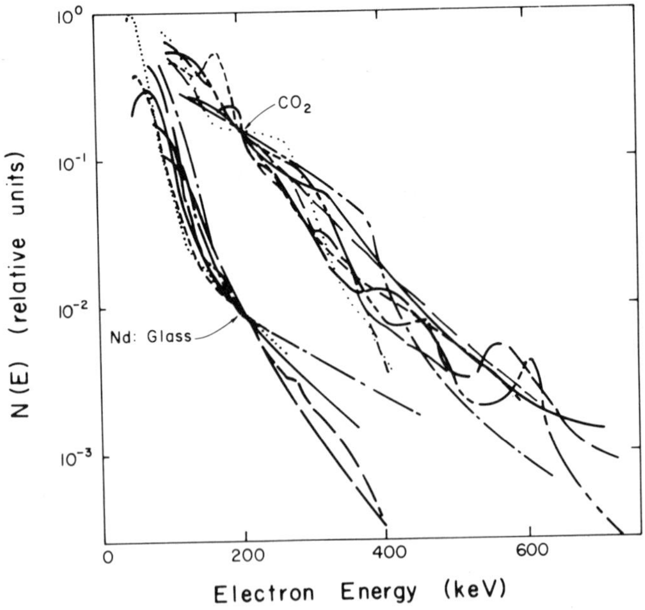

Fig. 6.
Hot electron energy distribution.

light is transmitted through. For the Nd:glass laser, 1000-Å-thick foil was found to be suitable for the 70 ps pulse irradiation while 2.5-µm-thick foil is needed for a 1.2 ns pulse. For a 1.2 ns CO_2 laser pulse, a 500 Å or thinner foil can be used. However, we also found 1000 Å to be suitable. The second foil is generally chosen to be 3.75 µm thick so that hydrogen ions of greater than 300 keV in energy can successfully penetrate the foil and still be energetic enough to allow detection. In the same figure, data from CO_2 and Nd:glass experiments are shown. The pulse lengths for both lasers were 1.2 ns FWHM. However, the Nd:glass laser intensity was at least a factor of 2 less than the CO_2 intensity. Both the front and the rear emitted ions were measured as a function of the foil separation. Again, by considering only the fastest ion components, the ratio of the velocity and the ion number are plotted in Figs. 7(b) and 7(c), respectively.

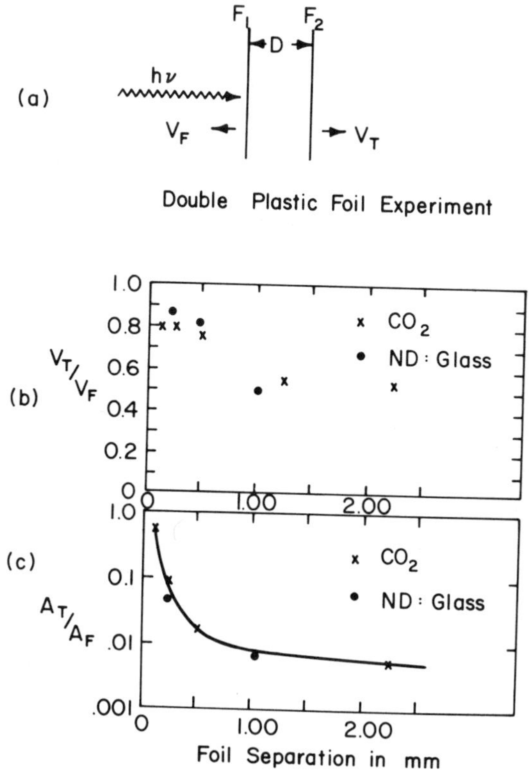

Fig. 7.
(a) Double foil target configuration. (b) and (c) data of the ratio of velocity and ion number, respectively, as a function of foil separation from CO_2 and Nd:glass experiments.

The velocity ratio is expected to be greater than .9 when the foil separation is zero. However, we observed this value to be slightly less as soon as the separations become non-zero and interesting enough it remains fairly constant even as the distance of separation increases. These behaviors seem to imply plasma expansion into the vacuum, but the steady current is maintained by continuous laser irradiation. As the distance of separation approaches 1.0 mm the rear ion velocity drops abruptly and approaches asymptotically to about 0.55 the value of the initial velocity immediately after the ions left the back side of the first thin foil. The corresponding energy of those back emitted ions appears to be equal to or lower than the value that can be accounted for by the simple stopping power of the cold material. The corresponding distribution for the ratio of the rear-front ion number shows the number of ions observed behind the back of the second foil begins to decrease sharply as soon as the separations between the foils is non-zero. Again, this number ratio levels off rapidly as the distance of separation approaches 1.0 mm. These data strongly imply that the vacuum provides fairly effective insulation against the electrons in the 10 keV region.

If we take the effective separation distance to be 1.0 mm, then for a 1.2 ns laser pulse length, the effective electron shorting velocity can be estimated to be about 8×10^7 cm/s. It is worth pointing out that we have previously estimated the electron temperature to be a few keV from the single foil experiments in order to explain the ion measurement. If the ion-temperature is assumed to be equal to the electron temperature, then the implied hydrogen plasma ion velocity agrees with the shorting velocity estimated here.

CONCLUSION

Several important conclusions can be made from the results presented in this paper:

(1) The sharp rising ion velocity distribution and the distinct cut off velocity are inherent features of high power laser-matter interactions.

(2) No obvious laser wavelength effects can be detected from the observed interactions.

(3) The bulk of the energy transported into the interior of the target is probably via the fairly energetic electrons with temperature in the 10 keV region.

(4) Vacuum insulation may effectively reduce the amount of suprathermal electron preheating.

ACKNOWLEDGEMENT

Material presented in this paper represent efforts contributed by many people at LASL. This work performed under the auspices of the U. S. Energy Research and Development Administration.

REFERENCES

1. Tai Ho Tan, A. H. Williams, and G. H. McCall, Nuclear Instruments and methods 131, 425 (1975).

2. G. H. McCall, Tai Ho Tan, A. H. Williams, "Measurements of Alpha Particles and Protons from a Laser-Imploded Target," Los Alamos Report No. LA-UR-76-2138.

3. C. E. Thomas, Appl. Opt. 14, 1267 (1975).

4. G. S. Waley R. Mason, Phys. Rev. Lett. 35, 520 (1975).

5. R. R. Goforth, F. Mayer, H. Brysh, R. Cover, 1976 IEEE Int. Conference in Plasma Physics, Austin, Texas, Paper 3A4 (1976).

6. J. E. Crow, P. L. Aner, J. E. Allen, J. Plasmas Phy. 14, 65 (1975).

7. J. Kephart, R. P. Godwin, and G. H. McCall, Appl. Phys. Lett. 25, 108 (1974).

8. Tai Ho Tan, G. H. McCall, A. Williams, D. Giovanielli, R. Godwin, Bull. Am. Phys. Soc. 21, 2A-13, 1047 (1976); Bull. Am. Phys. Soc. 21, 2A7, 1047 (1976).

9. D. Giovanielli, J. F. Kephart, and A. H. Williams, Journal of Applied Physics 47, 2907 (1976).

10. G. H. McCall, T. H. Tan, A. Williams, Bull. Am. Phys. Soc. 21, 2A-14, 1047 (1976); J. Kindel, D. Forslund, S. Gitomer, E. Lindman, Bull. Am. Phys. Soc. 21, 3A-12, 1067 (1976).

*Presented at the Fourth Workshop on "Laser Interaction and Related Plasma Phenomena," held at the RPI, Troy, New York, November 8-12, 1976.

ON THE DIAGNOSTIC OF NON-LINEAR EFFECTS *

J.L. Bobin

C.E.A. - Centre d'Etudes de Limeil
Boite Postale n° 27
94190 Villeneuve-Saint-Georges

ABSTRACT

Various methods capable of evidencing possible instabilities in laser plasma interaction are reviewed : direct absorption measurements, density profile modification, X-ray emission, forbidden and satellite lines, harmonic generation. A special emphasis is put on the last topic. The connexion of 3/2 ω_L emission with the two plasmon decay is investigated. The effects of various coupling schemes on the line profile of this line and the 2nd harmonic are presented. The available experimental data are reviewed : observed harmonic lines, spatial origin of such lines, intensity dependence, line shapes.

1. INTRODUCTION

Applications of laser interaction with plasmas such as inertially confined thermonuclear devices, require high intensities. The thermonuclear efficiency (fraction of burned T in a DT sphere) is proportional to the product of the density ρ times the radius R of the compressed sphere[1]. Furthermore one has[2]

(1.1) $$(\rho R)^2 \propto P_L \eta_A \eta_I$$

where P_L is the laser power, η_A the absorption efficiency and η_I

*Presented at the Fourth Workshop on "Laser Interaction and Related Plasma Phenomena," held at Rensselaer Polytechnic Institute, Troy, New York, November 12-18, 1976.

the implosion efficiency, ratio of the absorbed power to the mechanical power effectively applied on the compressed core. It can be shown that[3]

(1.2) $$\eta_I \propto R_c^{-2}$$

in which R_c is the critical radius where the light is actually absorbed (cut-off conditions : the density is such that the laser frequency ω_L is about equal to the plasma frequency ω_p). Now the intensity is

(1.3) $$I_L \propto \frac{P^2}{R_c^2}.$$

Substituting in (1.1) yields

(1.4) $$(\rho R)^2 \propto I_L \ \eta_A.$$

Putting numbers in the formulae, it turns out that for a 30% thermonuclear efficiency, ($\rho R \sim 3$ g/cm^2) and a 30% absorption efficiency, I_L should be around 10^{18} W/cm^2.

This fairly high value greatly exceeds the thresholds predicted for a number of instabilities in the laser plasma interaction 10^{14} W/cm^2 is a popular value in the case of 1.06 μ light (Nd)[4]. Interaction experiments have been performed so far with intensities up to 10^{17} W/cm^2 (Nd glass lasers) i.e. almost in the expected range. As usual it is very difficult to isolate a given mechanism in such brute fore experiments. The aim of the present review is to discuss some non linear effects and ways to evidence them with a special emphasis on harmonic generation.

2. EVIDENCE OF "NON CLASSICAL" ABSORPTION

For intensities lower than 10^{13} W/cm^2 (Nd) a model for interaction was set up long ago[5]. It implies absorption by inverse Bremsstrahlung in the under dense region and a structure of the flow dominated by electron thermal conductivity. The fractional absorption ϕ is readily calculated as a function of the incoming intensity, in such a model. Consider for instance an exponential density profile with scale-length L.

Now for a plasma with temperature T, the absorption coefficient is[6]

(2.1) $$K = \frac{\omega_p^2 \ \nu}{c\omega_L^2(1-\omega_p^2/\omega_L^2)^{1/2}} = \frac{a\rho^2}{T^{3/2}(1-\rho/\rho_c)^{1/2}}$$

where ν is an electron-ion collision frequency, a is a constant and $\omega_L >> \omega_p, \nu$. Close to the critical surface K has a different (non diverging) expression. However (2.1) can be used to calculate the absorption along a roundtrip in a density gradient, with an acceptable accuracy. Now, after the well known analysis of the laser driven flame, the absorbed intensity is

(2.2) $$I_A = b\ T^{3/2}$$

where b is another constant. The fractionnal absorption

(2.3) $$\phi = I_A/I_o$$

is then readily calculated in an exponential density profile. With scale-length L. It turns out that :

(2.4) $$\rho(L)/I_o = -\phi\, \mathrm{Ln}(1-\phi).$$

Putting number in the formula, ϕ is close to 1 when

(2.5) $$I_o \simeq 10^{11}\ L/\lambda_L \quad W/cm^2.$$

λ_L being the wave-length of the incoming light. Equation (2.4) gives the correct qualitative behavior of the fractional absorption. Figure 1 is a more accurate plot grounded on numerical simulations.

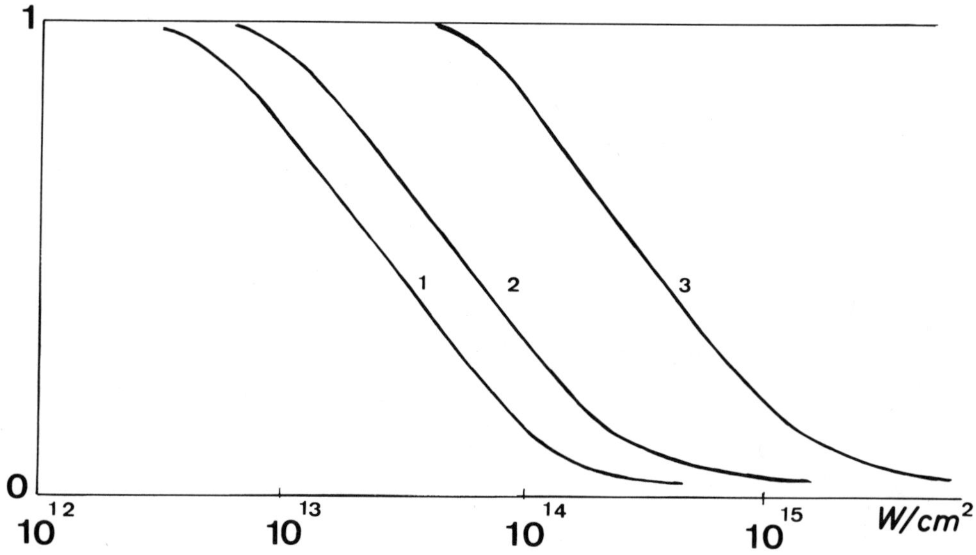

Figure 1

Fractional absorption of 1.06 μm light (Nd) in a CH_2 plasma, versus incoming intensity. The scale lengths L are : 1, 50 μm; 2, 100 μm; 3, 500 μm.

The same kind of curve also applies to the fractional absorption in a slab. In both cases at high intensities, the measured fractional absorption does not follow the predicted dependance[7].

3. NON LINEAR LAYERS

When non linear effects are expected to occur, they originate from resonances between the incoming frequency ω_L and the frequency of plasma waves[8]. Two such resonances* are of great importance :

$$\omega_L = \omega_p \qquad \text{or} \qquad \omega_L = 2\omega_p \tag{3.1}$$

the former taking place at the critical density, the latter at the quarter critical density. Thus, in the density gradient induced by the interaction of the laser beam with a solid surface, the non linear effects are expected to happen in two non linear layers having these densities. The situation is shown on figure 2.

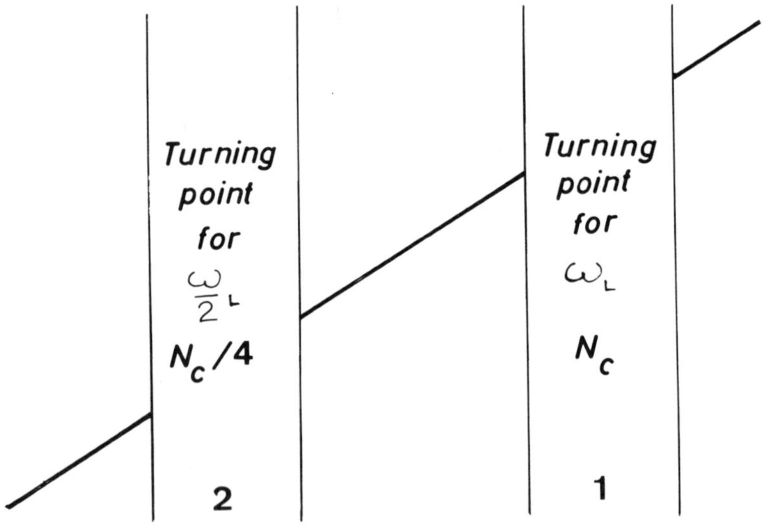

Figure 2

Non-linear layers 1 - N_c
2 - $N_c/4$

* more may appear. They will be mentioned in section 11.

The non linear effects of interest involve density profile modification under the influence of the ponderomotive force and wave-wave coupling between the incoming light and the plasma.

4. EFFECT OF THE PONDEROMOTIVE FORCE ON THE DENSITY PROFILE

The non linear ponderomotive force has been extensively dealt with in the previous workshops[9]. It causes the electrons to drift and through the self-consistent electrostatic field, act onto the plasma as a whole. Hydrodynamical effects are thus to be expected at high laser intensity. Let for the sake of brevity

$$\epsilon = \langle E^2 \rangle / 8\pi \qquad (4.1)$$

the average being taken over one period. The ponderomotive force can then be expressed as

$$-(\rho/\rho_c)\, d\epsilon/dx = df/dx \qquad (4.2)$$

where f is by definition the ponderomotive pressure. Note that

$$f = (\rho/\rho_c)\epsilon + (1/\rho_c)\int_\rho^{\rho_1} \epsilon\, d\rho \qquad (4.3)$$

the subscript 1 denoting a high density subsonic reference state.

Assume the plasma is isothermal. Disregarding the energy conservation, the equations of motion read

$$\partial n/\partial t + \vec{\nabla}.n\vec{v} = 0 \qquad (4.4)$$

$$nM\left[\partial \vec{v}/\partial t + (\vec{v}\cdot\vec{\nabla})\vec{v}\right] = -(n/n_c)\partial\epsilon/\partial x - T\,\partial n/\partial x. \qquad (4.5)$$

They should be coupled for completeness to the wave propagation equation (normal incidence) :

$$\frac{d^2E}{dx^2} + \frac{\omega^2}{c^2}(1-n/n_c)\,E = 0. \qquad (4.6)$$

As it is well known[10] this set of equations : (4.1) (4.4)-(4.6) admits solitary wave solutions which can be analytically calculated in the case of the weak cubic non linearity[11]. A density dip is associated with the maximum value of ϵ thus inducing density profile modulations. Any arbitrary density profile is to tend to such a shape. Indeed let the initial conditions be in 1D :

$$N(x) \quad , \quad 0 \text{ (motion less plasma)}$$

and the perturbed quantities be

$$N + n \quad , \quad v \ ;$$

linearizing (4.4) and (4.5) yields

(4.7) $$\partial n/\partial t + N\,\partial v/\partial x + v\,\partial N/\partial x = 0$$

(4.8) $$nM\,\partial v/\partial t = -(N/N_c)\,\partial \epsilon/\partial x - T\,\partial N/\partial x.$$

Differentiating (4.7) with respect to t and (4.8) with respect to x and combining the equations, one gets

(4.9) $$M\frac{\partial^2 n}{\partial t^2} = \frac{1}{N_c}\left(\frac{\partial \epsilon}{\partial x}\frac{\partial N}{\partial x} + N\frac{\partial^2 \epsilon}{\partial x^2}\right) + T\,\partial^2 N/\partial x^2$$

which expresses a zeroth order instability. At the critical density, one has in a linear gradient ($\partial N/\partial x$ = ct>0)

$$\partial \epsilon/\partial x < 0, \qquad \partial^2 \epsilon/\partial x^2 < 0$$

which imply $\partial^2 n/\partial t^2 < 0$ i.e. a depression is appearing in the density profile. This behavior has been seen in numerical simulations[12]. Due to the ρ/ρ_c dependance of the force, it is more important at cut-off.

5. INFLUENCE OF THE PLASMA FLOW

In the above derivations, the plasma is assumed to be as a whole, motionless. The influence of an overall flow is readily investigated in steady state[13]. Let J be the mass flow in a frame moving with a discontinuity in the plasma flow. The equations of motion are

(5.1) $$\rho u = J$$

(5.2) $$p + f + J^2/\rho - p_1 + J^2/\rho_1$$

(5.3) $$p/\rho = p_1/\rho_1 \qquad \text{(isothermal)}$$

where u is the particle velocity. The solution for f as a function of ρ is readily calculated. It reads in terms of $V = 1/\rho$

(5.4) $$f = -J^2 V - \frac{p_1 V_1}{V} + p_1 + J^2 V_1$$

and is represented on figure 3 for various situations according to the value of J. The maximum value of f decreases from p_1 to zero as J^2 increases from zero to p_1/V_1.

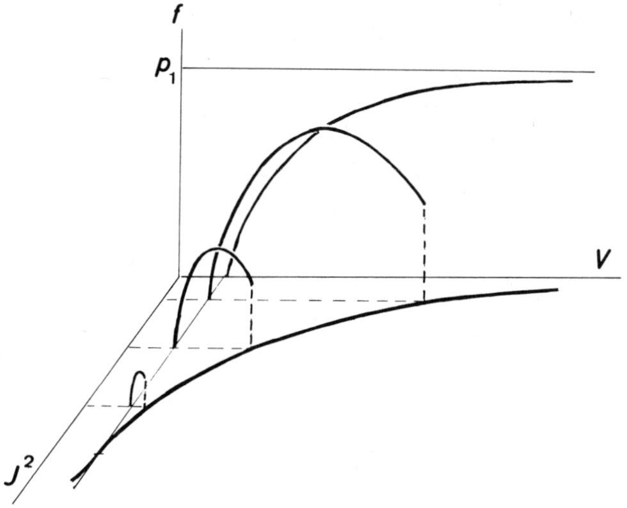

Figure 3

Ponderomotive pressure in various flow conditions

The case of zero flow is special. Then f is a monotonous function of V which tends asymptotically to p_1 for $V \to \infty$. A bell shaped solitary radiation wave structure of the type described in the previous section, is represented by an arc of this curve from $V = V_1$ to any larger value (infinitely large V correspond to negligibly small densities actually observed in numerical simulations). The structure is described by going back and forth along this arc. The integral in (4.3) is now zero.

When there is a flow the behavior of the solution is different. Only small amplitude solitons are allowed, and since the solution is bivalued in V, larger amplitude density modification are to result in a steep gradient connecting the subsonic state 1 to a low density supersonic state. The scale-length of this gradient is half the incoming light wave-length or even smaller. A low density plateau with modulated density is also expected and was actually observed in numerical simulations[14].

Such density profile modification can be diagnosed by interferometric probing of plasma flows resulting from CO_2 laser interaction[15]. However the measured density dip or steepening has a scale-length larger than expected. It is rather representative of the enveloppe of the swelling E field pattern.

6. PARAMETRIC DECAY, RESONANCE ABSORPTION AND HOT ELECTRONS

Non linearities also take place in the absorption of the incoming laser beam by the plasma. An abundant litterature, mainly theoretical has been devoted to the subject. The review by T.W. Johnston[16] contains a pertinent bibliography. Things are actually quite complicated since the plasma is non uniform, of finite size and undergoes convective motion.

Three wave absorptive processes are expected in both non linear layers. Representing by ∿∿ , ——— , – – –, a photon, a plasmon and a phonon respectively, they are :

- the ion acoustic decay instability

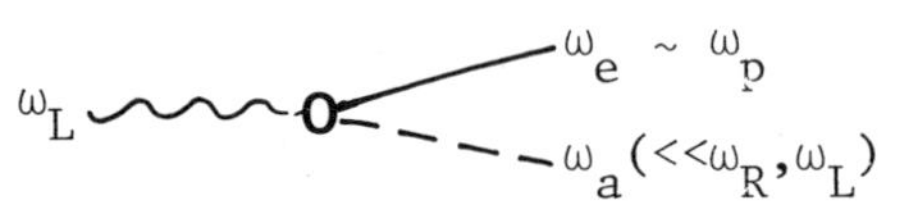

$$\left\{ \begin{array}{l} \omega_L = \omega_R + \omega_a \\ \vec{k}_L = \vec{k}_e + \vec{k}_a \end{array} \right.$$

which is resonant close to the critical density

- the two plasmon decay

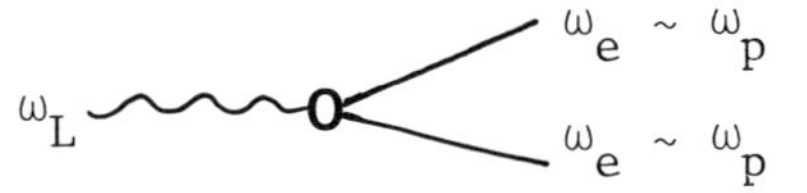

$$\left\{ \begin{array}{l} \omega_L = 2\omega_e \\ \vec{k}_L = \vec{k}_e + \vec{k}_e \end{array} \right.$$

which is resonant close to the quarter critical density.

The plasma inhomogeneity raises the threshold value of the incoming intensity for the occurence of such processes. It has another consequence. When the laser light has an oblique incidence with the electric field vector in the plane of incidence, the longitudinal component of the field tunnels beyond the refractive turning point and induces a resonance at the critical density[17]. The height and width of the resonance peak are determined by the absorption (resonance absorption).

When the above processes are present, the difference with the case of classical (i.e. inverse Bremsstrahlung) absorption is expected to show up in the electron distribution. Electron plasma waves trap electrons and create a high energy tail : suprathermal electrons. The resonant absorption locally strongly heat the plasma giving rise to a population of hotter electrons. In both cases the distribution is no longer maxwellian.

7. X-RAY DIAGNOSTICS

Electrons in a high temperature plasma radiate X-rays whose spectrum is representative of that of the electrons. Many ways of recording X-ray spectra have been used. Time resolution is as important as high spectral resolution. It is seldom possible to have both in a laser plasma interaction experiment.

For a maxwellian distribution of electrons in a hydrogen (or DT) plasma, the X-ray spectrum is a known function of the temperature. The amount of radiation recorded through different calibrated absorbing channels can also be accurately calculated in this case. Experiments with intensities over a few 10^{13} W/cm^2 (Nd) yield results unconsistent with the asumption. They evidence a larger emitted energy than expected, in the high energy (1-100 keV) region[18].

The correspondence between observed X-ray emission and electron distributions as mentionned in the previous section is displayed on the table 1 below.

Note that inhibited thermal conductivity would also cause the temperature of part of the electron population, to increase. Attempts have been made to calculate the X-ray spectrum through numerical simulations in which a modeling of the influence of non linear interaction is used. However the results are not fully satisfactory[19]. Since in any case quite a sophisticated unfolding is needed, continuum X-ray spectroscopy does not appear as a specific diagnostic.

TABLE 1
ELECTRON DISTRIBUTION AND ABSORPTION MECHANISMS

	THERMAL		SUPRATHERMAL
	Low T	High T	
Range of X-rays	0.1-5 keV	5-50 keV	up to 1 MeV
"Classical" collisional absorption	X		
Parametric decay instability			X
Resonance		X	
Inhibited transport		X	

8. FORBIDEN LINES

In high Z targets, lines are emitted in the far U.V. and X-ray parts of the E.M. spectrum, even at electron temperatures in the 1-10 keV range. Their intensity and broadening carry a lot of information as in the case of lower density plasmas. A single example will be dealt with here in connection with parametric decay instability which produces electron plasma waves.

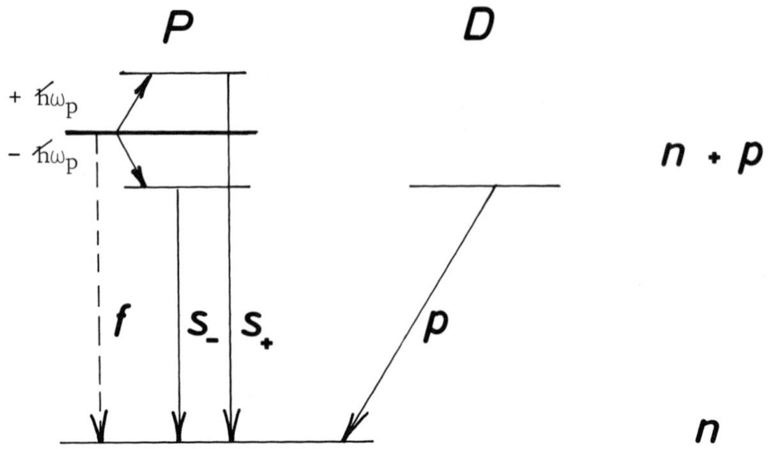

Figure 4

Transitions between excited states of Heliumlike ions. f is forbidden, s_+ s_- p are permitted.

As shown on figure 4, forbidden transitions between excited states of helium like ions can be made possible following a Stark splitting due to the electron plasma waves. Satellites on either side of the forbidden line are thus caused to appear. Such a behavior was seen in laser breakdown sparks in He in connection with Langhuir turbulence[20]. In CO_2 interaction with a low density column, the forbidden transition was also evidenced[21].

Helium like aluminium transitions with the ground state have been observed. The occurence of strong electron plasma waves due to parametric decay instability should show up through forbidden transitions between excited states.

9. RAMAN UP-CONVERSION AND HARMONIC GENERATION

Besides absorptive 3 wave processes, reflective (i.e. stimulated scattering) instabilities should also be considered. They are :

- Brillouin scattering

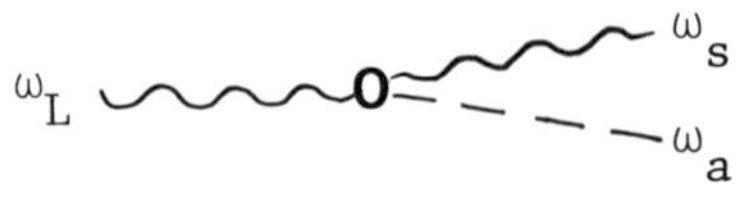

$\omega_a \ll \omega_L\ \omega_s$

$$\left\{ \begin{array}{l} \omega_L = \omega_s + \omega_a \\ \vec{k}_L = \vec{k}_s + \vec{k}_a \end{array} \right.$$

which occurs preferentially when photons travel in opposite directions.

- Raman scattering

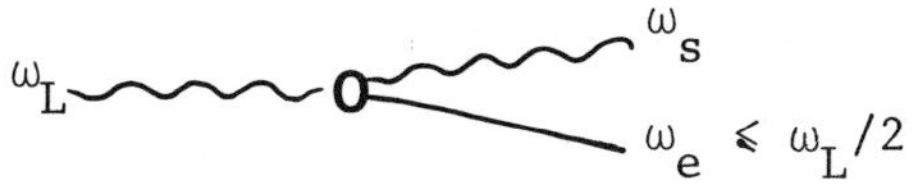

$$\left\{\begin{array}{l} \omega_L = \omega_s + \omega_e \\ \vec{k}_L = \vec{k}_s + \vec{k}_e \end{array}\right.$$

and its reverse process, Raman up conversion (or antistokes scattering) :

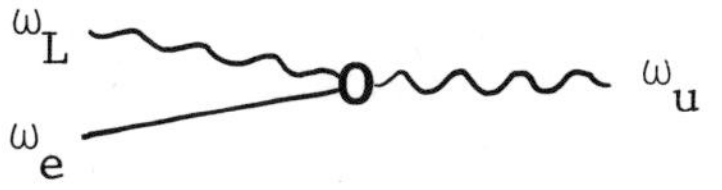

$$\left\{\begin{array}{l} \omega_u = \omega_L + \omega_e \\ \vec{k}_u = \vec{k}_L + \vec{k}_e \end{array}\right.$$

which is likely to appear when a large number of plasmons are present i.e. in both non linear layers where the parametric decays take place.

Thus, at the critical density where electron plasma waves with frequency $\omega_L-\omega_o$ are produced, the Raman up-conversion creates a second harmonic line

$$2\omega_L-\omega_a$$

i.e. with a frequency shift (red shift only). Since the parametric instability generates a wide spectrum of ion acoustic waves, the harmonic line should be fairly broad. Now in the resonant absorption, longitudinal oscillations are propagated to the critical surface. The incoming light may also couple with these oscillations and produce second harmonic generation without a frequency shift.

At the quarter critical density, the electron plasma waves have half the incident frequency. Thus the Raman up-conversion is expected to produce the frequency

$$\omega_L+\omega_e = 3\omega_L/2$$

without frequency shift.

In both non linear layers, cascade processes may generate higher order harmonics.

Thus, harmonic generation with integer or half integer order can results from the coupling of Raman up-conversion with any other 3 wave process creating plasmons. Density and angular (from the $\vec{k}$ vector matching selection rule) conditions are readily set up[22]. They happen to be quite restrictive. Growth rates for the harmonic line have also been calculated.

10. 4 WAVE UP-CONVERSION AND OTHER PROCESSES

The just mentioned restrictive angular conditions can be somewhat relaxed by taking into account the line widths. However not all possible situations can be dealt with this way. For instance the electron-ion parametric decay produces more likely daughter waves with k-vectors long compared with those of the involved E.M. waves. Figure 5 shows that in this case side anti-stokes Raman scattering is favoured.

In order to completely relax the condition, some extra momentum has to be carried away without perturbing too much the frequency matching. 4 wave processes implying a generated phonon can fulfil this requirement. The Raman up-conversion is now

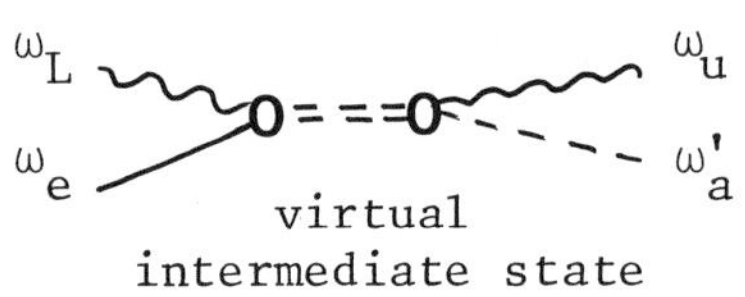

$$\begin{cases} \omega_L + \omega_e = \omega_u + \omega'_a \\ \vec{k}_L + \vec{k}_e = \vec{k}_u + \vec{k}_a \end{cases}$$

The generated frequency is accordingly

$$\omega_u = 2\omega_L - \omega_a - \omega'_a$$

implying a broader red shift and as shown on figure 6 favouring back scattering.

Since ion acoustic waves created at cut-off are convected down to the quarter critical density by the plasma flow, the 4 wave process may also contribute to the $3\omega_2/2$ harmonic generation, the line being red shifted by ω_a. The corresponding growth rates have been investigated[23].

Some other processes might also be involved in harmonic generation such as the usual 3 photon recombination of non linear optics

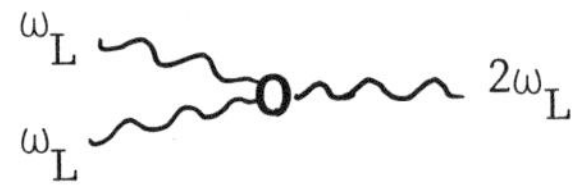

which implies a steep density gradient in order to provide the non linear coupling and be unharmed by the dispersion relations. This scheme can be coupled to the Brillouin scattering :

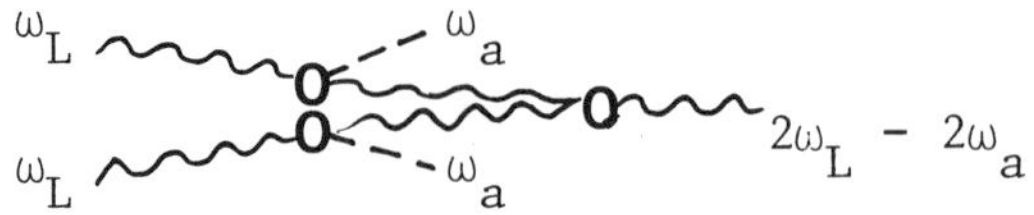

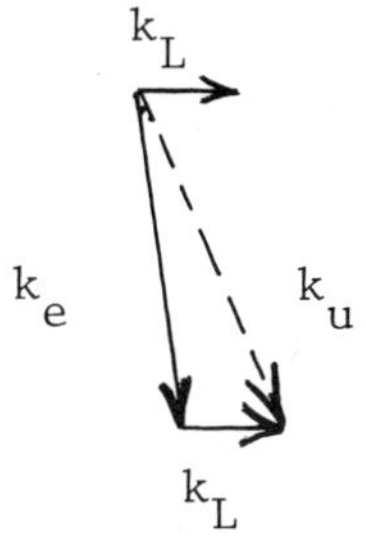

Figure 5

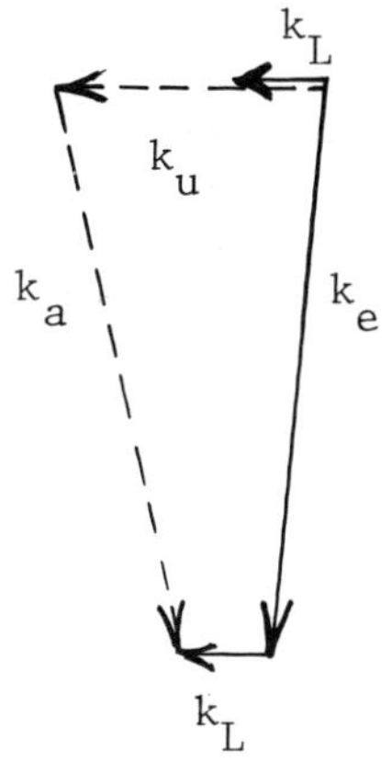

Figure 6

Finally $3\omega_L/2$ generation may also occur at cut-off according to the scheme :

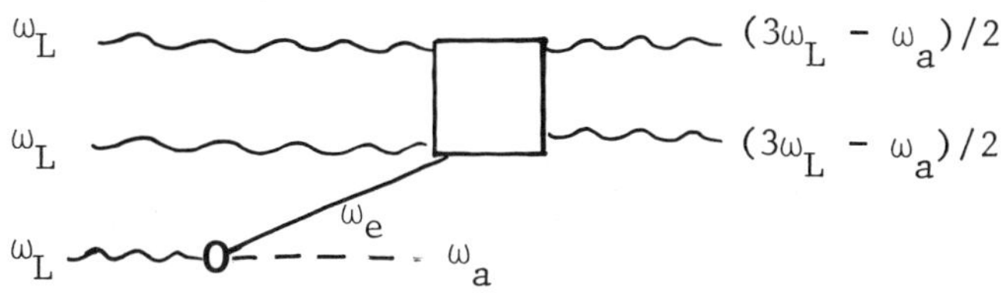

which was originally proposed in order to explain the occurence of a $3\omega_L/2$ harmonic with asymmetric broadening[24].

11. OBSERVATION OF HARMONIC LINES

The first question is of course : which lines were observed ? According to the previous discussion, harmonic lines are expected to appear with frequencies around

$$\omega_L/2^{\star}, 3\omega_L/2, 2\omega_L, 5\omega_L/2, 3\omega_L, \ldots$$

⋆ - This frequency would result from the absolute Raman instability at the quarter critical density (cut-off for $\omega_L/2$)

However, the resonances at the densities critical and quarter critical are not the only possible ones. Recent studies of E.M. propagation with fixed phase velocities in relativistic plasmas show that more resonant coupling may occur[25]. They all involve interaction between longitudinal and transverse waves. The following table 2 lists the resonances predicted by a perturbative theory. The actually observed frequencies are denoted by a star. Most of them result from Nd laser interaction with solid surfaces[24-26-27].

The $3\omega_L/2$, $2\omega_L/2$ and $3\omega_L$ lines were also obtained in CO_2 laser interaction[28]. More recently, $3\omega_L/2$ emission was also found in gas breakdown experiments with a CO_2 laser (N.R.C Group in Ottawa).

TABLE 2

Perturbative order	2	3	4	5	6	7	8
Resonance for ω_L/ω_p =	1	3/2*	2*	5/2*	{ 1 3*	{ 7/2 5/4*	{ 4 3/2*

(harmonics denoted by a star were actually observed)

12. SPATIAL ORIGIN OF HARMONIC LINES

In Nd laser interaction, the $3\omega_L/2$ and $2\omega_L$ lines can be so intense (about 1% of incident) that their spatial origin is easily investigated using usual filtered photographic imaging. Of course, the recording of second harmonic lines readily tells where the critical density is and whether or not self focusing or refraction spreading of the incoming beam occur. It has been shown that with nanosecond pulses and intensities up to $5.^{14}$ W/cm^2 the location and size of the $2\omega_L$ emitting zone exactly fits the focusing lens caustic as measured in vacuo. The intensity at the critical surface which is the relevant parameter for all fluid dynamical calculations is thus known with a fairly good accuracy : 50%[29].

In the same kind of experiments, the $3\omega_L/2$ light was found to be mostly emitted in a lower density region, farther from the target surface indicating an origin at the quarter critical density and therefore providing a straightforward measurement of the density gradient. Streaking both green and red images during the time of interaction has provided accurate information on the plasma flow in the corona while imploding gas filled glass shells. Evidence was then found of rather steep density gradients with scale-length of the order 1μm or less[30].

13. INTENSITY DEPENDENCE

When recording the harmonic emission through filtered photo-cathode one gets time resolution and relative intensity measurements. Preliminary studies with nanosecond Nd pulses showed that harmonic generation takes place mainly at the begining of the interaction process. Further more the peak intensity of the harmonic light can be plotted versus the incoming intensity. Two kinds of information can be derived from such a plot : a threshold value if any, and, through the power law dependance above threshold, some hints about the mechanisms involved.

Threshold measurements for the $3\omega_L/2$ line were reported in 27 (Nd lasers) and 28 (CO_2 laser). The values agree within order of magnitude with the theoretical predictions concerning the two plasmon decay in an inhomogeneous plasma : $\sim 2.10^{13}$ W/cm^2 (Nd), $\sim 2.10^{11}$ W/cm^2 (CO_2)[31].

Time-resolved intensity measurements were also performed[32]. They allow one to set up accurate plots with respect to the incoming flux density. In experiments with nanosecond Nd laser it was observed that the harmonic generation mainly takes place during the begining of the pulse, the peak being obtained when the incoming intensity reaches its plateau value. The maximum intensity was varied by changing the focusing conditions so that the energy flux on the target is kept constant.

One then readily gets the plot of figure 7 in which the

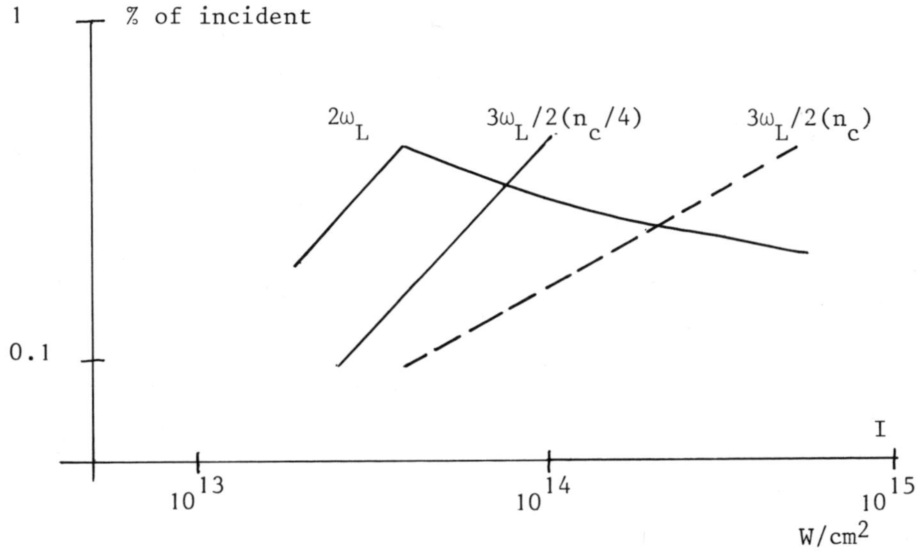

Figure 7

Intensity dependance of harmonic lines

maximum intensities of harmonic lines are displayed versus the maximum intensity at cut-off. However, the $3\omega_L/2$ line comes mainly from a zone outside the target where the flux density is smaller. Assuming that the scale-length of the density gradient is 100 μm and correcting accordingly for the geometry, one gets the solid line of figure 6 which as for the second harmonic before saturation has a unit slope. Since the energy flux on the target is a constant, this behavior is typical of mechanisms in which two impinging photons combine. Note that the process mentionned at the end of section 10 implies a power 0.5 dependence on the intensity at cut-off, whereas the actual dependance is a 0.6 power law : dotted line on figure 7.

Thus, these measurements (obviously above threshold) show that the $3\omega_L/2$ line is connected with the two plasmon decay. Also, the second harmonic undergoes a saturation for intensities greater than 4.10^3 W/cm^2. A possible candidate for this effect could be the density profile modification at cut-off. Indeed a steep gradient reduces the growth rate of parametric instabilities[33].

14. BROADENING OF HARMONIC LINES

In the earlier experiments[24], it was already noticed that the harmonic lines $3\omega_L/2$ and $2\omega_L$ exhibit an asymmetric broadening: on the Stokes side only. Refined measurements with a better spectral resolution were performed[34] using a narrow line (0.5 Å) as the pump. A more complex structure thus appeared. Similar results were also obtained with CO_2 lasers[35].

First look at the second harmonic line in the case of a 1.06 μm beam. The time resolved line shape (resolution ~ 0.1 ns) is reproduced on figure 8 with an ordinate scale linear instead of logarithmic. The main line (unshifted) and the different components are then more easily distinguishable. The rather broad sideband peaks at $\omega_{aM} \sim 0.2\ \omega_{ac}$ (ion plasma frequency at cut-off). According to Nishikawa[36], the highest growth rates of the parametric decay are obtained for $k_a \sim 0.1$-$0.2\ k_D$. This sideband is consistent with 4 wave up-conversion although it is 135° rather than back (180°) scattering. The same conclusion can be infered from the experiments with CO_2 laser.

Two satellites appear conspicuously : one blue shifted by $\Delta\omega_B \sim 0.25\ \omega_{ac}$ to which no mechanism was related so far, one red shifted by $\Delta\omega_R \sim 1.25\ \omega_{ac} \sim 5.10^{13}\ s^{-1}$. This frequency is about the cyclotron frequency for electrons in a 2.5 Megagauss magnetic field, recent theory predict in connection with radiation pressure effects[37]. The point have been investigated in more details by numerical simulation[38].

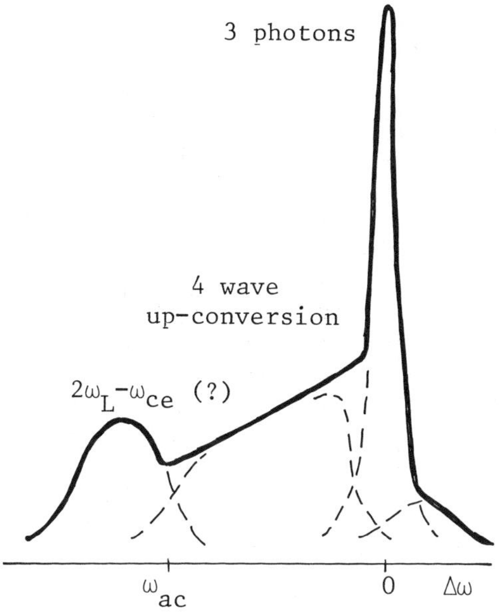

Figure 8 ($2\omega_L$)

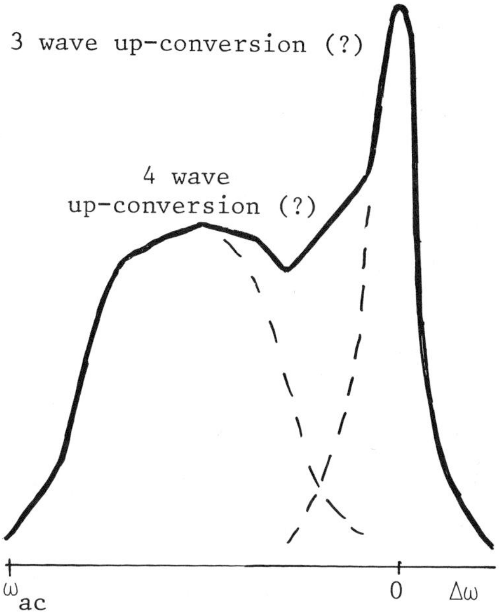

Figure 9 ($3\omega_L/2$)

The case of the $3\omega_L/2$ line is even more puzzling. The line structure was recorded in different laboratories using different targets[34-27-39]. An example is given in figure 9. The results are summarized in table 3. Two components were observed whose width and spacing seem to be weakly dependant on the target material. Shifts due to the 4 wave up-conversion obviously cannot account for the whole shape. The 3 wave up-conversion would yield an almost unshifted line.

15. CONCLUSION

Non linear effects are dominant in high intensity laser plasma interactions. Phenomena predicted long ago by theoreticians are now within the range of experimental investigations. The main draw back of the experiments however comes from the difficulty of isolating a given mechanism. Among the available diagnostics, the interferometric probing of the density profile and the continuum X-ray emission spectroscopy have not so far proven to be specific enough. On the contrary line spectroscopy with high frequency resolution can provide detailed informations about non linear processes. For instance, the sideband on the Stokes side of the second harmonic is clearly related to the parametric decay instability at cut-off. A satellite could be related to the generation of a D.C. magnetic field. However, since many features are still unexplained, the method should be considered only promising at this stage.

TABLE 3 [40]

	Intensity W/cm^2	line width Å	pulse duration ns	target	observations	δ_B Å	δ_R Å	Δ_B Å	Δ_R Å	$\frac{I_B}{I_R}$	I threshold W/cm^2
Lebedev Institute	10^{14}	10	2.5	glass shells	backscatter through lens	7-13	15-32	24	33	0.35-0.95	
I.P.P. Garching	10^{14}	30	5	flat surface D_2, C, Cu CH_2	backscatter through lens	10	15-50	30	30-50	0.2 -0.9	2-4 10^{13}
C.E.A. Limeil	10^{14}	0.5	3	flat surface H_2	135°	(*) $\delta_B+\delta_R=60$ Å		20	30-40	2	

Observation of $3\omega_L/2$ structure

(*) no reference line in this experiment

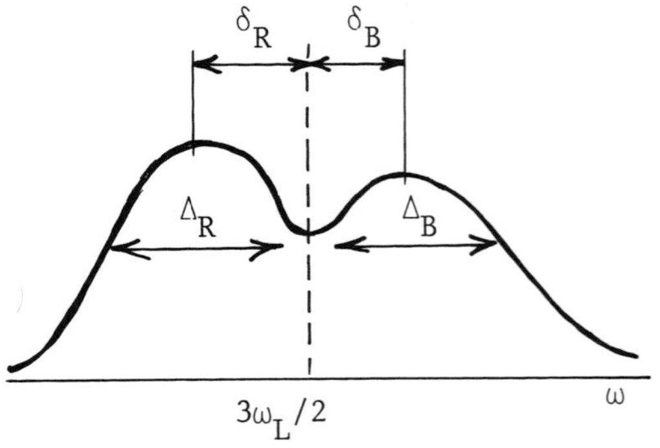

REFERENCES

1. J. NUCKOLLS, L. WOOD, A. THIESSEN, G.ZIMMERMAN, Nature 239 (1972) 135
2. R. KIDDER, Nucl. Fus. 14 (1974) 797
   J.L. BOBIN, C.R. Acad. Sc. Paris, 283B (1976) 309
3. J.L. BOBIN, J.M. REISSE, Rev. Phys. Appl. 11 (1976) 497
4. e.g. D.F. DUBOIS, Laser Interaction and Related Plasma Phenomena, Schwarz-Hora ed. Vol.3a (1974) 267
5. C. FAUQUIGNON, F. FLOUX, Phys. Fluids 13 (1970) 386
   J.L. BOBIN, Phys. Fluids 14 (1971) 2341
6. J. DAWSON, C. OBERMAN, Phys. Fluids 5 (1962) 517
7. E. FABRE, C. POPOVICS, C. STENZ, J. VIRMONT, Vth Plasma Physics and Controlled Fusion, IAEA, Tokyo, Vol 2 (1974) 435
   M. BOYLE et al., These proceedings
8. Yu. M. ALIEV, V.P. SILIN, Soviet Phys. JETP 21 (1965) 601
9. H. HORA, Laser Interaction and Related Plasma Phenomena, Schwarz-Hora ed. vol.1 (1971) 383
   F.F. CHEN, Laser Interaction and Related Plasma Phenomena, Schwarz-Hora ed. vol.3a (1974) 291
10. P. KAW, G. SCHMIDT, T. WILCOX, Phys. Fluids 16 (1973) 1522
11. R.Y. CHIAO, E. GARMIRE, CH. TOWNES, Phys. Rev. Lett. 13 (1964) 479
12. R. FAEHL, Thesis U. of California, 1976
    K. ESTABROOK, Private Communication
    H. ZEEGERS, B. DEGOVE, Private Communication
13. J.L. BOBIN, Rept. PRG.13, U.C. Davis, 1976
14. J.S. de GROOT, J.E. TULL, Phys. Fluids 18 (1975) 672
    K. ESTABROOK, E.J. VALEO, W.L. KRUER, Phys. Lett. 49A (1974) 109 - Phys. Fluids 18 (1975) 1151
    K. LEE, D. FORSLUND, J. KINDEL, E. LINDMAN, Report LA UR-75-2097 (1976)
15. T.P. DONALDSON, I.S. SPALDING, Phys. Rev. Lett. 36 (1976) 467
    J. MARTINEAU, These proceedings
16. T.W. JOHNSTON, Canadian Summer School on Plasmas, Banff (1975)
17. N.G. DENISOV, Soviet Physics JETP 4 (1957) 544
18. F. FLOUX, D. COGNARD, A. SALERES, D. REDON, Phys. Lett. 45A (1973) 483
    C. YAMANAKA et al., Phys. Rev. A6 (1972) 2335
19. G. DAHLABACKA, Conference on Transport in Laser Plasmas, Mount Lassen, 1976
20. J. BRETAGNE, Thesis, U.Paris-Sud, Orsay (1973)
21. J.J. TURECHECK, F.F. CHEN, Phys. Rev. Lett. 36 (1976) 720
22. J.L. BOBIN, Optics Communications, 14 (1975) 339
23. C. YAMANAKA et al., Phys. Rev. Lett. 32 (1974) 1038
    NGUYEN DUC LONG, K.J. PARBHAKAR, T.W. JOHNSTON, Proceedings of the 7th European Conference on Plasma Physics and Nuclear Fusion, Lausanne 1 (1975) 73

24. J.L. BOBIN, M. DECROISETTE, B. MEYER, Y. VITEL, Phys. Rev. Lett. 30 (1973) 594
J.L. BOBIN, Laser Interaction and Related Plasma Phenomena, Schwarz-Hora ed. Vol.3b (1974) 591
25. A. DECOSTER, To be published
26. PING LEE, D.V. GIOVANIELLI, R.P. GODWIN, G.H. Mc. CALL, Appl. Phys. Lett. 24 (1974) 406
27. H.C. PANT, K. EIDMANN, P. SACHSENMAIER, R. SIGEL, Optics Comm. 16 (1976) 396
28. E. FABRE et al., VIth Conf. on Plasma Physics and Nuclear Fusion, IAEA, Berchtesgaden (1976)
29. A. SALERES, M. DECROISETTE, C. PATOU, Optics Comm. 13 (1975) 321
30. S. JACKEL, B. PERRY, M. LUBIN, Phys. Rev. Lett. 37 (1976) 95
31. M.N. ROSENBLUTH, C.S. LIU, Phys. Rev. Lett. 29 (1972) 701
32. A. SALERES et al., J. de Phys.34 Colloque C2 (1973) 9
33. F.W. PERKINS, J. FLICK, Phys. Fluids 14 (1971) 2012
34. M. DECROISETTE, B. MEYER, Y. VITEL, Phys. Lett. 45A (1973) 443
35. E. FABRE, C. POPOVICS, C. STENZ, Proceedings of the 7th European Conference on Plasma Physics and Nuclear Fusion, Lausanne 1 (1975) 80
36. K. NISHIKAWA, J. Phys. Soc. Japan 24 (1968) 1152
37. J.J. THOMSON, C.E. MAX, K. ESTABROOK, Phys. Rev. Lett. 35 (1975) 663
J.A. STAMPER, Phys. Fluids 19 (1976) 7581
WEE WOO, J.S. de GROOT, Report PRG. 12, U.C. Davis, 1976
38. W.L. KRUER, These Proceedings pp. 709-720
39. N.G. BASOV et al., Nobel Symposium on Non Linear Plasma Physics, Aspenasgarden (1976)
40. after R. SIGEL, Nobel Symposium on Non Linear Plasma Physics, Aspenasgarden (1976)

LASER LIGHT ABSORPTION AND HARMONIC GENERATION DUE TO SELF-GENERATED MAGNETIC FIELDS*†

W. L. Kruer and K. G. Estabrook

University of California, Lawrence Livermore Laboratory

Livermore, California 94550 U.S.A.

ABSTRACT

It is shown that self-generated magnetic fields can play a significant role in laser light absorption. Even normally incident light will then be resonantly absorbed. Computer simulations and theoretical estimates for this absorption and the concomitant harmonic generation are given for parameters characteristic of some recent experiments.

I. INTRODUCTION

It has been shown that self-generated dc magnetic fields[1-8] can strongly inhibit the electron heat flow in recent laser plasma experiments. In this paper we will show that these magnetic fields can also introduce an interesting amount of light absorption and harmonic generation. Physically the absorption and harmonic generation occurs due to electrostatic waves resonantly generated by the oscillation of electrons driven by the $\underline{v}_{os} \times \underline{B}_o/c$ force, where $\underline{v}_{os}$ is the quiver velocity of an electron in the laser light

---

*Presented at the Fourth Workshop on "Laser Interaction and Related Plasma Phenomena" held at RPI, Troy, New York, Nov. 8-12, 1976.

†Research performed under the auspices of the USERDA, Contract No. W-7405-Eng-48.

field and $\underline{B}_o$ is the self-generated dc magnetic field. In contrast to the resonance absorption[9-11] usually considered in laser-target applications, this mechanism is operative even for normally incident light. As we will show, the magnitude of the dc magnetic field which has been experimentally measured[2] ($B \sim 4$ MG) is large enough to make the light conversion to plasma waves sizeable, but not so large as to make the critical surface inaccessible to the light. In contrast, this mechanism is negligible for the case of radio waves in the ionosphere.[9]

Although the basic physical mechanism is of course well-known,[12] its application to laser-plasma targets is novel and potentially important. We here investigate this mode conversion process in a regime complementary to that studied in recent magnetic fusion applications.[13] In these latter applications, the electron cyclotron frequency (ω_{ce}) is > the electron plasma frequency (ω_{pe}), whereas in our applications $\omega_{ce} \ll \omega_{pe}$. For example, for B = 4 MG and a Nd laser, $\omega_{ce}/\omega_{pe} \simeq .04$ at the critical density. Theoretical estimates for the coupling, saturation amplitudes, absorption, and harmonic generation will be given and explored in computer simulations for parameters characteristic of recent laser-plasma experiments. The calculations also show that the second harmonic generation may provide information both about the local scale length near the critical density and the size of the self-generated dc magnetic fields.

II. THEORETICAL ESTIMATES FOR THE MODE CONVERSION

The basic process will be investigated in a simple geometry. Let $\underline{B}_o = B_o\,\hat{x}$, and consider laser light normally incident ($\underline{k} = k(z)\hat{z}$) on a plasma with density $n = n(z)$. The electric vector of the light is orthogonal to $\underline{B}_o$ as it enters the plasma from vacuum. From Maxwell's equations and the linearized cold plasma equations, we can readily obtain coupled equations for E_y and E_z. Assuming a time-dependence of $e^{-i\omega_o t}$, we obtain

$$\frac{d^2E_y}{dz^2} = -\frac{\omega_o^2}{c^2}\,\varepsilon_1\,E_y \tag{1}$$

and

$$E_z = -\frac{\omega_{pe}^2}{\omega_o^2 - \omega_{ce}^2}\,\frac{i\omega_{ce}}{\omega_o}\,\frac{E_y}{\varepsilon_2}\,, \tag{2}$$

where ε_1 and ε_2 are the dielectric functions:

$$\varepsilon_1 = 1 - \frac{\omega_{pe}^2}{\omega_o^2} \frac{(\omega_o^2 - \omega_{pe}^2)}{(\omega_o^2 - \omega_{UH}^2)} \tag{3}$$

and

$$\varepsilon_2 = 1 - \frac{\omega_{pe}^2}{\omega_o^2 - \omega_{ce}^2} . \tag{4}$$

Here ω_o is the light frequency, ω_{pe} the local electron plasma frequency,

$$\omega_{ce} = \frac{eB_o}{mc} , \text{ and } \omega_{UH} = \sqrt{\omega_{pe}^2 + \omega_{ce}^2} .$$

Eq. (1) gives the well-known dispersion relation for the extraordinary wave. For laser pellet applications, $\omega_{ce} << \omega_o$, and the cut-off density n_{cut} (the density at which propagation cuts off for light waves incident from low density) is determined by $\omega_o^2 = \omega_{pe}^2 + \omega_{pe}\,\omega_{ce}$. Since $\omega_{ce} << \omega_o$, this cut-off density is only slightly less than the critical density n_{cr} (defined by $\omega_o = \omega_{pe}$). For example, for $\omega_{ce}/\omega_o = .04$, $n_{cut} = .96\ n_{cr}$.

As is apparent from Eq. (2), the extraordinary wave is partly longitudinal. This electrostatic field is simply the plasma response to the oscillation of electrons in the $\underline{v}_{os} \times \underline{B}_o$ force. Note that this field becomes infinite when $\varepsilon_2 = 0$, since the plasma response is then resonant. The resonance occurs at $n_R = n_{cr}[1 - (\omega_{ce}/\omega_o)^2]$, which is essentially the critical density for $\omega_{ce} << \omega_o$. Physically, the light is reflected at the cut-off density ($n_{cut} < n_R$), but its electric field tunnels into the higher density region, where upper hybrid waves are resonantly driven. Of course, collisions, thermal corrections or non-linear effects keep the resonantly-driven field finite. For light intensities commonly used, E_z is limited by strong wave-breaking as will be shown in the simulations.

The principal features of this absorption mechanism can be obtained rather simply. Since $E_z \equiv E_d/\varepsilon_2$, the behavior of the electrostatic fields is determined by the value of the effective driver field E_d at the resonance point. To determine this driver field we need to estimate E_y at the resonance point. Assuming a locally linear density profile,[9] we first estimate E_y at the cut-off density by analogy to the well-known Airy function solution for the $B_o = 0$ case and then multiply by the exponential decay of the field to the resonance point. The result is

$$E_d \simeq 1.3(k_oL)^{1/6} E_L \frac{\omega_{ce}}{\omega_o} e^{-\frac{\pi}{2} k_oL\left(\frac{\omega_{ce}}{\omega_o}\right)^{3/2}} , \tag{5}$$

where L is the density scale length and k_o and E_L are the wave number and electric field of the light in vacuum. Although an improved value of E_d can be obtained by a detailed solution of Eq. (1) (as is done in the simulations), this simple estimate will suffice to illustrate some of the scalings which is our principal goal in this section.

Several results are now readily obtained. First, the maximum plasma wave amplitude as determined by wave breaking is

$$E_d \simeq E_d/\left(\sqrt{\frac{X_d}{L}} + \frac{\sqrt{3}\,\lambda_{De}}{L}\right) , \tag{6}$$

where $X_d = \frac{eE_d}{m\omega_o^2}$ and λ_{De} is the electron Debye length. The second term in the denominator approximates a thermal correction to the cold plasma prediction. In addition, the absorption efficiency f is obtained by invoking energy balance

$$f \frac{cE_L^2}{8\pi} = \int dz \frac{\nu E_p^2(z)}{8\pi} ,$$

where ν is the magnitude of the plasma wave damping (either linear or nonlinear). As long as $\nu \ll \omega_o$, the integral is independent of the detailed value of ν (the height of the resonance is $\alpha\ 1/\nu$, and the width of the resonant region is $\alpha\nu$). Hence, we obtain

$$f \simeq 5(k_oL)^{4/3} \left(\frac{\omega_{ce}}{\omega_o}\right)^2 e^{-\pi(k_oL)\left(\frac{\omega_{ce}}{\omega_o}\right)^{3/2}} . \tag{7}$$

Defining $\beta \equiv (k_o L)^{1/3} \left(\frac{\omega_{ce}}{\omega_o}\right)^{1/2}$, we have $f \propto \beta^4 e^{-\pi\beta^3}$, and so the estimated absorption peaks at $\tilde{\sim}$ 40% for $\beta \simeq .8$. This behavior is easy to understand physically. As $\beta \rightarrow 0$ (for example, for $B_o = 0$), the coupling goes to zero. For large β, the resonance point becomes inaccessible. For laser pellet applications, we will see that β can be O(1) based on recent measurements of the magnitude of the self-generated magnetic fields.

It should be noted that the basic mechanism is linear, although the size of the self-generated magnetic field will depend on both the intensity and the uniformity of the irradiation. We also note that this mode conversion is operative whenever $\underline{E}_L \times \underline{B}_o \neq 0$. In the toroidal fields due to $\underline{\nabla}n \times \underline{\nabla}T$ sources, the coupling then occurs at preferred locations in the focal spot. However, we emphasize that there are a number of other sources of magnetic fields which may change their spatial structure.

III. COMPUTER SIMULATIONS OF THE MODE CONVERSION

To test this coupling, we have carried out simulations using an electromagnetic, relativistic particle code.[14] In choosing the parameters of a representative simulation, we are principally motivated by the recent NRL work[1,2]; in particular, we choose B_o = 4 MG, I = 10^{16} W/cm^2 (Nd light), and a rather short density gradient length (L/λ_o = 1.5). Such a short gradient is also consistent with some recent measurements of the density scale lengths in laser-produced plasmas.[15] For simplicity we use a fixed, neutralizing ion background, since the basic mechanism is independent of ion motion. In the code, the light is normally incident on an inhomogeneous plasma slab. To model the dc magnetic fields shown by the fluid calculations, we simply impose a constant magnetic field orthogonal both to the density gradient and to the electric field vector of the incident light.

The strong plasma wave generation near the critical density is clearly shown in Fig. 1, which is an electron phase space plot from the simulation. The electrostatic field is clearly limited by wave breaking and reaches a peak value of $eE_p/m\omega_o c \simeq .15$. (The simple estimate in Eq. (6) predicts $eE_p/m\omega_o c \simeq .17$). The time-averaged laser light absorption measured in the simulation is ≃ 15%, also in reasonable agreement with the estimate given in Eq. (7) which is 16%. Although modest, this absorption does not require oblique incidence or p-polarization and hence will reduce the sensitivity of mode conversion to angle of incidence and polarization.

Finally, we have carried out a number of simulations varying both the density gradient and the size of the magnetic field.

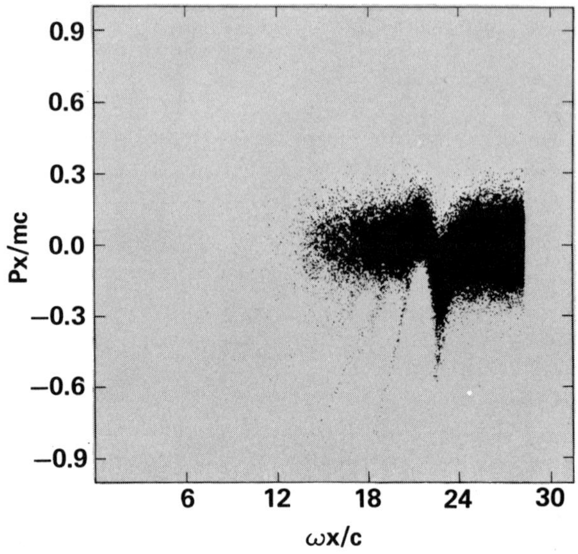

Fig. 1: A plot of electron phase space (p_x vs x) from the simulation at $\omega_o t = 125.2$. The parameters are $eE_L/m\omega_o c = .1$, $v_e/c = .1$, $k_oL = 10$, $\omega_{ce}/\omega_o = .04$ and fixed ions.

As shown in Fig. 2, the calculated electrostatic field is quite large and scales in reasonable agreement with the theoretical estimates. Such close agreement is probably fortuitous, since our theoretical estimates are rather crude. The absorptions measured in these short-time simulations also scale in reasonable agreement with the theoretical estimate. For example, when $\beta = .6$ we find an absorption of $\tilde{\sim}$ 30% (the estimated value is ≃ 35%). We emphasize that both the actual scale lengths and the values of B_o in experiments are not well-known and depend on such features as energy transport inhibition and the radial intensity profile of the light beam. The principal point is that for parameters characteristic of those recently measured, the self-generated magnetic field can introduce an interesting amount of absorption.

To further emphasize this point, let us compare the mode conversion introduced by a magnetic field with that introduced by oblique incidence. Comparing our results in Section II with similar estimates for the case of no magnetic field (but obliquely

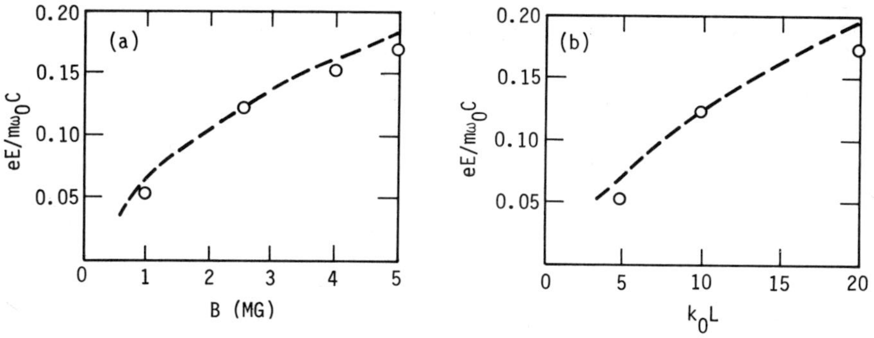

Fig. 2: The peak electrostatic field from the simulations as a function of a) the magnitude of the magnetic field ($k_oL=10$) and b) the density gradient scale length (B=2.5MG). The dashed line is our theoretical estimate. The parameters of the simulations are $eE_L/m\omega_oc=.1$, $v_e/c=.1$ (v_e is the electron thermal velocity), and fixed ions.

incident p-polarized light), we readily see that the presence of the magnetic field corresponds to an angle of incidence of $\theta \simeq \sin^{-1} (\omega_{ce}/\omega_o)^{1/2}$. For example, for $\omega_{ce}/\omega_o = .04$, $\theta \simeq 11°$. This is approximately the mean angle of incidence (averaged over energy) for light focused by an f/2 lens onto a planar target. Hence mode conversion due to a self-generated field can dominate that due to oblique incidence when large f number lenses are used (or when clam shells are used to focus light nearly normally onto spherical targets).

IV. SECOND HARMONIC GENERATION

Since the mode conversion gives rise to large electrostatic fields near the critical density, we can expect a sizeable level of second harmonic emission to occur. This emission can be simply computed by iterating the wave equation to obtain

$$\left(\frac{d^2}{dz^2} + \frac{4\omega_o^2}{c^2}\right) E_y^{(2)} = \frac{-4\omega_o^2}{c^2} \frac{4\pi}{2i\omega_o} J_y^{(2)}$$

where the superscripts refer to the component at frequency $2\omega_o$.

$$J_y^{(2)} = - n_o e v_y^{(2)} - n^{(1)} e v_y^{(1)} ,$$

where n_o is the background density, $n^{(1)}$ the density fluctuation, and $v_y^{(1)}$ is the oscillation velocity of electrons in the transverse field. Using our crude estimate (Equation 5) for the effective driver, noting that $J_y^{(2)}$ is very peaked near the critical density, and taking the magnetic field as small, we obtain

$$\frac{E_y^{(2)}}{E_L} \sim 2.7 \left(\frac{eE_L}{m\omega_o c}\right)\left(k_o L\right)^{2/3} \beta^2 e^{-\pi\beta^3}$$

Here $E_y^{(2)}$ (E_L) is the amplitude of the backreflected second harmonic light (the incident laser light) in vacuum.

This simple estimate is compared with 1½ dimensional simulation results in Fig. 3. The dependence on both magnetic field

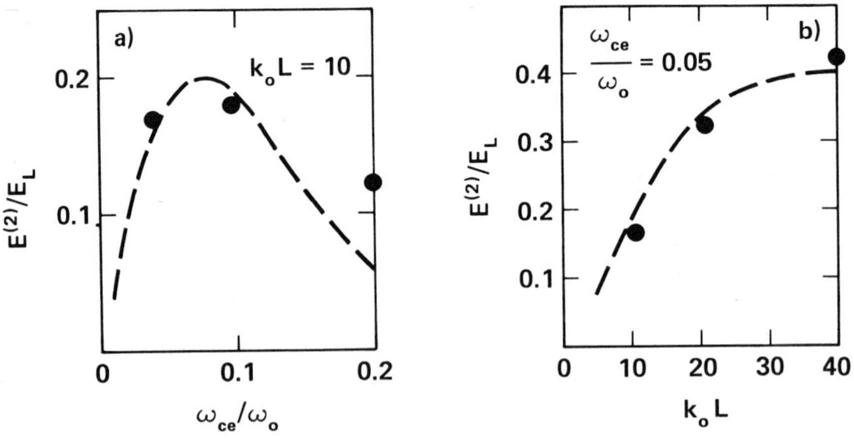

Fig. 3: The second harmonic field as a function of a) the magnetic field and b) the density scale length. The dashed line is a theoretical estimate. For these results a 1.5 dimensional code was used, $v_{os}/c=.1$, and $v_e/c = .09$.

and density scale length is in reasonable agreement. Note that the second harmonic emission can be quite large when the local scale length is long. However, with scale lengths more characteristic of steepened profiles, the energy in the second harmonic is of order 1% of the incident light energy. We note that a similar strong dependence of the magnitude of the second harmonic emission on local scale length has also been found[16] for the emission which accompanies resonance absorption of obliquely incident light. Hence the magnitude of the second harmonic emission provides an indirect indication of the local scale length near the critical density.

The reduction of the $2\omega_o$ emission by profile steepening suggests a simple explanation for some recent measurements at Limeil.[17] At low intensities the second harmonic emission was observed to increase in magnitude as the square of the incident light intensity. However, at a critical intensity of $\sim 5 \times 10^{13}$ W/cm^2 (Nd laser light), the dependence of this emission on intensity abruptly changed. Simple estimates show that this is just the intensity for which significant nonlinear profile steepening near the critical density is expected to onset and reduce the emission.

Finally the simulations also show that frequency structure in the second harmonic light may provide a diagnostic for the presence of large self-generated magnetic fields. As shown in Fig. 4, we observe a satellite line shifted down in frequency from $2\omega_o$ by an amount which is $\sim \omega_{ce}$. When the size of the magnetic field was doubled, the frequency shift became twice as large, confirming its dependence on B_o. Such frequency structure may arise, for example, from the coupling together of different Bernstein modes. An analytic theory of the structure should take into account finite larmor radius effects.

A magnetic field-induced line structure in the $2\omega_o$ light may indeed have been observed in the Limeil experiments.[17] A satellite line was observed shifted in frequency from the main line at $2\omega_o$ by an amount which was too large to be accounted for by an ion acoustic wave. If attributed to a magnetic field, the magnitude of the shift would correspond to a field of $\sim$ .5MG, which is a plausible value at the light intensity used in these experiments.

It is worth noting that self-generated magnetic fields can also account for a frequency splitting of the 3/2 ω_o light, which has been observed.[17,18] In simulations of stimulated scattering in underdense plasma with $n \gtrsim .25\ n_{cr}$, we found that the incident light decayed into scattered light waves plus several different Bernstein modes. The result was a spectrum of electrostatic waves near $\omega_o/2$ but differing in frequency by an amount proportional to

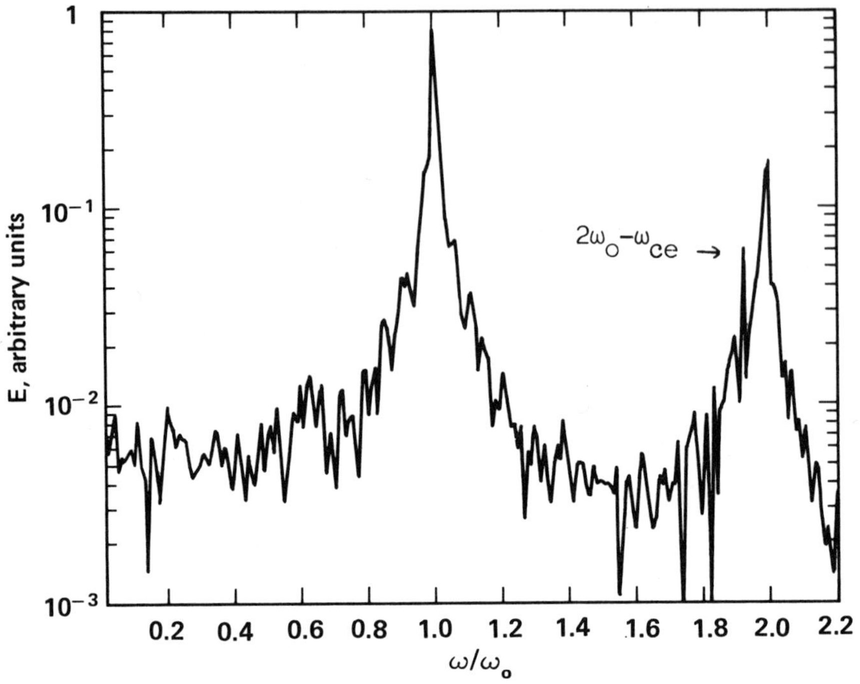

Fig. 4: The fourier-analyzed electric field of the incident and reflected light from a simulation using a 1½-D code. The parameters are v_{os}/c = .1, v_e/c=.09, ω_{ce}/ω_o =.1, and k_oL = 20.

ω_{ce}. These electrostatic waves in turn can scatter the incident light to give a split line near 3/2 ω_o.

V. SUMMARY

In conclusion, our calculations show that self-generated magnetic fields can play a significant role in laser plasma interactions. These fields introduce resonance absorption of even normally incident light and so can strongly reduce the dependence of linear mode conversion on angle of incidence and polarization. In addition, these magnetic fields provide an additional mechanism for second harmonic generation. Several features of this harmonic emission may provide indirect information about both the local density gradients and the size of the magnetic fields. More detailed calculations await a better understanding of the size and structure of the self-generated fields.

REFERENCES

1. B. H. Ripin et al., Phys. Rev. Letters 34, 1313 (1975).
2. J. A. Stamper and B. H. Ripin, Phys. Rev. Letters 34, 138 (1975); J. A. Stamper, K. Papadapoulous; R. N. Sudan, S. O. Dean, E. A. McClean, and J. M. Dawson, ibid 26, 1012 (1971).
3. J. B. Chase, J. M. LeBlanc, and J. R. Wilson, Phys. Fluids 16, 1142 (1973).
4. R. Haas, W. Mead, W. Kruer, D. Phillion, H. Kornblum, J. Lindl, D. MacQuigg, V. Rupert, and K. Tirsell, Phys. Fluids (in press).
5. N. K. Winsor and D. A. Tidman, Phys. Rev. Letters 31, 1044 (1973); D. A. Tidman and R. A. Shany, Phys. Fluids 17, 1207 (1974).
6. J. J. Thomson, C. E. Max, and K. G. Estabrook, Phys. Rev. Letters 35, 663 (1975).
7. E. Lindman, D. Forslund, J. Kindel, K. Lee, and W. Shanahan, Bull. Am. Phys. Soc. 20, 1378 (1975).
8. G. H. Dahlbacka, W. C. Mead, C. E. Max, and J. J. Thomson, Lawrence Livermore Laboratory Report UCRL-77150 (1975).
9. V. L. Ginzburg, The Propagation of Electromagnetic Waves in Plasmas (Pergamon Press, London, 1965), Chs. 4 and 5.
10. J. P. Friedberg, R. W. Mitchell, R. L. Morse, and L. I. Rudsinski, Phys. Rev. Letters 28, 795 (1972); D. Forslund, J. M. Kindel, and E. L. Lindman, Phys. Rev. A 11, 679 (1975); K. G. Estabrook, E. J. Valeo, and W. L. Kruer, Phys. Fluids 18, 1151 (1975).
11. J. S. DeGroot and J. Tull, Phys. Fluids 18, 672 (1975); P. Koch and J. Albritton, Phys. Rev. Letters 32, 1420 (1975).
12. T. Stix, Phys. Rev. Letters 6, 878 (1965).
13. J. J. Schuss and J. C. Hosea, Phys. Fluids 18, 727 (1945).

14. A. B. Langdon and B. F. Lasinski, in Methods in Computational Physics, edited by J. Killeen, B. Alder, S. Fernbach, and M. Rotenberg (Academic, New York), Vol. 16, 327-365 (1976).

15. T. P. Donaldson and I. J. Spalding, Phys. Rev. Letters 36, 467 (1976).

16. N. S. Erokhin, S. S. Moiseev, V. V. Mukhin, Nuclear Fusion 14, 333 (1974); K. Eidmann and R. Sigel, Physical Review Letters 34, 799 (1975).

17. J. L. Bobin, in Proceedings of the Nobel Symposium, edited by H. Wilhelmsson (Plenum Press, London) in press.

18. H. C. Pant, K. Eidmann, P. Sachsenmaier, and R. Sigel, Optics Comm. 16, 396 (1976).

"Reference to a company or product name does not imply approval or recommendation of the product by the University of California or the U.S. Energy Research & Development Administration to the exclusion of others that may be suitable."

NOTICE

"This report was prepared as an account of work sponsored by the United States Government. Neither the United States nor the United States Energy Research & Development Administration, nor any of their employees, nor any of their contractors, subcontractors, or their employees, makes any warranty, express or implied, or assumes any legal liability or responsibility for the accuracy, completeness or usefulness of any information, apparatus, product or process disclosed, or represents that its use would not infringe privately-owned rights."

LASER RADIATION FORCES: GENERATION OF AND DEPENDENCE ON MAGNETIC FIELDS†*

J. A. Stamper

Naval Research Laboratory

Washington, D.C.

ABSTRACT

Mechanisms are described through which the high frequency fields of the laser beam can directly exert time-averaged forces on the plasma. The generation of magnetic fields by these radiative forces is discussed and shown to depend on the field momentum transferred to the plasma during absorption. Refraction is used to illustrate the dangers of approximation. The magnetic field itself is shown to be responsible for a large, polarization dependent radiative force.

†Work sponsored by ERDA.

*Presented at the Fourth Workshop on "Laser Interaction and Related Plasma Phenomena," held at RPI, Troy, New York, November 12-18, 1976.

INTRODUCTION

The power densities obtained in a focused, high-power laser beam exceed the power densities attainable by any other method. Focused intensities of 10^{17} W/cm^2 have been achieved at several laboratories throughout the world. The radiative forces, i.e., the forces on the target plasma due directly to the high frequency electromagnetic fields which make up the laser beam, are of basic interest under such conditions. The period of the electromagnetic oscillations is in the femtosecond (10^{-15} sec) time scale. However, the phenomena of importance in laser-fusion applications or in experiments occur on a much longer time scale-picosecond or nanosecond (10^{-12} or 10^{-9} sec) time scale. We are thus interested in a time-averaged radiative force and this time average is denoted by enclosing in angular brackets $\langle\ \rangle$.

Radiative forces have been of particular interest recently because of their role in magnetic field generation and density profile modification. The steep density gradient resulting from radiative forces reduces backscatter, reduces the strength of the absorptive decay modes and enhances resonance absorption. The magnetic fields have a pronounced effect on the distribution of thermal energy due to the inhibition of conduction across the field. It is shown here that the magnetic fields are also directly responsible for a polarization dependent radiative force.

The velocity of an isolated electron subject only to the fields of an electromagnetic wave is out of phase with the electric and magnetic fields. Radiative forces thus require mechanisms whereby the oscillating electron velocity and magnetic field have components which are in phase so that there is a time-averaged $\underset{\sim}{v} \times \underset{\sim}{B}$ product. Plasma inhomogeneity, collisions and a time-averaged magnetic field are three such mechanisms. The slowly varying (spontaneous) magnetic fields can be generated by thermal forces or radiative forces. A background discussion of the inhomogeneity and collisional mechanisms is given before describing magnetic field generation and the magnetic field itself as the mechanism producing a magnetic radiative force.

PONDEROMOTIVE FORCE

The term ponderomotive force will be used for the radiative force on a cold, collision-free, unmagnetized plasma. The mechanism is then plasma inhomogeneity. The ponderomotive force is essentially an electrostatic type of radiative force such as that responsible for forces on the wall of a cavity containing radiation. The electromagnetic fields for a plasma varying only in the x-direction are given in the WKB approximation[1] by

$$\underset{\sim}{E} = \hat{e}(E_v/\bar{n}^{\frac{1}{2}})\cos\theta \tag{1}$$

$$\underset{\sim}{B} = \hat{b}(E_v/\bar{n}^{\frac{1}{2}})(\bar{n}\cos\theta - \frac{c}{2\omega\bar{n}}\frac{d\bar{n}}{dx}\sin\theta) \tag{2}$$

where $\bar{n}$ is the refractive index, E_v is the vacuum electric field, $\hat{e}$, $\hat{b}$ are unit vectors and $\theta = \omega t + (\omega/c)\int \bar{n}dx$. The non-relativistic electron (quiver) velocity is then $\underset{\sim}{v} = \hat{e}(eE/m\omega)\sin\theta$. Note, because of the plasma inhomogeneity (i.e., $d\bar{n}/dx \neq 0$), that $\underset{\sim}{B}$ has been phase shifted so that it has a component which is in phase with $\underset{\sim}{v}$. There is thus a time-averaged force $\underset{\sim}{f}_p = -(ne/c)\langle \underset{\sim}{v} \times \underset{\sim}{B}\rangle$ which can be written as

$$\underset{\sim}{f}_p = \frac{E_v^2}{16\pi}\frac{\omega_p^2}{\omega^2\bar{n}^2}\frac{d\bar{n}}{dx}(\hat{e}\times\hat{b}) \tag{3}$$

Since $d\bar{n}/dx < 0$ when x increases into the target, this is a de-confining force as has been discussed by Hora.[1] A more general derivation and expression is given in a later section.

COLLISIONAL FORCE

A cold, homogeneous, unmagnetized plasma is now considered. The mechanism responsible for a radiative force is collisions and an effective collision frequency ν is assumed. The electric and magnetic fields are assumed to be in phase so that they can be expressed as

$$(\underset{\sim}{E}, \underset{\sim}{B}) = (\underset{\sim}{E}_o, \underset{\sim}{B}_o)\sin\omega t \tag{4}$$

The electron velocity is determined by $m\dot{\underset{\sim}{v}} = -e\underset{\sim}{E} - m\nu\underset{\sim}{v}$ so that it is given by

$$\underset{\sim}{v} = \frac{e\underset{\sim}{E}_o}{m\sqrt{\omega^2 + \nu^2}}\cos(\omega t + \phi) \tag{5}$$

where the phase ϕ is given by $\tan\phi = \nu/\omega$. Note that, in the collisional case, it is the velocity which is phase-shifted so that it has a component which is in phase with $\underset{\sim}{B}$. There is thus a collisional, radiative force $\underset{\sim}{f}_c = -(ne/c)\langle \underset{\sim}{v} \times \underset{\sim}{B}\rangle$ which can be written as

$$\underset{\sim}{f}_c = \frac{KI\bar{n}}{c}(\hat{e}\times\hat{b}) \tag{6}$$

where $I = (c/4\pi)|\langle \underset{\sim}{E} \times \underset{\sim}{B}\rangle|$ is the energy (Poynting) flux and the absorption coefficient K is defined by $\bar{n}[1 + (\nu/\omega)^2]\, Kc = (\omega_p/\omega)^2\nu$. The collisional force is in the propagation direction since it is due to laser field momentum transferred to the plasma during absorption. A more general derivation of the radiative force is given next-including both the ponderomotive force and the collisional force. This radiative force is shown to generate slowly varying magnetic fields.

MAGNETIC FIELD GENERATION[2]

Theory

A cold, collisional, unmagnetized plasma is considered in this section. The ponderomotive and collisional forces can then be expressed simply in terms of the real and imaginary parts of a dielectric function. Thus, in this section, complex notation is used. The only time variations assumed are high frequency sinusoidal variations ($\partial/\partial t \rightarrow -i\omega$). The complex representation of a real high frequency vector $\underset{\sim}{A}$ (field or current) is denoted by $\underset{\sim}{A}'$. Thus $\underset{\sim}{A} = \mathrm{Re}(\underset{\sim}{A}')$, where $\underset{\sim}{A}' = \underset{\sim}{A}_o e^{-i\omega t}$. The time average $\langle \underset{\sim}{A} \circ \underset{\sim}{B}\rangle$ of a general (scalar, cross or tensor) product of two such real vectors can be expressed as $\frac{1}{2}\mathrm{Re}(\underset{\sim}{A}_o \circ \underset{\sim}{B}_o^*)$.

In order to define the dielectric function ϵ, the high frequency current density and electric field are related by a complex conductivity: $\underset{\sim}{J}' = \sigma \underset{\sim}{E}'$. Thus, $c\nabla \times \underset{\sim}{B}' = \partial \underset{\sim}{E}'/\partial t + 4\pi\sigma \underset{\sim}{E}'$ and can be written as $-i\omega\epsilon \underset{\sim}{E}'$ where $\epsilon = 1 + i4\pi\sigma/\omega$. It is stressed that the dielectric function ϵ is a convenient parameter only (not a constitutive relation) and that the vacuum form of Maxwell's equations are used. Note, however, that $\nabla \cdot (\epsilon \underset{\sim}{E}') = \nabla \cdot (\nabla \times \underset{\sim}{B}') = 0$ where $\epsilon = \epsilon_r + i\epsilon_i$.

The radiation pressure $\underset{\approx}{P}^r$ is defined in Ref. (2) except that, when collisions are present, the symbol ϵ represents the real part, $1 - \omega_p^2/(\omega^2 + \nu^2)$, of the dielectric function.[3,4] When $\nu/\omega \ll 1$, the presence of ν in ϵ has little effect on the radiative force and, as will be shown, has little effect on the magnetic field generation associated with the radiative force. It is, in fact, the imaginary part ϵ_i which is of primary importance in magnetic field generation. Nevertheless, the correct general form of the radiation pressure is of current interest and is used here in the derivation of the radiative force. Thus, a general derivation is given - not requiring the explicit introduction of collision frequency.

The radiation pressure[2,3,5,6] is the sum of a pure field pressure $\underset{\approx}{P}^f$ and a quiver pressure $\underset{\approx}{P}^q$. Field pressure (negative of Maxwell stress tensor) is $U\underset{\approx}{I} - \langle \underset{\sim}{EE} + \underset{\sim}{BB}\rangle/4\pi$, where $U = \langle(\underset{\sim}{E})^2 + (\underset{\sim}{B})^2\rangle/8\pi$ is field energy density. The quiver pressure is $mn\langle\underset{\sim}{vv}\rangle$ where the electron quiver velocity $\underset{\sim}{v}$ is related to the oscillating current density by $\underset{\sim}{J} = n(-e)\underset{\sim}{v}$. In complex notation, the oscillating current $\underset{\sim}{J}'$ and electric field $\underset{\sim}{E}'$ are related by a resistivity $\rho = \rho_r + i\rho_i$ or a conductivity $\sigma = \sigma_r + i\sigma_i$ according as one writes $\underset{\sim}{E}' = \rho\underset{\sim}{J}'$ or $\underset{\sim}{J}' = \sigma\underset{\sim}{E}'$. The real resistivity represents, as in the low frequency case, the frictional (due to collisions or plasma waves) dissipation of energy while the imaginary resistivity represents electron inertial effects. Since σ_r and σ_i each depend on both ρ_r and ρ_i, this separation of frictional and inertial effects applies only for resistivity and not for conductivity. One can thus evaluate the imaginary resistivity independently of the dissipative effects. Balancing the oscillating electric force $-e\underset{\sim}{E}$ with the linearized inertial force $m\partial\underset{\sim}{v}'/\partial t = -i\omega m\underset{\sim}{v}'$ gives the imaginary part of the resistivity as

$$\rho_i = \frac{-4\pi\omega}{\omega_p^2} = \frac{-\sigma_i}{|\sigma|^2} \tag{7}$$

where $\omega_p = \sqrt{4\pi e^2/m}$ is the plasma frequency. This relation is used to express the quiver pressure in terms of the electric field. Using $\underset{\sim}{J}' = \sigma\underset{\sim}{E}'$ gives the quiver pressure as $\underset{\approx}{P}^q = (m/2ne^2)$ times $\mathrm{Re}(\sigma\underset{\sim}{E}_0\sigma^*\underset{\sim}{E}_0^*)$ or $\underset{\approx}{P}^q = (m|\sigma|^2/ne^2)\langle\underset{\sim}{EE}\rangle$. Then using Eq. (7) and $4\pi\sigma_i/\omega = 1-\epsilon_r$ gives $\underset{\approx}{P}^q = (1-\epsilon_r)\langle\underset{\sim}{EE}\rangle/4\pi$. The total radiation pressure can thus be written as

$$\underset{\approx}{P}^r = \frac{1}{4\pi}\left[\tfrac{1}{2}\langle\underset{\sim}{E}^2 + \underset{\sim}{B}^2\rangle\underset{\approx}{I} - \langle\epsilon_r\underset{\sim}{EE} + \underset{\sim}{BB}\rangle\right] \tag{8}$$

The desired expression for the radiative force is derived, for sinusoidal time variations, from[6] $f^r = -\langle\nabla\cdot\underset{\approx}{P}^r + \partial\underset{\sim}{g}/\partial t\rangle$. The field momentum density is defined as $\underset{\sim}{g} = \langle\underset{\sim}{E}\times\underset{\sim}{B}\rangle/4\pi c$. Expanding $\nabla\cdot\underset{\approx}{P}^r$ with the usual vector-tensor indentities, using the vacuum Maxwell's equations with $\nabla\cdot(\epsilon\underset{\sim}{E}') = 0$ (as noted previously), and time-averaging the real expressions gives

$$\underset{\sim}{f}^r = \frac{\epsilon_r - 1}{8\pi}\nabla\langle\underset{\sim}{E}^2\rangle + \epsilon_i\omega\underset{\sim}{g} \tag{9}$$

plus a term $(1/8\pi)\,\mathrm{Im}[(\epsilon_i\underset{\sim}{E}')(\nabla\cdot\underset{\sim}{E}^*)]$ depending on spatial variations in the phase of $\underset{\sim}{E}'$. This term is not considered here but could be important when the phase changes appreciably over a laser wavelength.

Equation (9) shows that ϵ in the usual ponderomotive force (first term) should be replaced by its real part and that, through the imaginary part of ϵ, there is a term depending on the net field momentum $\underset{\sim}{g}$. The first term is in the direction of the laser intensity gradient while the second term is in the propagation direction. When there is no net field mementum (as with standing waves) or when absorption is negligible ($\epsilon_i = 0$), then the force is primarily in the direction of the intensity gradient. This was the case considered by Lindl and Kaw.[7]

If an effective collision frequency ν is defined by $4\pi\sigma = \omega_p^2/(\nu - i\omega)$, the field momentum term is the expected force associated with inverse bremsstrahlung absorption. Thus, $\epsilon_i\omega\underset{\sim}{g} = K\underset{\sim}{I}n/c$ where $\underset{\sim}{I} = c^2\underset{\sim}{g}$ is the laser intensity or time-averaged energy flux and the absorption coefficient K is defined by $\bar{n}[1 + (\nu/\omega^2]K = (\omega_p/\omega)^2(\nu/c)$. The refractive index is denoted by $\bar{n}$.

The magnetic source term $\underset{\sim}{S}$ (initial time growth rate of the field) can be expressed[5] in terms of the radiation-induced electric field $\underset{\sim}{E}^r = \underset{\sim}{f}^r/ne$ as $\underset{\sim}{S} = -c\nabla \times \underset{\sim}{E}^r$. Contributions to $\underset{\sim}{S}$ from the two terms in Eq. (9) depending on the real and imaginary parts of ϵ are denoted respectively by $\underset{\sim}{S}_1$ and $\underset{\sim}{S}_2$. For $\nu/\omega \ll 1$ we have

$$\underset{\sim}{S}_1 \cong -(c/8\pi n^* e\omega^2)\ \nabla\nu^2 \times \nabla\langle \underset{\sim}{E}^2\rangle \tag{10}$$

$$\underset{\sim}{S}_2 \cong -(1/n^* ec)\ \nabla \times (\nu\underset{\sim}{I}) \tag{11}$$

where n^* is the critical electron density.

The ratio of S_1 to S_2 when $\nu/\omega \ll 1$, is roughly (assuming similar angular variations and a common scale length L) $(|\underset{\sim}{E}|/|\underset{\sim}{B}|)(\nu/\omega)\ /2k_oL$ where $k_o = \omega/c$. Adiabatically, we would have $|\underset{\sim}{E}|/|\underset{\sim}{B}| \sim \omega/\nu$ at the critical density so that $S_1/S_2 \sim 1/kL = \lambda_o/4\pi L$ and is thus generally small. The fields are non-adiabatic near the critical region so that there is some uncertainty in this estimate. However, for oblique incidence, we have probably over estimated S_1/S_2 since the gradient scale length for $\langle E^2\rangle$ in S_1 is likely to be larger than the density gradient scale length.

Discussion

The simplicity of Eq. (11) compared to radiative source terms depending directly on a tensor formulation [3,6,8] makes this form useful for checking field geometries and for use in computer modelling. For example, the field due to resonant absorption[3] of a vertically polarized laser beam on a spherical pellet (assuming ∇ is radial and $\underset{\sim}{I}$ tangential at the turning point) is primarily above and below the center and in a horizontal direction.

A comparison of the magnetic source terms derived here with those depending explicitly on $\underset{\approx}{P}^r$ requires understanding the role of collisions. The primary effect of collisions on magnetic field generation is through the transfer of field momentum during absorption. This is treated explicitly in Eq. (11). However, the explicit dependence of $\underset{\approx}{P}^r$ on collision frequency only comes through its rather small effect on ϵ_r. The much larger effect is through ϵ_i (which is proportional to collision frequency). When working directly with $\underset{\approx}{P}^r$ this dependence on ϵ_i is contained implicitly in the spatial variations of fields or intensity. When the gradients are consistent with absorption and reflection, then operating on $\underset{\approx}{P}^r$ (even ignoring the dependence of ϵ_r on collision frequency) includes the major effects of absorption and reflection, i.e., field momentum transfer.

A connection between the two types of source terms can be made by noting that the laser intensity $\underset{\sim}{I}$ in Eq. (11) is proportional to the difference in the radiation pressure $\underset{\approx}{P}^r$ components parallel and perpendicular to the propagation direction. For a plane wave, we have $I = (c/\bar{n})(P^r_{\parallel} - P^r_{\perp})$. Thus, Eq. (11) is seen to represent the field generation associated with an anisotropy in $\underset{\approx}{P}^r$. This source dependence on the anisotropy in $\underset{\approx}{P}^r$ is explicitly given when evaluated directly from the tensor formalism and when spatial variations are assumed [6,8] which rule out the scalar source (Eq. (10)).

One can make a rough estimate of the radiative source without a detailed knowledge of the fields. The source is estimated at the turning point where resonance absorption is important. It is assumed that all quantities vary along the target normal (x- direction) and that (due to absorption) the intensity $\underset{\sim}{I}$ also varies along the propagation direction (y-direction). Then $\underset{\sim}{S}_2 = -\hat{z}\,(1/n^*ec)\partial(\nu I)/\partial x$ so that one can estimate the source as $S_2 \sim (I\omega/n^*ec)\;(\nu/\omega) \times (1/L)$ where L is the gradient scale length. For this purpose, it is useful to note that $I\omega/n^*ec$ in MG -μm/psec is 1.254×10^{-14} times I in W/cm^2 at the 1.06 μm wavelength.

For example, in the resonant absorption case considered in Ref. (3), the density scale length L is 3.18 μm and the effective collision frequency associated with resonant absorption is $\nu/\omega = .08$ (observed in the simulation in Ref. (3)). Then, using the incident intensity of $I = 2.2 \times 10^{16}$ W/cm^2, gives $S_2 = 6.9$ MG/psec, in good agreement with the values obtained in Ref. (3) by the analytic treatment (6 MG/psec) and by the simulation (7.4 MG/psec). The calculation using Eq. (11) shows that magnetic field generation results from field momentum transfer during absorption and not from absorption itself.

Growth of the magnetic field will occur until it is limited by convection, diffusion or the duration of the generation process. The field development equation is [5]

$$\frac{\partial \underset{\sim}{B}}{\partial t} = \nabla \times (\underset{\sim}{V}_e \times \underset{\sim}{B}) + \frac{c^2}{4\pi\sigma} \nabla^2 \underset{\sim}{B} + \underset{\sim}{S} \tag{12}$$

where only the radiative contribution to the magnetic source term, S is considered and a constant, isotropic conductivity $\sigma = \omega_p^2/4\pi\nu'$ is used to define an effective collision frequency ν'.

Since $\underset{\sim}{V}_e = \underset{\sim}{V} - \underset{\sim}{J}/ne$ (where $\underset{\sim}{V}_e$, $\underset{\sim}{V}$ are electron and plasma velocities) there is a plasma convective $(\nabla \times (\underset{\sim}{V} \times \underset{\sim}{B}))$ as well as a current convective $(e^{-1} \nabla \times (\underset{\sim}{J} \times \underset{\sim}{B}/n))$ contribution to $\partial B/\partial t$. The current convective contribution is present even with fixed ions and is small compared to diffusion when the Hall field is small compared to the Joule (resistive drag) field. This occurs when the electron cyclotron frequency $\omega_c < \nu'$, the effective collision frequency. The plasma convective contribution is small compared to field diffusion when the field diffusion time $\tau_d < \tau_c$, the plasma convection time. When we can assume a common scale length L, $\tau_d = L^2/D_m = (L\omega_p/c)^2/\nu'$ and $\tau_c = L/V$. Thus, diffusion dominates plasma convection when $L < (c/\omega_p)^2(\nu'/V)$.

If plasma convection (first term) limits $(\partial \underset{\sim}{B}/\partial t \rightarrow 0)$ the radiative field growth $(\underset{\sim}{S}_2)$, then, at normal incidence, the cylindrical θ component of Eq. (12) is $(n^*ec)\partial(VB)/\partial Z = \partial(\nu I)/\partial r$. Thus, the convectively limited field is

$$B_c(\text{MG}) \sim 10^{-6} \frac{\nu}{\omega} \frac{\delta}{r} \frac{I(\text{W/cm}^2)}{V(\text{cm/sec})} \tag{13}$$

when δ and r are, respectively, axial and radial gradient scale lengths and $\underset{\sim}{V}$ is plasma flow velocity.

However, if diffusion (second term) limits the field growth $\underset{\sim}{S}_2$ then Eq. (12) becomes $(\nu/\nu')(c/\omega_p)^2\nabla^2\underset{\sim}{B} = (n^*ec)^{-1} \nabla \times \underset{\sim}{I}$ when a fixed collision frequency is assumed. The diffusively limited field is then

$$B_d(\text{MG}) \sim 10^{-17}\left(\frac{\ell\omega_p}{c}\right)^2 \frac{I(\text{W/cm}^2)}{r(\mu\text{m})} \frac{\nu}{\nu'} \tag{14}$$

where ℓ is the Laplacian scale length for $\underset{\sim}{B}$ and r is the gradient scale length for I. Note that, if the same effective collision frequency applies both to absorption (ν) and field diffusion (ν') then the collision frequency cancels. This would more likely apply to ion turbulence than resonance absorption.

The smaller of B_c or B_d is the better approximation to the actual field since the more rapid saturation mechanism will limit the growth. Note that the collision-free skin depth c/ω_p at the 1.06 μm critical density is 0.17 μm. Thus, for ℓ less than a few microns, diffusion could dominate and limit the fields to less than a few megagauss.

REFRACTION

Field momentum transferred to the plasma during absorption gives a solenoidal character to the force per electron. The solenoidal character is important since it leads to magnetic field generation. This suggests that other mechanisms leading to field momentum transfer could give a solenoidal character to the radiative force. Refraction is such a mechanism and is considered next in the ususal adiabatic approximation. It is shown that the approximate theory does describe a solenoidal radiative force per particle. However, the more exact theory (Eq. 9) shows, for a collision free plasma, that the radiative (ponderomotive) force is irrotational.[9] It is shown further that refractive force can be derived from the ponderomotive force by considering a localized beam.[10] This section gives insight into the limitations of approximate descriptions of radiative forces and shows in particular that refraction does not generate magnetic fields.

Consider first the direction of the radiative force implied by the usual description of refraction.[11] A ray undergoes a directional along its path given by $d\theta/ds = |\hat{k} \times \nabla \bar{n}|$. The change in direction of the $\underset{\sim}{k}$ - vector can be expressed in terms of the unit vector $\hat{k}$ (along the propagation direction) as $d\hat{k}/ds = (\hat{k}\cdot\nabla) = -\hat{k} \times (\hat{k} \times \nabla \bar{n})$. This directional change is usually thought of as a bending of the wave front due to phase velocity ($c/\bar{n}$, in a plasma) variations normal to the propagation direction. However, the corresponding group velocity ($c\bar{n}$) variations normal to $\hat{k}$ imply a change in the direction of field momentum density and energy flux.

Thus, thinking of $\hat{k}$ as a unit vector along the field momentum shows that the field momentum density changes in the $-\hat{k} \times (\hat{k} \times \nabla\bar{n})$ direction. The resulting radiative force on the plasma is in the opposite or $+\hat{k} \times (\hat{k} \times \nabla n) = -(\underset{\approx}{I}-\hat{k}\hat{k})\cdot\nabla\bar{n}$ direction. It is shown next that the total collision-free radiative force has, in the adiabatic approximation, a contribution along the $\hat{k} \times (\hat{k} \times \nabla\bar{n})$ direction and hence can be identified as resulting from refraction.

Adiabatically, $\epsilon E^2 = B^2$ and $\underset{\sim}{v}_g = c\bar{n}\hat{k}$. Thus using $\underset{\sim}{B}\underset{\sim}{B} = B^2$ times $[\underset{\approx}{I} - (\hat{e}\hat{e} + \hat{k}\hat{k})]$ in the general expression (Eq. (8)) for the radiation pressures gives

$$\underset{\approx}{P}^r \cong \frac{I(1-\epsilon)}{2\bar{n}c}\,\underset{\approx}{I} + \underset{\sim}{v}_g\underset{\sim}{g} \tag{15}$$

where the I is the magnitude of the energy (Poynting) flux $\underset{\sim}{I} = c^2\underset{\sim}{g}$. The directed radiation pressure $\underset{\sim}{v}_g\underset{\sim}{g}$ is what one would expect physically.

The steady-state radiative force $f \cong -\nabla\cdot\underset{\approx}{P}^r$ can now be evaluated. Using energy conservation $(\nabla\cdot\underset{\sim}{I} = 0)$, $I/\bar{n} = (c/4\pi)\langle\underset{\sim}{E}^2\rangle$ and noting that $(\hat{k}\cdot\nabla)\bar{n}\hat{k} = (\hat{k}\hat{k}\cdot\nabla)\bar{n}$ (since $\hat{k}$ has no variation along $\hat{k}$) gives

$$\underset{\sim}{f} \cong \frac{\epsilon - 1}{8\pi}\,\nabla\langle\underset{\sim}{E}^2\rangle - \frac{I}{c}(\underset{\approx}{I} - \hat{k}\hat{k})\cdot\nabla\bar{n} \tag{16}$$

This approximate radiative force thus has the usual ponderomotive force plus a contribution due to refraction. Note, however, that more exact expression for the collision free radiative force (the first term in Eq. (9)) does not contain the refractive force explicitly. The use of a localized (adiabatic) description has not preserved the spatial (irrotational) character of the radiative force per particle. Thus, the approximate theory does not show that refraction does not generate time-averaged magnetic fields. In fact, the field momentum transfer during refraction and reflection (as a limiting case of refraction) does not generate magnetic fields.

The refractive force is contained within the ponderomotive force. This can be seen most easily by considering a beam of limited lateral extent and no axial variation. The beam axis is taken to be the z-axis with $\hat{k}$ a unit vector in the +z direction. It is assumed that $\langle\underset{\sim}{E}^2\rangle$ is a function of (x,y) only and vanishes on the surface of a bounding cylinder - as in Fig. 1. Consider the force on a volume element in Fig. 1. The total force $\underset{\sim}{F} = (8\pi)^{-1}\int(\epsilon-1)\nabla\langle\underset{\sim}{E}^2\rangle dV$ is expressed as the sum of a volume integral $\underset{\sim}{F}_V$ and surface integral $\underset{\sim}{F}_S$ by integrating the equation

$$(\epsilon-1)\nabla\langle E^2\rangle = \nabla[(\epsilon-1)\langle\underset{\sim}{E}^2\rangle] - \langle E^2\rangle\nabla(\epsilon-1) \tag{17}$$

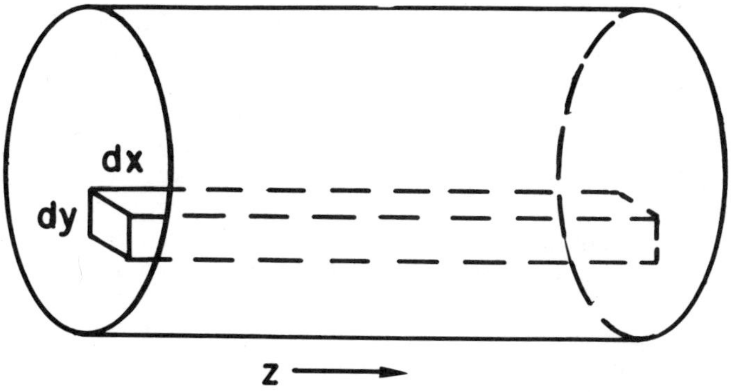

Figure 1. Cylinder bounding a laser beam. $\langle \underset{\sim}{E}^2\rangle$ is a function of x, y only and vanishes on the cylindrical surface.

The volume force is $-(8\pi)^{-1}\int \langle \underset{\sim}{E}^2\rangle \nabla\epsilon dV$ and the surface force is $-(8\pi)^{-1}\oint (\epsilon-1)\langle \underset{\sim}{E}^2\rangle d\underset{\sim}{S}$. Consider the surface force on the volume element in Fig. 1. The surface forces on the sides of the element will cancel in pairs with adjacent elements so that the total force on the sides is just that of the total beam and vanishes. The surface force on the ends of the element is $(8\pi)^{-1}\langle E^2\rangle[\epsilon(z+dz)\hat{k}dxdy + \epsilon(z)(-\hat{k})dxdy]$, where $dxdy[\epsilon(z+dz) - \epsilon(z)] = dxdy\, \nabla\epsilon\cdot(\hat{k}dz) = (\hat{k}\cdot\nabla\epsilon)dV$. Thus the surface integral, for this case, can be converted to the volume integral $-(8\pi)^{-1}\int_V \langle \underset{\sim}{E}^2\rangle\hat{k}(\hat{k}\cdot\nabla\epsilon)dV$. Expressing the magnitude I of the Poynting flux as $(\bar{n}c/4\pi)\langle E^2\rangle$ and noting $\nabla\epsilon=\nabla\bar{n}^2=2\bar{n}\nabla\bar{n}$ then gives the total force on the beam as $-c^{-1}\int_V I(\underset{\approx}{I} - \hat{k}\hat{k})\cdot\nabla\bar{n}dV$. Thus, the effective force per unit volume on a beam of finite lateral extent and having no axial variation is the refractive force (second term in eq. (16)).

Although there is a net field momentum transferred to the plasma during refraction or reflection, the refractive force is included in the ponderomotive force and hence does not lead to magnetic field generation. Also note that reflection as a limiting case of refraction does not generate magnetic fields. These conclusions are based on the Ohm's law, for a cold, unmagnetized plasma and could be modified somewhat when the current and electric field are related by a more general electron equation of motion. However, the general conclusion is that radiative forces generate magnetic fields primarily through absorption and not through refraction or reflection.

RADIATIVE FORCES IN A MAGNETIZED PLASMA

Introduction

Plasma inhomogeneity and collisions were discussed earlier as mechanisms leading to a time-averaged radiative force. The slowly varying (spontaneous) magnetic fields, generated either by thermal or radiative forces, are now shown to be a third mechanism for producing radiative forces.[12] This magnetic radiative force is polarization dependent, lying along the oscillating electric field, and can be comparable in magnitude to the ponderomotive force. It can thus introduce an appreciable azimuthal asymmetry about the propagation direction.

The zeroth order motion of an electron is a linear oscillation along the oscillating electric field. Due to the oscillating magnetic field there are also excursions along the propagation direction so that the orbit is a figure 8. For an isolated electron in an electromagnetic field the orbit is closed and there is no time-averaged force. In presence of plasma inhomogeneity, the orbit is open and drifts toward regions of lower laser intensity.[1] This is the ponderomotive force. In the presence of collisions, the orbit is open and drifts in the propagation direction.[2] This is the absorptive force. The magnetic force can also be qualitatively understood in terms of a modification to the figure 8 orbit.

On the left in Fig. 2 is pictured the orbit of an electron in the presence of purely oscillating fields ($\underset{\sim}{E}$, $\underset{\sim}{B}$). On the right is shown how the orbit is modified by a steady magnetic field $\underset{\sim}{B}_1$

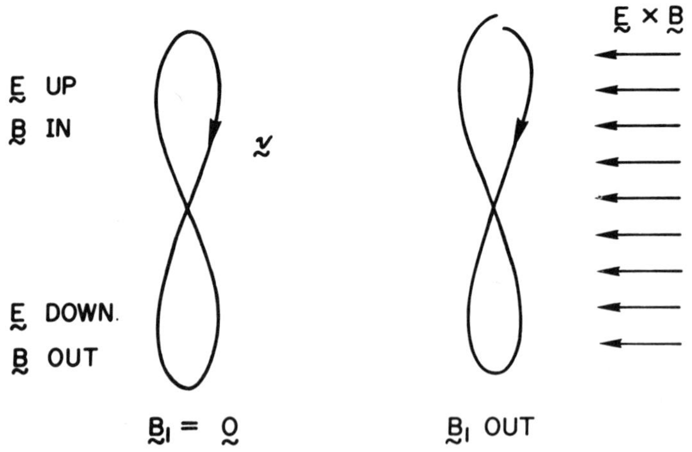

Figure 2. Modification of electron orbits by a slowly varying magnetic field.

out of the page. The primary velocity is vertical due to the vertically oscillating electric field. The oscillating magnetic field $\underset{\sim}{B}$ is in and out of the page. The $\underset{\sim}{v} \times \underset{\sim}{B}$ force introduces excursions along the propagation direction to form the figure 8 orbit.

The $\underset{\sim}{v} \times \underset{\sim}{B}_1$ force at the bottom aids the restoring $\underset{\sim}{v} \times \underset{\sim}{B}$ force so that the orbit is tighter. The $\underset{\sim}{v} \times \underset{\sim}{B}_1$ force at the top opposes the $\underset{\sim}{v} \times \underset{\sim}{B}$ force so that the orbit balloons out. The $\underset{\sim}{v} \times \underset{\sim}{B}$ force is inward at both orbit extremes whereas the $\underset{\sim}{v} \times \underset{\sim}{B}_1$ force is upward at both extremes. Thus the steady magnetic field has introduced an orbit asymmetry in the vertical direction, i.e., along the oscillating electric figure. The orbit is no longer closed and there is a time-averaged motion in the vertical direction.

Theory

To show the result quantitatively, consider the Lorentz equation for one electron in the presence of a uniform electromagnetic wave $(\underset{\sim}{E}, \underset{\sim}{B}) = (\underset{\sim}{E}_o, \underset{\sim}{B}_o) \sin(\omega t-kz)$ and a uniform dc magnetic field $\underset{\sim}{B}_1$.

$$m \frac{d\underset{\sim}{v}}{dt} = -e(\underset{\sim}{E} + \frac{1}{c} \underset{\sim}{v} \times \underset{\sim}{B} + \frac{1}{c} \underset{\sim}{v} \times \underset{\sim}{B}_1) \tag{18}$$

First, consider the ordering of the terms. Non-relativistically, the primary oscillatory velocity is $\underset{\sim}{v}_e = e\underset{\sim}{E}/m\omega)\cos(\omega t-kz)$. The higher orders $\underset{\sim}{v}_e \times \underset{\sim}{B}$ term produces an oscillating kz(t) variation of order $O(v_e/c)O(\tilde{B}_o/B^*)$, where $B^* = \omega mc/e$ is about 100 MG. For B_1 fields less than 10 MG we can ignore the quivering kz(t) variation in the $\tilde{E}$ and $\tilde{B}$ fields. The terms are ordered such that $E = O(B) = O(B_1)$ and $v_e/\tilde{c} \ll 1$. It can be shown that the initial thermal component of $\underset{\sim}{v}$ is decoupled from the effect of B_1 so that, for simplicity, the thermal component of $\underset{\sim}{v}$ is ignored.

There are two first order corrections to the oscillatory velocity $\underset{\sim}{v}_e$. The magnetic force $\underset{\sim}{v}_e \times \underset{\sim}{B}$ produces the standard second harmonic perturbation corresponding to the figure 8 motion. The $\underset{\sim}{v}_e \times \underset{\sim}{B}_1$ term produces an oscillatory velocity

$$\underset{\sim}{v}_1 = (B_1/B^*)\, v_e \sin \omega t\; \hat{b}_1 \times \hat{e} \tag{19}$$

where $v_e = eE_o/m\omega$ and $\hat{b}_1$, $\hat{e}$ are unit vectors along $\underset{\sim}{B}_1$ and $\underset{\sim}{E}_o$.

This magnetic perturbation velocity $\underset{\sim}{v}_1$ is in phase with the oscillating fields and thus produces a second-order, low frequency force $\underset{\sim}{f}_b = -(ne/c)\langle \underset{\sim}{v}_1 \times \underset{\sim}{B}\rangle$ which can be written as

$$\underset{\sim}{f}_b = \left(\frac{\omega_p}{\omega}\right)^2 \frac{\omega_c}{c} \frac{I}{c} \cos\theta\, \hat{e} \tag{20}$$

where $\omega_p^2 = 4\pi n e^2/m$, $\omega_c = eB_1/mc$, $I = (c/4\pi)|\langle E \times B\rangle|$ is the laser intensity, and θ is the angle between B and B_1.

Discussion

Equation (20) shows that the magnetic part of the radiative force is primarily along the laser electric field-as noted earlier. This laser polarization dependence of the radiative force directly affects the plasma dynamics and indirectly (through its effect on magnetic field generation) affects the heat flux. The problem is basically three-dimensional since there is no longer azimuthal symmetry about the laser beam. The importance of the magnetic force depends on its magnitude relative to the radiative forces on an unmagnetized plasma. As shown earlier, these can be written as the sum of a collisional force f_c representing field momentum transfer during absorption and a (deconfining) ponderomotive force f_d due to field inhomogeneity. The collisional force is $f_c = (\omega_p/\omega)^2(\nu/c)I/c$, where I is laser intensity. The ponderomotive force is $f_d \cong -(8\pi)^{-1}(\omega_p/\omega)^2\nabla\langle E^2\rangle$. On can thus roughly compare the magnitudes of the contributions to the radiative force according to the characteristic lengths: $f_b:f_c:f_d = (\omega_c/c):(\nu/c):|\nabla|$. For some conditions of interest to laser-induced fusion ($n = 10^{21}$ cm^{-3}, $kT_e = 1$ keV, $B = 5$ MG, $\bar{Z} = 10$, $\ell n\Lambda = 10$) we have $c/\nu = 30$ μm and $c/\bar{\omega}_c = 3$μm. Gradient[1] scale lengths on the order of microns are seen in numerical studies. Thus, the magnetic force may be of the same order as the other radiative forces and should be considered in laser-fusion applications.

Momentum conservation considerations give some insight into the radiative forces. Since f_b is perpendicular to the laser field momentum, it does not directly involve a laser field momentum transfer to the plasma. The radiative force acts on the plasma through the slowly varying magnetic field. The component of f_b perpendicular to B_1 produces a net drift along $f_b \times B$ and there is a net acceleration along B_1.

The electric field or force per particle associated with the absorptive force and magnetic force are, in contrast with the ponderomotive force, nonpotential. This is essentially due to their dependence on the laser propagation direction and polarization. They can thus generate magnetic fields and the problem of a field dependent force generating magnetic fields must be solved self-consistently.

REFERENCES

1. H. Hora in Laser Interaction and Related Plasma Phenomena, (H. Schwarz and H. Hora, Eds.), Plenum, New York 1971, p. 387.

2. J. A. Stamper, Phys. Fluids 19, 758 (1976).

3. J. J. Thomson, Claire Ellen Max and Kent Estabrook, Phys. Rev. Lett. 35, 663 (1975).

4. J. A. Stamper. NRL Memo Report 2775, April, 1974, p. 7.

5. J. A. Stamper and D. A. Tidman, Phys. Fluids 16, 2024 (1973).

6. J. A. Stamper, O. C. Barr, J. Davis, G. A. Doschek, C. M. Dozier, U. Feldman, B. M. Klein, W. M. Manheimer, E. A. McLean J. M. McMahon, D. J. Nagel, D. A. Tidman, R. R. Whitlock, K. Whitney, N. K. Winsor, and F. C. Young, in Laser Interaction and Related Plasma Phenomena, Edited by H. Schwarz and H. Hora, (Plenum, New York, 1973), Vol 3B, p. 750.

7. John D. Lindl and Predhiman Kaw, Phys. Fluids 14, 371 (1971).

8. J. A. Stamper, NRL Report 7411 (1972).

9. This was pointed out to the author by S. E. Bodner.

10. R. H. Lehmberg, private communication.

11. M. Born and E. Wolf, Principles of Optics, 2nd Edition (Pergamon, New York 1964), p. 121

12. J. A. Stamper and S. E. Bodner, Phys. Rev. Lett. 37, 435 (1976).

MEASUREMENTS OF SPONTANEOUS CURRENTS AND MAGNETIC FIELDS IN LASER-PRODUCED PLASMAS*

Michel G. Drouet

Direction Sciences de base
Institut de recherche de l'Hydro-Québec
Varennes, Québec, Canada J0L 2P0

An extensive experimental study of the associated electrical currents and magnetic fields generated in plasmas produced by nanosecond CO_2 and Nd-glass laser irradiation of a copper target in a background of air, N_2, Ar and He at pressure ranging from 10^{-5} to atmospheric was conducted. Space and time resolved measurements of current, magnetic field, UV radiation and laser pulse structure were performed. The results show correlations between a) the distribution of laser target irradiation and the distribution of emitted current, and b) the dependences, on the background pressure, of both the UV radiation and the magnetic field, and reveal a two-phase expansion - diffusion and convection - of the magnetic field.

I. INTRODUCTION

The spontaneous generation of a large magnetic field in a laser-produced plasma has, recently, been the subject of a large number of experimental and theoretical studies. These spontaneous magnetic fields are of importance in laser fusion if they are large enough to affect electron thermal conduction and plasma dynamics[1-7]. Apart from the recent measurements of Stamper and Ripin[8], who obtain direct evidence for the existence of megagauss

*Presented at the Fourth Workshop on "Laser Interaction and Related Plasma Phenomena," held at Rensselaer Polytechnic Institute, Troy, New York, November 12-18, 1976.

fields in the focal region of the laser, all measurements of the magnetic field[9-19] have been performed outside of the immediate focal region where, in fusion experiments, the effect of the field should be the most detrimental to a successful implosion of the laser irradiated pellet. These measurements were used to determine how the magnetic field scales with radius[11-18,20,21] as it expands from the focal region where it is generated. The magnetic field maximum was found to vary as $R^{-\nu}$ where R is the radius measured from the focal spot. Using these scaling parameters it was inferred, by extrapolation, that fields up to megagauss levels were possible at the focus. Such an extrapolation has been hampered, not only by the apparent lack of consistent measurements, ν varying between 1 and 4, but also by the lack of understanding of the

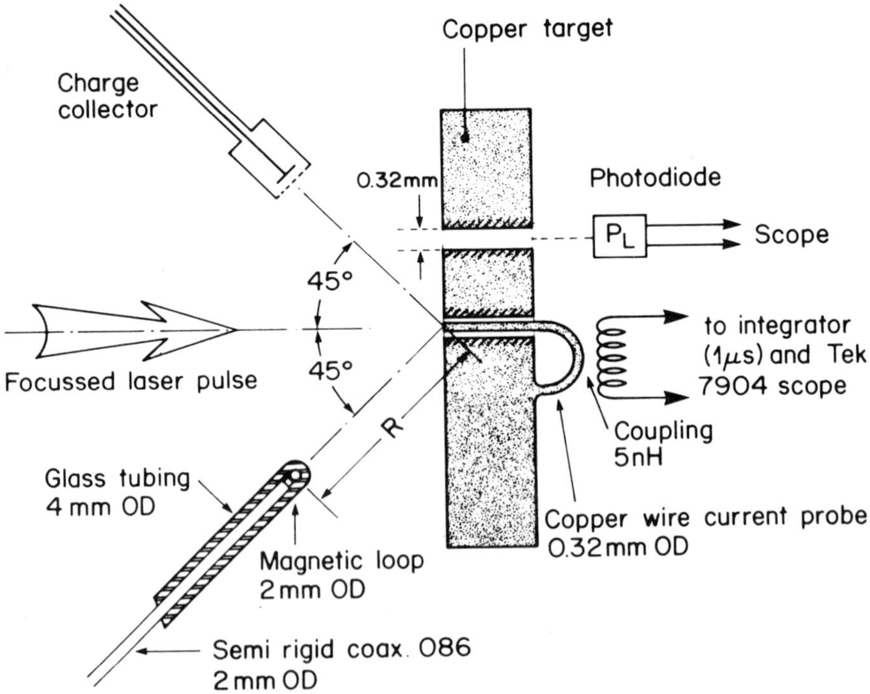

Fig. 1. Illustration of the experimental arrangement.

field topology at the plasma-target interface and of the development of the field, in the plasma, in front of the target.

In our experimental study, reviewed here, different diagnostics and conditions of plasma production were used. Two different lasers, working at 1.06 and 10.6 μm, produced the plasma by irradiation of a target in different background gases over a range of pressures. Extensive interdependent measurements, resolved in both space and time, of electric current, magnetic field, laser pulse intensity and UV radiation were obtained which, we believe, lead to a better understanding of the origin and development of the magnetic field in laser-produced plasmas. Although obtained with a conducting target, the present results should, to a large extent, be pertinent also for insulating target. It is believed that these results are of interest in laser fusion experiments because it now appears possible to infer the value of the magnetic field in the interaction region from measurements obtained far from the laser target thus corroborating direct measurements of the field by methods such as proposed by Stamper and Ripin[8], which, although representing a significant advance, are still difficult to use.

II. EXPERIMENTAL ARRANGEMENT

The plasma was produced in an experimental chamber by laser irradiation of a thick plane copper target at an ambient pressure ranging from 10^{-5} to 10^3 torr of air, argon, helium, CO_2 or N_2. The beam from a Nd-glass laser (0.1 - 4 J, 20 ns, 3 mrad) or a CO_2 laser (Valcartier[22], 1.7 ns, 2.5 Gw, 1.3 mrad) was focused at normal incidence on the target. The focal spot diameter in the absence of plasma was, in both cases, 250 μm at half intensity.

Four diagnostic techniques were used. They are shown in Fig. 1. Two of the diagnostic methods, the wire probe and the photodiode, are shown mounted behind the target. To determine the radial dependence of the physical quantity measured by either method, at the target surface, the target has to be moved transversely with respect to the laser axis. In contrast, the charge collector and the magnetic loop were inserted radially through ports of the experimental chamber. Their radial distance R, from the point of laser impact on the target surface, could be varied along a line making an angle of 45° with the incident laser beam.

III. MEASUREMENT OF THE CURRENT EMITTED FROM THE TARGET

The current flow, through the plasma-target interface, was

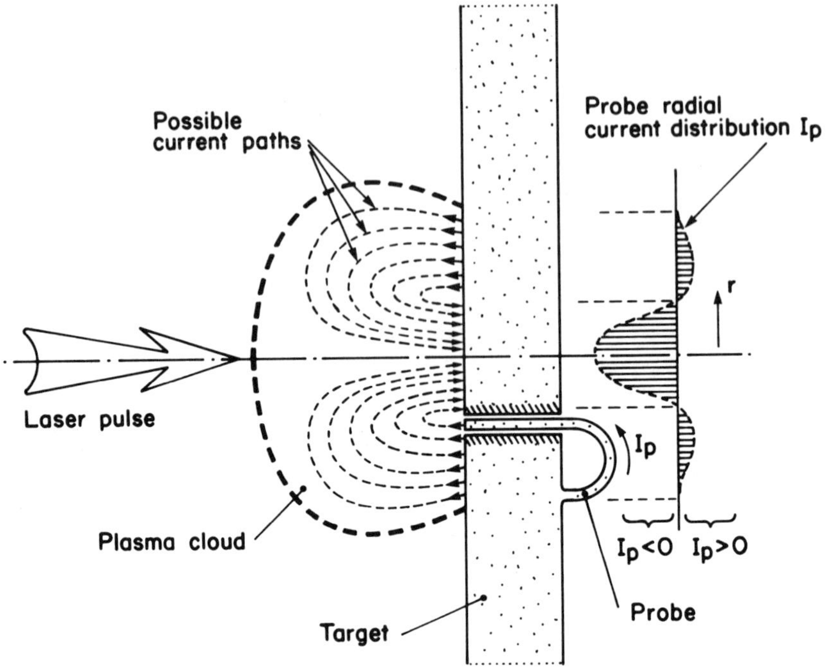

Fig. 2 Diagram showing the target with the wire probe sampling part of the current flow through the plasma-target interface.

monitored using a small wire probe (0.32 mm OD) embedded in the target[23] as shown in Fig. 1. The wire probe, insulated from the target surface, inductively couples the current element it carries with a secondary coil. The probe coil mutual inductance was 4×10^{-9} H and the time response less than 4 ns. The induced voltage in the coil, proportional to dI/dt, was integrated to give a signal corresponding to the current I in the probe. This signal was displayed on an oscilloscope. The current distribution at the surface of the target was obtained[24,25] by plotting the measured value of the current corresponding to the same delay with respect to the laser pulse for different positions of the probe relative to the laser axis. Assuming that the current paths are as illustrated in Fig. 2, then the distribution of the probe current may correspond to that also given in Fig. 2, as a function of r, with a central circular cathode region, with current flowing in the target and centered with respect to the laser axis, surrounded by an annular anode region with current flowing out of the target.

A typical oscilloscope recording of the voltage pulse induced in the secondary coil during an experiment, the probe being located on the laser axis, is presented in Fig. 3, with, for reference, the intensity profile of the Nd-glass laser pulse as monitored by a photodiode. We note in Fig. 3 that: a) the current flow appears simultaneously with the beginning of the laser irradiation of the target; b) it changes direction during the laser pulse build-up; c) it reaches a maximum, in absolute value, near the end of the laser pulse and then, d) it slowly decays. Except for the initial short lived precursor, which will be explained later, the shape of the dI/dt pulse, in Fig. 3, is comparable to that of the dB/dt variation published in the literature[11,13,26]. In addition, the direction of the axial current flow is the same as the laser-pulse propagation direction; this is in agreement with the direction of the azimuthal magnetic field reported previously by many authors[11,13,17,18,19,26].

A typical current distribution as measured by the probe, as a function of its position relative to the laser axis, is presented in Fig. 4. The distribution obtained shows a central cathode surrounded by an annular anode as discussed earlier in relation to Fig. 2. This current distribution is topologically consistent with that proposed by McKee et al[15] which they obtained by taking the curl of their measured magnetic field distribution. However, our results demonstrate the active participation of the target in the current flow process. This concept is illustrated schematically in Fig. 5b which can be contrasted with the previously suggested scheme of Fig. 5a.

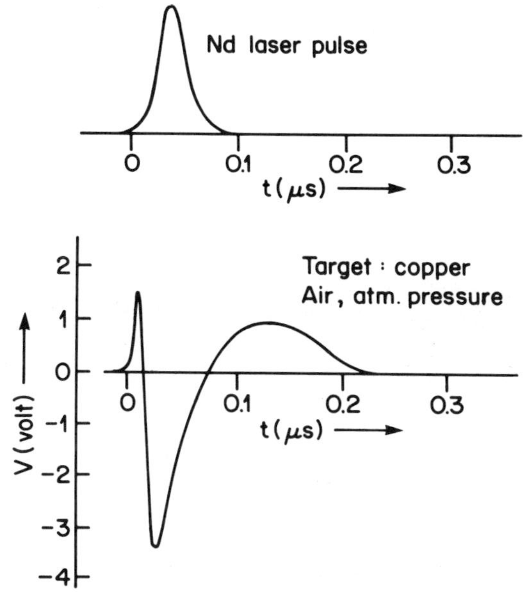

Fig. 3. Typical voltage pulse V corresponding to the probe dI/dt.

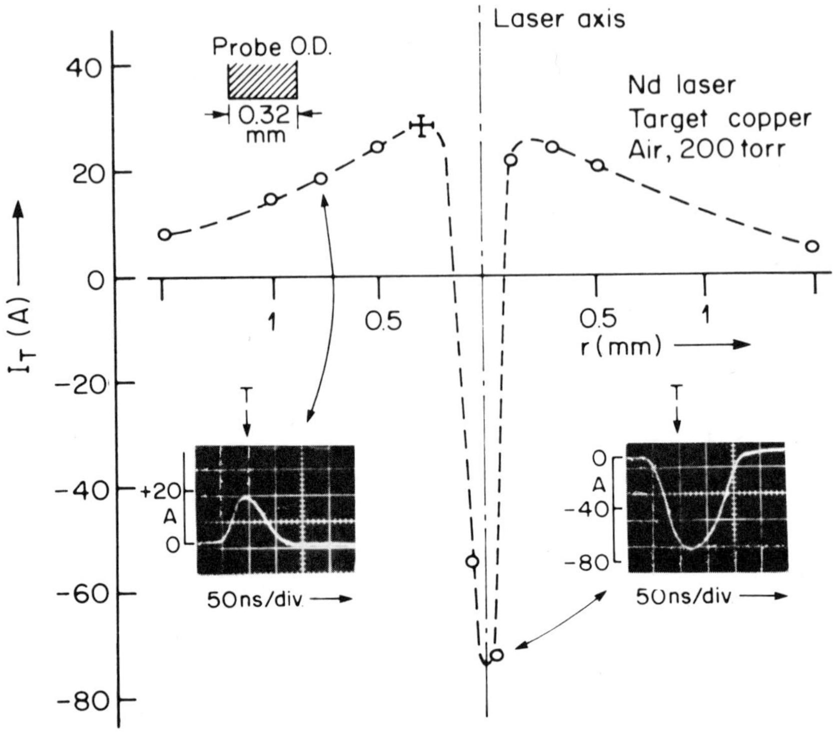

Fig. 4 Typical probe current distribution. A positive value implies conventional current flow across the target surface into the plasma.

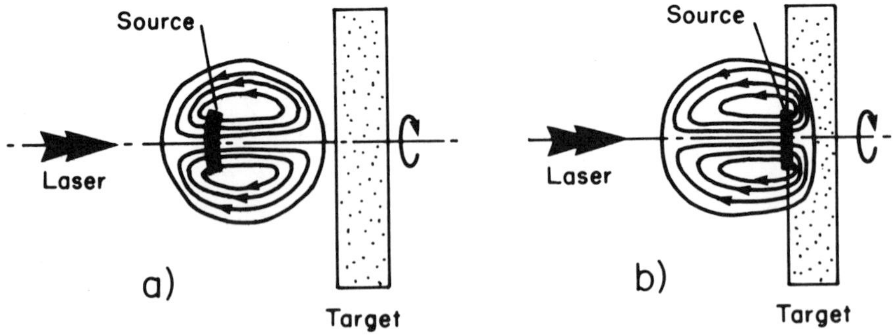

Fig. 5 Two different locations for the current flow paths. The source are located at the front of the expanding plasma (a) or at the plasma-target interface (b).

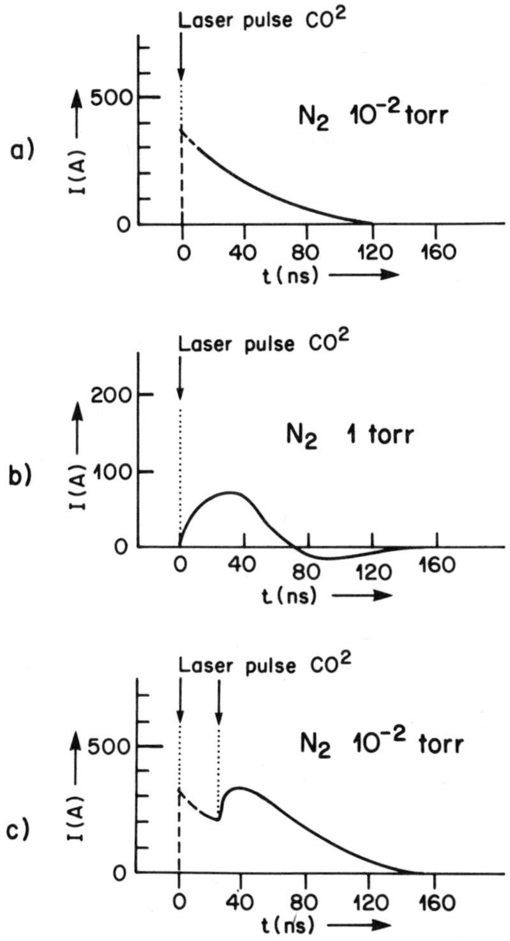

Fig. 6 Typical oscilloscope recordings of the current pulse in a CO_2 laser plasma. A positive value implies conventional current flow into the target.

Typical oscilloscope recordings of the current pulse monitored by the probe for three different experimental conditions in a CO_2 laser plasma are shown in Fig. 6. The shape of the pulse varies with pressure. In the case (Fig. 6c) of double-pulse irradiation, at 10^{-2} torr, the shape of the current pulse induced by the second laser pulse is different from that resulting from the first laser pulse; it can be more easity compared with the current pulse at higher pressure (b). This can be attributed to the fact that the gas pressure in the vicinity of the target must be much

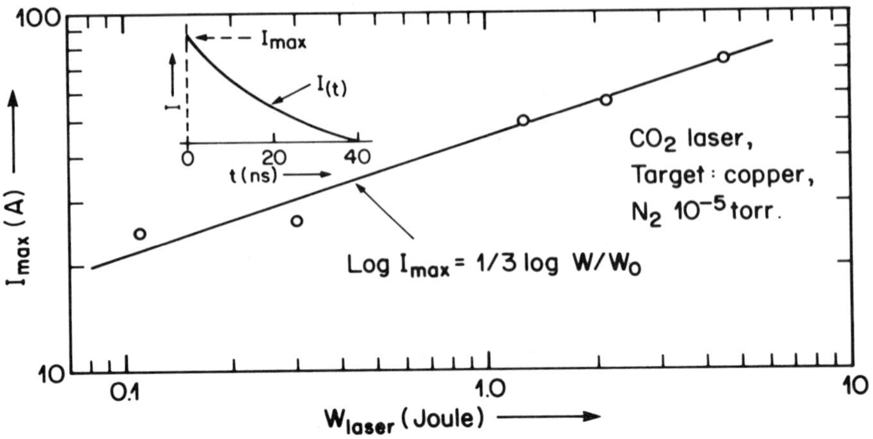

Fig. 7 Measured maximum value in the current pulse collected by the probe as a function of the laser energy.

higher than 10^{-2} torr when the second laser pulse is fired only 23 ns after the first pulse.

The relationship between the amplitude of the current collected by the probe and the laser energy is shown in Fig. 7, at a pressure of 10^{-5} torr nitrogen. If one assumes that the mechanism of conversion of the laser energy W, incident on the target, into the magnetic energy $\frac{1}{2}\ LI^2_{max}$ associated with the current I_{max}, is independent of laser power then, $\frac{1}{2}\ LI^2$ should be proportional to W. The third power dependence, shown in Fig. 7, suggests that the equivalent inductance L, which characterizes the volume of the whole current distribution, may be, itself, proportional to the current I_{max}. This possibility could be checked if the current flow pattern could be determined[13,15] by mapping the magnetic field in front of the target. Results of this type published to date[13,15] do not, unfortunately, allow such verification. However, it is interesting to note that in a free jet irradiated by laser, the dependence on the laser energy W of the magnetic field B, measured recently[27], was found to be $B \propto W^{1/3}$.

The relationship between the nitrogen background pressure and the current collected by the probe, at constant CO_2 laser power (2.5 GW, 1.7 ns), is shown in Fig. 8. Arifov et al[28], using a ruby laser, found a similar pressure dependence for the target net current (I_n in Fig. 9) with a maximum at 10^{-2} torr. We believe that the net target current, which has been used as a diagnostic by many authors[28,29,30] is only a small part of the total current emitted by the target. As illustrated in Fig. 9, it corresponds to that small part of the current which does not flow directly back to the target via closed loops through the photoionized gas in front of the target. Instead, the net current returns via the walls of the experimental chamber and is measured as a voltage drop through the resistor R.

The resonance-like behavior shown in Fig. 8 at 10^{-2} torr is much more evident than the variation with pressure of the net current (defined above) as measured by Arifov[28] and the variations with pressure of the magnetic field induced in laser-produced plasma, as measured for example by Schwirzke[13] and shown, for comparison, in Fig. 8.

At a pressure corresponding to the resonance (10^{-2} torr), the radial distribution of the current was measured by the probe. The results are presented in Fig. 10. Measurements of the current distribution at lower (10^{-5} and 10^{-3} torr) and higher (10^{-1} torr) pressures were also made. At these pressures, the current distributions were also characterized by the existence of a "virtual" cathode surface, centered on the laser axis, with conventional current flow into the target, surrounded by an annular anode surface with current flowing out of the target. The size of the

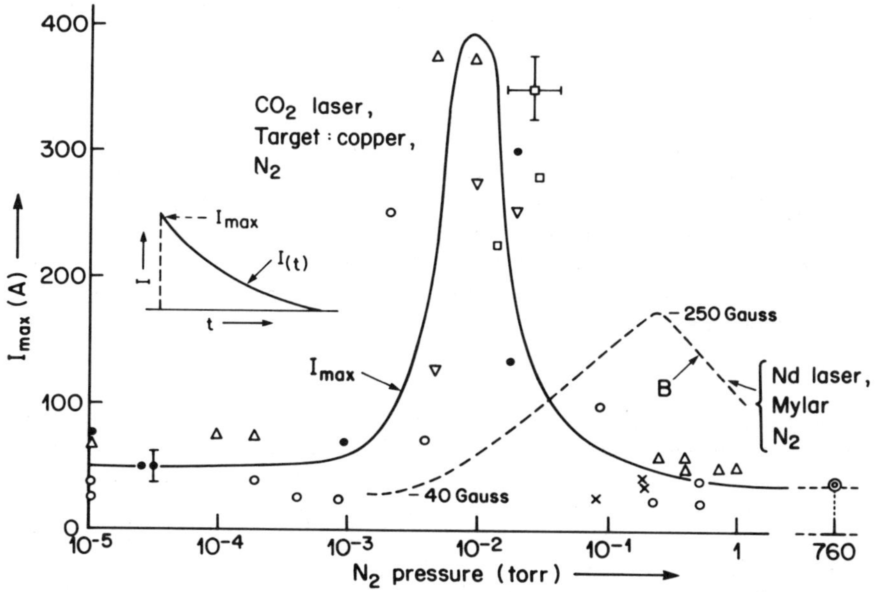

Fig. 8 Dependence of the probe current on the Nitrogen background pressure. The different symbols used correspond to different sets of experiments.

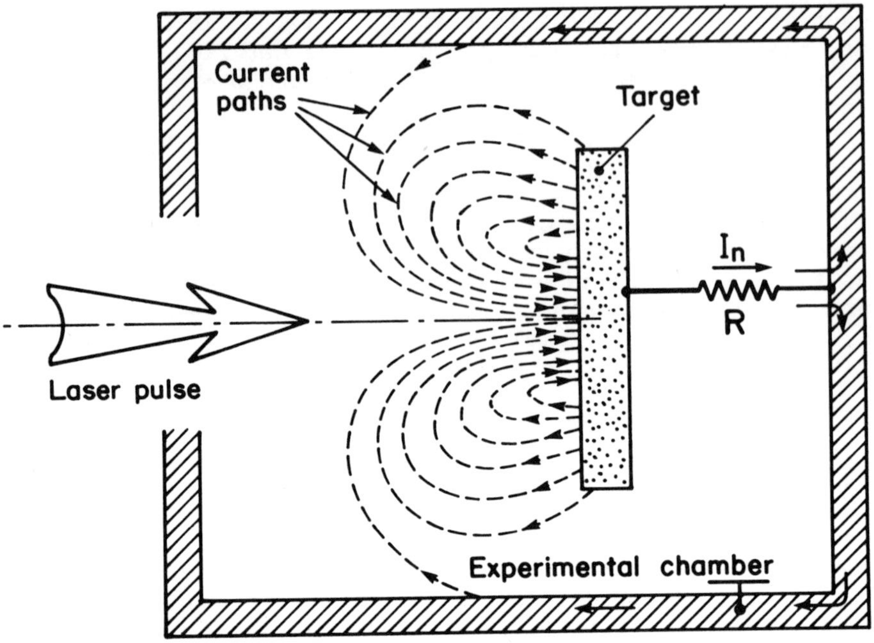

Fig. 9 A small part of the current, I_n, returns to the target via the walls of the chamber and is measured as a voltage drop across the resistor R.

cathode surface was found to be approximately 1 mm in diameter, except at about 10^{-2} torr, where it is 20 mm. The resonance effect at 10^{-2} torr in nitrogen manifests itself by both a larger current density and a larger cathode surface, in such a way that the total cathode current flow, estimated at approximately 60 A over the range of nitrogen pressure studied, reaches a value of 250 kA at the resonance point.

Two interpretations, illustrated in Fig. 11, have been offered[25] to explain the fact that the cathode surface is, in general, larger than the focal spot. The laser plasma produced in front of the target would be the source of intense radiation of energetic electrons which, by bombardment, would create, at the plasma-target interface, the necessary conditions[31] leading to secondary electron emission from the target surface. If the source is isotropic, the bombardment will vary as $\cos^3\theta$ with θ defined as shown in Fig. 11a. Such a non-uniform bombardment could produce the bell-shaped current distribution having the value given by the expression $I = [A \cos^3\theta - B]$ as shown in Fig. 10.

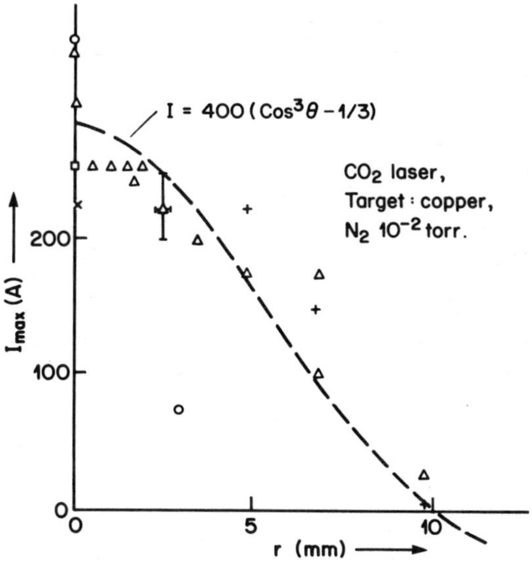

Fig. 10 Radial distribution of the probe current. A positive value implies conventional current flow into the target.

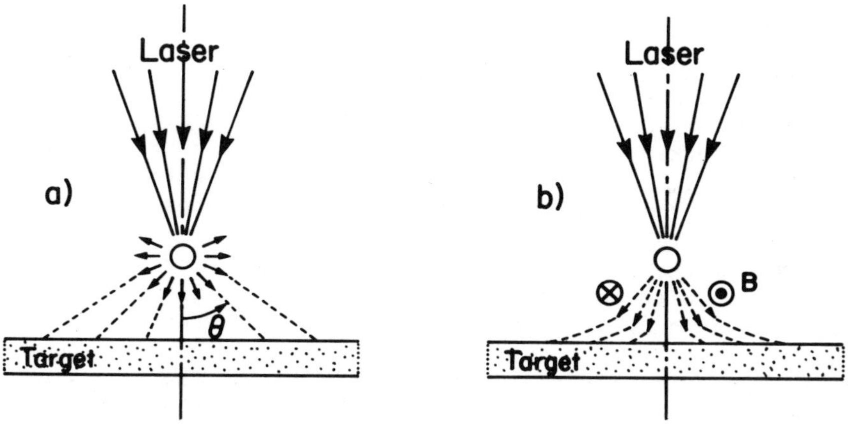

Fig. 11 a) Isotropic source producing a bombardment of the target proportional to $\cos^3\theta$.

b) The primary electrons produced by the source, are deflected by the magnetic field B of the higher secondary electron current.

IV. DISTRIBUTIONS OF THE LASER IRRADIATION AND THE EMITTED CURRENT

The existence of a large electrical current flow in the plasma produced by laser irradiation of a conducting target has been discussed above. In this section we discuss the relation between the distribution of the 1.06 μm laser irradiation, and the distribution of the electrical current flowing through the plasma-target interface for a plane copper target. The experiment is illustrated in the lower right corner of Fig. 12. The Nd-glass laser (1.06 μm, 4 J, 30 ns, 3 mrad) was partially focused by a lens of focal length 150 mm, the focal point being 15 mm below the target surface. A light probe, P_L, located behind the solid copper target, was used to monitor the intensity of the laser irradiation through a small orifice in the target. By varying the position of the probe-target assembly, different recordings of the intensity of irradiation are obtained which correspond to different regions in the laser beam.

Typical laser irradiation distribution across a diameter of the impact footprint are shown in the lower left corner, Fig. 12. For these measurements, performed in a nitrogen atmosphere of 10^{-4} torr, the total laser energy was reduced to 0.1 J in order to avoid breakdown at the target surface and the production of a plasma which might have affected the results. The measured profiles are interpreted as the envelopes of the contributions from three hot laser spots which are labelled (a), (b), and (c) on the diagram. The total duration of the laser pulse is about 80 ns.

The electrical current flow through the interface between the laser plasma and the target was monitored, (Fig. 12, lower right corner) using the wire probe technique discussed previously. Typical probe current distribution, obtained in a nitrogen atmosphere of 0.3 torr, across the diameter of the laser footprint are shown in Fig. 12 for a laser energy of 4 J. These results correspond to the same region of the laser footprint explored by the light probe. These profiles confirm the existence of a central cathode region, with conventional current flowing into the target, surrounded by an annular anode region with the current flowing out the target. We note that, a) the maximum cathode current (≃ 65 A) is reached approximately at the time of 80 ns; b) the high contrast observed at early times, in the stucture of the cathode profile, decreases and disappears in later distribution (130, 155, 180 ns); c) the anode profile at an early time varies inversely with the distance (1/x) from an arbitrary reference position; d) as the size of the cathode increases, the region on the outside of the cathode edge which is initially part of the anode, with current flow outwards, becomes part of the cathode, with current flow inwards - this explains the current reversal noted earlier, Fig. 3.

A comparison of irradiation and current distributions shown in Fig. 12 reveals that the cathode region corresponds to the region irradiated and, more specifically, that the structure iden-

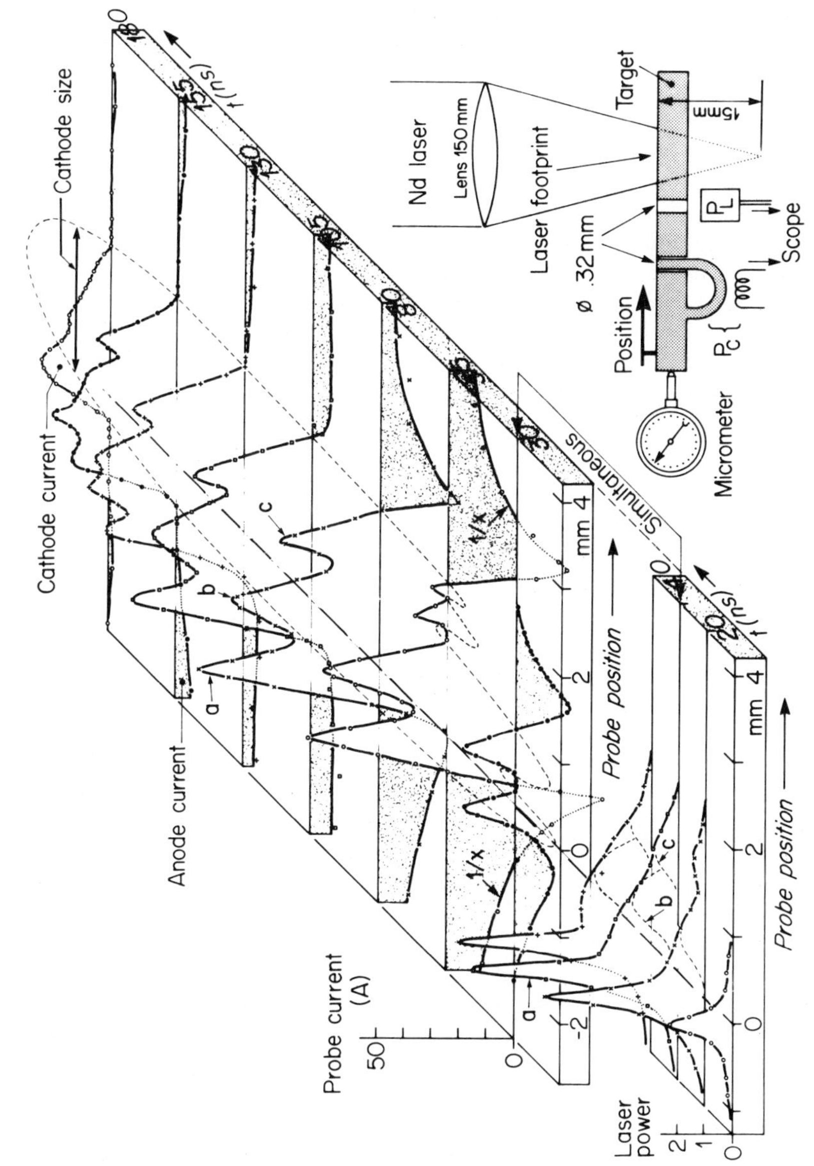

Fig. 12 Probe-target assembly (lower right corner) and profiles, at different times, of the laser power (relative units, lower left corner) and of the probe current as a function of position in the laser footprint.

tified in the laser profiles with hot spots (a), (b), and (c), in Fig. 12, is translated into a structure in the cathode current profile with maxima (a), (b), and (c) in the same positions. In agreement with earlier observations (Fig. 3), we note that the total build-up of the current is coincident with the laser-pulse duration with the maximum current value reached near the end (80 ns) of the laser pulse. A detailed interpretation of these results has been published[45].

The results presented here show that localized deposition of the laser energy on a target leads to a highly correlated non uniform emission of current. We believe that this observation is of the same nature as that reported by Haas et al[32,33]. They found that the laser plasma exhibits a structure of intense localized X-ray emission hot spots which correlate with the structure of the energy distribution in the focal spot. A possible interpretation for their results, based on our measurements, is that the non uniform irradiation leads to a structured emission of current, which results in a non uniform magnetic field distribution and the formation of hot spots[6,7] by magnetic field inhibition of the electron thermal conduction in these regions.

V. SIMULTANEOUS MEASUREMENTS OF CURRENT AND MAGNETIC FIELD

The magnetic field in the plasma, in front of the target, was monitored using a small inductive probe[9,11,13,17,19,34]. The probe consisted of a semi-rigid coax cable with its central wire terminated in the form of a double turn loop and soldered to the shield. As shown in Fig. 1, it was covered by a thin glass tubing. The probe was calibrated and its response time was less than 1 ns. In the experiment, the signal was accurately reversed when the probe was rotated through 180° insuring that the measured signal was in fact that resulting from a transient magnetic field and that spurious electrostatic or photoelectric signals were negligible.

Figure 13 displays typical target current, measured with the wire probe, and magnetic field pulses recorded simultaneously in various gases for a CO_2 laser plasma. As can be seen in Fig. 13, the initial rise of the magnetic signal and that of the current signal were always found to be coincident, regardless of the distance of the magnetic probe from the focal point. The above mentioned signals coincided with the laser pulse to within 1 ns using the short pulse 1.7 ns CO_2 laser. In the case of the long pulse 30 ns Nd-glass laser pulse, these signals were both delayed with respect to the beginning of the laser pulse, by up to 30 ns, depending on the laser power; the delay increases as the laser power decreases.

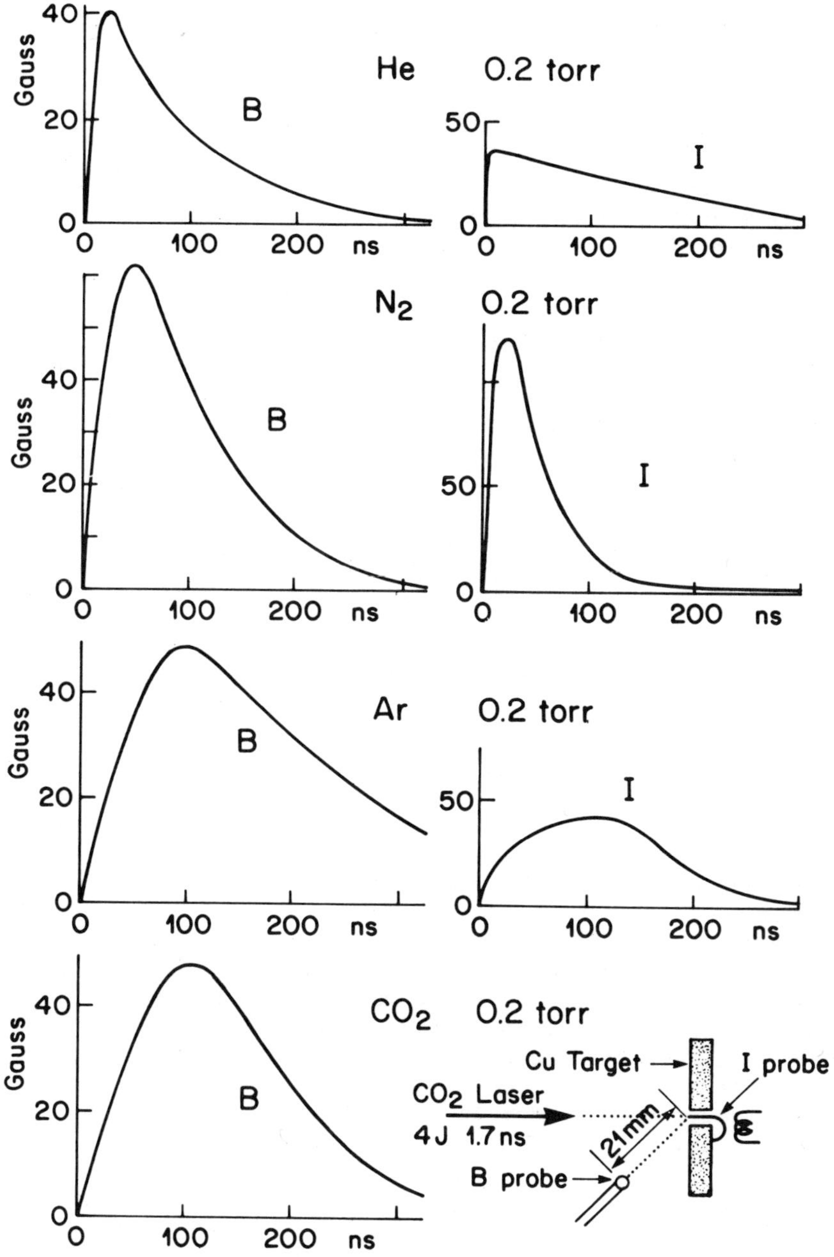

Fig. 13 Typical magnetic field and current signals in Helium, Argon, Nitrogen and CO_2 at 10^{-2} torr.

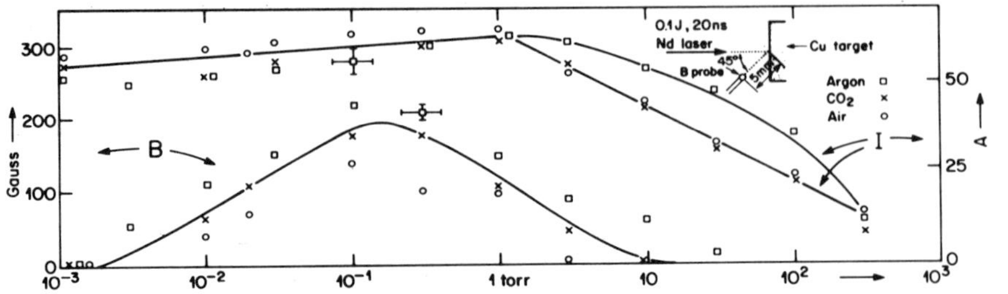

Fig. 14 Effect of background pressure on the current peak and azimuthal magnetic field for Nd-glass laser irradiation of a copper target in Air, N_2, Ar and CO_2.

The simultaneous rise of the magnetic field signal irrespective of the probe position is in contrast with earlier reported observations[11,13] in which the signals were delayed as the magnetic probe was moved to larger distances from the target. This delay was attributed to the propagation of a magnetic front at the velocity of the expanding laser-produced plasma[11,13]. However, in more recent studies[16,18], the magnetic field was found, to temporally coincide with the laser pulse, regardless of the distance from the target, in agreement with our own observation[19]. A calculation was made of the value of the axial electric current which would generate, at the probe location, the measured magnetic field shown, for example, on Fig. 13. The value of the calculated current was approximately that measured by the current probe. This favorable agreement supports the current flow pattern proposed earlier in relation to Fig. 2.

The background pressure dependence of the maximum self-generated current and azimuthal magnetic field is illustrated in Fig. 14. The effect of gas pressure on the magnetic field is in agreement with previous investigations[12,13,15,18]. One will note that the magnetic field attains its maximum value at 21 mm from a target irradiated by the CO_2 laser pulse at a pressure of 10^{-1} torr argon. In comparable conditions[18] the maximum was observed with the probe located 8 mm from the target at a pressure of 6 10^{-1} torr argon. This confirms an earlier report[12] that the pressure at which the magnetic field attains its maximum value decreases as the probe is moved away from the target; this is also well documented, later in this paper, by the results of Fig. 18 and 19.

Various hypotheses to explain the effect of background pressure on the value of the peak magnetic field have been proposed. According to Bird et al[12] and McKee et al[15] enhanced field generation should occur in the optimum pressure range. For Case et al[16] the reason is that a larger electric current would flow through the ionized background gas at the pressure corresponding to the maximum B. The dependence of the current on pressure shown in Fig. 14 does not support either of these hypothesis. Our own interpretation for the dependence of the magnetic field, on the background pressure, is based on the experimental observation, discussed in the next section, of a similar dependence on pressure of the "UV" radiated from the laser plasma.

VI. ULTRAVIOLET RADIATION AND MAGNETIC FIELD DEPENDENCE ON BACKGROUND PRESSURE

An ion collector constructed using a commercial BNC connector was used for the determination of the dependence of the photoelectron yield on the background pressure, the type of gas used and the distance from the laser impact point on the target surface. A small disk was soldered to the center conductor of the connector and a wire mesh was spot welded to the cylindrical ground shield of the connector. The electrode to mesh spacing was 4.5 mm. A potential of -200 V was applied on the photo-cathode. The device was connected to the oscilloscope through a blocking capacitor.

The measured collector signals were consistent with previously published data[30,35,36]. Two components may be distinguished in

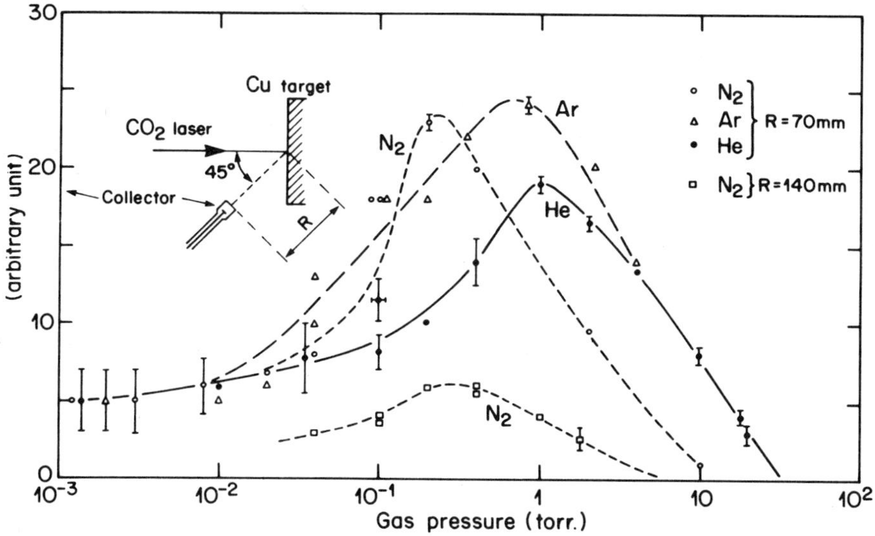

Fig. 15 Effect of background pressure on the amplitude of the collector UV signal for CO_2 laser focused irradiation (4 J, 1.7 ns), in Argon, Nitrogen and Helium.

the signals: a large spike which has been attributed to photo-electrons ejected from the collector by soft X-rays and extreme ultraviolet emitted from the plasma; the second component coming later in time corresponds to ions collected from the expanding plasma. The first component of the collector signal appears almost simultaneously with the laser pulse and for this reason it has often been used as a time reference for the determination of the time of flight of the ions.

While, at high vacuum, energetic electrons could contribute[30] to this early component of the collector signal, we expect that, in the pressure range studied, there must be some photoionization inside the collector and the conduction current between the cathode and the grid may also contribute. Therefore, without further studies, it is not possible to attribute the first signal to the sole effect of the UV irradiating the photo-cathode. However, we considered that, in our range of pressures, this signal - being mainly photo-electrons from the collector cathode and conduction current through the photoionized gas in the collector chamber - is related to the intensity, at the position of the collector, of the UV radiation which, as supported by the $1/R^2$ dependence observed in both laser experiments, should originate from the laser impact region.

The dependence of the height of the UV pulse on the value of the helium, argon or nitrogen atmospheric pressure is shown in Fig. 15 in the case of a 4 J, 1.7 ns, CO_2 laser irradiation focused on a copper target. We have no interpretation for the increase of the UV signal with increasing pressure from 10^{-3} torr to the optimum pressure of 2.10^{-1} torr in N_2, 8.10^{-1} in argon and 1 torr in helium. However, its decrease for larger values of pressure must result from increased absorption of the radiation at higher pressures. In this regard it is interesting to note that the values of the optimum pressures for the three gases studied are approximatively in the inverse ratio to their respective average value of photoionization cross section[37]. Furthermore, the slope of the curve, representing the amplitude of the UV as a function of pressure for pressure above the optimum, calculated using these values of the photoionization cross section, is steeper than the slope shown in Fig. 15. This suggests that the source of radiation is intense. The $1/R^2$ dependence, observed in Fig. 15 for the two nitrogen curves, is also found in Fig. 17 which gives the value of the height of the UV pulse as a function of the distance from the Nd-glass laser impact on the target. In Fig. 16, we present the time evolution, obtained at relatively high pressure, of the collector signals corresponding to radial distances R from 60 to 240 mm. A preliminary computer calculation has indicated that these results correspond to a diffusion of radiation from a source located at the origin and varying with time as the laser pulse.

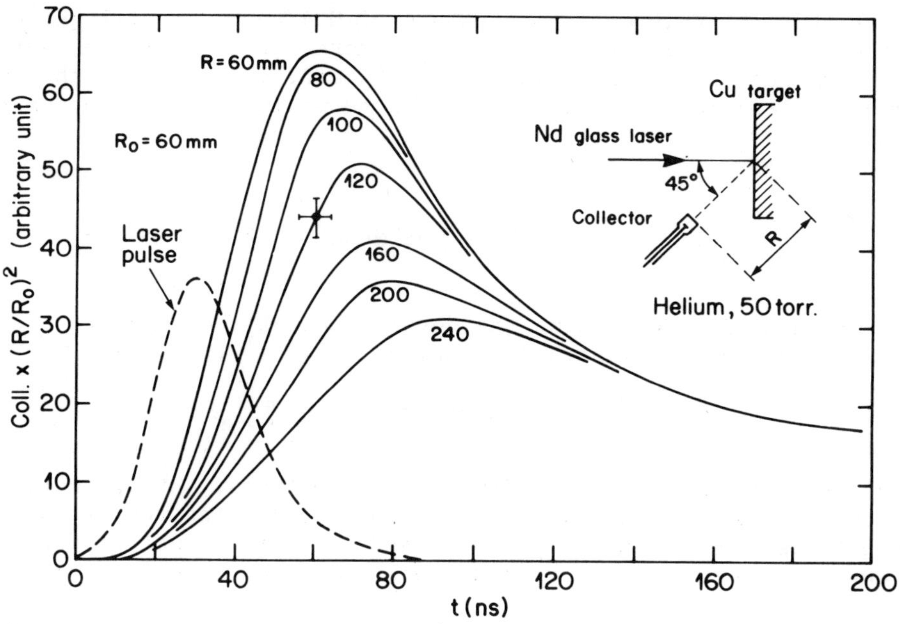

Fig. 16 Collector UV signal as a function of time for increasing values of the target-collector distance R in Helium (Nd-glass laser, 4 J, 30 ns).

The usefulness of these UV measurements becomes evident when one observes that both the magnetic field and the UV pulse from the collector have a similar dependence on the background gas pressure. This is demonstrated by a comparison of the curves of Fig. 15, 18 and 19. In particular, the helium pressure, at which the maximum UV signal occurs, is the same as the pressure at which the maximum magnetic field occurs, see Fig. 15 and 19. In nitrogen, Fig. 15 and 18, the maxima occur at a lower pressure than in helium. Our interpretation for the similarity between the behavior of the magnetic field and the collector signal as a function of gas pressure is that the current/magnetic field toroidal configuration - illustrated in Fig. 2 - is determined by the existence of a conducting region, in front of the target, produced by photoionization of the background gas by irradiation from the laser plasma at the focal point. As an example, we observe, in Fig. 15, that the collector signal, at the position R_0 = 70 mm, is maximum at a pressure of 0.4 torr in nitrogen. Assuming that the relative spectral distribution of the UV radiation does not vary with pressure we may expect that the maximum extent of photoionization in region $R > R_0$ will also correspond to that pressure of 0.4 torr. If, in addition, we assume that the spatial configuration of the current paths is as shown in Fig. 2, then we may expect that, at this optimum pressure of 0.4 torr, it will extend further, i.e. to larger R. Therefore, more current will be carried in the region corresponding to $R > R_o$ and, as a result, the magnetic field measured at $R = R_o$ will be maximum at this pressure.

VII. EXPANSION OF THE MAGNETIC FIELD

The equation describing the development of the magnetic field is obtained from the generalized ohm's law[7,11,13],

$$\vec{E} = \frac{J}{\sigma} - \vec{V}_e \times \vec{B} - \frac{1}{n_e e} \nabla (n_e k T_e) \tag{1}$$

where $\vec{V}_e$ is the fluid velocity of the electrons, σ the electrical conductivity, n_e and T_e the electron density and temperature respectively, J the current density and where the electric field, $\vec{E}$, must be solenoidal in order to generate a magnetic field $\vec{B}$:

$$\frac{\partial B}{\partial t} = - \nabla \times \vec{E} \tag{2}$$

i.e.

$$\frac{\partial \vec{B}}{\partial t} = \frac{1}{\mu_o \sigma} \nabla^2 \vec{B} + \nabla \times (\vec{V}_e \times \vec{B}) + \frac{k}{e} \nabla T_e \times \nabla \ln n_e \tag{3}$$

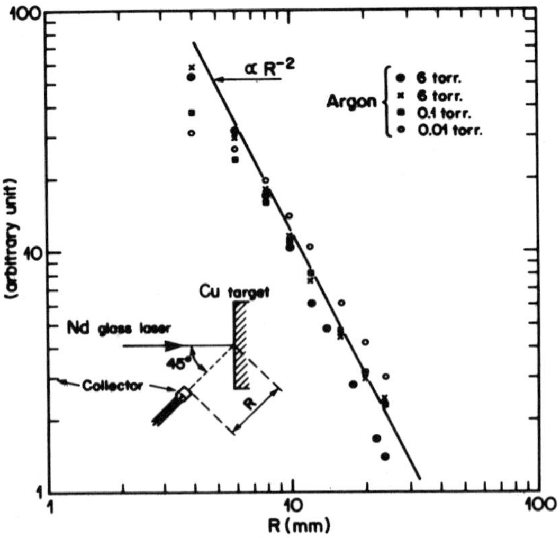

Fig. 17 Peak amplitude of the collector UV signal as a function of the target-collector distance R in Argon (Nd-glass laser, 4 J, 30 ns).

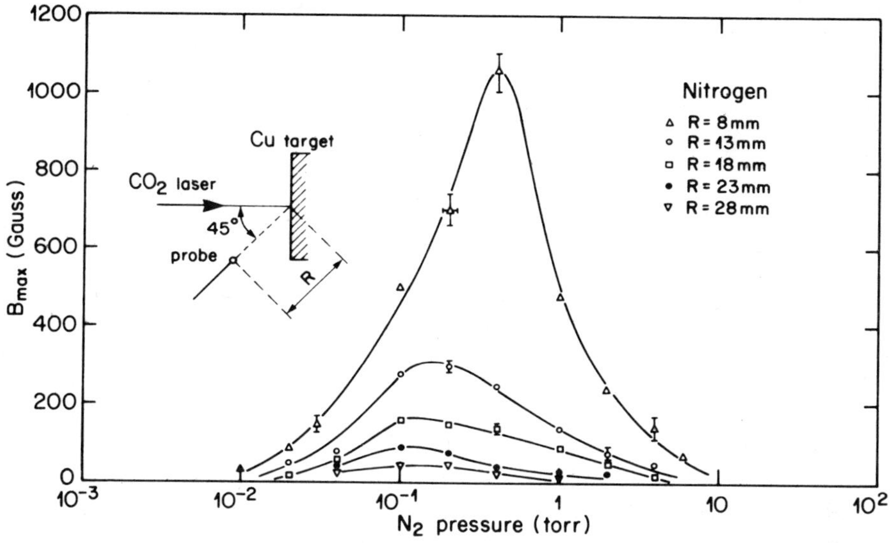

Fig. 18 Peak amplitude of the azimuthal magnetic field as a function of the Nitrogen pressure for different probe-target distances R (CO_2 laser, 4 J, 1.7 ns).

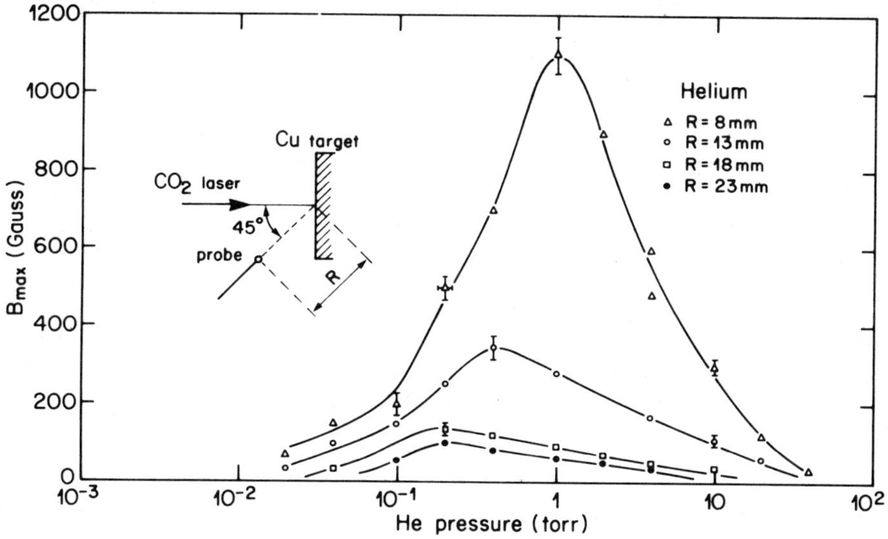

Fig. 19 Peak amplitude of the azimuthal magnetic field as a function of the Helium pressure for different probe-target distances R (CO_2 laser, 4 J, 1.7 ns).

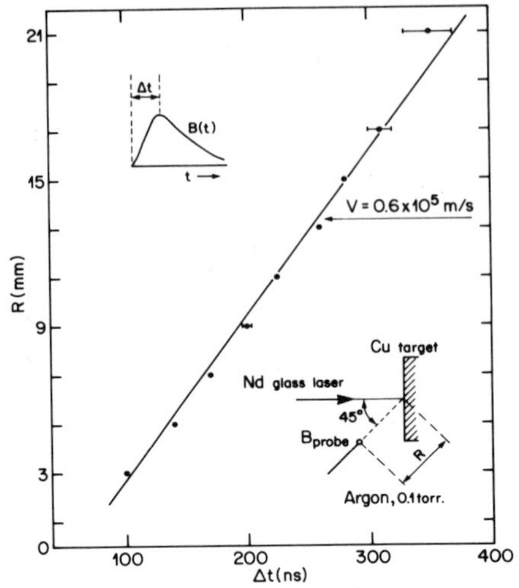

Fig. 20 Radial position of the peak amplitude of the azimuthal magnetic field as a function of time in Argon (Nd-glass laser, 4 J, 30 ns).

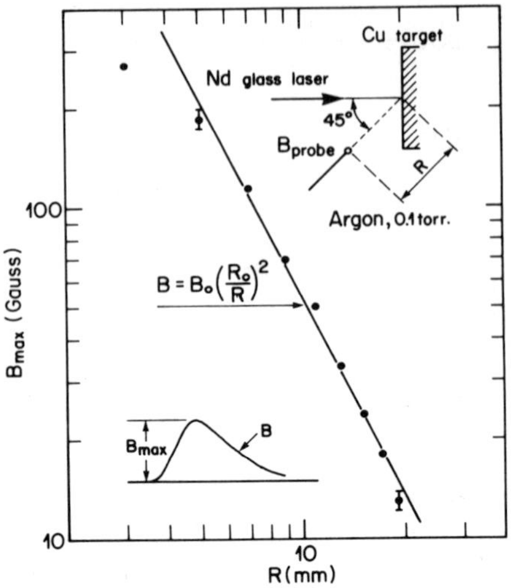

Fig. 21 Peak amplitude of the azimuthal magnetic field as a function of the probe target distance in Argon (Nd-glass laser, 4 J, 30 ns).

The first two terms on the right hand side represent the well-known diffusion and convection of the magnetic field. The last term is the source term.

It has been considered in the past[11,13] that convection of the B field should largely dominate over diffusion. The magnetic field, generated in the focal region, would be carried outwards by the expanding plasma. This assumption of "frozen in" field lines is supported by the $1/R^2$ dependence of the peak field observed by several authors[13,17,18,20]. This is also confirmed here by the results obtained at low laser power, Fig. 20, where we observed that the peak of the B field occurs with a delay which increase linearly with the distance from the focal spot - i.e. it propagates at constant velocity - and, in addition, the amplitude of the peak B field is found, Fig. 21, to vary as $1/R^2$. The velocity of propagation is of the order of 10^5 m/sec as observed previously for the velocity of expansion of the laser plasma[11,13,15,16,20,28,29,38,39].

At high laser power, we find that the behavior of the field is quite different. As shown in Fig. 22 and 23, in both gases studied, N_2 and He, respectively, at the optimum pressure, 0.4 torr in N_2, 1 torr in He, the magnitude of the peak field varies as $1/R^3$. If the magnetic field expands in the photoionized region in front of the target in a manner analogous to the diffusion of a scalar quantity from an impulsive point source[40,41] this relation

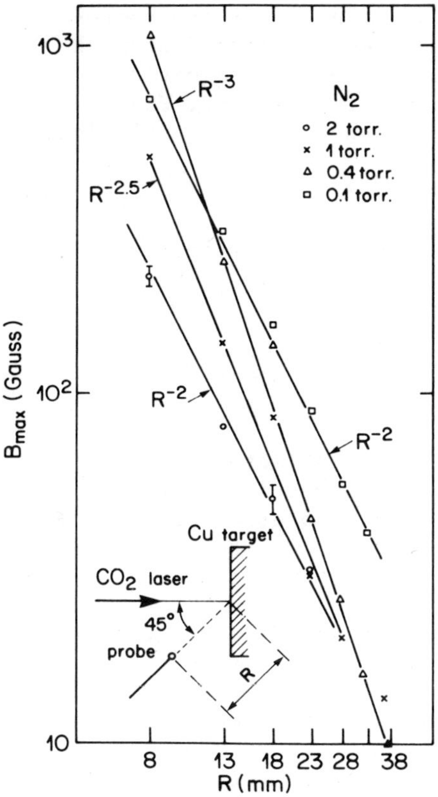

Fig. 22 Peak amplitude of the azimuthal magnetic field as a function of the probe target distance R for different values of the Nitrogen pressure (CO_2 laser, 4 J, 1.7 ns).

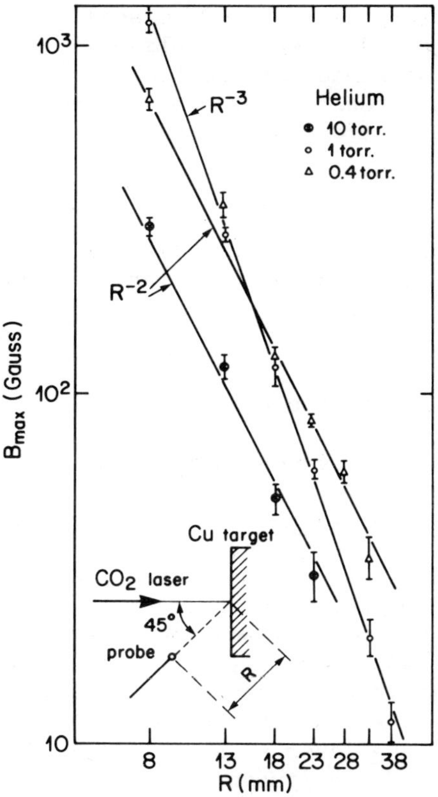

Fig. 23 Peak amplitude of the azimuthal magnetic field as a function of the probe target distance R for different values of the Helium pressure (CO_2 laser, 4 J, 1.7 ns).

would be expected for the time occurrence of the maximum value of this quantity at a given radius. At gas pressures, both half an order of magnitude higher and lower than the optimum, the $1/R^2$ dependence, observed in Fig. 22 and 23, suggests that the expansion of the peak field is controlled by convection.

The relative importance of the terms of convection and diffusion in equation (3) is normally given by the Reynold's magnetic number (e.g. [13])

$$R = \mu_o \sigma LV \tag{4}$$

where L is a characteristic length for the field extension ($\sim 10^{-2}$ m), V would correspond to the expansion velocity of the laser plasma ($\sim 10^5$ m/sec). The value of $\mu_o\ \sigma$ for the photoionized background plasma in argon has been estimated from the measurement of the time occurrence t of the peak magnetic field as a function of the distance R from the focal point, Fig. 24. It was found that the results satisfied the following relation,

$$R^2 = 8 \times 10^3\ t. \tag{5}$$

Knowing that the scalar diffusion with spherical symmetry is characterized by a similar relation[40,41]:

$$R^2 = 6\ \mu_o\ \sigma\ t \tag{6}$$

leads to the evaluation of $\mu_o\ \sigma$.

These considerations lead to a wrong estimate of R of the order of 10^6, which would indicate that convection of the B field, by the plasma, dominates over diffusion. The error comes from neglecting to properly take into account the existence of the two different plasmas: the photoionized background plasma, the extent of which, as discussed above, is considered maximum at the optimum pressure, and the laser plasma which expands, from the origin, at approximately 10^5 m/sec. The background plasma is created within the laser-pulse duration, which is very short - especially in the case of the CO_2 laser - compared to the time of expansion of the laser plasma (10^{-7} sec). As shown earlier the magnetic field configuration is also established within the laser pulse duration and it extends beyond the laser plasma well into the photoionized region. Then equation (3) should be applied independently to describe the behavior of the field in both plasma regions. In the photoionized region, in the time preceeding the arrival of the expanding laser plasma, there is no convection and the expansion of the field should correspond to diffusion. In the laser plasma, the conductivity is expected to be much larger than that for the photoionized plasma and the magnetic Reynold's number should be much higher than the value above, therefore conduction should dominate. The source term has not been taken into consideration here for the source should be located near the plasma-target interface and does not affect the rate of change of the magnetic field in the region studied ($R > 8$ mm).

Therefore, the expansion of the field is characterized by two successive phases: a phase of diffusion followed by one of convection. The phase of diffusion lasts until the arrival of the expanding laser plasma and, therefore, lasts longer at points further from the target. In addition, the duration of the diffusion phase decreases with decreasing background gas pressure due to an increase in velocity of the expanding laser plasma[13,38]. At high pressure ($\sim$ 10 torr range) the diffusion phase may be absent due to the limited spatial extent of the photoionized region resulting from a very large absorption of the UV radiation at this pressure. The experimental results indicate that, for both gases studied, it is at the optimum pressure that the peak field satisfies the "diffusion relation", $1/R^3$. These conditions correspond to an optimum between a higher velocity of the expanding laser plasma at lower pressure and a smaller photoionized region at higher pressure.

We note that the diffusion of the magnetic field has been treated theoretically by Tidman[42] and by Widner[20,21]. Our results, which suggest the existence of an initial phase of diffusive expansion in the photoionized background gas, are in agreement with the analysis of Widner, according to which diffusion should dominate over convection for electron temperature less than 100 eV. This must certainly be the case for the photoionized background plasma. In fact, a temperature of 1 eV has been estimated using the value of conductivity determined above, eqn. (5), (6) and Fig. 24.

It is worth adding that the diffusion process is characterized by a $1/R^3$ dependence of the peak amplitude of the signal only in the case of an impulsive source of short duration with respect to the time of propagation from the origin to the location of recording[40,41]. For example, diffusion was proposed earlier in relation to Fig. 16 as the mean of propagation of the UV radiation in the background gas; however in this particular case the amplitude of the maximum does not satisfy the $1/R^3$ relation because of the long duration of the source. Conversely, the fact that the $1/R^3$ relation is satisfied by the expansion of the magnetic field indicates that the source of magnetic field is located at the origin and acts only for a very short time - comparable to the duration of the laser pulse. This is in agreement with the fact, illustrated in Fig. 3 and 12, that the target current is found to be at a maximum at the end of the laser pulse. We will, therefore, consider that the source of current/magnetic field is located at the origin and lasts approximately for the duration of the laser pulse - 1.7 ns in the case of the CO_2 laser. We also know that the conducting background plasma is created within the laser pulse duration as evidenced by the build-up of the B field simultaneously with the occurrence of the laser pulse. In addition, at the end of the laser pulse, the size of the laser plasma, expanding at 10^5 ms^{-1}, should still be very small (< 1 mm) with the photoionized region, where the diffusion regime dominates, extending beyond (R > 1 mm). Thus it appears justified to extrapolate the $1/R^3$ dependence of the magnetic field shown in Fig. 22 and 23 down at least to 1 mm from the origin. In our conditions, the extrapolation yields a value of the magnetic field in excess of 0.5 MG. Concurrent measurement of the field near the focal point by methods such as proposed by Stamper and Ripin[8] would be of great interest to check this extrapolation and to confirm the nature of the expansion of the field close to the focal region.

VIII. THE MODEL

On the basis of this experimental study, a complete model of

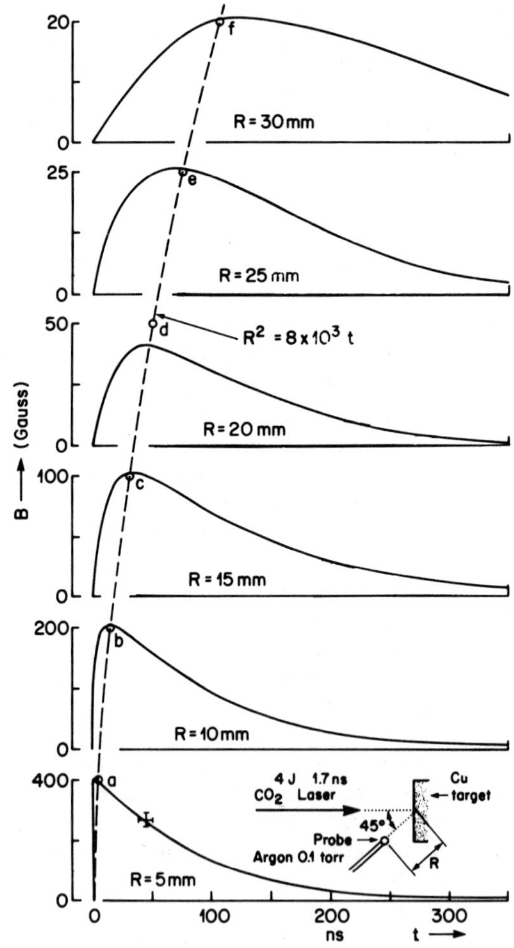

Fig. 24 Magnetic field as a function of time for increasing values of the target-probe distance R. The dashed line drawn through the points a, b, c, d, e and f satisfies the relation $R^2 = k\ t$ where $k = 8 \times 10^3\ m^2\ s^{-1}$.

the interaction for magnetic field generation in a plasma produced by laser irradiation of a conducting target is now proposed.

a) First, we consider that the laser plasma produced by laser irradiation of the target is an intense source of energetic ions and UV radiation.

b) The UV radiation produces a second plasma, in front of the target as shown in Fig. 25, by photoionization of the background gas. The characteristics of this second plasma are pressure-dependent. Tenuous at low pressure ($< 10^{-2}$ torr), this plasma is limited, in its volume, at high pressure ($>$ 1 torr) due to large absorption of the UV radiation in the gas near the source.

c) As a result of the intense UV and ionic bombardment of the target surface, and of the presence of a very large electric field generated in the plasma by thermoelectric process[3,14,43,44], a very large current is emitted by the target. The structure in the distribution of the emitted current is related to the spatial structure in the laser pulse. Thus any laser pulse inho-

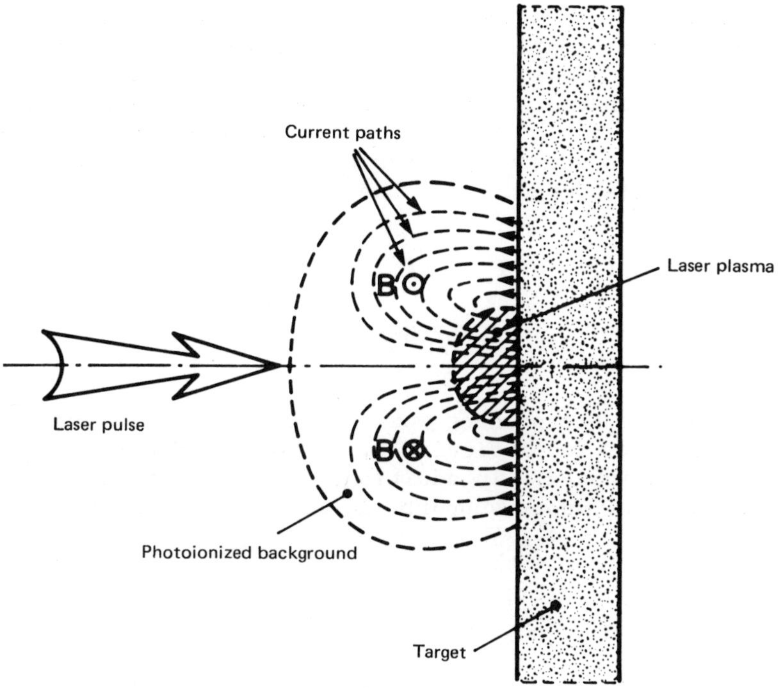

Fig. 25 Illustration of two plasma concept. The current paths "diffuse" in the photoionized background; those contain in the laser plasma are carried "frozen". Both mecanisms bring a change in the local magnetic field B.

mogeneity will be translated into an inhomogeneity in the current/magnetic field configuration which should lead to the creation of hot spots[6,7]. This should explain the correlation which has been observed[32,33] between the laser and X-rays spatial structures.

d) Current flows and, owing to the cylindrical symmetry of the interaction about the laser axis and to charge conservation, the distribution of current is toroidal. At the target surface, there exists a cathode region centered on the laser axis, with conventional current flowing into the target, surrounded by an annular anode region with current flowing out of the target. In general, the current builds up during the laser pulse and decays subsequently.

e) The interface current I and the magnetic field B occur simultaneously regardless of the position in the photoionized background plasma. The extent of the current/magnetic field configuration, in the background gas, is that of the photoionized background plasma and is therefore pressure dependent.

f) As a result of both the non-uniformity of the field and of the expansion of the laser plasma into the photoionized background plasma, the spatial configuration of the magnetic field/current distribution will change with time. The development of the configuration is characterized by two successive phases: a phase of diffusion in the background plasma followed by one of convection with the expanding laser plasma. The occurrence of these two phases is dependent on the laser power and the background pressure.

ACKNOWLEDGMENTS

Contributions to the present work by colleagues at IREQ, INRS-Energie and CRDV, in particular R. Bolton, P. Kieffer and H. Pépin, are acknowledged. In addition, the author is grateful to C. Richard for continuous support and to C.H. Leigh for reading the manuscript.

REFERENCES

1. J.B. Chase, J.M. Leblanc and J.R. Wilson, Phys. Fluids, 16, 1142 (1973).

2. N.K. Winsor and D.A. Tidman, Phys. Rev. Letters, 31, 1044 (1973).

3. D.A. Tidman and J.A. Stamper, Appl. Phys. Letters, 22, 498 (1973).

4. D.A. Tidman and R.A. Shanny, Phys. Fluids, 17, 1207 (1974).

5. B.H. Ripin et al, Phys. Rev. Letters, 34, 1313 (1975).

6. R.S. Craxton and M.G. Haines, Phys. Rev. Letters, 35, 1336 (1975).

7. R.S. Craxton, "A Numerical Investigation of Magnetic Fields in Laser-Produced Plasmas", Ph.D. Thesis, Imperial College, London, August 1976.

8. J.A. Stamper and B.H. Ripin, Phys. Rev. Lett. 34, 138 (1975).

9. V.V. Korobkin and R.V. Serov, JETP. Letters, 4, 70 (1966).

10. G.A. Askaryan, M.S. Rabinovich, A.D. Smirnova and V.D. Studenov, JETP Letters, 5, 93 (1967).

11. J.A. Stamper, K. Papadopoulos, R.N. Sudan, S.O. Dean, E.A. McLean and J.M. Dawson, Phys. Rev. Letters, 26, 1012 (1971).

12. R.S. Bird, L.L. McKee, F. Schwirzke and A.W. Cooper, Phys. Rev. A, 7, 1328, 1973.

13. F. Schwirzke, in Laser Interaction and Related Plasma Phenomena, edited by H. Schwarz and H. Hora, Plenum, New York, vol. 3A, 213 (1974).

14. J.A. Stamper et al, in Laser Interaction and Related Plasma Phenomena, edited by H. Schwarz and H. Hora, Plenum, New York vol. 3B, 713 (1974).

15. L.L. McKee, R.S. Bird and F. Schwirzke, Phys. Rev. A, 9, 1305 (1974).

16. R.S. Case Jr. and F. Schwirzke, J. Appl. Phys. 46, 1493 (1975).

17. J.N. Olsen and C.W. Mendel, J. Appl. Phys. 46, 4407 (1975).

18. R. Serov and M.C. Richardson, Appl. Phys. Lett. 28, 115 (1976).

19. M.G. Drouet, R. Bolton, G. Saint-Hilaire, P. Kieffer, Z. Szili, H. Pépin, G. Grek, A. Thibaudeau and K. Trépanier, Appl. Phys. Letters 29, 469 (1976).

20. M.M. Widner, Sandia Laboratories Pulbication no. SLA-73-0500, 1973 (unpublished).

21. M.M. Widner, Phys. Fluids, 16, 1778 (1973).

22. F. Rheault, J.L. Lachambre, P. Lavigne, H. Pépin, H.A. Baldis, Rev. Sci. Instrum. 46, 1244 (1975).

23. M.G. Drouet and S. Gruber, IEEE Trans. PAS-95, 105 (1976).

24. M.G. Drouet and R. Bolton, Phys. Rev. Letters, 36, 591 (1976).

25. M.G. Drouet, and H. Pépin, Appl. Phys. Letters, 28, 426 (1976).

26. R.P. Godwin, in Laser Interaction and Related Plasma Phenomena, edited by H. Schwarz and H. Hora, Plenum, New York, vol. 35, 691 (1974).

27. H.M. Thompson and J.W. Daiber, Bull. Am. Phys. Soc. 20, 1284 (1975).

28. T.U. Arifov, G.A. Askaryan, I.M. Raevskii and N.M. Tarasova, Sov. Phys. JETP, 28, 201 (1969).

29. S.I. Andreev, Yu. I. Dymshits, L.N. Kaporskii and G.S. Musatova, Sov. Phys. Tech. Phys. 13, 657 (1968).

30. W. Demtröder and W. Jantz, Plasma Phys. 12, 691 (1970).

31. J.A. Stamper, Phys. Fluids, 18, 735 (1975).

32. R.A. Haas, M.J. Boyle, K.R. Manes and J.E. Swain, J. Appl. Phys. 47, 1318 (1976).

33. A.A. Haas, D.W. Phillion, V.C. Rupert, H.N. Kornblum, L.N. Koppel and K.G. Tirsell, Proceedings of the IEEE Conference on Plasma Science, Austin, May 1975, paper 3A8.

34. R. Serov, M.G. Richardson and P. Burtyn, Rev. Scient. Instrum. 46, 886 (1975).

35. K. Dick. H. Pépin, J. Martineau, K. Parbhakar and A. Thibaudeau, J. Appl. Phys. 44, 3284 (1973).

36. A.W. Ehler, J. Appl. Phys. 46, 2464 (1975).

37. A.N. Zaidel' and E. Ya. Shreider, Vacuum Ultraviolet Spectroscopy, Ann Arbor - Humphrey Science Publishers, Ann Arbor 1970.

38. J.L. Bobin, Y.A. Durand, Ph.P. Langer and G. Tonson, J. Appl. Phys. 39, 4184 (1968).

39. J.F. Reintjes, T.N. Lee, R.C. Eckardt and R.A. Andrews, J. Appl. Phys. 47, 4457 (1976).

40. H.S. Carslaw and J.C. Jaeger, "Conduction of Heat in Solids", 2nd Edition, Oxford University Press, London, p. 257, 1959.

41. R.V. Churchill, "Operational Mathematics", 2nd Edition, McGraw-Hill, New York, 1958.

42. D.A. Tidman, Phys. Rev. Lett. 32, 1179 (1974).

43. J.A. Stamper, U.S. Naval Research Laboratory Report No. 7411, 1972 (unpublished).

44. J.A. Stamper and D.A. Tidman, Phys. Fluids, 16, 2024 (1973).

45. M.G. Drouet, R. Bolton, P. Kieffer, G. Saint-Hilaire, and Z. Szili, Phys. Letters 59A, 210 (1976).

HEAT TRANSPORT IN SPATIALLY FLUCTUATING LASER-GENERATED D.C. MAGNETIC FIELDS*§

Claire Ellen Max, Wallace M. Manheimer,† and Jeffrey J. Thomson

University of California, Lawrence Livermore Laboratory

Livermore, California 94550 U.S.A.

ABSTRACT

In this paper, we investigate anomalous electron thermal conduction perpendicular to d.c. laser-generated B fields. We find that d.c. spatial fluctuations δB about a mean field B_0 can produce large enhancements in the cross-field diffusion coefficient, over its $\delta B = 0$ value. There are by now a multitude of suggested sources for laser-generated d.c. magnetic fields. When all these processes act at once, the likely result is an overall "average" field B_0 which is probably toroidal, due to $\nabla n \times \nabla T$ sources, plus a spatially fluctuating component δB due to the sum of all the other field sources. For parameters appropriate to laser fusion experiments, we use a quasilinear model to describe the effect of particle scattering by fluctuations δB. We find that if $\delta B \lessapprox B_0$, the diffusion coefficient can be comparable to the Bohm value: $D \sim (v_t^2/\Omega_c)(|\delta B|^2/B_0^2)$. Next we discuss the effect of field-line wandering on the flux. Finally, we investigate an analogue of neoclassical diffusion appropriate to laser-plasma conditions.

---

*Presented at the Fourth Workshop on "Laser Interaction and Related Plasma Phenomena" held at RPI, Troy, New York, Nov. 8-12, 1976.

§Research performed under the auspices of the USERDA, Contract No. W-7405-Eng-48.

†Permanent address: U.S. Naval Research Lab., Washington, D. C.

I. INTRODUCTION

It is becoming apparent that there are a number of plausible mechanisms for producing laser-generated d.c. magnetic fields. Many of these mechanisms yield fields which are megagauss in magnitude, but whose geometric configurations differ markedly from each other. To date, the interaction of several different field-generation processes with each other has not been examined in detail. It seems qualitatively clear, however, that many of these mechanisms are capable of occurring simultaneously.

In the present work, we take seriously the possibility that several field-generating processes can occur at once, yielding a total $\underset{\sim}{B}$ whose geometry and magnitude may be quite complex. We are interested in describing electron heat transport in such a realistic and messy configuration. Rather than confronting a very difficult exact calculation of the field geometry, we choose to model the total field $\underset{\sim}{B}$ as an average part $<\underset{\sim}{B}>$, plus a part $\delta\underset{\sim}{B}$ representing static spatial fluctuations on a variety of length scales. We then examine the effects of these spatial fluctuations on transport.

In Section II, we briefly review some major field generation mechanisms, with an eye towards delineating which spatial scales and magnitudes are favored by each one. In the remainder of the paper, we consider three different mechanisms for transport across fluctuating B fields: resonant scattering, field line wandering, and neoclassical transport. We shall show that the latter applies in the extremely collisionless regime, and yields a modification to the classical scaling for transport coefficients. The first two mechanisms, resonant scattering and field-line wandering, apply under less rigidly collisionless conditions. These yield diffusion coefficients independent of the collision frequency, giving a Bohm-like scaling. We conclude by discussing the implications of these scalings for laser-plasma calculations and experiments.

II. REVIEW OF FIELD GENERATION MECHANISMS

The aim in this section is to motivate our model for the structure of complex B fields. To do so, we briefly derive the spatial scales and magnitudes of fields produced by several major mechanisms. We use these to construct our model composite field structure, $\underset{\sim}{B}(\underset{\sim}{x}) = <\underset{\sim}{B}> + \delta\underset{\sim}{B}$.

At present, the best candidate for providing the average component of $\underset{\sim}{B}$ is the so-called "$\nabla n \times \nabla T$ field" which surrounds the laser spot. Non-colinear density and temperature gradients at the edge of a large laser spot can produce a d.c. magnetic field.[1] In very simplified form, the MHD equations give

$$\frac{\partial \underset{\sim}{B}}{\partial t} = -\frac{c}{en}\left(\underset{\sim}{\nabla} n \times \underset{\sim}{\nabla} kT\right) + \underset{\sim}{\nabla} \times \left(\underset{\sim}{V} \times \underset{\sim}{B}\right) \quad . \tag{1}$$

Here $\underset{\sim}{V}$ is the fluid velocity, n the number density, and T the temperature. Initial growth of $\underset{\sim}{B}$ is dominated by the first term on the right. Thus if ∇n points axially into the surface of a slab target, and $\underset{\sim}{\nabla} T$ points radially inwards toward the spot center, the resulting field is toroidal in shape and has a scale size typical of the whole laser spot.

The saturated magnitude of this "average" field $\langle B \rangle$ can be crudely estimated by setting $\partial/\partial t$ equal to zero in (1):

$$\left(\frac{\langle B_{sat}\rangle}{1\ \mathrm{MG}}\right) \cong 5\left(\frac{\theta_e}{1\ \mathrm{keV}}\right)^{\frac{1}{2}}\left(\frac{10\ \mu\mathrm{m}}{L}\right)\left(\frac{\overline{A}/\overline{Z}}{2.5}\right)^{\frac{1}{2}} F(\theta) \quad . \tag{2}$$

We have taken $V \sim c_s \cong (\overline{Z}kT_e/\overline{A}m_p)^{\frac{1}{2}}$; $F(\theta)$ represents an angular factor due to the cross products in Eq. (1). For the parameters indicated in (2), $\langle B_{sat}\rangle$ is expected to be several megagauss. (In reality, these estimates will be modified[2] by inclusion of "thermal force" and "resistive" terms in Eq. (1).)

Next we consider several mechanisms which contribute smaller-spatial-scale components, $\delta\underset{\sim}{B}$, to the total magnetic field structure. First, the $\underset{\sim}{\nabla} n \times \underset{\sim}{\nabla} T$ process can generate small-scale structures as well as large ones. For example, at discontinuities which occur in shocks, ionization fronts, or composition jumps,[3] $\underset{\sim}{\nabla} n$ can be locally very large. If the discontinuity lies at an angle to the background temperature gradient, sizeable magnetic fields can be produced in a small region. The analysis of Ref. 3 shows that the scale sizes of these fields are determined by resistive diffusion, and are typically several microns after 100 psec, for $n_e = 10^{21}\ \mathrm{cm}^{-3}$, $\theta_e = 1$ keV. The field magnitudes are limited by pressure balance, $B_{max} = (8\pi nkT)^{\frac{1}{2}}$, for times longer than an Alfven wave transit time across the discontinuity. Fields of many megagauss, localized to regions a few microns across, are produced by this mechanism.

Self-focused filaments or hot-spots in the laser heated plasma provide a different source for small scale $\underset{\sim}{\nabla} n \times \underset{\sim}{\nabla} T$ fields. Inside a filament, the plasma density is depressed due to the ponderomotive force.[4] The density gradient points radially away from the axis of the cylindrical filament. The background temperature gradient, when crossed with this localized density gradient, yields a magnetic field which wraps around the cylindrical filament. If the laser light pressure $E_0^{\,2}/8\pi$ is roughly equal to the plasma pressure nkT, self-focused structures typically have transverse dimensions of only one or two vacuum wavelengths. Thus for a Nd laser, scale sizes of a few microns for the resulting B field are once again expected. Typical magnitudes of several megagauss are easily attained. Such structures would occur predominantly toward the center of the laser spot, where the laser intensity is high.

Small scale magnetic fields can be produced by mechanisms other than the $\underset{\sim}{\nabla} n \times \underset{\sim}{\nabla} T$ source. Radiation pressure during resonance absorption of laser light has been shown[5] to yield megagauss fields, localized to within a few vacuum wavelengths of the critical density surface. Typically these fields have complex structure: they change sign within distances of less than a vacuum wavelength, as one goes from the underdense to the overdense plasma. Megagauss fields associated with ripples and bubbles in the critical density surface have been seen in computer simulations.[6] These occur at high laser intensities, and are typically a few vacuum wavelengths in extent.

In more collisional plasmas, a field-generating thermal instability may be an important source.[7] For the parameters $\bar{Z} = 10$, $T_e = 500$ eV, $n_e = 10^{21}$ cm^{-3}, and background scale lengths $L = 10$ μm, the typical scale size for the fastest growing mode of this thermal instability is about 10 μm.

Let us summarize these results. Large-scale toroidal $\underset{\sim}{\nabla} n \times \underset{\sim}{\nabla} T$ fields are expected to exist around the edge of the laser spot, and to have the large sizes typical of the spot itself. Most of the other mechanisms discussed here yield small-scale fields with sizes of a few microns to fractions of a micron, for Nd lasers. These include fields due to discontinuities, fields around laser light filaments, and fields due to resonance absorption and critical surface nonuniformities. In high-Z plasmas, the thermal instability can produce fields with sizes on the order of 10 μm. Furthermore, each of these processes seems capable of yielding megagauss levels for the magnitude of B. The interested reader may wish to add other field generation mechanisms as well; the present list is intended to be illustrative rather than exhaustive.

With this motivation, we take the total magnetic field to be

$$\underset{\sim}{B}(\underset{\sim}{x}) = \langle \underset{\sim}{B} \rangle + \delta \underset{\sim}{B} \quad , \tag{3}$$

with $\delta \underset{\sim}{B}$ containing significant components having spatial scales of a fraction of a micron to 10 microns, for a Nd laser. The averaged part, $\langle \underset{\sim}{B} \rangle$, is presumed to represent the large-scale toroidal $\underset{\sim}{\nabla} n \times \underset{\sim}{\nabla} T$ field. In the remainder of this work, we shall need to assume in addition that $|\delta B|$ is somewhat smaller than $|\langle B \rangle|$, in order for our analytic tools to be valid.

III. RESONANT SCATTERING FROM FIELD FLUCTUATIONS

In this section and the next, we shall further model the local magnetic field structure as

$$\underset{\sim}{B} = B_o \hat{z} + \hat{y} \sum_k \delta B_k \exp(ik_z z) \quad , \tag{4}$$

with $\delta B\,(k = 0) = 0$, and $|\delta B_k| \ll |B_o|$.

We now iteratively solve the Lorentz force equation for an electron, so that the electron's trajectory is $\underset{\sim}{x}(t) = \underset{\sim}{x}^o(t) + \delta \mathbf{x}(t)$. Here x^o is the trajectory calculated using only the zero order field $B_o\hat{z}$, while the two transverse components of $\delta \underset{\sim}{x}$ satisfy

$$\delta \ddot{x} + \Omega \delta \dot{y} = v_z \sum_k \delta \Omega_k \exp ik_z(z_o + v_z t) \quad , \tag{5a}$$

$$\delta \ddot{y} - \Omega \delta \dot{x} = 0 \quad . \tag{5b}$$

In these expressions, $\Omega = eB_o/m_e c$, $\delta\Omega_k = e\delta B_k/m_e c$, and z_o, v_z represent the initial z position and z velocity of the electron.

The solution of Eq. (5a) which satisfies $\delta x = 0$ and $\delta \dot{x} = 0$ at $\Delta t = 0$ is

$$\delta x = v_z \sum_k \left(\frac{\delta\Omega_k \exp(ik_z z_o)}{\Omega^2 - k_z^2 v_z^2} \right) [e^{ik_z v_z \Delta t} - (\tfrac{1}{2})(1 + k_z v_z/\Omega) e^{i\Omega\Delta t}$$

$$- (\tfrac{1}{2})(1 - k_z v_z/\Omega) e^{-i\Omega\Delta t}] \quad . \tag{6}$$

We wish to find the spatial diffusion coefficient in the x-direction,

$$D_{xx} \equiv \lim_{\Delta t \to \infty} \langle|\delta x|^2\rangle / 2\Delta t \ .$$

To do this, we multiply (6) by its complex conjugate, and then average over initial positions z_0, assuming that all initial positions are equally likely. We then use the identity

$$\pi\delta(x) = \lim_{\Delta t \to \infty} \frac{\sin^2 x\,\Delta t}{x^2\,\Delta t} \tag{7}$$

to obtain

$$D_{xx} = \frac{\pi}{4} v_z^2 \sum_k \frac{|\delta\Omega_k|^2}{\Omega^2} \left\{ \delta(k_z v_z + \Omega) + \delta(k_z v_z - \Omega) \right\} \ . \tag{8}$$

If $v_t = (kT_e/m_e)^{\frac{1}{2}}$ is a typical value for v_z, this result shows that the magnetic field fluctuations responsible for diffusion in this case have $|k_z|\rho_L = 1$, where $\rho_L = v_t/\Omega$ is the electron larmor radius in the background field. Hence the name "resonant scattering". A crude estimate for the size of D_{xx} is

$$D_{xx} \sim \frac{\pi}{4} \left(\frac{\delta B}{B}\right)^2 \left(\frac{v_t^2}{\Omega}\right) \ . \tag{9}$$

In general, $\underset{\sim}{k}$ will have components perpendicular to the direction of $\langle \underset{\sim}{B} \rangle$, in addition to the z component considered here. The result[8] for a general direction of $\underset{\sim}{k}$ reduces to that given in Eq. (9), when $k_\perp v_\perp \ll \Omega$. Results similar to Eq. (8) are well known in the theory of cosmic ray propagation in galactic magnetic fields.

IV. EFFECTIVE DIFFUSION DUE TO FIELD LINE WANDERING

The fact that fluctuations $\delta\underset{\sim}{B}$ exist leads to another type of cross-field transport, qualitatively different from resonant scattering. In this second process, an electron remains rigorously tied to the total field line. But because of the fluctuating component $\delta\underset{\sim}{B}$, the total field line can wander sideways relative to the average field $\langle\underset{\sim}{B}\rangle$. Thus the electron's trajectory carries it away from the position it would have had in $\langle\underset{\sim}{B}\rangle$ alone.

We again model the field by Eq. (4). The trajectory of a field line, and thus of the electron, is given by

$$\frac{dy}{dz} = \frac{B_y}{B_z} = \sum_{k_z} \left(\frac{\delta B_k}{B_o}\right) e^{ik_z z} \quad . \tag{10}$$

The zero-order orbit $z = z_o + v_z t$ is substituted into Eq. (10), which can then be integrated to give the y deflection in a time Δt,

$$\delta y = v_z \sum_k \left(\frac{\delta\Omega_k}{\Omega}\right) \exp(ik_z z_o) \left\{\frac{\exp[ik_z v_z \Delta t]-1}{ik_z v_z}\right\} \quad . \tag{11}$$

To obtain the y diffusion coefficient

$$D_{yy} \equiv \lim_{\Delta t \to \infty} \frac{<|\delta y|^2>}{2\Delta t} \quad , \tag{12}$$

we multiply (11) by its complex conjugate, and average over initial positions z_o. All initial positions are assumed equally likely. After using Eq. (7) once again, we obtain

$$D_{yy} = \pi v_z \sum_k \left(\frac{|\delta\Omega_k|^2}{\Omega^2}\right) \delta(k_z) \quad . \tag{13}$$

Had we retained the x and y components of $\underset{\sim}{k}$, the full expression for D_{yy} would have reduced to Eq. (13), for $\tilde{k}\, v \ll \Omega$. Perturbations $\delta B_{\underset{\sim}{k}}$ having $\underset{\sim}{k} \cdot <\underset{\sim}{B}> = 0$ are the ones which contribute to this effective cross-field diffusion process.

One can define a typical length scale by

$$\ell \equiv \frac{\sum_k |\delta\Omega_k|^2 \, \delta(k_z)}{\sum_k |\delta\Omega_k|^2} \quad . \tag{14}$$

In terms of ℓ, the size of the diffusion coefficient is estimated as

$$D_{yy} \sim \pi \left(\frac{\delta B}{B_o}\right)^2 \left(\frac{v_t^2}{\Omega}\right) \left(\frac{\ell}{\rho_L}\right) , \tag{15}$$

if a typical value for v_z is $v_t \equiv (kT_e/m_e)^{\frac{1}{2}}$.

V. NEOCLASSICAL DIFFUSION

When the plasma is sufficiently collisionless that an electron can be mirrored back and forth between local maxima of δB before suffering a collision, a type of diffusion occurs which is analogous to neoclassical transport in toroidal confinement devices. As the trapped electron bounces, it undergoes drifts transverse to $<\underset{\sim}{B}>$. The magnitude of the sideways displacement due to these drifts becomes the effective step-size for the cross field diffusion process. The neoclassical diffusion coefficient is approximately given by[9]

$$D_{NC} \cong \nu_{eff} \, |\Delta r|^2 \, \mathcal{F} , \tag{16}$$

where ν_{eff} is an effective collision frequency defined below, $\mathcal{F}$ is the fraction of trapped particles, and $|\Delta r|$ is the width of the bounce orbit. We calculate these quantities as follows.

The average toroidal field is "unwrapped" into a straight cylinder, so that

$$\underset{\sim}{B} = B_o(r)\hat{z} + \delta\underset{\sim}{B}(r,\theta,z), \qquad |\delta B| << B_o \quad . \tag{17}$$

This is a valid approximation if the laser spot size is much greater than local gradient scale lengths. The fluctuating field δB is derived from a radial vector potential $A_r = A(r)\cos\eta$, with $\eta \equiv m\theta + kz$:

$$\begin{aligned} \delta B_r &= 0 \\ \delta B_\theta &= -kA\sin\eta \\ \delta B_z &= (mA/r)\sin\eta \; . \end{aligned} \tag{18}$$

For δB's which vary slowly compared to ρ_L, the bounce motion is given by

$$m_e \frac{d^2 s}{dt^2} = -\mu \frac{\partial B(s)}{\partial s} , \qquad (19)$$

where $\mu = m_e v_\perp^2/2B$ is the magnetic moment. If one uses (18) for B(s), and expands cos η about the point $\eta \cong \pi/2$, one obtains (for $B_o > 0$) the equation for the bounce motion along $\hat{z}$:

$$\ddot{z} + \omega_B^2 z = 0, \; \omega_B^2 = \frac{\mu}{m_e}\left(\frac{k^2 m|A|}{r}\right) = \frac{k^2 v_\perp^2}{2}\left(\frac{|\delta B_z|_{max}}{B_o}\right) . \qquad (20)$$

Since the maximum value of $|B|$ is $|B|_{max} = B_o + |mA/r|$, an electron is trapped if its velocity components satisfy

$$\frac{v_\parallel^2}{v_o^2} < \frac{m|A|}{rB_o} = \frac{|\delta B_z|_{max}}{B_o} . \qquad (21)$$

Thus the angular width of the trapped-particle region of velocity space is

$$(\Delta\Theta)^2 \cong \frac{|\delta B_z|_{max}}{B_o} , \qquad (22)$$

and the fraction of trapped particles is

$$\mathcal{F} \cong \Delta\Theta = \left(\frac{|\delta B_z|_{max}}{B_o}\right)^{\frac{1}{2}} . \qquad (23)$$

As in any random walk process, one expects $\Delta\Theta \propto t^{\frac{1}{2}}$. Inverting the latter relation gives an effective collision frequency

$$\nu_{eff} \cong \frac{\nu}{(\Delta\Theta)^2} = \nu \left(\frac{B_o}{|\delta B_z|_{max}} \right) , \tag{24}$$

where ν is the classical collision frequency.

To complete the calculation of the diffusion coefficient, the effective step-size $|\Delta r|$ is needed. The radial component of the ∇B and curvature drift velocity is approximately

$$\frac{dr}{dt} \cong - \left(\frac{m_e c}{qB_o^2} \right) \left(\frac{v_\perp^2 m^2}{2r^2} \right) A \cos \eta , \tag{25}$$

for trapped particles with $v_{/\!/} \ll v_\perp$. Expanding $\cos\eta$ about $\eta \cong \pi/2$, using the periodic solution of Eq. (20), and assuming that $|\Delta r| \ll r$, we integrate Eq. (25) to obtain the radial excursion during a bounce orbit:

$$|\Delta r| \cong \rho_L \left(\frac{|\delta B_z|_{max}}{B_o} \right) \left(\frac{3\pi \sqrt{2}\, m}{kr} \right) . \tag{26}$$

Here we have taken a typical amplitude of the z bounce motion to be π/k.

Finally, we re-assemble the factors needed to calculate the neoclassical diffusion coefficient using Eq. (16), and obtain

$$D_{NC} \cong \nu\rho_L^2 \left\{ \left(\frac{|\delta B_z|_{max}}{B_o} \right)^{\frac{1}{2}} \left(\frac{18\pi^2 m^2}{k^2 r^2} \right) \right\} . \tag{27}$$

The expression in brackets represents the amount by which D_{NC} is enhanced above the classical value $\nu\rho_L^2$.

In order for this analysis to be valid, it is necessary that $\omega_B \tau \gg 1$, where $\tau = \nu^{-1}$, or

$$(k\rho_L)\,(\frac{\delta B}{B_o})^{1/2}\,(\Omega\tau) \gg 1 \quad . \tag{28}$$

It is also necessary that

$$k, \frac{m}{r} \ll \rho_L^{-1} \quad . \tag{29}$$

Eq. (28) is a rather stringent condition on $\Omega\tau$, since each of the other factors in the inequality is less than unity. In essence, this neoclassical analysis applies to a more collisionless regime than the previous two sections do. For neoclassical transport, an electron must survive many bounce periods without a collision. In contrast, the quasilinear treatments of Sections III and IV require only that an electron survive many cyclotron periods without a collision.

VI. CONCLUSIONS

Each of the diffusion mechanisms described above occurs at a rate significantly faster than the classical value. Yet we shall show below that they all represent significant transport inhibition relative to the $B_o = 0$ case.

It is appropriate to compare diffusion due to resonant scattering and field-line wandering with the Bohm diffusion coefficient $D_B = v_t^2/16\Omega$, since the scaling is similar. From Eqs. (9) and (15) we obtain

$$\frac{D_{xx}}{D_B} \sim 4\pi\,(\frac{\delta B}{B_o})^2 \quad , \tag{30}$$

$$\frac{D_{yy}}{D_B} \sim 16\pi\,(\frac{\delta B}{B_o})^2\,(\frac{\ell}{\rho_L}) \quad . \tag{31}$$

For plausible numbers such as $(\delta B/B_o)^2 \sim 10^{-1}$, $\ell/\rho_L \sim 1$, both of these processes occur at rates faster than the Bohm value. These rates can be compared to the classical one by multiplying by the factor $(\Omega\tau/16)$. The neoclassical result given in Eq. (27) has a scaling similar to the classical rate, $D_{c\ell} = \nu\rho_L^{\,2}$. If we write $m/r \approx k_\perp$, then

$$\frac{D_{NC}}{D_{c\ell}} \simeq \left(\frac{\delta B}{B_o}\right)^{\frac{1}{2}} \left(200\,\frac{\lambda_{/\!/}^{\,2}}{\lambda_\perp^{\,2}}\right) \quad . \tag{32}$$

For $\lambda_{/\!/} \sim \lambda_\perp$, neoclassical diffusion is faster than classical if $\delta B/B_o > 3 \times 10^{-5}$. This is highly likely to be true, since each of the field generation mechanisms of Section II yields megagauss level fields.

Recent experiments[10] have indicated that transport may occur at rates significantly below the zero magnetic field classical value. The present results are quite consistent with this hypothesis. If we compare our diffusion coefficients with the classical rate $D_o = v_t^{\,2}\tau$ appropriate for a field-free plasma, we find

$$\frac{D_{xx}}{D_o} \sim (\Omega\tau)^{-1} \left\{ \frac{\pi}{4} \left(\frac{\delta B}{B_o}\right)^2 \right\} \tag{33}$$

$$\frac{D_{yy}}{D_o} \sim (\Omega\tau)^{-1} \left\{ \pi \left(\frac{\delta B}{B_o}\right)^2 \frac{\ell}{\rho_L} \right\} \tag{34}$$

$$\frac{D_{NC}}{D_o} \sim (\Omega\tau)^{-2} \left\{ \left(\frac{\delta B}{B_o}\right)^{\frac{1}{2}} \left(200\,\frac{\lambda_{/\!/}^{\,2}}{\lambda_\perp^{\,2}}\right) \right\} \quad . \tag{35}$$

For $\Omega\tau >> 1$, all of these expressions represent transport slower than the classical value for a field-free plasma.

From the discussion of Section II, it seems likely that field fluctuations with the scale sizes necessary for these three processes will in fact be present in laser plasmas. The required space scales are summarized as follows. Resonant scattering requires $\lambda_{\parallel} \approx 2\pi\rho_L$. Since $\rho_L = .15$ μm for 5 MG and 1 keV, parallel wavelengths of a few microns are thus necessary. Field-line wandering needs $\underset{\sim}{k}\cdot\langle\underset{\sim}{B}\rangle = 0$, which is a geometrical condition easy to satisfy locally. Our crude approximations for both these above cases require $\lambda_{\perp} >$ a few microns. Neoclassical diffusion requires δB's with scale lengths $\gg \rho_L$, or $L >$ a few microns once again. Since we found in Section II that space scales of 1-10 μm are likely to be well represented in the δB spectrum, we conclude that all of these conditions can be met, since δB's of the correct spatial scale will be present.

Finally, we have not considered here the effect of electric field fluctuations on cross-field transport. If laser-driven heat flow causes ion-acoustic or other instabilities to occur, the resulting electric field perturbations may have a very strong influence on diffusion processes. This possibility raises a host of new issues, and deserves careful consideration in a separate paper.

REFERENCES

1. K. A. Brueckner and S. Jorna, Rev. Mod. Phys. 46, 325 (1974), and references therein.

2. D. G. Colombant and N. K. Winsor, "The Thermal-Force Terms and Self-Generated Magnetic Fields in Laser-Produced Plasmas", U.S. Naval Research Laboratory Report 3362, 1976.

3. D. A. Tidman, Phys. Rev. Lett. 32, 1179 (1974); D. A. Tidman and L. L. Burton, Phys. Rev. Lett. 37, 1397 (1976).

4. M. S. Sodha, A. K. Ghatak, and V. K. Tripathi, "Self Focusing of Laser Beams in Plasmas and Semiconductors", in Progress in Optics XIII (North-Holland, Amsterdam, 1976), edited by E. Wolf, and references therein; C. E. Max, Phys. Fluids 19, 74 (1976).

5. J. A. Stamper, Phys. Fluids 18, 735 (1975) and Phys. Fluids 19, 758 (1976); J. J. Thomson, C. E. Max, and K. G. Estabrook, Phys. Rev. Lett. 35, 663 (1975); B. Bezzerides, D. F. DuBois, and D. W. Forslund, to be published.

6. K. G. Estabrook, Phys. Fluids 19, 1733 (1976); D. W. Forslund, Bull. Am. Phys. Soc. 21, 1066 (1976).

7. D. A. Tidman and R. A. Shanny, Phys. Fluids 17, 1207 (1974); B. A. Al'terkop et al., JETP Lett. 19, 170 (1974); B. A. Al'terkop and E. V. Mishin, Phys. Lett. 46A, 319 (1974); L. A. Bol'shov et al., JETP Lett. 19, 168 (1974).

8. C. E. Max, W. M. Manheimer, and J. J. Thomson, to be published.

9. R. D. Hazeltine, "Review of Neoclassical Transport Theory", in Advances in Plasma Physics vol. 6 (Wiley, N.Y., 1976), edited by A. Simon and W. B. Thompson.

10. R. C. Malone, R. L. McCrory, and R. L. Morse, Phys. Rev. Lett. 34, 721 (1975); W. C. Mead et al., Phys. Rev. Lett. 37, 489 (1976).

NOTICE

"This report was prepared as an account of work sponsored by the United States Government. Neither the United States nor the United States Energy Research & Development Administration, nor any of their employees, nor any of their contractors, subcontractors, or their employees, makes any warranty, express or implied, or assumes any legal liability or responsibility for the accuracy, completeness or usefulness of any information, apparatus, product or process disclosed, or represents that its use would not infringe privately-owned rights."

ELECTROMAGNETIC WAVE INTERACTIONS WITH INHOMOGENEOUS PLASMAS*

A. Y. Wong

University of California, Department of Physics
Los Angeles, CA 90024 and TRW Defense and Space Systems,
Redondo Beach, CA 90278

Abstract

Experimental observations of electromagnetic wave propagation and interactions with a large inhomogeneous laboratory plasma are summarized. The dominant process is the conversion of electromagnetic waves into electrostatic waves whose field strengths are enhanced by several orders of magnitude in the vicinity of the resonant layer where the incident frequency matches the local plasma frequency. Strong local accelerations of electrons and ions by these localized resonant fields are described. Density cavities produced by the digging actions of these ponderomotive forces are called "cavitons" which coexist with the rf fields. Parametric decay instabilities are found to be a much weaker process compared with the caviton formation at the resonant layer. A double resonance technique of controlling ion dynamics at a specific location in an inhomogeneous plasma is demonstrated. The correlations between the present experiments and laser-plasma interactions, as well as radio wave interactions with the ionosphere, are discussed.

*This paper has been included in these Proceedings, since it was scheduled for presentation at the 4th Workshop on "Laser Interaction and Related Plasma Phenomena" held at Rensselaer Polytechnic Institute, 8-12 November 1976; however, the author could not attend.

INTRODUCTION

In recent years, laser-pellet experiments [1] and the active interaction with the ionosphere by radio waves [2] have focused attention on the physics underlying the electromagnetic wave interactions with inhomogeneous plasmas. Very interesting though puzzling results have been observed, but in either case the experimental situation is not easily accessible to detailed diagnostics. Comparisons between experimental findings and the vast body of theoretical work and computer simulations [3,4] have not always been completely satisfying, especially in regimes where large excitation fields are used.

With the advance of large-scale laboratory plasmas [5], it is now possible to study such interactions and to investigate their temporal and spatial developments from linear to nonlinear regimes. It is the purpose of this paper to follow the propagation of electromagnetic waves along a gentle density gradient and examine the linear and nonlinear interactions in the vicinity of the resonant region where the wave frequency is close to the local plasma frequency. The analysis is considerably simplified by the gentle density gradient whose scale length is many times the electromagnetic free-space wavelength and by the absence of any external confining magnetic field. In the linear regime, the three dimensional propagation characteristics are followed in a time-space resolved experimental study (Section 1). The large plasma dimension actually enables us to observe the propagation of electromagnetic wave packets at the speed of light. The linear conversion between electromagnetic (hereafter abbreviated as EM) waves and electrostatic (ES) waves in the critical region is presented (Section 2). Nonlinear effects are readily observed even though the external field might be weak because the plasma resonance enhances this driving field by two to three orders of magnitude. The conversion from long wavelength EM field to short wavelength ES field traveling at much reduced group velocities can take place in many nonuniform plasmas with and without magnetic fields. Since this conversion takes place with an accompanying large increase of electrostatic fields by several orders of magnitude, this process is of great importance in understanding the source of nonlinear phenomena.

In the nonlinear regime, the development of an instability leads to a new state — a density cavity created by large electric field intensity gradients, hereafter referred to as a "caviton" (Section 3). This new state can be easily generated in an inhomogeneous plasma and has a very low threshold. In Section 4 the accelerations of electrons and ions by these localized cavitons are then discussed. The parametric decay instabilities, which could grow temporally in a nonuniform electric field region near the EM cutoff, is described. They are shown to be unimportant in comparison with the generation of cavitons. A double resonance technique with two separate pump frequencies, which permits an external remote control of the ion dynamics in the plasma, is illustrated (Section 5). Finally, in Section 6 the correlations with laser-pellet interactions and the excitation of the ionosphere by radio waves are outlined.

1. Electromagnetic Propagation

A large steady-state magnetic field-free plasma [5] of 2M dia. and 4M length is specially developed for this experiment as shown in Fig. 1. The plasma is generated by dc discharges between 100 hot filaments and the chamber wall in argon at pressures between 10^{-4} and 10^{-3} torr. The plasma is contained by multimirror confinement with 10,000 permanent magnets at the interior chamber walls. The high efficiency of plasma production is a result of the containment of the ionizing electrons and the long path of travel between reflections. An axial density gradient ($1M < n/\nabla n < 10M$) is produced by generating the plasma preferentially near one end of the device and by controlling the ionizing mean-free-path through the neutral pressures. Radial gradients are avoided by azimuthally symmetric plasma generation and gas feeds. In the range of microwave

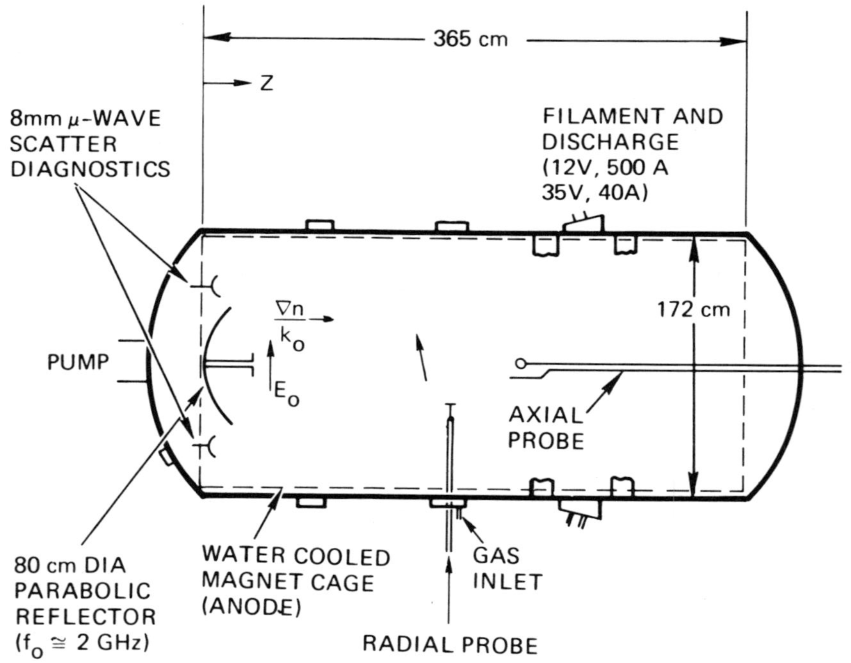

FIG. 1 - Schematic of the large TRW QUIPS (Quiescent Plasma Studies) chamber used in the microwave interaction with plasmas. The radiating antenna, 80 cm in diameter, was used in the 2 GHz experiment. Langmuir, rf and plasma wave probes are mounted on the axial and radial probes. The inside surface is lined with 10,000 permanent magnets to provide surface-cusp confinement of ionizing electrons.

frequencies chosen for our experiment (1 GHz $\leq f_o \leq$ 3 GHz, 10 cm $\leq \lambda_o \leq$ 30 cm), there are 10 — 100 free space wavelengths over a density gradient scale length. Up to 40KW of pulsed microwave power has been used in our experiments.

The first experiment on the EM propagation by Stenzel, Wong and Kim [6] was in the simplest possible arrangement: a normal incidence with the EM propagation vector $\underline{k}_o || \nabla n_o$ in the far field zone, Fig. 2. A low-aperture and low gain (G = 2.7dB wide-angle H-plane 3dB beam width $\theta \simeq 56°$) rectangular, open wave-guide antenna has been used. The EM signal is detected by a coaxially-fed short wire dipole antenna or a shielded magnetic loop constructed from coaxial wires which could be moved axially or radially. The normally incident wave $\underline{k}_o || \nabla n_o$ is detected near the axis of our plasma. In order to achieve effective free-space propagation measurements, fast-rise ($\tau_{rise} \approx$ 5 - 6 cycles) phase-coherent EM wave bursts are produced by using pin-diode gating. The large plasma size enables us to distinguish the straight line EM propagation along the axis from the longer paths taken by EM reflections from surrounding chamber walls. A fast-sampling scope is used for this purpose with the sampling gate width set in the subnanosecond range. The wall is lined with microwave absorbers (fine-wire steel wool) to minimize reflections. The changes in phase and group velocities along the density gradient (Fig. 2), are measured from the time-of-flight measurements, Fig. 3. Our results agree essentially with what is expected from the homogeneous EM dispersion relation $\omega^2 = \omega_p^{\,2}(z) + k^2c^2$ applied locally over the gentle density gradient. The sampling scope also allows us to make a polar plot of the radial EM field during its propagation. As a result, we could observe the transition from a propagating wave to a standing wave when the EM burst reaches the EM cutoff z_c, $\omega_o = \omega_p(z_c)$, for normal incidence of the EM wave. We wish to point out that for oblique propagation of EM waves at an angle θ with respect to the density gradient, the EM cutoff location, z_c, is given by $\omega_p(z_c) = \omega_o \cos\theta$ as readily derived from the wave

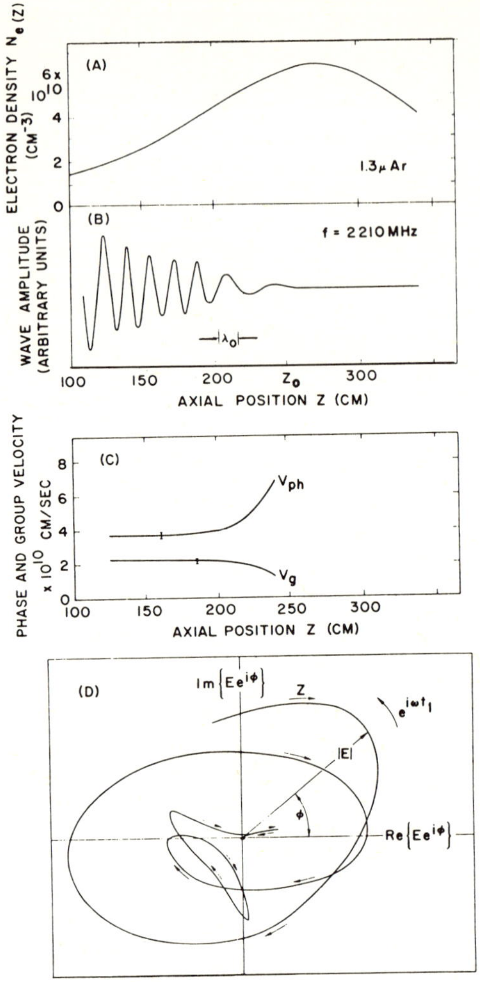

FIG. 2 - Propagation of EM waves radiated by a dipole antenna for normal incidence $\underline{k}||\nabla n_o$: a) Axial electron density profile. b) Axial variation of radial electromagnetic field, $E_r(z) \cos[\omega t_1 - k(z)z]$, ωt_1 = constant. λ_o is the free space wave-length and z_o the resonant location $\omega_p(z_o) = \omega_o$. c) Measured spatial dependence of phase velocity [from Fig. 2(b)] and group velocity [from Fig. 3(b)]. d) Amplitude and phase of the radial electric field sampled at the same time t_1 at different axial positions. Phase behavoir shows the transition from a propagating pattern ($z \ll z_c$, where z_c is the location for the EM cutoff.) to a standing wave pattern at $z \approx z_c$. [After Stenzel, Wong and Kim, Ref. 6]

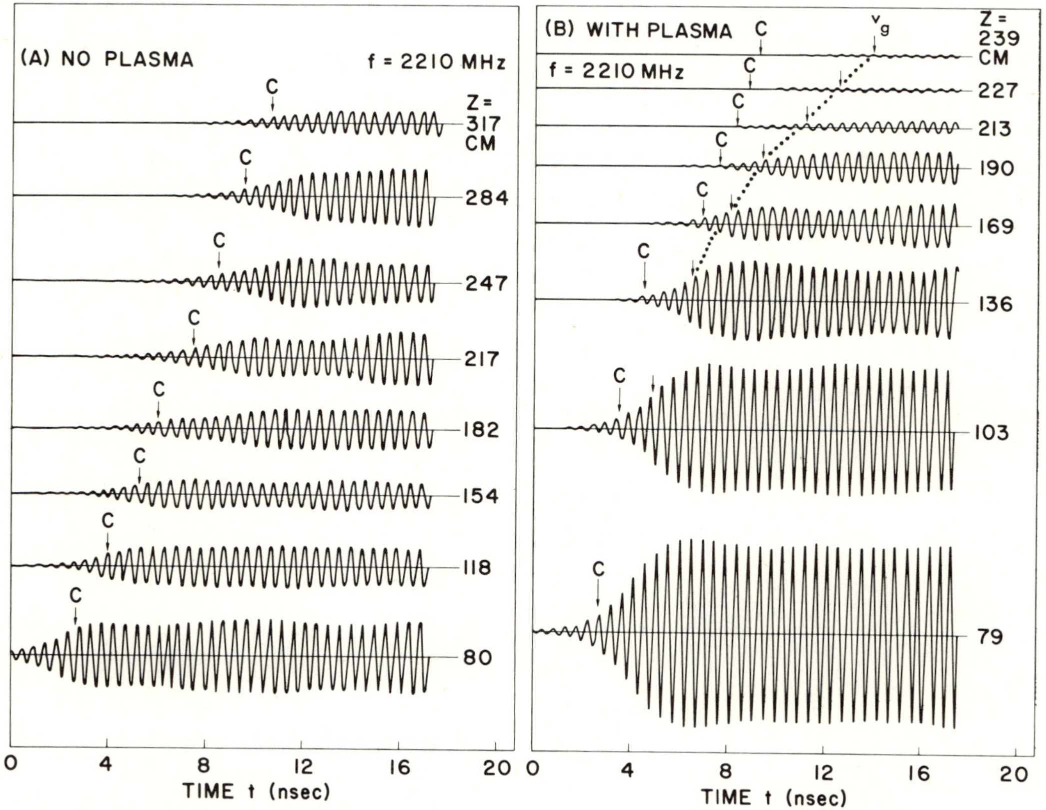

FIG. 3 - Time-of flight measurements with fast-rise phase-coherent microwave bursts. Wave packet envelope propagates with speed of light (arrow marked "c") in vacuum, (a), with group velocity $v_g < c$ (arrow marked "v_g") in the plasma, (b). The applied signal in (b) is larger than in (a). Note in (b) the strong amplitude decrease near cutoff due to refraction. [After Stenzel, Wong, and Kim, op. cit.]

momentum conservation in a direction perpendicular to the density gradient. This location, z_c, is to be distinguished from the resonant z_o [$\omega_p(z_o) = \omega_o$] further up on the density profile where the plasma enhancement of the incident field occurs.

The theoretical consideration of the conservation of energy flux ($v_g E^2 A$ = const, where E is the electric field and A is the illuminated area) in one dimension implies that the reduction in group velocity in the vicinity of the reflection point gives rise to amplitude swelling as in the usual one dimensional treatment. However, three dimensional effects are important in our experimental situations. As a result of the decreasing refractive index of the plasma with distance $\eta(z) = [1 - \{\omega_p^2(z)/\omega^2\}]^{1/2}$, an initially divergent beam now rapidly diverges as it travels along the density gradient, resulting in a much larger illuminated area A. The net result is an overall reduction in EM field amplitude near the plasma cutoff as supported by a three dimensional ray-tracing calculation [7] as shown in Fig. 4. This divergent pattern is also supported by direct experimental measurement of the polarization vector of the EM wave taken off the axis at half the radius of the chamber. Using a nonlinear scattering dipole [8] which gives the minimum perturbation to the field pattern, the electric field is found to point in the radial direction near the radiating source but rotated by $\Delta\theta > 40°$ toward the axis near the cutoff.

The same refractive effects of an inhomogeneous plasma also significantly enhance the convergence of an initially convergent beam at the source, giving rise to an effective shorter focal length. The ray-tracing diagram in Fig. 4 essentially confirms this focusing property of the plasma.

2. Linear Conversion from Electromagnetic to Electrostatic Waves

Except for regions very close to the axis, the EM waves from our antenna are obliquely incident with respect to the density gradient (along $\hat{z}$), and thus possess an electric field component E_o

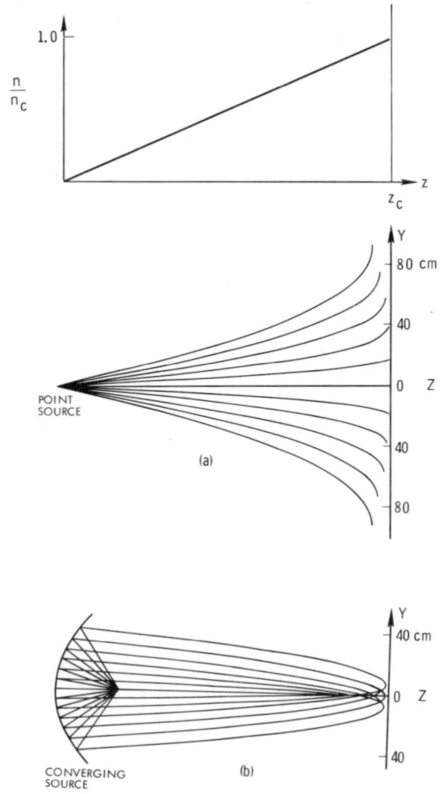

FIG. 4 - (a) Two dimensional trajectories of initially divergent beams in a plasma with a linear density gradient. The rays are strongly refracted near the critical layer.

(b) Ray trajectories of initially convergent beams. [After W. DiVergelio and A. Y. Wong.]

parallel to the density gradient: $E_o = E_{em} \sin\theta$ where θ is the angle between $\underset{\sim}{k}_o$ and the density gradient $(\partial n_o/\partial z)\hat{z}$. This axial field E_o is enhanced by the plasma resonance at the location $\omega_p(z_o)=\omega_o$. In the presence of a density gradient this enhanced field moves electrons from a high density region to a low density region, exciting one dimensional electron density perturbations propagating along the axis. Even though the incident EM field is three dimensional, the behavior of the plasma driven by the resonant field aligned mainly along the axis can be described by the following one dimension equations. As a first approximation ions are assumed to be stationary

$$\frac{\partial n}{\partial t} = -\frac{\partial}{\partial z}(nv) \qquad \text{continuity equation} \tag{2.1}$$

$$\frac{\partial v}{\partial t} + v\frac{\partial v}{\partial z} + \frac{eE}{m_e} + \frac{1}{nm_e}\frac{\partial p_e}{\partial z} + \nu_e v = 0 \qquad \text{equation of motion} \tag{2.2}$$

$$\frac{\partial E}{\partial z} = 4\pi e(n_i - n) + \frac{\partial E_o}{\partial z} \qquad \text{Poisson's equation} \tag{2.3}$$

where E is the total electric field inside the plasma, E_o is the external rf electric field of frequency ω_o, p_e is the electron kinetic pressure, and ν_e is the electron collision rate. Differentiating Eq. (2.3) with respect to t and integrating the resulting equation with respect to z, we have

$$\frac{\partial E}{\partial t} = 4\pi\ env + \frac{\partial E_o}{\partial t} \tag{2.4}$$

where Eq. (2.1) has been used to replace $\partial n/\partial t$. Differentiating the last equation once again with respect to t we have

$$\ddot{E} - \ddot{E}_o = 4\pi e \frac{\partial}{\partial t}(nv) \quad . \tag{2.5}$$

Substituting the equation of motion of electrons on the right-hand side and Fourier analyzing in time we obtain:

$$-\omega^2 E + \frac{\omega}{(\omega+i\nu)}\left[\omega_p^2 E - v_{th}^2\frac{\partial^2 E}{\partial z^2} + v_{th}^2\left(\frac{1}{n_o}\frac{\partial n_o}{\partial z}\right)\frac{\partial E}{\partial z}\right] = -\omega^2 E_o \tag{2.6}$$

The plasma response expressed in terms of the total electric field E can also be written as

$$\varepsilon(z)\ E = E_o \tag{2.7}$$

where

$$\varepsilon(z) = \frac{1}{k_{D_o}^2}\left[\frac{\partial^2}{\partial z^2} - \frac{1}{n_o}\frac{\partial n_o}{\partial z}\ \frac{\partial}{\partial z}\right] + \left[1 - \frac{\omega_p^2(z)}{\omega_o^2} + i\ \frac{\nu}{\omega_o}\right]$$

is the differential operator representing the plasma resonance in a nonuniform plasma in analogy to the usual plasma dielectric function in a uniform plasma $k_{D_o}^2 = \omega_o^2 m_e/3kT_e$. The first term of $\varepsilon(z)$ represents the propagation of the ES field; the second term originating from the ambipolar field $E_A = -(kT_e/en_o)\nabla n_o$ represents the wave convection in the inhomogeneous plasma and the spatial damping in the direction of $-\nabla n_o$; the terms inside the second brackets stand for the cold plasma dielectric effect, with ν being the electron-ion collision frequency. The density gradient allows charge accumulations to occur under this enhanced driving field E and density oscillations n_1 are excited according to the following equation derived from Eq. (2.7) by differentiating it with respect to z, neglecting second order terms and the spatial dependence of E_o:

$$\varepsilon(z)n_1 = \frac{-eE}{m\omega_o^2}\frac{\partial n_o}{\partial z} = d\,\frac{\partial n_o}{\partial z} \tag{2.8}$$

where $d = -eE/m\omega_o^2$ is the excursion distance of the electron. The source term on the right-hand side describes the high frequency density perturbation created by the oscillating field E in the presence of the density gradient. Of course E and n_1 are related to each other through the Poisson's equation.

The solution of Eq. (2.7) can be expressed in terms of well known functions if the ambipolar field E and the collisional term ν_e/ω_o are small; these approximations are good in the presence of a gentle linear density gradient and nearly collisionless environments. Equation (2.7) then becomes

$$\left[\frac{1}{k_D^2}\frac{\partial^2}{\partial z^2} + 1 - \frac{\omega_{pe}^2(z)}{\omega_o^2}\right] E = E_o \tag{2.9}$$

where $\omega_{pe}^2(z) = \omega_o^2[1 + (z-z_o)/L]$, L = density gradient length. The general solution E/E_o is expressible in terms of the Airy functions [9]

$$\frac{E}{E_o} = \pi(k_D L)^{2/3}\,[A_i^2(x) + G_i^2(x)]^{\frac{1}{2}} \tag{2.10}$$

where $x = (k_D L)^{2/3}\ (z-z_o)/L$. The main predictions of Eq. (2.9) are (for $k_D L > 1$):

1) The electric field is enhanced near $z = z_o$ where $\omega_{pe}(z_o) = \omega_o$ and the enhancement factor $E/E_o \approx (k_D L)^{2/3}$. In typical experiment $10 \lesssim k_D L \lesssim 10^4$, the enhancement is very significant. It is this large enhancement which makes nonlinear effects easily observable even for very modest incident fields $E_o^2/4\pi nkT \approx 10^{-4}$. A simple and qualitative way of looking at the enhancement is that the energy is carried to the resonant region by EM waves at nearly the speed of light. This energy is convected away at a much lower velocity, namely the electron thermal velocity. The enhancement factor can be estimated by $E/E_o \propto (c/v_e)^{1/2} \approx (\lambda_o/\lambda_D)^{1/2} \approx (k_D L)^{1/2}$ since $c = \omega/k_D$, and $\lambda_o \approx L$. In order to compare with the experimental dependence on the density gradient L we must include the dependence of the incident field E_o on L which can be obtained from our ray-tracing calculation. As mentioned previously, E_o in an initially divergent beam tends to decrease with gentler density gradient.

2) The plasma electric field changes its phase by 180° across the resonant layer as the dielectric constant ε changes sign. This phase reversal is important in understanding the generation of density perturbations by the oppositely directed velocities across the resonant layer according to the following equation:

$$\frac{\partial n}{\partial t} = -v\frac{\partial n}{\partial z} + n\frac{\partial v}{\partial z} \approx -n_o\frac{\partial v}{\partial z} \quad \text{where} \quad v \approx \frac{-eE(z)}{m}$$

3) Electrostatic plasma waves are excited near this resonance region z_o and propagate down the density gradient $-\nabla n_o$ direction. In the linear regime the propagation up the density gradient is prohibited by the plasma cutoff. The propagation of these waves accounts for the convective loss which limits the enhancement of the electric field.

4) These ES plasma waves should have shorter and shorter wavelengths as they propagate down the density gradient as given by the

approximate relation $k^2 = k_{D_o}^2 = k_{D_o}^2 [1-\omega_p{}^2(z)/\omega^2]$.

These theoretical predictions have been borne out by close examination of the region between the EM cutoff and the resonant location. Figure 5(a) shows that for oblique incidence, the EM waves are converted to short-wavelength ES oscillations which propagate down the density gradient ($-\nabla n_o$) with decreasing wavelengths. By lowering the plasma density such that there is no resonant location anywhere in the plasma, no ES waves are observed propagating down the density gradient.

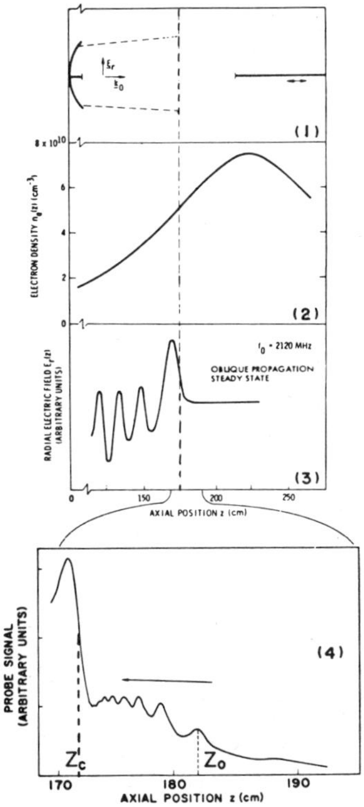

FIG. 5 - (a) 1. Experimental arrangement of the radiating antenna and the parobolic dish, which produces oblique incident EM waves. Up to 10W CW power is used. 2. Axial density profile. 3. Spatial dependence of the radial electric field $E_r(z)$. 4. Linear converted plasma waves propagating down the density gradient towards the EM cutoff after being excited at resonant location z_o by the resonantly enhanced field E_z. The wavelength is obtained by interferometry technique.

This experimental fact further confirmed the resonant location as the origin of the ES waves. The theoretical and experimental agreement on wavelength measurements is good away from the resonant region, but becomes poor near $\omega_p(z) = \omega_o$ since the wavelength is very sensitive to density perturbations by probes near the critical region. The conversion from EM to ES waves is found to be more efficient at steeper density gradients. We believe that this is caused by the reduced divergence of the incident beam in steeper gradients. Using a coherent EM pulse train as shown in Fig. 3, we have measured the time lag between the appearance of the short wavelength ES waves and the first arrival of the EM pulse at the cutoff location. We have found that the linear conversion from EM to ES waves takes place in a time scale $[10(2\pi/\omega_{pe})^{-1}]$ determined essentially by electron inertia.

In order to avoid perturbations by the probe on the ES wave, a remote method is devised for estimating the resonant ES wave amplitudes. The method is based on the ponderomotive force, $\nabla(\langle E^2\rangle/8\pi)$, which expels electrons from the resonant region. Ions follow so as to preserve space-charge neutrality, resulting in a density perturbation $\Delta n/n_o = E^2/8\pi n_o kT_e$ [see Eq. (3.2) in Section 3]. After turnoff of a pulsed incident EM wave, the associated density perturbation ($\Delta n/n \simeq 4\%$ at 10W incident power) propagates out from the high-field strength region and is measured with a Langmuir probe a few centimeters away from the resonance region in the absence of high-frequency signals. We have verified that $\Delta n/n_o$ varies linearly with E^2 and that no ionization takes place. This observation yields local resonant electric field strengths $E_{es} \gtrsim 100$ V/cm which are 3 orders of magnitude above the local EM fields. The conversion efficiency $E_{es}/E_{em} = 10^3$ agrees with the enhancement factor computed exactly from Eq. (2.7) using $\nu/\omega \simeq 10^{-4}$ and $L = 100$cm.

When much higher pump power ($E_o^2/4\pi nT_e \simeq 10^{-2}$) is used, the perturbations on the spatial density profile can be more easily measured by a sampling technique at various stages of development.

The measurements presented in Fig. 5(b) show a series of cavities starting with the deepest cavity of the largest width at the resonant region to progressively shallower and narrower cavities down the density gradient. Both the separation and the depth of cavities follow approximately an Airy function pattern and are consistent with the fact that the envelope of rf fields which create these cavities obey the Airy pattern predicted from the inhomogeneous density profile [Eq. (2.10)].

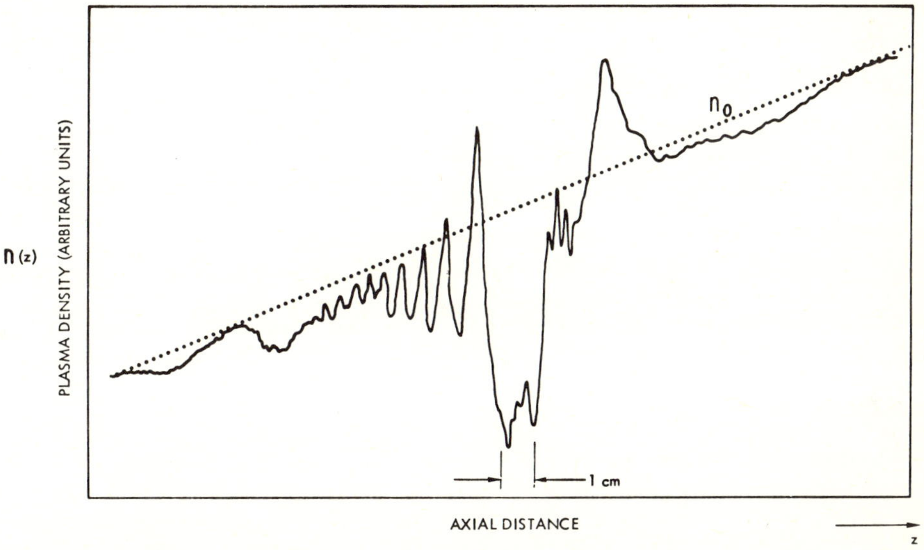

FIG. 5 - (b) Axial density profile sampled at 5 μsec after a high-power incident pulse (0.2 μsec, P_o = 10KW, f_o = 1 GHz) is terminated. Both the large density caviton and the smaller cavities are generated by the pondermotive force of electrostatic fields.

A special experimental arrangement has been designed [10] to confirm this large enhancement factor. The plasma region between cutoff and resonance is simulated by a linear density gradient provided by judiciously arranging the filaments inside a 30cm dia. plasma chamber as shown in Fig. 6. The driving field E_o along the density gradient is replaced by a quasistatic field at a lower frequency 360 MHz generated by an rf electrode.

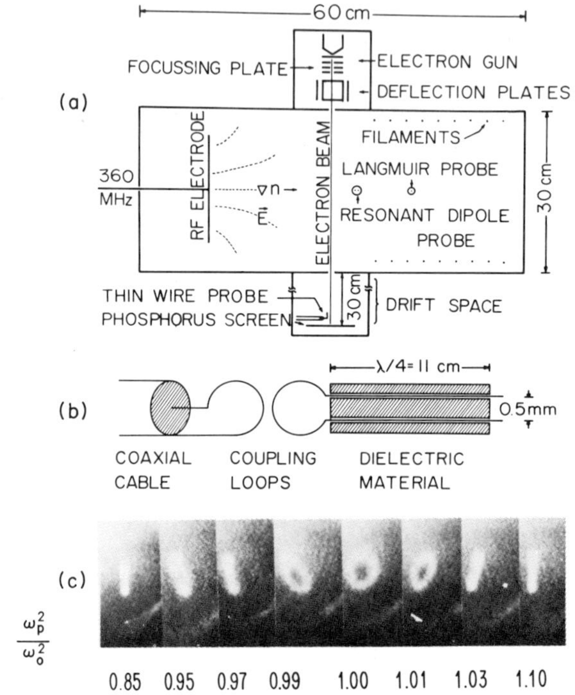

FIG. 6 - (a) UCLA DP device with electron beam diagnostics. (b) Resonant-dipole probe consisting of a $\lambda/4$ resonator in a ceramic shield. The potential difference measured between the exposed probe tips is magnetically coupled to a 50-Ω coaxial cable. (c) Lissajous patterns as function of increasing density (from left to right) $\omega^2{}_{pe}(z_o)/\omega_o{}^2 = 0.85, \ldots, 1.10$. $P_o = 10W$, $\omega_o/2\pi = 360$ MHz, steady state. The horizontal deflection is caused by the plasma rf field and the vertical deflection by a constant external field. The horizontal field peaks at the resonant location $\omega_p{}^2(z_o)/\omega_o{}^2 = 1$ and its phase changes by π across the layer. Ref. 10.

A very narrow test electron beam (0.5mm diameter, 5-9KeV, 0.1μA, $n_b = 10^4$ cm^{-3}) traverses the resonant region along the radial direction in approximately 1/2 — 1/4 the rf period.

Using the probe to map out the radial extent of the resonant region (Δr = 5cm) and taking into account the amplification of the beam deflection in the drift space, the correlation between the electric field and the beam spread is established (under typical experimental conditions of $\omega_o/2\pi$ = 360 MHz and drift region of 50cm, 1mm deflection of the 9KeV beam corresponds to 5 volts/cm rf field in the plasma). A large enhanced field of 50 volts/cm rf field using the electron beam diagnostics, which corresponds to E_T/E_o=100.

This electron beam method has enabled us to calibrate our probe readings and the comparison is given in Table I. In order to demonstrate that ES plasma waves originate from the resonant layer, we have reduced the peak density such that there is no resonant layer anywhere and no propagating ES waves are observed.

Table I

Comparisons of Electron Beam and Resonant Dipole Probe Diagnostic Techniques of Measuring RF Fields.

	Electron Beam	Resonant Dipole Probe
Location of Resonant Region	No Disturbance for Low Intensity Beams	Shifted up to 2cm Toward Higher Density Side ‡
Resonant Width	1 cm	Broadened to 2cm ‡
E_T/E_o at Resonance §	Steady State: 30 Peak: 100	Steady State: 3 Peak: 10 ‡
Frequency Response	dc to 1 GHz*	
Measured Growth Rate of RF Fields	1 — 2μs	10μs

‡ Due to the sheath near the probe. The sheath could produce high and low frequency noise. The probe also exhibits spurious resonant probe effects.

§ Ordinary single wire probes give much lower readings.

*Determined by transit time of beam across resonant region.

The phase relationship between the self-consistent field E_T and the applied field E_o is measured by the Lissajous method. The electron beam is modulated vertically by $E_o \cos\omega_o t$ through a set of plates at the source. After the horizontal modulation by the resonant field $E_T \cos(\omega_o t + \phi)$, it emerges on the phosphorous screen as a Lissajous figure whose ellipticity depends on E_T and ϕ, Fig. 6. This method indeed confirms that the plasma field E_T undergoes a change of phase of π across the resonant region.

This electron beam could also act an an rf probe for frequency measurements, as it possesses both good frequency response (400 MHz) and spatial resolution (0.5 mm) without perturbing the region of interest. A high impedance probe located outside the plasma column collects the beam after it traverses the interaction region and the frequencies can be monitored through a sensitive amplifier.

A word of caution against the use of Langmuir probes for high frequency measurements at the critical layer should be made. A Langmuir probe, be it floating or at any bias, collects plasma and causes density perturbations. Since the resonance at the critical layer has a Q $\gtrsim 10^3$ or more, any slight density perturbation ($\Delta n/n_o \gtrsim 10^{-3}$) could significantly modify the resonant behavior. Furthermore, a separate experiment on probe resonances [11] has shown that high rf fields are concentrated at the tip of a cylindrical probe seen by the deflection of a probing electron beam glazing by the probe tip. Together with the plasma sheath around a probe, this concentrated rf field induces resonant phenomena which are peculiar to the probe and not necessarily related to the plasma proper. The failure to observe stationary cavitons in an experiment of similar arrangement [12] is quite possibly due to the perturbing effects of the probes used in that experiment.

3. Nonlinear Regime

Formation of Localized Fields and Cavities (Cavitons). The highly localized linear enhancement of the driven oscillating field can lead to interesting nonlinear phenomena driven by the ponderomotive force $F = -\nabla\langle E^2\rangle/8\pi$. This force results from electron oscillations in a nonuniform field, through the $\underline{v}\cdot\nabla\underline{v}$ term in the electron momentum equation. The time-averaged drift motion of electrons causes the ions to move by means of the ambipolar field. Combining the electron and ion momentum equations we have

$$\rho \frac{\partial v_i}{\partial t} = -\nabla(P_e + P_i) - \left(\frac{\omega_p^2}{\omega^2} \nabla \frac{\langle E^2\rangle}{8\pi}\right) \qquad (3.1)$$

where $\rho = n_e m_e + n_i m_i$, $P_e = nkT_e$. Under quasistatic conditions, the above equations describes the balance between changes in the plasma pressure and the ponderomotive force

$$\nabla(P_e + P_i) = -\frac{\omega_p^2}{\omega^2} \nabla \frac{\langle E^2\rangle}{8\pi}$$

$$n = n_o \text{ Exp } - \frac{\omega_{po}^2}{\omega^2} \frac{\langle E^2\rangle}{8\pi n_o kT} \qquad (3.2)$$

We believe the following sequence of events describes the EM interaction with the inhomogeneous plasma. The applied EM field E_o is enhanced by the process of linear conversion to ES modes, whose field is maximum at the resonant location where $\omega_{pe} \approx \omega_o$. The maximum field enhancement is limited by the convective loss of electron plasma waves out of the resonant region and transit time damping. The density cavity (hereafter referred to as "caviton") created by the ponderomotive force of this enhanced rf field, is favorable to the trapping of rf field, reducing the convective loss. This results in a higher rf field which in turn creates a deeper density cavity and the process continues until saturation mechanisms set in. The particles expelled from this region of cavity formation will be studied in the next section.

This development of cavitons has been studied in the smaller inhomogeneous plasma in which an electron beam diagnostic could be used. The temporal and spatial development of the rf field E_T and the corresponding density perturbation δn are studied by pulsing on

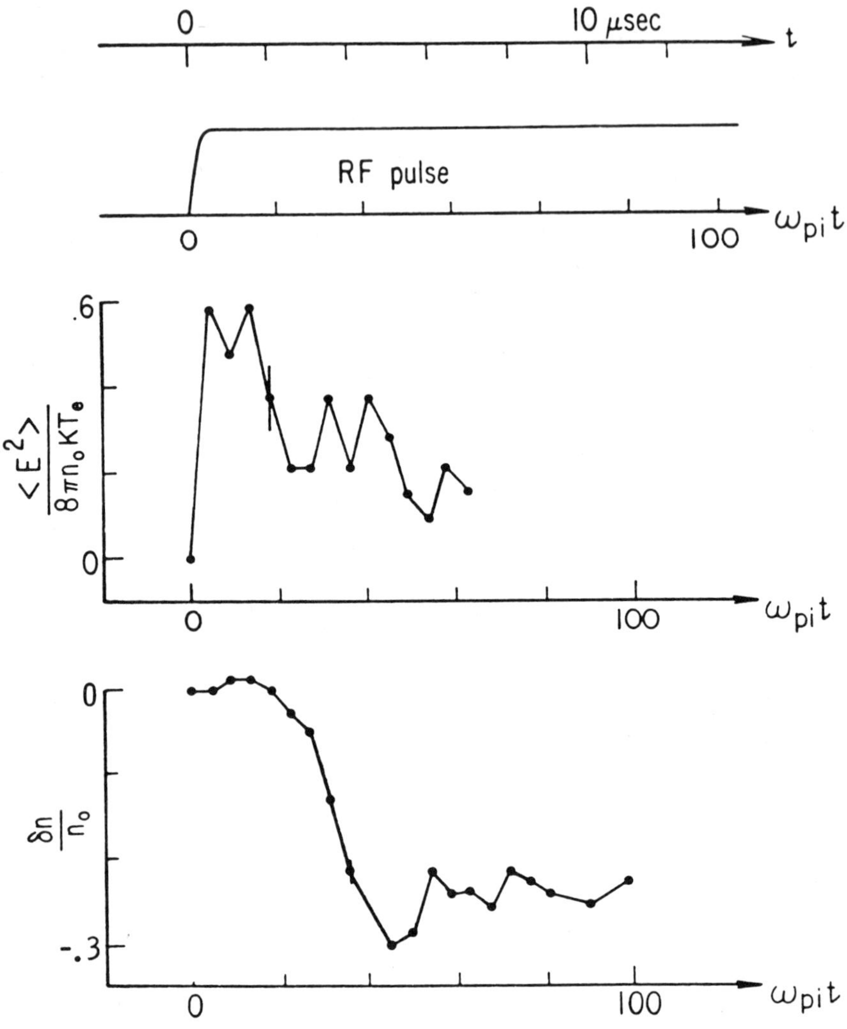

FIG. 7 - (a) Temporal variations of total rf fields $\langle E_T^2 \rangle / 4\pi n_o T_e$ and density perturbation $\delta n/n_o$ at the resonant location. Rf field is measured by an electron beam technique; $\delta n(\tau)/n_o$ is measured by Langmuir probe immediately after an external pulse of duration τ is terminated to avoid rf interference. Oscillations at ion acoustic frequencies ($0.6\ f_{pi}$) are observed to develop at later times.

[After H. Injeyan, P. Leung, and A. Y. Wong, UCLA DP device; $\frac{\omega_o}{2\pi} = 360$ MHz, $L = \frac{1}{n_o} \frac{\partial n_o}{\partial z}^{-1} = 20$ cm, Te = 2eV]

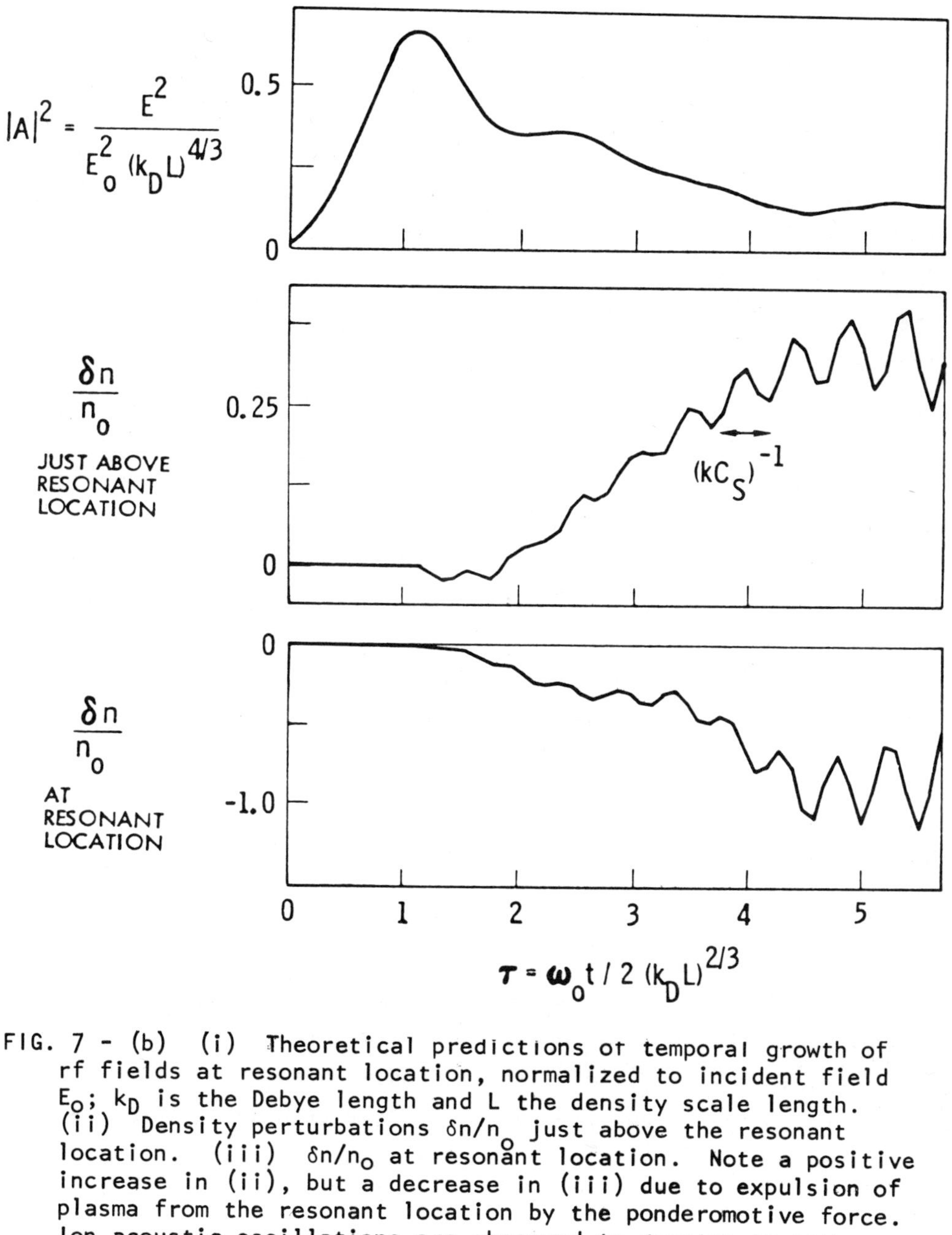

FIG. 7 - (b) (i) Theoretical predictions of temporal growth of rf fields at resonant location, normalized to incident field E_o; k_D is the Debye length and L the density scale length. (ii) Density perturbations $\delta n/n_o$ just above the resonant location. (iii) $\delta n/n_o$ at resonant location. Note a positive increase in (ii), but a decrease in (iii) due to expulsion of plasma from the resonant location by the ponderomotive force. Ion acoustic oscillations are observed to develop in both (ii) and (iii). [After G. Morales and Y. C. Lee.]

the externally imposed field E_o. The rf field is measured by the electron beam technique [13] while the corresponding density perturbation was monitored by Langmuir probes immediately after the rf field is turned off. The time and space resolved measurements made by Quon, et al, [13] show a rapid rise of the self-consistent rf field to $E^2/8\pi n_o T_e \approx 0.6$ within 0.5 μsec after the turn on of the external rf pulse, Fig. 7(a). This field readjusts itself to $E^2/8\pi \delta n T_e \approx 1$ as the density cavity $\delta n(\delta n/n_o \approx 0.3)$ develops. The electron beam diagnostic gives a faster response of the E field than earlier measurements by a high impedance rf probe [9], which show a gradual development of the rf field and density cavity. The electron beam method is regarded as much more reliable because it does not disturb the critical layer, whereas the probe response does. The initial enhancement of the high frequency field is explainable in terms of the vanishing of the linear dielectric constant at the critical layer. After the rf field is built up, the change in the density profile reduces the resonant enhancement, although the reduction is partially offset by the trapping of the rf field inside the developing density cavity. Measurements of the absolute rf field strength and density perturbation at saturation (after 10 μsec or 10 ion plasma periods) establish $\delta n \propto \langle E_o{}^2 \rangle$ qualitatively and $\langle E_T{}^2 \rangle / 8\pi n_o k T_e \approx \delta n/n_o$ quantitatively within 30% error. The spatial extent δz of the initially created caviton increases with the density gradient length L, as is expected from a larger resonant region.

The dynamics of cavitons such as splitting and propagation depends on both the pulse duration and the density gradient. For short pulses $\tau \lesssim 50\omega_{pi}^{-1}$ in a gentle density gradient ($k_D L > 500$), the $\langle E_\tau{}^2 \rangle$ and δn in the caviton break up after saturation into two or more peaks and propagate out from the resonant location towards regions of higher and lower densities, Fig. 8(a). The velocity of the uphill propagating caviton is less than the downhill propagating caviton by twice the plasma drift down the density gradient.

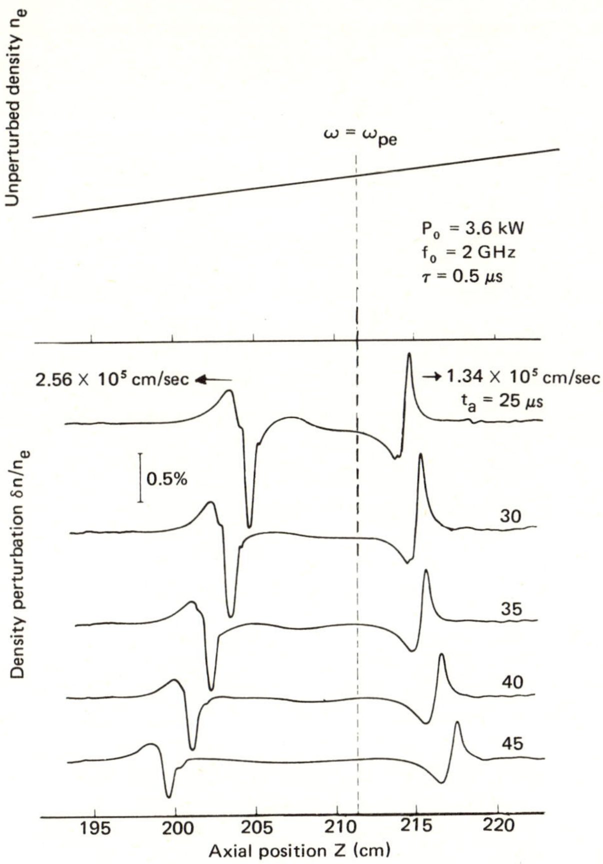

FIG. 8 - (a) When QUIPS is operated in a high density regime corresponding to the pump frequency f_o = 2 GHz, the density scale length, when normalized to the free space wavelength λ_o or the Debye length λ_D, is effectively longer than that in the low density regime (8b). Under a high power (3.6 KW) short pulse (0.5 μs) irradiation a caviton is created and then split up into two halves propagating up and down the density profile; the uphill propagation is different from the downhill propagation speed by the plasma drift speed of 0.6 x 10^5 cm/sec. The ion acoustic speed for argon plasma is 2 x 10^5 cm/sec. [After R. Stenzel and A. Y. Wong, QUIPS data.]

When the density gradient becomes steeper ($k_D L < 500$) the cavitons propagate predominantly down the density profile. If the rf pulse is applied for a longer time ($\tau > 50\ \omega_{pi}^{-1}$), the initial caviton is widened and additional cavitons are developed, Fig. 8(b). The formation of **shoulders** as a result of the expulsion of plasma from the cavity becomes more pronounced. The experimental observation, [Fig. 8(b)], corrobates well with the solutions of the nonlinear Schrödinger equation by Morales and Lee, Fig. 8(c) including ion dynamics. In both the experiment and theory, the appearance of a single caviton in earlier times is followed by the subsequent formation of additional cavitons.

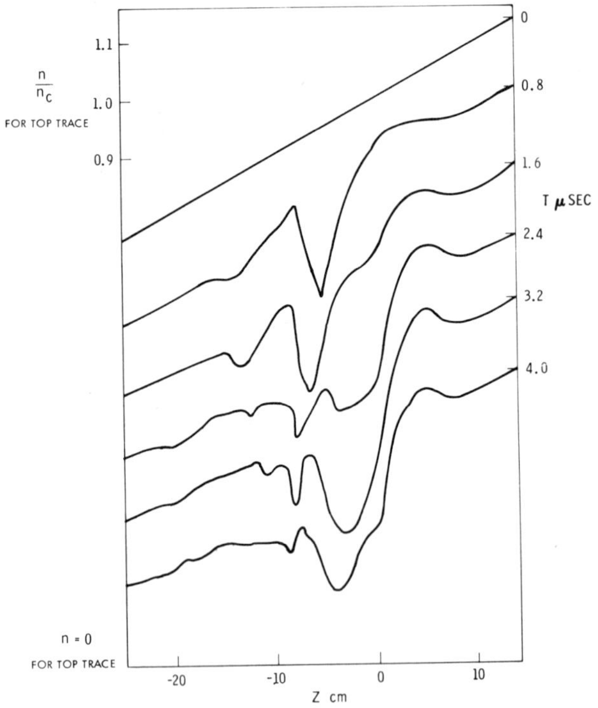

FIG. 8 - (b) When QUIPS is operated in a lower density regime corresponding to f_o = 1 GHz, the spatial profile at the end of increasing pulse durations (0-4 μs) shows the formation of multiple cavitons from the initial single caviton (P_o = 10 KW). [After W. DiVergelio and A.Y. Wong.]

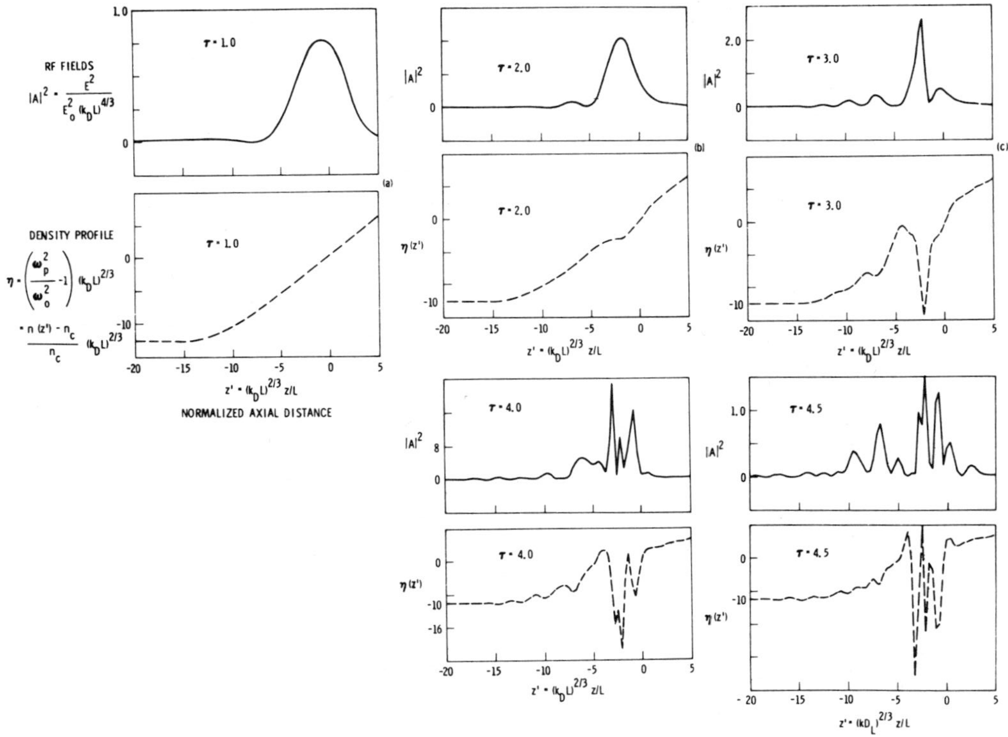

FIG. 8. - (c) Theoretical calculations of spatial profiles of both normalized rf fields A^2 and density profile $\eta(z')$ as function of normalized axial distance z' at successive times $\tau = \omega_o t/2k_D L)^{2/3}$ after rf excitation. Resonant location is at z'_c where $\eta(z'_c) = 0$. The evolution of the density profile is to be compared with the experimental date 8b. [After G. Morales and Y. C. Lee.]

Two significant results follow this observation. Particles can now interact with a series of cavitons in the vicinity of the resonant locations. Each passage through a caviton enhances the particle energy according to the maximum allowable electric field for each caviton. The cumulative gain in energy through a number (N) of sharply localized cavitons is roughly N times the energy gain through one caviton. Secondly, this series of cavitons is the origin of ion acoustic waves which convect energy out of the resonant location after the external rf field is terminated. We have verified that the caviton dynamics develops in the ion time scale by using two different plasmas, argon and xenon. The rise and decay times of cavitons are proportional to $(\text{ion mass})^{1/2}$.

For much longer rf pulses (approaching the steady state condition) more and more plasma is expelled away from the resonant region and a plateau is formed in front of the resonant location which now lies on a steep density gradient, Fig. (9). Probe measurements indicate a rf field profile which helps support the sharp density gradient. Using two different gases (Ar and Xe), it is observed that this profile develops in a time scale $\tau \propto M_i$.

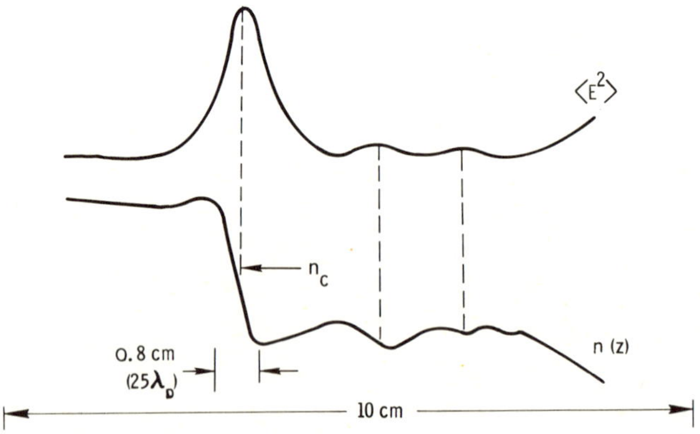

FIG. 9 - (a) Plateau formation in density profile (argon plasma) when a long pulse (f_o = 300 MHz, P_o = 10W, τ = 8 μsec) is applied. Rf fields are measured by a high impedance probe. [After H. Injeyan, P. Leung, and A. Y. Wong, UCLA DP device.]

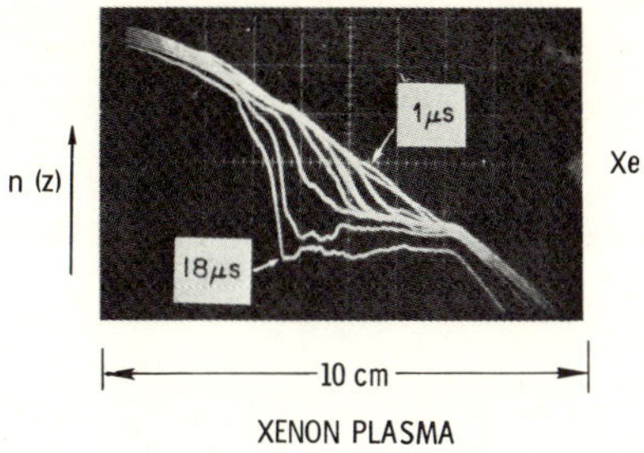

FIG. 9 - (b) Temporal development of the plateau formation in xenon plasmas. Density profiles are sampled after termination of applied pulses whose durations are 1, 2, 3, 4, 5, 10, 16, 18 μsec. Plateau formation takes a longer time in xenon than in argon by the mass ratio $\sqrt{M_{xenon}/M_{argon}}$. [After T. Yoshizumi, B. Quon, and A.Y. Wong.]

Trapping of RF Fields. A caviton is formed when a plasma is expelled from a region of intense rf fields. The rf fields are then surrounded by slightly higher densities on all sides. Such a situation favors the trapping of the rf fields. In order to verify this hypothesis, the rf pump is turned off abruptly [10] (0.3 μsec fall time) and the temporal decay of E inside the density cavity is studied by the electron beam technique. Because the smaller electron neutral collision rate at lower pressures favors this observation, the following parameters have been chosen: Ar pressure 5 x 10^{-5} torr, $\omega_o/2\pi$ = 100 MHz, $\nu_{en}/\omega_o \approx 10^{-3}$, and $L = n_o/(dn_o/dz)$ = 40cm. In the absence of the wave trapping, the rf field would convect out of the resonant region on the time scale $t = \int_{z_o}^{z_o-\Delta z} dz/v_g(z) = 200$ nsec, where $v_g(z) = [(3kT_e/m_e)(z_o-z)/L]^{1/2}$ is the local group velocity and Δz is the resonant width, while with trapping, the rf field decays in the time scale of the electron neutral collision (1.6μsec). A strong pump field (15 W) is pulsed with duration (1 μsec) just long enough to produce the enhanced field inside the density cavity. The subsequent slow decay of E_T (time constant $\tau = 1.5$ μsec) indicates collisional decay of the trapped plasma oscillations, Figs. 10(a) and 10(b). The density profile monitored immediately at the end of a short rf pulse shows a density barrier on both sides of the density cavity as expected from the expulsion of particles by the ponderomotive force.

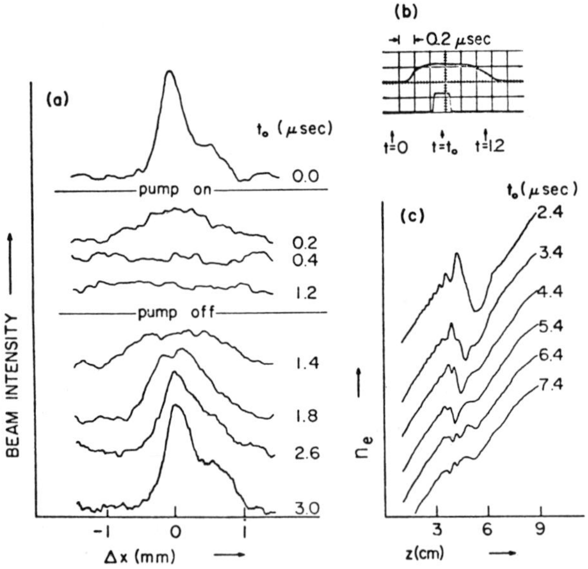

FIG. 10- (a) Temporal evolution of the electron beam deflection before, during, and after the rf excitation; the trapped rf field is confirmed by the persistent beam deflection after rf is turned off. (b) The time resolution is achieved by switching the 9 KeV electron beam on only during the desired time. Upper trace shows $\langle E_0^2 \rangle$ verses t. Lower trace shows the pulse during which the electron beam is switched on.
(c) Relaxation of the density cavity after the pump is turned off. [Kim et al., op. cit.]

After turning off the rf pulse, the density cavity propagates with an initial velocity approximately 4 times sound speed and slows down to the sound speed as the amplitude becomes smaller.

A combined numerical and analytical calculation of the temporal evolution of our experiments has been performed by Morales and Lee [14]. Assuming a slow and fast variation of the electric field $E = \tilde{E}(z,t)\, e^{-i\omega t+ikz}$ and using Eqs. (2.2) and (2.3), the following nonlinear Schrodinger equation for the variation of the amplitude of the electric field E was obtained

$$\frac{2i}{\omega_o}\frac{\partial\tilde{E}}{\partial t} + \frac{3}{k_D^2}\frac{\partial^2\tilde{E}}{\partial x^2} + \left[1 - \frac{\omega_p^2(|E|^2)}{\omega_o^2} + i\,\frac{\omega_p^2}{\omega_o^2}\frac{\nu}{\omega_o}\right]\tilde{E} = E_o \qquad (3.3)$$

where E_o is the externally applied electric field. Nonlinear perturbations δn on the zeroth-order density profile $n_o(z)$ due to ponderomotive forces $|E|^2$ were included in the ω_p^2/ω_o^2 term as follows using a static approximation

$$\omega_p^2 = \omega_{po}^2(z_o)\left[1 + \frac{z}{L} + \frac{\delta n}{n_o}\right]$$

where $\delta n/n_o = -|E|^2/[8\pi n_o(T_e+T_i)]$ and a linear spatial variation of the zeroth-order density has been assumed. Transit time damping by fast electrons passing through the localized region has not been included. Their findings in the weak, intermediate and strong field regimes correspond qualitatively to our observations. In the weak field regime, $(E^2/4\pi nT \ll 1)$, a propagating pattern of the electrostatic wave is observed with no significant perturbation of the density profile. In the intermediate to strong field regime, $(E^2/4\pi nT_e \approx 0.1)$, stationary density perturbations develop in the course of time. In the strong field limit, the spiky rf field breaks up into two isolated components at later times. The spike initially develops at the resonant location, then propagates down the density gradient carrying the density cavity with it.

In the strong field case, $(E^2/4\pi nT \approx 1)$, the time development of a caviton has been studied further with the ion dynamics described by the equation of motion, (3.1). In Fig. 8(c), a single caviton is shown to break up into two separate cavitons simultaneously with the development of individual rf peaks. The experimentally and theoretically obtained temporal development of cavitons as expressed in Fig. 8(b) and 8(c) is qualitatively similar. Unlike the static approximation mentioned above, the inclusion of ion dynamics reveals that density cavities are not necessarily aligned with envelopes of rf fields. The envelope of the linearly converted modes together with their associated density cavities along the density gradient are qualitatively similar to the experimental results in Fig. 5(b).

There is qualitative agreement between the experimental and theoretical results in the temporal developments of the rf fields and density profiles as shown in Fig. 7(a,b). The rf fields reach a peak in half an ion plasma period before any ion density perturbation can take place. The peak E^2 drops by a factor of 4 to the steady state level in both experiment and theory. The theory predicts a rise time of .05 μsec (or $\omega_{pi}t = 0.4$) for our experimental parameters, which cannot yet be verified by the experiment; both the rise time of the rf signal and the detector response limit the detected rise time of E^2 to 0.5 μsec. Even though the damping term in the theory is only phenomenological the theoretical and experimental values of the enhancement factor $\frac{E}{E_o}$ agree to within 30%: $\left(\frac{E}{E_o}\right)_{theory} = 70$, $\left(\frac{E}{E_o}\right)_{exp.} = 100$.

The rf field inside a caviton bears a resemblance to the wavefunction inside a quantum mechanical potential well, which corresponds to the density cavity in the present case. The splitting of the caviton is analogous to the transition from the ground state to the first excited state of a higher mode structure. The difference lies in the mutual interaction between the "wavefunction" and the potential well, which is a characteristic of the nonlinear Schrödinger equation.

The electron beam technique used in our experiment allows the "wavefunction" or the profile of the rf field to be measured without significant perturbations to the system.

4. Particle Acceleration. We now present experimental evidence which indicates that the creation of cavitons and the generation of fast electrons and ions are intimately related.

The large oscillating electric field inside the localized cavity accelerates plasma electrons in two ways. Those electrons which cross the localized layer of 20 λ_D thick along the density gradient, in less than half the period $T = 2\pi/\omega_p$ would gain an amount of energy $\Delta W \gtrsim (E^2/4\pi nT_e)^{1/2}\ T_e$; this requires fast electrons in the velocity range $v > 3v_{th}$. A second acceleration which affects the bulk electron distribution in the critical layer takes place because of the ponderomotive potential energy $\phi_p \gtrsim (\langle E^2\rangle/8\pi nT_e)\ T_e$. This acceleration can also be looked upon as the low frequency drift of electrons responding to a spatially inhomogeneous electric field.

The transit time damping is experimentally verified in our previously mentioned capacitor plate experiment [10] in which an injected fast electron beam traversed across the radial extent of the caviton region in less than 1/2 period in a direction perpendicular to $\underline{E}$, (Fig. 6). This beam, which acts like a group of fast test electrons, acquires a velocity component in the direction of the oscillating E field $v = \frac{eE\tau}{m}\hat{z}$ where τ is the transit time. This velocity component is observed by the deflection of the electron beam registered on the fluorescent screen in the direction of the oscillating field. Once the electrons are accelerated, they pull the ions with them by means of self-consistent fields.

Recently another verification of the transit time damping was put forth by Wong, Leung and Injeyan. By turning on a weak magnetic field (5G) perpendicular to the density gradient such that the electron cyclotron radius ρ_e is less than Δ, the width of the caviton, the transit time damping is suppressed; the electric field is enhanced by a factor of 3 corresponding to the reduced damping rate.

The process of ion acceleration is studied in an experiment [15] performed in QUIPS, when a 1 GHz 40 KW pulsed microwave source (Applied Microwaves) is used with a large horn antenna (14.4dB gain, linear polarization) at one end of the chamber. Short microwave pulses of typical duration (0.4 μsec) are radiated onto the plasma density profile which rises from the antenna end to a maximum density near the opposite end (3 M away). The argon plasma characteristics are $1.5\text{eV} < kT_e < 2\text{eV}$, $kT_i \simeq 0.2\text{eV}$, $5 \times 10^9 < n_e < 10^{11}$ cm^{-3}, and $100\ \text{cm} < [(1/n_o)(\partial n_o/\partial z)]^{-1} < 1000\text{cm}$. Using a boxcar integrator in connection with probe diagnostics, the temporal development of the spatial density profile, the electron and ion distributions, and the plasma potential can all be taken before and after the incident rf pulse. In other words, complete time-resolved information about the phase space is obtained experimentally.

The experimental arrangement and the main observations of the generation of fast ions can be best described by the simultaneous spatial and temporal descriptions of the process shown in Fig. (11). The short electromagnetic pulse (0.4 μsec) is converted to an electrostatic pulse of high field intensity near the resonant layer $\omega_o = \omega_p(z_o)$, as a result of the much smaller group velocity and the resonant enhancement at that layer. Observations with an ion energy analyzer at several microseconds after the EM pulse show the presence of a large ion current (twice the normal saturation value) just outside the resonant region. Using an energy discrimination grid, we found this enhanced current to consist of ions of much higher energy than the ambient, Fig. 12(a). Hotter ions are observed to arrive at the analyzer at earlier times as expected, Fig. 12(b). The time-of-flight measurements are consistent with the energy discrimination data. The ion current peak is followed by a density

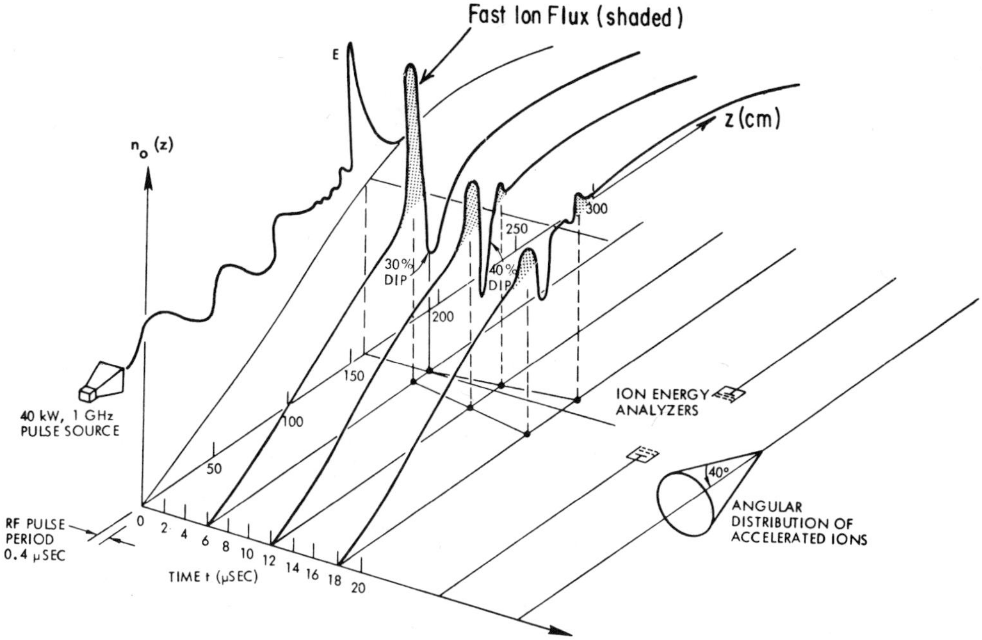

FIG. 11 - Space-time representation of fast ion flux (shaded) driven by the ponderomotive force superimposed on the background density profile. Density cavities are created as a result of ion expulsion. The polarization of ion trajectories is represented by the cone with a half-width of 40° with respect to the z-axis. The left hand side is a representative trace of the E field amplitude in EM wave and the linearly converted ES wave.

depression ($\Delta n/n = 30\%$), which is a result of the expulsion of ions from the resonant region. As this ion structure travels down the density profile, the peak consisting of a wide range of ion velocities quickly disperses, while the density "caviton" composed of background ions maintains its shape as a nonlinear ion perturbation for a much longer time.

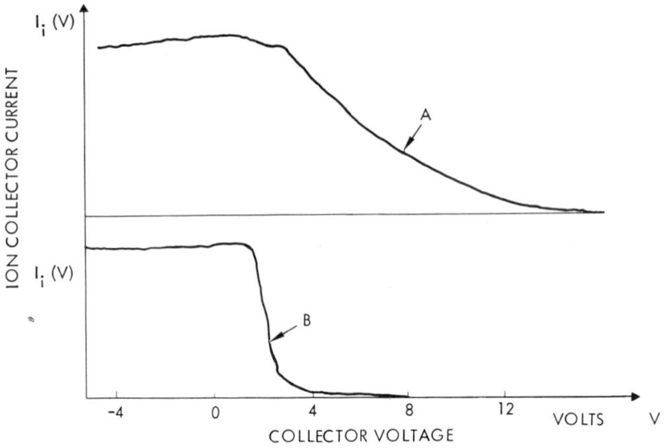

FIG. 12 - (a) Energetic ion flux generated by high power rf pulse. Ion flux (A) monitored on the collector of an ion energy analyzer sampled immediately after 0.4 μsec rf pulse, is plotted versus discriminator grid voltage. The curve (B) in the absence of rf is plotted as a reference. The ion distribution function is the derivative of this curve.

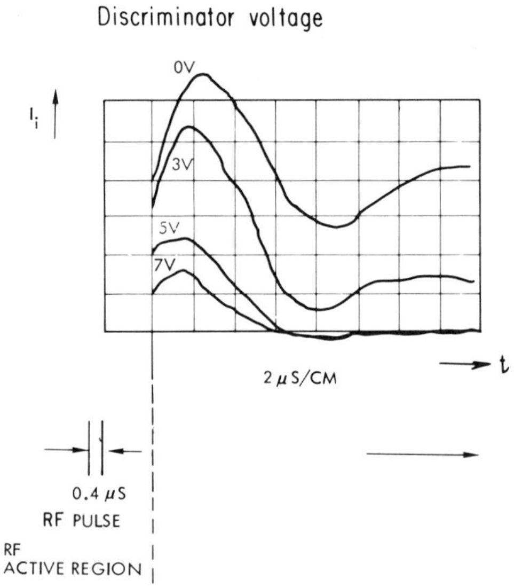

FIG. 12 - (b) Energy-discriminated ion fluxes arrive at the energy analyzer at various times dependent on their mean energy.

By rotating the energy analyzer at various angles with respect to the axis, we find that the ions move predominantly along the axial direction with a half-width angle of 40° as shown by the polar plot. The cross section over which ions are accelerated is as broad as the illuminated area of 7.5×10^3 cm^2. Using oppositely directed analyzers, we observed accelerated ion fluxes in both directions up and down the density profile with a much stronger ion flux down the density profile ($-\partial n_o/\partial z$ direction). This is most probably a result of the steeper profile of the intensity of the high frequency field (measured by the electron beam deflection techniques) in the same direction as the density gradient. Both the ponderomotive and plasma pressures drive the particles down the density gradient, whereas the ponderomotive force has to overcome the plasma pressure in order to drive ions up the density gradient.

The temporal behavior of the ion saturated current following the short 0.4 μsec rf pulse is displayed together with the ion temperature and plasma potential in Fig. 13. The diagnostics consist of radially and axially movable ion energy analyzers and Langmuir probes all located outside the resonant region, at 2cm down the density gradient from the resonant location. These locations not only avoid the interference of the probes with the development of the rf field and density cavities at the resonance, but also allow the ion signal to be well separated in time from the rf pulse. As shown in Fig. 13(b) and 13(c), the electron response first comes out of the resonant region followed by the ionic response as monitored by a Langmuir probe and an ion energy analyzer all mounted on the same movable shaft located 2cm away. The ions eventually catch up with the electrons and they travel together. The ion current is observed to increase to a peak value 1.7 times its normal saturated value in the first 8 μsec after the rf turn-on due to the presence of energetic ions. The energy spectrum of the ions was measured with an ion energy analyzer which discriminates against ions below a certain energy by a grid placed in front of the

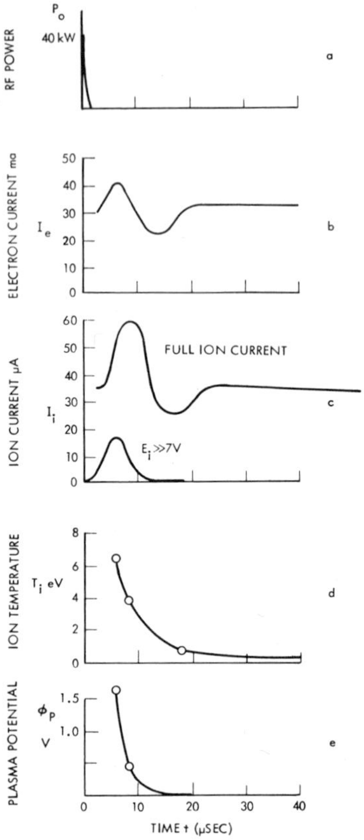

FIG. 13 - Temporal behavior of various characteristic parameters near the resonant layer.

(a) Rf excitation pulse.

(b) The electron saturation current collected by a Langmuir probe at the same location as the ion-energy analyzer. Note the earlier arrival time of the electron response with respect to the ion response in (c).

(c) Argon ions collected by energy analyzer at various times after the termination of the exciting rf pulse at 0.4 μsec. The analyzer is located at 2 cm below the resonant region, monitoring ions accelerated towards the lower-density side. Higher-energy ions observed with a discriminating grid arrive earlier.

(d) Ion temperatures and plasma potentials obtained from ion-energy analyzer. [After A.Y. Wong and R.L. Stenzel, Phys. Rev. Letters **34** (1975) 727].

collector (a third grid left at the floating potential is placed in front of the discriminator grid to reduce the disturbance of the analyzer on the ambient plasma). Two curves of ion collector currents are displayed in Fig. 12(a) to demonstrate the presence of energetic ion components when resonant acceleration takes place. This energy analysis corroborates well with the time-of-flight measurements as shown in Fig. 12(b). The faster ions with energy > 7 eV arrive earlier than the bulk of fast ions represented by the main peak. Ion fluctuations propagating at the ion acoustic velocity $C_s = (kT_e/M_i)^{1/2}$ ($\Delta n/n = 5\%$, $\omega \approx 0.1\ \omega_{pi}$) are also observed, [Fig. (14)], in the vicinity of the resonant layer where these energetic ions are generated. We believe that these fluctuations, which are observed after the energetic ions have passed through, are excited by interactions between the accelerated ions and the background.

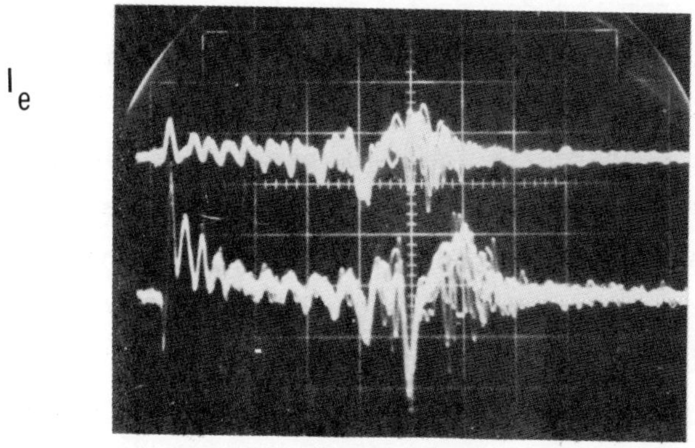

FIG. 14 - Two single exposures of ion wave instabilities excited by fast ions expelled from the density cavity, travelin through the background plasma. Observation probes biased at electron plasma potential are placed 2 cm downstream from the resonant location. The ordered oscillations starting from left are density perturbations created by ponderomotive forces of ES waves traveling toward the detection probe. The random oscillations in the middle are the ion fluctuations excited by fast ions ejected from the main density cavity.

The sampled results of the ion energy analyzer, Fig. 12(a), show a broader ion distribution or an apparent higher ion temperature, which is a consequence of the spatial distribution of ions on the ponderomotive potential curve. This higher temperature decays to the ambient value when the ion acceleration process is terminated. The curves in Fig. 13 also reveal the systematic increase in plasma potential at times when the ion temperature and density both increase. The increase in plasma pressure in the hot region just outside the density cavity causes an increase in the plasma potential which, together with the decay of the resonance field due to convective loss, ultimately limits the ion acceleration by the ponderomotive force.

The above process requires an rf pulse of sufficient duration ($\tau > \Delta z/v_e \approx 0.1$ sec) to set up the initial electron drift and displacement away from the resonant layer ($\Delta z \lesssim 2$ cm). Once this duration is exceeded, the net electron drift and the field necessary to accelerate the ions depend mainly on the peak rf power. As shown in Fig. (15), the energetic ion flux bears approximately the same linear relation to the peak power, even though the pulse duration changes by a factor of 5. The insensitivity to the total energy content of the applied pulse distinguishes this process from that of turbulent wave-particle interactions; the turbulent ion wave spectrum must depend on the energy input into the system and the pulse duration should be many times the ion plasma wave period for a turbulent wave spectrum to develop.

Based on our model, the magnitude of the transient electric field intensity or the ponderomotive potential $\langle E^2\rangle/8\pi$ can be directly estimated from the maximum energy gained by the ions since the change in ion energy $\Delta \frac{n}{2} mv_i^2 \approx E^2/8\pi$. The experimental observation of ion energy up to 7 kT_e indicates the presence of strong transient fields $E^2/4\pi nkT_e \approx 7$ which is 700 times larger than the incident field $E_o^2/4\pi nkT_e \approx 10^{-2}$. This enhancement factor is confirmed by our

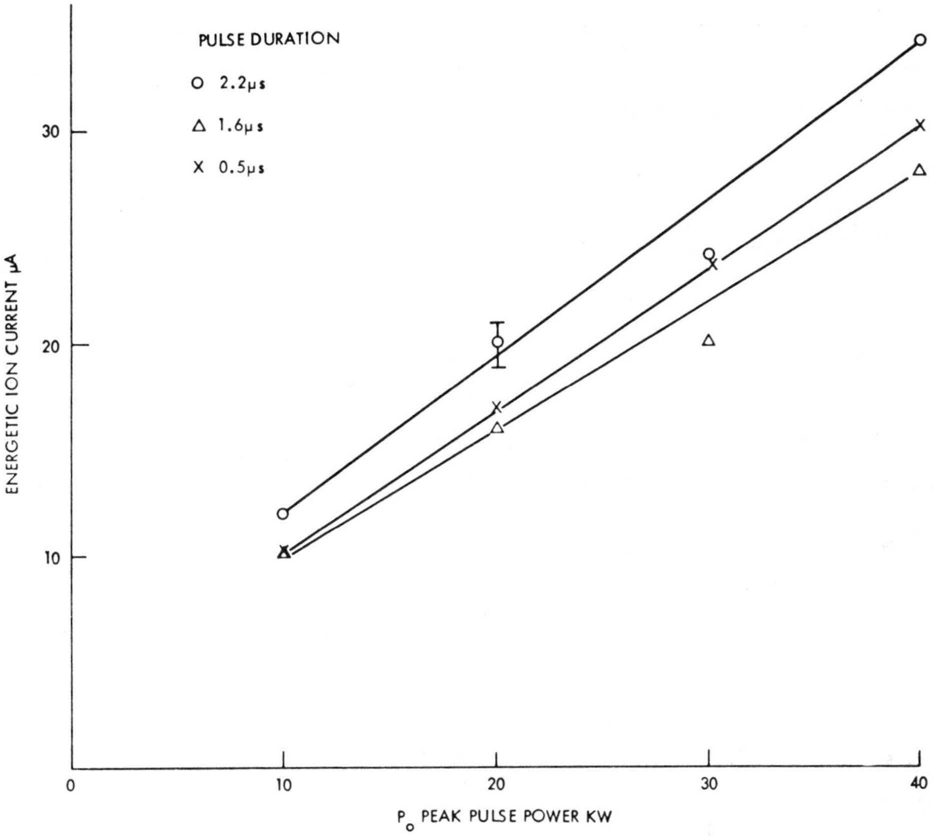

FIG. 15 - Energetic ion flux versus incident power P_o for three different pulse durations τ showing its dependence on P_o but not on τ.

separate electron beam measurements under similar situations. This strong field condition could only be maintained before the plasma ions have time to move and before the change in plasma density shifts the resonant location, since equilibrium conditions require $E^2/4\pi nkT_e \approx 1$. Even though the incident field is weak, resonant enhancement easily brings our experimental parameters into the strong field regime.

The energy gained by these fast ions is of the same order of magnitude as the energy carried by the electron plasma waves out from the resonant layer:

$$\frac{\text{Energy carried out by fast ions}}{\text{Energy convected out by electron waves}} = \frac{n\left(\frac{1}{2}\,mv_i^{\,2}\right)\,v_i A\tau}{(nkT_e)\;v_g\;A\tau} \cong \frac{v_i}{v_g} \cong 1$$

where $E^2/8\pi nkT_e \stackrel{\sim}{\sim} 1$ is used.

The measured group velocity v_g of electron plasma waves near the resonant layer $v_g \lesssim 10^6$ cm/sec is in the same range as the speed of the energetic ions. The resonant enhancement of external fields is therefore limited by the convective losses of both waves and particles.

This process of acceleration by localized high frequency fields can occur over a much shorter time scale and impart more energy to the ions than the linear process of acceleration by ion acoustic waves excited by the parametric process. This latter process could be eliminated by applying the pump field over a duration shorter than the growth rate of the ion acoustic wave. The plasma inhomogeneity which raises the thresholds in parametric processes and shortens the wave-particle interaction region actually favors this localized field, and its intractions with particles.

The ion acceleration by steep gradients of EM fields is to be distinguished from the ambipolar acceleration due to a steep density gradient. The latter case can only cause ion acceleration in the direction down the density gradient and not up the density gradient, as is permissible in the acceleration by the ponderomotive force.

It is interesting to point out that a continued acceleration of ions down the density gradient can be sustained by sweeping the excitation frequency towards a lower range. Ions accelerated by the first caviton continue to feel accelerated downward by the new caviton generated just below, resulting in a cumulative gain in the final ion energy. A rough estimate gives a final ion energy $W = N\phi_o$ where ϕ_o is the ponderomotive potential of a single caviton; of course the rate of sweep must be adjusted to maintain a proper phase between the accelerated ions and the newly created caviton.

Generation of Magnetic Fields in the Resonant Region. A quasistatic magnetic field as observed by DiVergilio, Wong, Kim and Lee [16] was generated in the resonance region by the incident electromagnetic wave. The magnetic field, B, lies in the plane perpendicular to the density gradient and to the incident wave polarization. Fig. (16) illustrates the spatial dependence of B_y in the x-z plane (y = 0), where the density gradient is along z and the incident wave is polarized in the x direction. The traces in Fig. 16(a) were obtained by sampling the integrated output of a magnetic probe near the end of a 2.5 μsec microwave burst of peak power 7kW at 19Hz, corresponding to $\frac{E_o^2}{4\pi nkT_e} = 1.7 \times 10^{-3}$.

As a function of z, for a given radial position, x, B_y is seen to peak to a value of .03 gauss near the critical density and reverse sign 5-10 cm below critical, indicating the presence of a current layer in the resonant region flowing in the x direction. B_y also changes sign in the radial direction, with $B_y \simeq 0$ at x = 0. This geometry is explained by the following physical picture. The incident electromagnetic wave undergoes strong absorption in the resonant region. The x component of the absorbed wave momentum drives an electron current along the critical layer, while space charge forces along the density gradient transfer the z component of momentum to the ions. Referring to Fig. 16(b), we note that the divergent incident wave drives a current in opposite directions on either side of x = 0, resulting in the observed radial sign change in B_y. The field geometry is essentially identical to that predicted by Thomson, Max and Estabrook [17], when their results are extended to a divergent source. The orthogonal relationship between the quasistatic magnetic field and the incident wave polarization was found to be preserved when the polarization was changed by rotating the microwave horn.

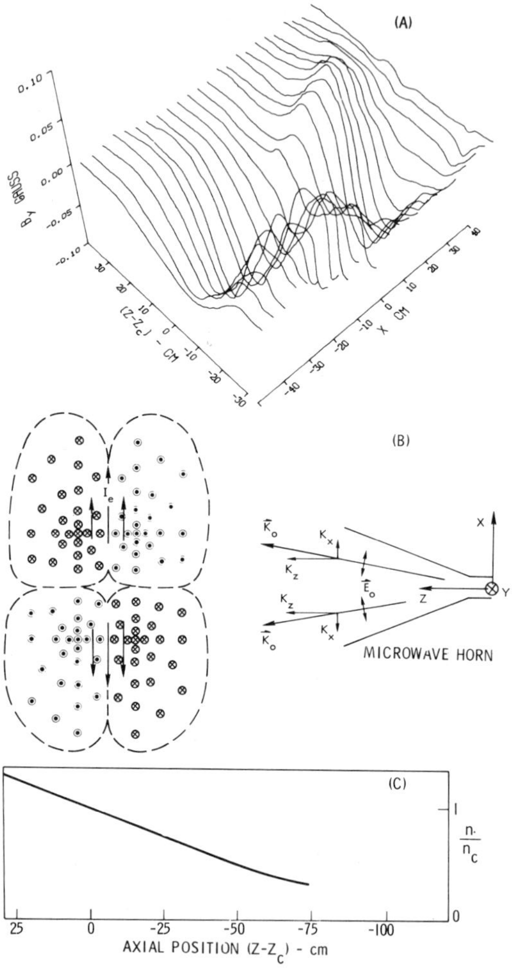

Fig. 16 - (a) Steady state B_y as a function of axial position, z, and radial position, x. Positive B_y corresponds to field in positive y direction, which is into paper in (b).

(b) Diagram of results shown in (a), indicating geometry of field with respect to microwave source (1 GHZ, 7KW, $E_o^2/4\pi nT_e \approx 1.7\times10^{-3}$). Inferred electron current paths are shown by solid lines and possible return paths by dotted lines.

(c) Axial density gradient before rf turn on. Position scale applies to (b) and (c). [After W. DiVergilio, A.Y. Wong, H.C. Kim and Y.C. Lee, Ref. 16].

Generation of Second Harmonics at the Critical Layer. The nonlinear term $v \cdot \nabla v$ should also produce a driving term at the second harmonic frequency by virtue of the anharmonic electron orbit which is the result of an inhomogeneous driving field. We now include the nonlinear term $v \cdot \nabla v$ in the equation of motion for the electron.

$$\frac{\partial}{\partial t} v = \frac{-eE}{m} - \frac{1}{n_o m} \frac{\partial}{\partial x} nT - \nu_e v - v \frac{\partial}{\partial x} v \quad . \tag{3.4}$$

Substitution into Eq. (2.5) yeilds the plasma response to the second harmonic driving term as:

$$E(2\omega_o) \cong \frac{\frac{e}{m} \frac{E(\omega_o)^2}{2\Delta}}{(2\omega_o)^2 - \omega_p^2 - 3k^2 v_{th} + i2\nu\omega_o} \tag{3.5}$$

where Δ is the steepest gradient length of the resonant layer.

$$\frac{E(2\omega_o)}{E(\omega_o)} \approx \frac{eE(\omega_o)/2m\Delta}{3\omega_o^2} = \frac{1}{6} \frac{eE(\omega_o)}{m\omega_o^2} \frac{1}{\Delta} = \frac{1}{6} \frac{d}{\Delta} \tag{3.6}$$

where d = distance of electron excursion under the resonant driving field = $eE(\omega_o)/m\omega_o^2$.
The strength of the second harmonic normalized with respect to the total field at the fundamental is proportional to the ratio of the electron excursion distance and the gradient length of the resonant layer.

This second harmonic field has been observed with the use of the electron beam diagnostics. Referring to Fig. 6, the image of the electron beam on the fluorescent screen describes the response of a test electron to the horizontal resonant field and the vertical reference field at the applied frequency ω_o. When a reference field is imposed on the vertical deflection plates at the electron beam source, a vertical figure "8" Lissajous figure, indicative of a horizontal second harmonic field, was observed at either side of the resonant location where the gradients of rf fields are steepest.

Parametric Decay Instability in an Inhomogeneous Medium. [18] The parametric process can be defined as the regenerative coupling between resonant modes. As a result of the regenerative nature of this process, it can be observed even for rather small amplitudes of driving fields. An example is given in Fig. 17(a), in which ω_o represents the pump frequency and ω_1, ω_2 are the initially stable fluctuations at two resonant frequencies of the plasma. If the frequencies and wave-numbers are such that relationships $\omega_o = \omega_1 + \omega_2$ and $\underline{k}_o = \underline{k}_1 + \underline{k}_2$ hold, then the coupled mode $\omega_o - \omega_2$ regeneratively enhances ω_1 and similarly for ω_2. For example, ω_o could be an electromagnetic pump, and ω_2 and ω_1 are the respective electron and ion resonant modes of the system; or the three modes could be electron modes or ion modes.

In the earlier days of theoretical treatments on the interaction of EM waves with plasmas, the pump field important for the parametric instability was thought to be the oscillating EM field which lies predominantly along the radial direction in the case of axial propagation. However, experiments have revealed that the electrostatic field of the linearly converted mode, enhanced by the plasma, plays a far more significant role in generating parametric instabilities in the axial direction. These two processes are distinguished experimentally by examining the directions of propagation of the decay products (ω_1, $\underline{k}_1$) and (ω_2, $\underline{k}_2$), with respect to the direction of the pump field. Earlier theoretical estimate implied that the maximum amplitude of this instability should occur at the location of maximum pump field, namely at the resonant location. However, as shown by experimental data below, nonlinear effects such as density perturbations limit the optimum pump field amplitude to a rather small value $E^2/4\pi nT \lesssim 10^{-2}$.

Experimentally the parametric decay instability was observed in the large QUIPS chamber to grow temporally within a limited region in the vicinity of the EM cutoff, which is several cm below the resonant location, Fig. 17(b). This region is bounded by an evanescence of the EM field on the one side and the rise of the electrostatic field on the other, as shown in Fig. 17(d). It is located near the minimum in the profile of the pump field $|E_{oz}|$. The non-uniform pump field $|E_{oz}|$ gives rise to an absolute decay instability situated near its minimum field strength which is above the parametric decay threshold. The ionic sideband ω_i monitored by a Langmuir probe biased above the plasma potential and the electronic sideband $E(\omega_o-\omega_i)$ monitored by a high impedance rf probe are observed to grow within a limited region without significant convection out of that region, Fig. 17(c). This threshold has been experimentally determined in terms of the externally applied field E_o as $E_o^2/4\pi nT_e \approx 10^{-6}$, Fig. 18. The spectra of the high frequency sidebands $(\omega_o-\omega_i)$, where ω_i is the low frequency ionic sideband, at various steady pump intensities are observed to evolve from discrete lines to a broad spectrum. The integrated power density $\int |E^2(\omega)|d\omega$ of the high frequency sideband is observed to saturate when the pump slightly exceeds the threshold.

In order to understand the development of the turbulent spectrum at high pump intensities, the external pump was pulsed on and off. The instability monitored by the amplitude of either of the decay waves grows in the ion time scale (1 - 10 μsec), as shown in Fig. 19. The instability starts with a coherent spectrum until saturation when the waves become much less coherent with a steady-state bandwidth of $\Delta\omega/\omega_o \simeq \frac{1}{3}\,\omega_{pi}/\omega_{pe} = \frac{1}{3}\,(m_e/m_i)^{1/2}$. Both the ion density oscillation $n_i(\omega_i)$ and the high frequency sideband $E(\omega_o-\omega_i)$ show a stationary phase region inside the minimum in the $|E_{oz}|$ profile, Fig. 17. The leakage of waves out from this region of absolute growth is slow, a necessary condition of the absolute instability. The group velocity of the electron sideband was measured by gating the incident EM field and monitoring the propagation of the wave packet out of the region of maximum temporal growth. The measured group velocity is three orders of magnitude smaller than the electron

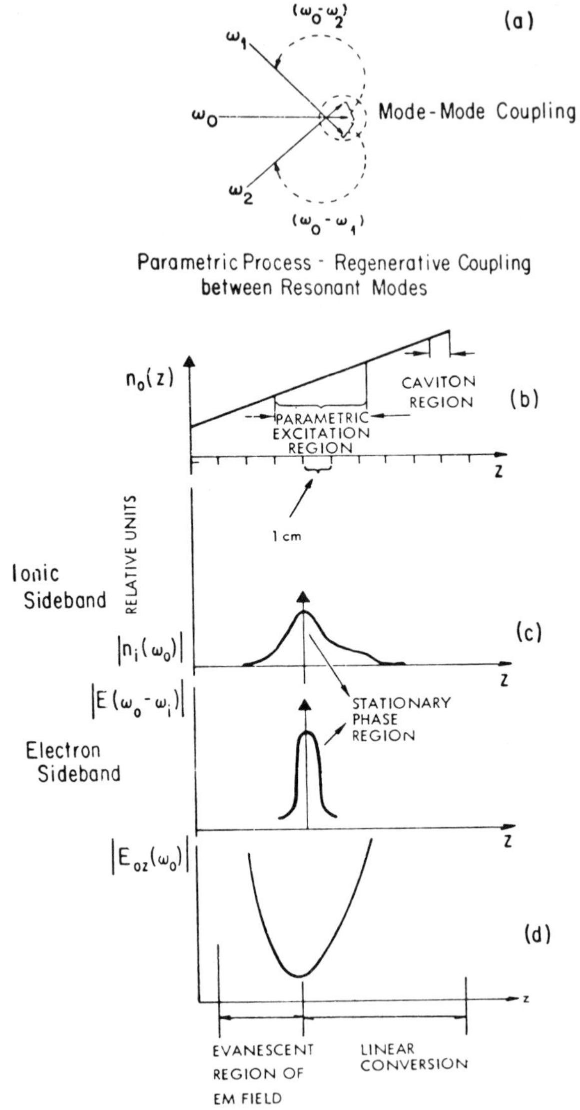

FIG. 17- (a) Schematic representation of parametric process, the modes couple to regeneratively enhance one another. (b) Location of parametric instabilities on the density profile with respect to the caviton region. (c) Amplitudes of ion and electron sidebands $n_i(\omega_i)$ and $E(\omega_0-\omega_i)$ are observed to grow in the parametric region with stationary phase. (d) The lower curve depicts the absolute amplitude of the axial pump field $|E_z|$ and the existence of a "well" in the pump profile.

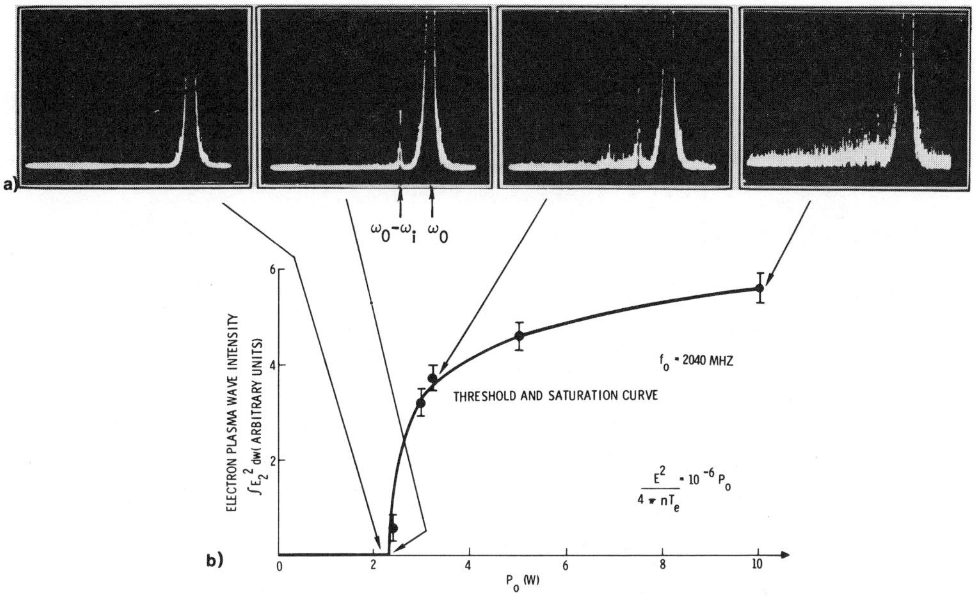

FIG. 18- (a) Spectra of high frequency pump and sidebands corresponding to the various pump power P_o (see arrows in b). (b) Intensity of the electron sideband integrated over frequencies versus pump intensity, showing the sharp threshold and the saturation effects. The normalized field intensity can be obtained from the incident power P_o as $E_o^2/4\pi n_o T_e = 10^{-6} P_o$.

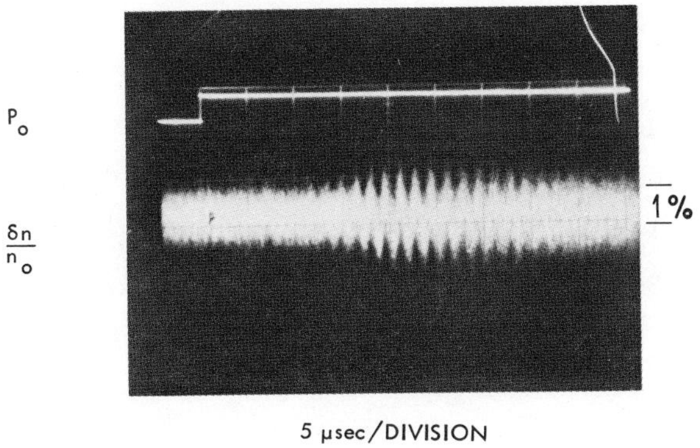

FIG. 19- Temporal growth of ion sideband $n_i(\omega_i)/n_o$ (lower trace) after the pump P_o (upper trace) is abruptly turned on. Note the transition from a well-defined ionic frequency into a turbulent saturation state at a small level of 1-2%.

thermal velocity. The amplitude of the electron sideband is at least an order of magnitude below the value of electric fields at the resonant layer. This is deduced from the fact that no observable density perturbations (< 5%) have been linked to the ponderomotive force of the high frequency sideband. The amplitude of the ion sideband is equally small $\delta n/n \lesssim 2\%$.

The temporal growth of waves within a particular location is qualitatively similar to a theoretical concept advanced by Arnush and Kennel [19]; they showed that an inhomogeneous pump profile can restrict spatially growing waves to a limited region. The magnitude of the pump field enters into the dispersion relation of the parametrically decaying waves such that at certain amplitudes of the pump, the waves are reflected at a certain location where the pump field reaches a certain amplitude. Standing waves are set up, giving rise to temporally growing waves in a region between reflection points. No detailed theory has been worked out for the case of an inhomogeneous pump profile composed of contributions from both the axial component of the incident EM waves and the linear converted waves.

Judging from measured wave fields and density perturbations, this decay instability is a much weaker nonlinear phenomenon than the one in the resonant region. There are two reasons why this decay instability is not observed at the resonant location where the field is largest. First, the perturbations of the rf field on the zeroth-order density and, therefore, on the dispersion relations are so large that the linear parametric analysis no longer applies. Secondly, the nonuniform density profile makes it difficult for the parametric waves to propagate with proper frequency and wave number matching over any distance more than one wavelength long.

The minimum incident EM field above which a parametric process takes place is measured to be $E_o^2/4\pi n T_e \approx 10^{-6}$. The theoretical threshold for the instability in a homogeneous plasma is

$$\left(\frac{E^2}{4\pi n T_e}\right)_{\text{Theory}} = 16 \frac{\Gamma_e}{\omega_e} \frac{\Gamma_i}{\omega_i}$$

where Γ_e, Γ_i are the damping rates of the electron and ion sidebands ω_e, ω_i respectively. For typical experimental parameters, the theoretical threshold $E^2/4\pi n T_e \sim 10^{-5}$ is larger than the experimentally observed threshold by an order of magnitude, a rather surprising result in view of convective losses in an inhomogeneous plasma. Our explanation is that the incident field E_o is enhanced near the critical region, resulting in a larger pump field. In making comparisons with theory, the enhanced self-consistent field intensity E^2 should be used, which, according to our estimate, is at least an order of magnitude above the external pump field $E_o{}^2$. The end result is that the experimental threshold is equal to or above the theoretical threshold obtained for homogeneous plasmas. Similar lower experimental thresholds have also been observed in parametric experiments in the ionosphere [2] and in other laboratory experiments. The same explanation using self-consistent fields should apply.

5. Double Resonance Excitation

The linear conversion of EM waves into ES waves of high field strengths near the critical layer produces many interesting mode coupling processes. Consider the plasma being radiated by two incident EM frequencies, ω_1, ω_2 separated by an ion acoustic wave frequency. The beating of their linearly converted ES waves produces $\omega_i = \omega_2 - \omega_1$ and $k_i = k_2 - k_1$, which could be either a forced ion oscillation or one that fits in with the ion wave dispersion relation. The latter process is called a double resonance excitation [20] since both the pump wave and the resulting coupled waves are resonant modes of the plasma. Other double resonance schemes using a combination of ES and EM pumps are of course possible.

The coupling process could be represented by the momentum

equation for both ions and electrons

$$\rho_o \frac{\partial v_i}{\partial t} = - \frac{\partial}{\partial z} (P_e + P_i) - \frac{\partial}{\partial z} \frac{<E^2>}{8\pi} \tag{4.1}$$

and the ion continuity equation

$$\frac{\partial \rho}{\partial t} + \rho_o \frac{\partial v_i}{\partial z} = 0 \quad . \tag{4.2}$$

The ponderomotive force term provides the coupling between the two pump waves. Combining (4.1) and (4.2) yields

$$\frac{\partial^2 n}{\partial t^2} - C_s^2 \frac{\partial^2 n}{\partial z^2} = \frac{1}{m_i} \frac{\partial^2}{\partial z^2} \frac{<E^2>}{8\pi} \tag{4.3}$$

where

$$C_s^2 = \frac{kT_e + kT_i}{m_i} \approx \frac{kT_e}{m_i} , \quad T_e >> T_i \quad .$$

We can invoke the quasistatic approximation to replace the source term $<E^2>/8\pi$ by $n_s kT_e$ in the right hand side and arrive at an equation for the ion density fluctuation:

$$\frac{\partial^2 n}{\partial z^2} - \frac{1}{C_s^2} \frac{\partial^2 n}{\partial t^2} = - \left(\frac{\partial^2}{\partial z^2} n_s\right)_{\text{source}} \quad . \tag{4.4}$$

The left hand side is the ion wave equation while the right hand side is the driving term, as a result of the coupling between the two linearly converted waves:

$$E = E_1(z) \cos \omega_1 t + E_2(z) \cos \omega_2 t$$

where $E_1(z)$ are given by the Airy function in the solution of Eq. (2.9). The low frequency term $<E^2>$ is obtained after averaging over the fast variation and retaining only $\omega_2 - \omega_1 = \Delta\omega$ term:

$$<E^2> = 2E_1(z) E_2(z) \cos (\omega_2 - \omega_1)t \quad .$$

If the source as represented by the right-hand side extends only over a finite region, the ion response in Eq. (4.3) could be solved by using the Green's function:

$$n(z,t) = \int_{\substack{\text{source}\\ \text{region}}} \frac{-\frac{\partial^2}{\partial z'^2} n_s(z') \exp\left[-i\Delta\omega\left(t - \frac{|z-z'|}{C_s}\right)\right]}{4\pi|z-z'|} d^3z' \quad . \tag{4.5}$$

For the steep density gradient case where the source is well localized, we have the approximate solution:

$$n(z,t) \cong n(0) \exp\left[-i\Delta\omega\left(t - \frac{|z|}{C_s}\right)\right] \quad .$$

Ion waves excited at the source region with amplitude n(0) propagate up and down the density gradient at the ion acoustic phase velocity. This is similar to placing a grid at x = 0 to excite ion acoustic waves by causing density perturbations. However, the present scheme has the advantage that no physical obstacles needs to be placed at the desired location, and is a very useful scheme for exciting ion acoustic waves or turbulence in the central region of a plasma.

In a gentle density gradient, the high frequency ES waves extend over a large region ($\approx$ 2-3 λ) and within this source region a simple beating between these two Airy patterns of slightly different frequencies results. Outside the source region the normal ion wave propagation pattern prevails. This is confirmed experimentally as shown in Fig. 20. The interference pattern in the density fluctuation is obtained by using the difference frequency $\Delta\omega = \omega_2 - \omega_1$ as a reference. The changing wavelength in the propagation pattern is a result of the beating of the ES waves and is in general a non-resonant ion mode. Outside the source region, a wavelength-frequency relation similar to the ion acoustic dispersion is observed. At lower frequencies ($\Delta\omega/\omega_{pi} \lessapprox 0.1$) the phase velocity is approximately C_s, the ion acoustic phase velocity. At higher frequencies ($\Delta\omega/\omega_{pi} > 0.1$), the axial phase velocity is higher than the ion acoustic wave velocity, $(v_{ph})_z = \omega/k_z \approx 2 - 3\ C_s$, because the excited ion waves also tend to propagate in an off-axis direction. The axial wave-number k_z becomes smaller, $k_z = k \cos\theta$, where θ is the propagation angle resulting in an apparently higher wave velocity.

The above measurements in the limit of small frequency separation yield directly the group velocity of the high frequency wave propagating in an inhomogeneous plasma. A knowledge of the group velocity is useful in the calculation of wave convective loss from the resonant location.

6. Correlations with Laser-Plasma Interactions and Ionospheric Plasmas

The microwave simulation experiment lies in a parameter range between laser-plasma interactions at the one end and radio wave interactions with the ionosphere at the other (Fig. 21). We shall attempt in this last section to make connections with these two kinds of experiments.

We shall first devote our attention to the laser experiments. Some of the significant observations in experiments involving laser irradition of solid targets are the following:

1. Fast ions stream out along the density gradient with an angular spread of 40°.

2. Up to 10^7 neutrons are observed per shot.

3. Radiation frequencies at 2 ω_o, 3/2 ω_o, 5/2 ω_o are observed although $\omega_o/2$ is not usually observed.

4. Hot spots are observed at the surface from which high energy electrons come out. These hot spots are indicative of the presence of sharp density gradients.

5. Even a small pre-pulse (orders of magnitude down from the mail pulse in power) could affect the subsequent behavior of the plasma under the irradiation by the main pulse; for example, the

characteristics mentioned under items 1, 2, 3, and 4.

Computer simulation experiments revealed further information on the structure near the critical surface:

6. A density plateau is observed to form just in front of the critical layer after the plasma is blown off.

7. The transverse dimension of the critical surface is unstable and is subject to the formation of smaller density cavities.

Our microwave modeling experiments have demonstrated that the significant enhancement near the critical layer can bring even a small incident field ($E^2/4\pi nT_e \gtrsim 10^{-3}$) to a sufficient large field, $E^2/4\pi nT_e \approx 10$. Strong nonlinear effects such as zeroth-order density perturbations and accelerations of particles occur over a wide range of incident powers and explain why a pre-pulse could have a significant effect on the subsequent laser-plasma interactions. The strong field at the critical layer and its gradient aligned along the density gradient are responsible for the strong directivity of particle acceleration along the density gradient. Particles gain energy up to 10 kT_e either by transit time damping through the critical region or acceleration down the pondermotive potential ($E^2/4\pi nq$).

We have identified this region of strong field and its associated density depression as a nonlinear state called "caviton." In a plasma with nearly uniform radial density profile, the caviton has a large radial extent and is subject to breaking up into smaller cavitons. These smaller cavitons are locations of enhanced fields and could be identified as the hot spots from which energetic particles are emitted. Once electrons are accelerated, ions follow through ambipolar fields. The maximum energy gained by the ions is of the order of ponderomotive potential or $\Delta W \approx (E^2/4\pi nkT_e)kT_e$. Taking $E^2/4\pi nkT_e \approx 10$ and $kT_e \approx 10$ KeV, ions could be accelerated up to 100 KeV. DT or DD reactions could take place between accelerated ions and background and might explain the production of neutrons under conditions in which no significant compression of the target takes place.

The unusual combination of observed radiation frequencies could also be explained by the presence of the caviton. Consider a caviton which is deep enough so that its lowest density is 1/4 the critical density. Oscillations at the half frequencies $\omega_o/2$ could be created and trapped inside the caviton, driven by the parametric decay of the enhanced field at the pump frequency. The sharp field gradients inside the caviton then provide the mixing of $\omega_o/2$ and ω_o frequencies to form higher harmonics 3 $\omega_o/2$, 5 $\omega_o/2$, etc. These frequencies, which are higher than the ambient plasma frequency, can escape and be detected, whereas the half frequency ω_o remains trapped.

The microwave experiments have confirmed the prediction by computer simulation of the development of a density plateau in front of the critical location. The evolution of this plateau was observed to be a result of the expulsion of plasma from the critical region by the pondermotive force.

Similarly in the ionosphere for rather weak incident power ($P_o \gtrsim 3\ \mu\omega/m^2$, $E_o^2/4\pi nT \approx 10^{-5}$) strong electrostatic waves in the direction of the density gradient were observed [2] with the Thomson radar (400 MHz). From the frequency spectra of the backscattered Thomson signal, ES waves at the incident pump frequency ω_o and the sideband frequencies $\omega_o - \omega_i$ are present. According to our laboratory experiments, this implies that both the linearly converted ES modes and ion oscillations which result from parametric instabilities are present. Fast electrons have also been observed by airglow measurements and could be the result of collisionless interactions between particles, ES waves, and cavitons. Our laboratory experiments suggest that the presence of fast ions should now be searched for.

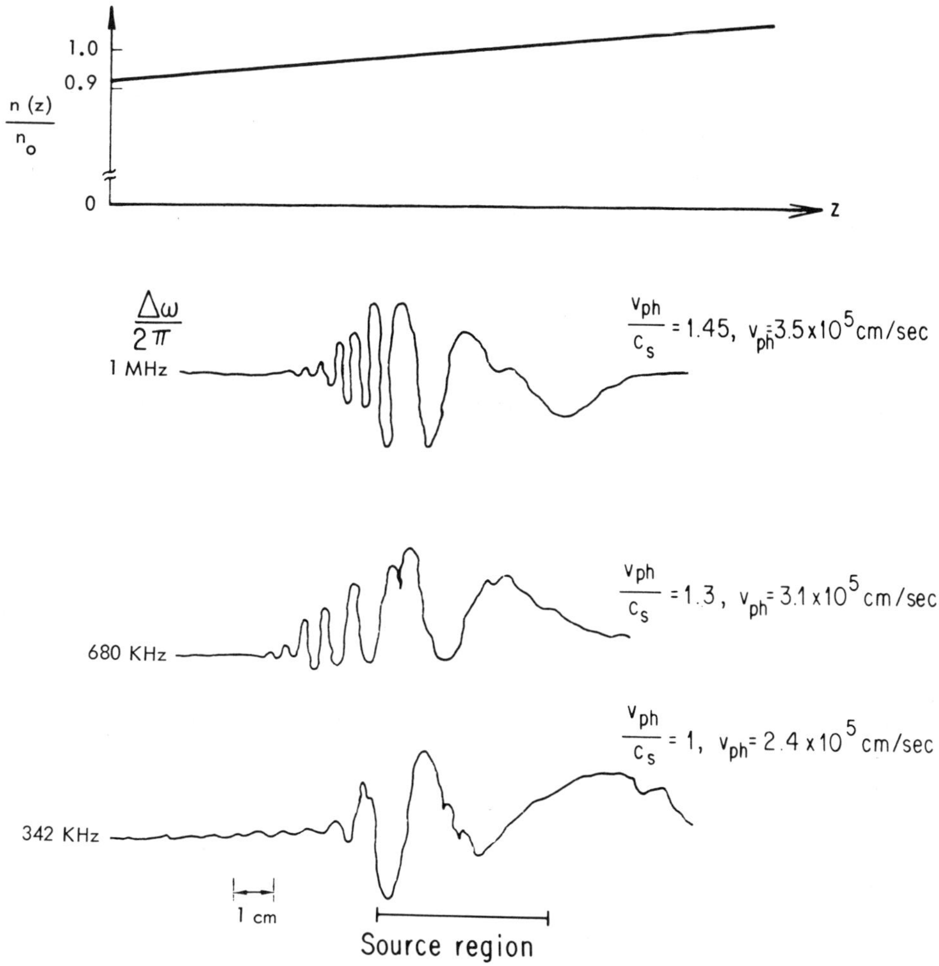

FIG. 20- Ion waves at ω_i excited by two incident EM waves ω_1. ω_2 where $\omega_i = \omega_2 - \omega_1$, in the vicinity of their resonant locations. Interferometric traces show a changing wavelength pattern inside the source region (where the two linearly converted ES waves ω_1, ω_2 interact) settles into a stable ion wavelength outside.

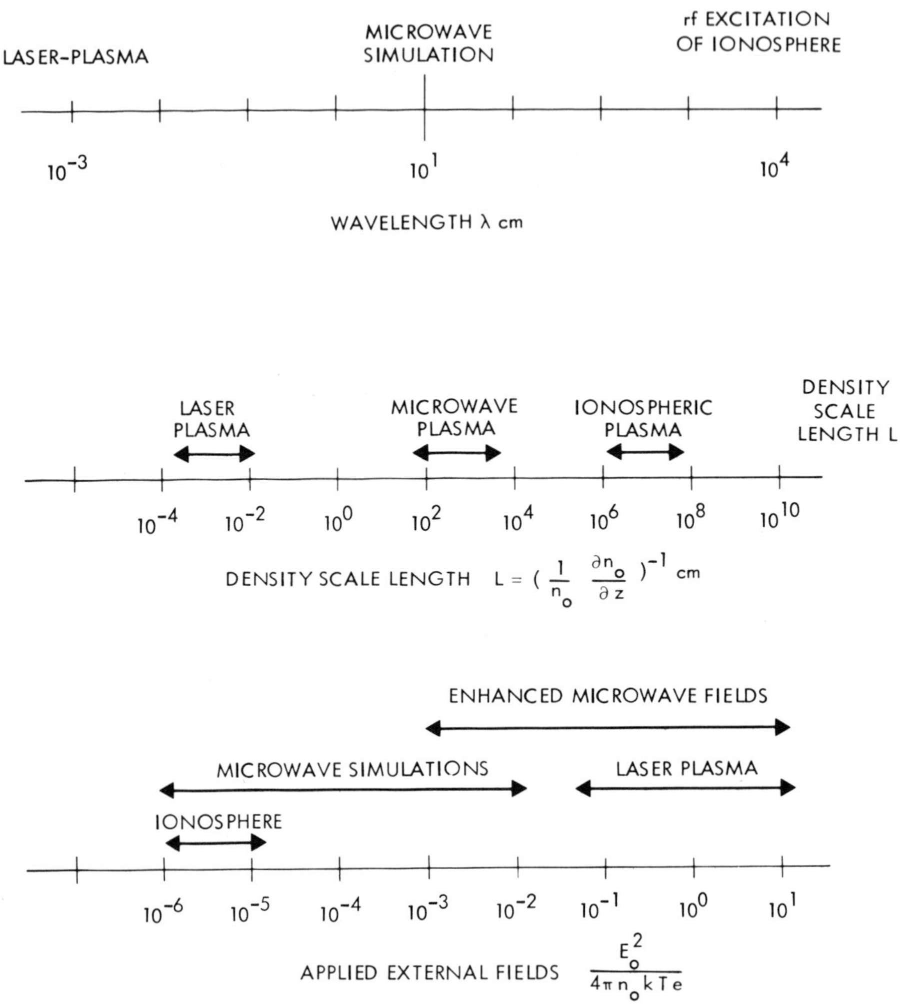

FIG. 21- A comparison of the experimental parameters, wavelength λ, density scale length L, and applied external fields E_o^2 for the three cases of interest--laser-plasma, microwave simulation, and rf excitation of ionosphere. The microwave parameter regime overlaps the laser and ionospheric regime in L/λ (number of free space wavelength in a density gradient length) and the normalized field strengths $E^2/4\pi n_o T$, where E is the self-consistent field in the plasma.

It is quite likely that the double resonance technique employing two pump frequencies could be used to excite ion waves or even couple to the ion cyclotron motion at a pre-determined height. Indeed, some preliminary experiments [21], performed at Boulder, Colorado, using a high frequency pump of 9.2 MHz whose modulation frequency can be swept continuously, indicated an enhancement of ionospheric fluctuations at the electrostatic ion cyclotron wave frequencies of O^+ and NO^+. If future ionsopheric experiments using Thomson radar continue to confirm such observations, an active control of the ionospheric dynamics is a distinct possibility using rather modest rf powers.

Acknowledgments

The author wishes to acknowledge his long association with Drs. R. L. Stenzel, H. C. Kim, G. Morales, Y. C. Lee, and D. Arnush, whose works form the basic references here. He is grateful to W. DiVergilio and D. Eggleston for assistance and critical reading of the manuscript. The experimental work at TRW Systems was supported by Lawrence Livermore Laboratory Contract #7950105 and at UCLA by AFSOR under grant #72-2332.

References

1. J. F. Kephart, R. P. Godwin, and G. H. McCall, Appl. Phys. Letters 25 (1974) 108; A. W. Ehler, J. Appl. Phys. 46 (1975); Ripin, et al., Phys. Rev. Letters 34 (1975) 1313.

2. W. F. Utlaut, J. Geophys. Res. 75 (1970) 6402; A. Y. Wong and R. J. Taylor, Phys. Rev. Letters 27 (1971) 644; W. E. Gordon, H. C. Carlson, and R. L. Showen, J. Geophys. Res. 77 (1972) 1242.

3. J. P. Friedberg, R. W. Mitchell, R. L. Morse, and L. I. Rudsinski, Phys. Rev. Letters 28 (1972) 795; N. G. Denisov, Soviet Phys. -JETP 4 (1957) 544; A. D. Piliya, Zh. Tekh. Fiz. 36 (1966) 818 [Soviet Phys. Tech. Phys. 11 (1966) 609]; M. M. Mueller, Phys. Rev. Lett. 30 (1973) 582; D. Kelley and A. Banos, Phys Fluids 17 (1974) 2275.

4. G. J. Morales, Y. C. Lee, and R. B. White, Phys. Rev. Letters 32 (1974) 457; Phys. Rev. Letters 33 (1974) 1016; E. Valeo and W. Kruer, Phys. Rev. Letters 33 (1974) 750; P. Koch and J. Albritton, Phys. Rev. Letters 32 (1974) 1420; K. E. Valeo, and W. Kruer, Phys. Fluids 18, 1151 (1975); D. W. Forslund, J. M. Kindel, K. Lee, L. Lindman, and R. L. Morse, Phys. Rev. Letters 11 (1975) 679; K. Nishikawa, H. Hojo, K. Mima, and H. Ikezi, Phys. Rev. Letters 33 (1974) 148; K. Mima, K. Nishikawa, and H. Ikezi, Phys. Rev. Letters 35 (1975) 726.

5. A. Y. Wong, R. L. Stenzel, D. Arnush, B. D. Fried, C. F. Kennel, and H. Reim, APS Bull. 14 (1972) 1017.

6. R. L. Stenzel, A. Y. Wong, and H. C. Kim, Phys. Rev. Letters 32 (1974) 654.

7. W. DiVergilio and A. Y. Wong, TRW Memo, 1976.

8. R. L. Stenzel, Rev. Sci. Instr. 45 (1974) 626.

9. H. C. Kim, A. Y. Wong, and R. L. Stenzel, UCLA PPG-177 (June 1974).

10. H. C. Kim, R. L. Stenzel, and A. Y. Wong, Phys. Rev. Letters 33 (1974) 886.

11. R. L. Stenzel, H. C. Kim, and A. Y. Wong, Radio Science 10 (1974) 485.

12. H. Ikezi **et al.**, J. Phys. Soc. Japan 37 (1974) 766 observed resonant effects but not stationary density cavities as in Kim et al.

13. B. Quon, T. Yoshizumi, H. Injeyan, P. Leung, and A. Y. Wong, UCLA Internal Memo (1975).

14. G. Morales and Y. C. Lee, Phys. Rev. Letters 33 (1974) 1016; UCLA PPG-211 (Feb. 1975).

15. A. Y. Wong and R. L. Stenzel, Phys. Rev. Letters 34 (1975) 727.

16. W. DiVergelio, A. Y. Wong, H. C. Kim and Y. C. Lee, Phys. Rev. Letters 38, to be published 1977.

17. J. J. Thomson, C. E. Max and K. Estabrook, Phys. Rev. Letters 35 (1975) 663.

18. A. Y. Wong, UCLA PPG-143 (1973); R. L. Stenzel, A. Y. Wong, D. Arnush, B. D. Fried, and C. F. Kennel, Proceedings of the AGARD-NATO Meeting, CTP-138, No. 4-1 Edinburgh (November 1973); A. Y. Wong, R. L. Stenzel, H. C. Kim, and F. F. Chen, Plasma Physics and Controlled Nuclear Fusion Research, v. II. 589, IAEA-CN-33/H 4-1 (1975).

19. D. Arnush and C. Kennel, Phys. Rev. Letters 30 (1973) 597.

20. A. Y. Wong, D. R. Baker, and N. Booth, Phys. Rev. Letters 24 (1970) 804; D. Arnush, K. Nishikawa, B. Fried, C. Kennel, and A. Y. Wong, Phys. Fluids 16 (1973) 2270.

21. A. Y. Wong, R. J. Taylor, D. Arnush, B. D. Fried, and C. F. Kennel, TRW Memo (December 1971).

THE NONLINEAR FORCE OF ELECTRODYNAMIC LASER-PLASMA INTERACTION[1,2]

Heinrich Hora[3]

Dept. of Theoretical Physics, The University of New South Wales

Kensington, NSW Australia

Abstract

At high laser intensities the interaction with plasmas generates electrodynamic forces of a nonlinear type, where the dielectric plasma properties can generate explosions with 100 and more times the radiation pressure. The complexity of the derivation of this force is described and analytical and numerical results are reviewed. Some problems are explained. From the numerical results, which are experimentally confirmed (by Wong and Stenzel) of low entropy transfer of optical energy into fast moving thick plasma layers without remarkable increase of the temperature. The use of high efficiency compression of laser irradiated spherical plasmas for exothermal nuclear reactions is described, showing a great improvement on the reaction gains compared with the thermokinetic compression schemes.

I INTRODUCTION

With the advent of the laser, the various types of forces generated in materials irradiated by electromagnetic radiation had to be analysed. One must distinguish from purely thermokinetic forces, due to gasdynamic pressures of the generated plasma, the forces of electrodynamic nature as expressed by the Poynting vector, gradients of quadratic expressions of the electric or magnetic laser field strength $\underline{E}$ and $\underline{H}$, including dielectric properties which later are dependent on the gasdynamic properties of the generated plasma or (nonlinearly) on the actual laser

[1] Presented at the Fourth International Workshop Conference on "Laser Interaction and Related Plasma Phenomena" at Rensselaer Polytechnic Institute, Troy, New York, 8-12 November, 1976.
[2] Supported by ARGC grant No. B75/15538.
[3] Also from Hartford Graduate Center (Rensselaer Polytech. Inst.), Hartford, Conn.

fields. These electrodynamic and essentially non-linear forces will be discussed in this review.

The necessity of including the nonlinear forces in the discussion of laser plasma interaction, was triggered by the first publications on lasers for nuclear fusion.[1-4] It was indicated by the various anomalous experimental results, such as the generation of keV-ions [5], the quadratic increase of the ion energy with the laser intensity [6,7] and the measurement of electron emission currents greater than permitted by space charge laws for classical electron emission [8,9]. Further indications are the superlinear momentum transfer at the onset of the plasmatization [10], the confusing scattering of reflectivity measurements ([11] see Fig. 1.1), the analysis of two groups of generated plasmas from laser irradiated spheres (one with thermokinetic and one with nonlinear properties) [12] and the measurement of high-energy x-ray emission [13] first analysed by Eidmann [14] as a phenomenon differing from the thermal properties. An analysis of these phenomena by the action of the nonlinear force [11, 15, 16, 17, 18] was not easily convincing as complicated thermokinetic processes were involved.

The success of the last years, to show the transfer of 60% of the laser energy for spherically irradiated targets without self-focusing [19] into a block of fast plasma with more than 60 keV ion energy, or the comparable microwave experiments of Wong and Stenzel [20,21] accelerating plasma of 2 eV temperature to cold blocks of 7 keV ion energy demonstrated directly the action of the nonlinear forces. It has to be mentioned that the self-focusing in plasmas as measured first by Korobkin and Alcock [22] could be straightforwardly explained by a dynamic nonlinear force mechanism to derive the threshold and the beam diameter [23] as had been reproduced by several authors [24 - 26] generalized for thermal effects [27,28] and numerically studied [29]. The measurement of the generation of MeV ions [30-33] could be explained by an acceleration by the nonlinear force, including a wave bending relativistic self-focusing [34] which causes a shrinking down of the beam diameter to a diameter of one wavelength within a very short distance.[35,36]

The importance of the nonlinear force is its use for compression of plasma for exothermal nuclear reactions. Following the result [4] for the gain in inertial confinement, that an increase of the initial density causes a quadratic decrease of the necessary energy input for some gains, Nuckolls [37] and other authors [38,39] had the idea of compressing the laser irradiated plasma by this interaction where extensive numerical calculations [37,40] used the compression from recoil due to the gasdynamic ablation based on thermal processes. These calculations have not included the time delay for the thermalization which causes at least a restriction to low intensities [41]. In contrast to this scheme, the use of the nonlinear force for high intensity direct transfer of optical energy into kinetic energy of fast (cold) plasma blocks [11,42] is a low entropy, and therefore highly efficient process causing a plasma compression with 1000 times less laser energy (50 times smaller lasers) for the same reaction gains [43,44].

The development of the derivation of the nonlinear force of laser plasma interaction and of the several analytical and numerical applications is given in numerous papers with different aims. Based on the preceding reviews [17,18] the development of the last few years will be described here. As the derivation of electrodynamic forces is connected with developments over more than 100 years, including electrostriction, magnetostriction, radiation pressure, and ponderomotive forces, and as the present terminology is not unique, some appropriate remarks on these points have to be included. Nevertheless, the subject of this paper is a well defined phenomenon and it was the advent of the laser showing curious interaction effects which caused a unique new development of this force during the last ten years. The roots for the theory are the electron theory of Lorentz, a product of this century. As could be seen in deriving the nonlinear force, the former Maxwellian theory can lead to wrong results if nonlinear phenomena -- all electrodynamic forces are of such kind -- are treated. A textbook example of such errors with the ponderomotive force described by the Maxwell theory was described by Weizel [45] in a very effective way. Another important source for the development of the nonlinear force of laser plasma interaction was the two-fluid theory of the plasma, derived by Schluter [46] at the beginning of the modern plasma theory.

II BASIC DERIVATION OF THE NONLINEAR FORCE

Assuming that the plasma can be described by the two-fluid model [46], where these are, e.g. the fluids of the electrons and of the ions of a fully ionized hydrogen plasma for dimensions larger than the Debye radius, the force density $\underline{f}$ in the plasma of ion density n, temperature T, current density $\underline{j}$ and with oscillating fields $\underline{E}$ and $\underline{H}$ of a frequency ω is given by (first Schluter equation) [16,46]

$$\underline{f} = - \underline{\nabla} nKT + \frac{1}{c} \underline{j} \times \underline{H} \tag{1}$$

where c is the vacuum velocity of light, K the Boltzmann constant, and the plasma frequency ω_p is given by

$$\omega_p{}^2 = \frac{4\pi e n_e}{m_e} \tag{2}$$

and e is the charge, n_e the density and m_e the mass of the electrons. From the second Schluter equation [46], (Ohm's law)

$$\frac{4\pi}{\omega_p{}^2}\left(\frac{\partial \underline{j}}{\partial t} + \nu \underline{j}\right) = \underline{E} + \frac{1}{c} \underline{v} \times \underline{B} - \frac{1}{e n_e c} \underline{j} \times \underline{B} \tag{3}$$

neglecting the second (Hall) and third (Lorentz) terms on the right hand side, one derives a dispersion relation for the oscillating $\underline{E}$ and $\underline{H}$ from the Maxwellian equations resulting in a (complex) refractive index $\tilde{n}$

$$\tilde{n}^2 = 1 - \frac{\omega_p{}^2}{\omega^2 + \nu^2}\left(1 + i \frac{\nu}{\omega}\right) \tag{4}$$

where the collision frequency ν

$$\nu = \frac{\omega_p^2 \, \pi^{3/2} \, m_e^{1/2} \, Ze^2 \, \ell n \, \Lambda}{8\pi \, (2\, KT)^{3/2} \, \gamma_\varepsilon(\mathbf{Z})} \tag{5}$$

In Eq. (5),$\ln\Lambda$ is the Coulomb logarithm, Z the ion charge, and $\gamma_E(Z)$ is Spitzer's correction (a factor between 0.5 and 1) for including electron-electron collisions. The possibility of using the d-c collision frequency ν also for the high-frequency field has been proved by comparision with the quantum mechanical derivation of inverse bremsstrahlung [47] which analogy even holds [48-50] for the nonlinear dependence of the optical constants at very high laser fields $\underline{E}$, where the effective temperature T is given by that of thermal motion T_{th}

$$T = T_{th} + \varepsilon^e_{osc}/K \tag{6}$$

and the oscillation energy of the electrons

$$\varepsilon^e_{osc} = \frac{e^2 \, \underline{E}^2}{m_o \, \omega^2} \tag{7}$$

If one substitutes the WBK approximation for $\underline{E}$ and $\underline{H}$ of a laser frequency ω in striated inhomogeneous plasma for perpendiculary incident linearly polarized laser radiation (propagation into x-direction, E is parallel to y-direction)

$$\underline{E} = \underline{i}_2 \frac{E_v}{\tilde{n}^{1/2}} \exp(iF) \exp(\mp k(x)\, x/2) \tag{8}$$

$$\underline{H} = \underline{i}_3 \, \underline{E}_v \, \tilde{n}^{1/2} \exp(i\, F) \exp(\mp k(x)\, x/2) \tag{9}$$

$$F = \omega(t \mp \int^x \mathrm{Re}(\tilde{n}(\xi) d\xi/c)\,; \quad k(x) = \omega/(xc) \int^x \mathrm{Im}(\tilde{n}(\xi) d\xi \tag{10}$$

into Eq. (1), one arrives after averaging over one period $1/\omega$ at

$$\overline{\underline{f}} = - \underline{\nabla}\, nKT - \underline{i}_1 \frac{E_v^2}{16\pi} \frac{\omega_p^2}{\omega^2} \frac{\partial}{\partial x} \frac{1}{|\tilde{n}|} \tag{11}$$

This derivation of the nonlinear force $\underline{f}_{NL}$

$$\underline{f}_{NL} = \underline{f} + \underline{\nabla}\, n\, K\, T \tag{12}$$

led to the impossible result if obliquely incident laser radiation was used, as a net force in the plasma surface resulted for general polarization. This problem was overcome by comparing the nonlinear force as derived from a formula [49] based on the Lorentz theory which should hold for nondispersive media only,

$$\underline{f}_{NL} = +\underline{\nabla} \cdot (\underline{\underline{U}} - (\tilde{n}^2 - 1)\overline{\underline{E}\,\underline{E}}) - \frac{\partial}{\partial t}\frac{\overline{\underline{E} \times \underline{H}}}{4\pi c}, \tag{13}$$

where $\underline{\underline{U}}$ is the Maxwellian stress tensor. The justification for extending the validity of the formula to dispersive media was concluded from the fact that for perpendicular incidence of linearly polarized laser light on a stratified plasma (with any degree of inhomogeneity) the force f_{NL} is [16] averaged on time

$$f_{NL} = \frac{1}{8\pi} \nabla (\overline{\underline{E}^2} + \overline{\underline{H}^2}) \tag{14}$$

in agreement with the more intuitively derived value [15] resulting for the restriction of the inhomogeneity to the WBK case

$$0 < \frac{c}{2\omega}\frac{1}{|\tilde{n}|^2}\frac{\partial}{\partial x}|\tilde{n}| << 1 \tag{15}$$

in the same value of Eq. (11)

$$f_{NL} = -\underline{i}\, \frac{E_v^2}{16\pi}\frac{\omega_p^2}{\omega^2}\frac{\partial}{\partial x}\frac{1}{|\tilde{n}|} \tag{16}$$

This can be seen either by direct use of Eqs. (8) and (9) in the general expression (14) for perpendicular incidence or by the WBK-like substitution $\overline{E^2} = E_v^2/|\tilde{n}|$; $\overline{H^2} = E_v^2|\tilde{n}|$ with the vacuum values E_v of the amplitude [41]

$$f_{NL} = \frac{1}{8\pi}\nabla(\overline{E^2} + \overline{H^2}) = \frac{E_v^2}{16\pi}\nabla\left(\frac{1}{|\tilde{n}|} + |\tilde{n}|\right) = \frac{|\tilde{n}|^2 - 1}{16\pi}\nabla\overline{E^2} \tag{17}$$

The last expression was known for the forces in a collisionless WBK-like striated plasma at microwave irradiation. [50]

From this indirect and partial justification of the Landau-Lifshitz-formula, Eq. (13), as being applicable for the dispersive plasma, it could be derived [16] by algebraic rewriting of Eq. (13) plus the Maxwellian equations, that the two-fluid equation of motion, Eq. (1) should read

$$\underline{f} = -\underline{\nabla}\, nKT + \frac{1}{c}\,\underline{j} \times \underline{H} + \frac{1}{4\pi}\,\underline{E}\,\nabla\cdot\underline{E} - \frac{1}{4\pi}\underline{\nabla}\cdot\underline{E}\,\underline{E}(|n|^2 - 1) \tag{18}$$

It was remarkable that Schluter's derivation, Eq. (1), reproduced one of the last nonlinear terms of Eq. (18) while Spitzer's derivation [51] from the microscopic theory (Boltzmann equation) did not show this term. With this generalized equation of motion (18), the difficulties for oblique incidence of laser radiation were avoided.

A third derivation of the nonlinear force was that using the quivering motion of the electrons in the laser field. Following the $\underline{E}$-field the electrons oscillate and due to the Lorentz force due to this motion and the $\underline{H}$-field, a longitudinal oscillation with the frequency 2ω results, causing an eightlike motion in the plane of $\underline{E}$ and the propagation direction at linear polarization. In vacuum this 8 is standing, but in a medium, the phase between $\underline{E}$ and $\underline{H}$ causes a net penetration of the electron. In an inhomogeneous medium a net plasma motions with a nonlinear force given by Eq. (16) results. The detailed description of this derivation is given in a preceding review [17] to which nothing has to be added. It is a further confirmation for the correctness of the derivation of the formula (16) and a further justification of the use of the first Schluter equation (1) for the net motion of plasma due to high frequency fields and for the justification of the necessary generalization of the Landau-Lifshitz formula, Eq. (13) for the dispersive plasmas. Only under these aspects of consistency can we consider the derivation of Eq. (14) or Eq. (16) as correct.

The predominance of the nonlinear force $\underline{f}_{NL}$ over the thermokinetic force $\underline{f}_{NL} = -\underline{\nabla} nKT$ has been discussed before extensively [17]. We review here that a necessary condition is that the excess of the electromagnetic force density in one region over the other region has to be larger than the excess of the thermokinetic energy density. For the case of perpendicularly incident radiation on a stratified plasma,

$$(E^2 + H^2)/8\pi - (E_v{}^2 + H_v{}^2)/8\pi \geq nKT \tag{19}$$

This is the case, if the laser intensity I is larger than I*, which is for the approximation $\omega/\nu >> 1$

$$E_v{}^2 \geq E_v{}^2 = 8\pi\, n_{ec} \left\{ \frac{\omega_p{}^2 \pi^{3/2} m_e{}^{\frac{1}{2}}\, Z\, e\, \ell n\Lambda}{8\pi\omega\gamma_E\,(Z)\, Z^{3/2}} K^{\frac{1}{2}} \right\}^{\frac{1}{2}} T_{th}{}^{\frac{1}{4}} \tag{20}$$

or for

$$I_v \geq I_v \simeq \begin{Bmatrix} 7.5 \times 10^{13} \\ 7.5 \times 10^{10} \end{Bmatrix} T_{th}^{\frac{1}{4}} \ W/cm^2 \begin{Bmatrix} \text{Nd glass} \\ CO_2 \text{ laser} \end{Bmatrix} \qquad (21)$$

where T_{th} is in electron volts. The exact value has been calculated by iteration and does not deviate essentially [17]. The calculation of Steinhauer and Ahlstrom [52], could be used to confirm Eq. (21) if Eq. (6) was used into their treatment [17]. It has not yet been understood on the basis of a comparison of Eq. (19), why the mentioned authors [52] could arrive at a predominance of the nonlinear force even for less intensities $I < I^*$ if $T > 10^4$ eV.

The picture of the quivering motion explains again the nonlinear nature of the force as well as its "semi-relativistic" property. While the nonlinearity in Eqs, (1) and (13) can be seen from the fact of the binary products of $\underline{E}$ and/or $\underline{H}$-components ($\underline{j}$ being considered as equivalent to $\underline{E}$ by Ohm's law), the net plasma acceleration by the quivering motion is mainly given by the longitudinal motion of the electrons with twice the laser frequency 2ω. This component of motion is [17] of a v/c_o magnitude, therefore a typical relativistic effect, however, c_o is there the group velocity of light, so that for strong dielectric effects ($|\tilde{n}| << 1$), we have a relativistic process without the relativistic mass corrections.

III GENERAL RESULTS

The generation of the nonlinear force for oblique incidence of radiation had been treated in an early stage [16] on the basis of the Landau-Lifshitz formula to check whether a polarization dependence of the forces exists. It was found that there are no net forces in the plane of the plasma surface and the magnitude of the acceleration perpendicular to the surface did not show a strong polarization dependence for an angle of incidence of less than 20°. For larger angles of incidence the approximations used became invalid.

For oblique incidence it was found that where patterns of nodes and antinodes of a standing wave have been generated, a net motion in each striation can result in high plasma velocities, which net momentum is compensated by the inverse motion in the neighbouring layer [53].

It is to the credit of F. F. Chen [26] that he could derive the existing parametric instabilities by interaction of an electromagnetic wave with a plasma from the action of the nonlinear force in the plasma. The thresholds and the growth rates of the instabilities have been derived generally from the ansatz of the nonlinear force different from the special assumptions which were necessary in the preceding treatments of these instabilities. A further application was the derivation of the threshold of the laser power for self-focusing in low density ($\omega_p << \omega$) plasma in agreement with numbers derived before [23-25].

The first quantitative theory of self-focusing of a laser beam [23] was based on the model suggested by Schluter [54] that the lateral gradient of the intensity of the laser beam ($\nabla(E^2 + H^2)/8\pi$) expels plasma from the center of the beam. The generated density gradient causes a gasdynamic counterpressure ∇nKT. In equilibrium one can derive [23] density gradients comparable with the condition of total reflection of particular beams within an angular region fulfilling the diffraction conditions. From this calculation a minimum laser power P* follows if the self-focusing is calculated for a slab

$$P \geq P^* = \begin{cases} 1.46 \times 10^6/T_{th}^{5/4} \text{ Watts}; \ \omega_p \leq \omega; \ T > 30\text{eV} \\ 11 \times 10^3 \ T \ \text{ Watts}; \ \omega_p << \omega \end{cases} \tag{22}$$

which gives the values for a Gaussian beam [24-26] for $\omega_p << \omega$

$$P^*_{Beam} = 8 \times 10^3 T \quad \text{Watts} \tag{23}$$

(T in eV).

From the fact that the self-focusing diameter is limited to such value where the electromagnetic energy density is equal to nKT, what can be evaluated from a relation of Eq. (21), it could be calculated for the experiments of Korobkin and Alcock [22] that their laser beam of a power of 10^7 Watts (in agreement with Eq. (22)) must shrink down to a 3μ diameter in agreement with the experimental result. The generation of Brillouin scattering at self-focusing in plasmas [24] has been concluded in the same way as known from self-focusing in solids or liquids.

The results of the momentum transfered to the plasma being pushed to lower density (ablation) can be understood by a recoil due to the increase of the momentum of the photons in the plasma. Using the momentum of the photons in vacuum

$$P_o = \varepsilon_n/c \tag{24}$$

where ϵ_L is the energy of the acting laser photons, the momentum transferred to the inhomogeneous plasma corona between the vacuum and a certain depth x_2 at perpendicular incidence of the radiation and before the reaching minimum of $|\tilde{n}|$ is (Eq. (66b of Ref. [16])

$$P_{inh} = P_o(1 - |1\tilde{n}(x_2)| \ )^2/(2 \ |\tilde{n}(x_2)|), \tag{25}$$

if a WBK-like reflectionless penetration of the light can be assumed. As P_o correspond to the usual radiation pressure, P_{inh} is increased, e.g. by 100, if $|\tilde{n}(x_2)|$ is 1/200 to result in a compressing pressure of 101 in towards the plasma interior. To understand this process one must conclude that instantly at every point $x < x_2$, to the momentum P_ϕ of each photon has been increased to a value

$$p_\psi = \frac{\hbar\omega}{c|\tilde{n}|}\left(1 - \frac{1}{2}\frac{\omega_p^2}{\omega^2}\right) \tag{26}$$

$$= \frac{\hbar\omega}{2c|\tilde{n}|}\left(1 + |\tilde{n}|\right) \tag{27}$$

where $\hbar$ is Planck's constant divided by 2π. The fact that such a momentum agrees neither with Abraham's nor Minkowski's value of the photon momentum has been pointed out by Klima and Petrzilka [55,56]. They arrived at the same value of the momentum by a procedure of calculating the momentum transfer to a plasma completely independent from our case of a stationary penetration of light into an inhomogeneous plasma. Their case was a homogeneous plasma where the light was a non-stationary, increasing and decreasing wave pocket. In this case the plasma was moved with the light at the beginning of the wave pocket (due to $\nabla(E^2 + H^2)/8\pi$) and was pushed back at the other end of the wave pocket so that finally no momentum was transferred to collisionless plasma; however a shift of the plasma had occurred. The momentum of the photons in the plasma had the value of Eq. (26) where the ω_p^2/ω - term expresses the mechanism of a less energy transfer from the field to the electrons and back. Without a final solution of the Abraham-Minkowski problem having been found [57], its connection with the nonlinear forces had been pointed out by Basov [58], explaining that Tamm's theory of the Cerenkov radiation needed a concept following the Abraham description and in some parts the Minkowski description.

It should be remarked, that Peierls et. al. [59], found the same photon momentum as in Eq. (26) for dielectric solids, however arrived finally [60] at a differing value. The result is then that an exchange of momentum occurs at the surface of a solid, while it was remarkable that a slab of plasma between vacuum would not receive a momentum by the light at stationary conditions, as the increase of the photon momentum in the plasma due to Eq. [26] is just compensated by the reflected photons at the boundaries. The forceless compensation of the photon momentum resulted straight forward in the Fresnel formulae of reflectivity [61]. That there exists a difference to the case of solids, is evident as forces are acting on transparent spheres in laser beams [62-64]. It has to be noted however, that the direction of motion of the pellets in a medium intensity laser beam can be inverse to that of Ashkin et. al. [62] if small dust particles of smoke are used, as Riedmuller [65] has measured.

The basic consideration of the nonlinear force for several cases of calculations was done by several authors, as the connection with gas dynamic calculations [66], numerical evaluations [66-76], for self focusing [21-23, 77]. A very remarkable application was the success of Stamper et. al. [79, 80] to apply the model of the quivering motion to explain the spontaneously generated magnetic fields in a laser produced plasma, discovered by Stamper and reconfirmed by many groups [81,82]. The results were discussed at the preceding conference.

A discrepency of results using the quivering motion to the $\nabla(E^2 + H^2)/8\pi$ description should be explained here to some larger extent. M. J. Hollis [83] has measured gasbreakdown at low pressures, where the Debye length was larger than the focal diameter. Self-focusing should then not occur following the nonlinear force theory [23-29] but it can be expected that the electrons are being expelled from the center of the beam by the nonlinear force as proved by the nonlinear force self-focusing [22,23].

Hollis measured energies of emitted electrons up to 180 eV which value corresponded to the maximum oscillation energy in the beam focus. However, he observed a polarization dependence with a lower electron energy perpendicular to the direction of E at linear polarization and a maximum parallel to E. This result is not understandable from a force given by $\nabla(E^2 + H^2)/8\pi$, as the directions are cancelled by quadratures. The quivering model (see Fig. 1) results in this polarization dependence, as the eightlike motion at the area [1] causes a net acceleration for a monotonically increasing profile of the external beam intensity while the electron is not being moved at [2].

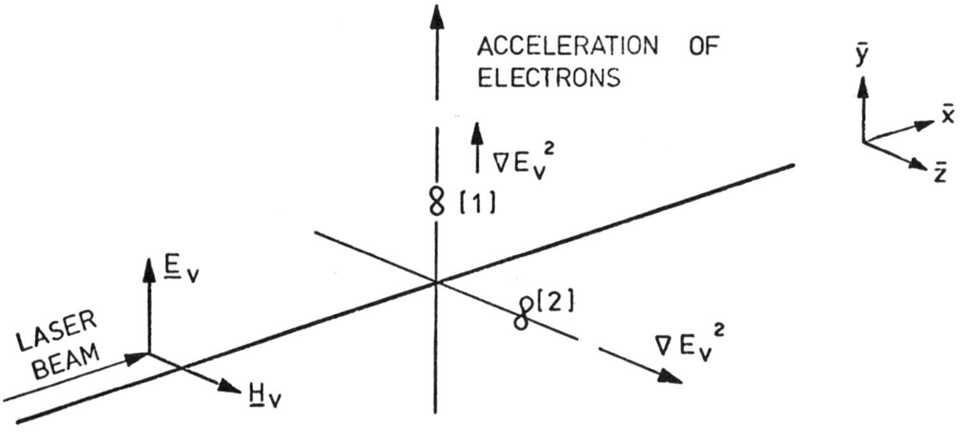

Figure 1
Laser beam focused into low pressure gas accelerates the generated electrons preferentially parallel to the E-vector of the laser beam. [83,84]

The motion of the electron is an eight along $\underline{E}$ in the plane perpendicular to the magnetic laser field $\underline{H}$. At [2], the eight is locally fixed in space, as $\underline{E}$ and $\underline{E} \times \underline{H}$ are to first order constant along the direction of these vectors as no phase between $\underline{E}$ and $\underline{H}$ can be assumed because of $\tilde{n} = 1$. No net acceleration should then occur into the z-direction.

At [1], there is a y-dependence of the intensity, or

$$E_v = E_o \; (1 + \alpha_1 y + \alpha_2 y^2 + \ldots) \qquad (28)$$

The linear motion of the electron along E is given by

$$\ddot{y} = \frac{e}{m} E_o \; (1 + \alpha_1 y + \alpha_2 y^2 + \ldots)\cos \omega t \qquad (29)$$

Using

$$y = y_1 + y_2 + \ldots \qquad (30)$$

where y_1 is the solution for $a_\nu = 0$ ($\nu = 1 \ldots$) and y_2 is the second order iteration with $a_\mu = 0$ ($\mu = 2 \ldots$) etc., we find the second order approximation of

$$\dot{y} = \dot{y}_1 + \dot{y}_2 = -\frac{eE_o}{m\omega} \sin \omega t - \left(\frac{eE_o}{m\omega}\right)^2 \alpha_1 \left[\frac{t}{2} + \frac{1}{4\omega} \sin 2\omega t\right]$$

and

$$y = y_1 + y_2 = -\frac{eE_o}{m\omega^2} \cos \omega t - \left(\frac{eE_o}{m\omega}\right)^2 \left[\frac{t^2}{4} - \frac{1}{8\omega^2} \cos 2\omega t\right] \qquad (31)$$

where the very last term is of importance only for $t < 1/\omega$. For longer times we find a time averaged drift length

$$\bar{y} = -\left(\frac{eE_o}{m\omega}\right)^2 \alpha_1 \frac{t^2}{4} \qquad (32)$$

penetrated by the quivering electron.

If the lateral intensity dependence of the beam is approximated as quadratically decreasing from the centre of the beam to zero at a distance $d_o/2$, the function of E_v is given by $a_1 = -2/d_o$; $a_\mu = 0$ ($\mu = 2 \ldots$), the end velocity is then exactly $v = eE_o/(m\omega)$, and the translational kinetic energy ϵ_{trans} of the electron (being not seriously influenced by ions because of the large Debye length l_D) after leaving the beam area is

$$\varepsilon_{trans} = \frac{m}{2} \overline{v^2} = \frac{e^2 E_o^2}{2m\omega^2} = e^e_{osc} \qquad (33)$$

This corresponds to the well known result for very high density plasmas ($l_D << d_o$) where the translational energy reached by the ions after acceleration is Z-times ($Z \sim$ ion charge) of the maximum oscillation energy of the electrons [17,41]. The fact that there is no drift acceleration of the electrons along the z-axis, see [2] Fig. 1, is a complete interpretation of the measured polarization dependence. [83]

It is noted, that there is a similar interpretation of the result of Yablontovich [85] possible, where an electron acceleration parallel to E in laser induced gas-breakdown has been observed, however, at higher plasma densities, highly pronounced at the beginning of the cut-off density. If no other interpretations are relevant (as e.g. resonance absorption with a vector effect of the nonlinear force) at least the dielectric swelling has to be added to a mechanism of the kind described.

The mentioned discrepancy of the result of the quivering model from the other derivations of the nonlinear force $\nabla(E^2 + H^2)/8\pi$ which does not depend on the polarization, may be due to the insufficiency of our assumption of no phase difference between E and H near [2] in Fig. 1. It is being suggested that the exact Maxwellian description of an optical wave bundle results in such a phase, that the higher order nonlinear terms in the force density of the stress tensor description compensates to zero near [2]. This example demonstrates once more that the nonlinear radiation force of the laser-plasma interaction is more complicated than a simple ponderomotive electrostriction or a $\underline{j} \times \underline{H}$-mechanism, as it has been demonstrated from the beginning. [16]

Another general analytical result has to be mentioned, where the approximate solution of the transfer of the maximum oscillation energy of electrons ϵ_e in the laser field of an intensity I is

$$\Sigma^C_{cx} = m\, c^2 (\sqrt{1 - A(I) I/I_{rec}} - 1); \quad I_{rec} = \frac{2mc^3\omega^2}{8\pi\, e^2} \tag{34}$$

$$(e^2 E_v^2)/(m\, \omega^2) \qquad ; \text{ if } I << J_{rec} \tag{35}$$

with a factor A between 1 and 1.1 depending on polarization and intensity as discussed by Schwarz, et. al. [86], is being transfered into translation energy ϵ^i_{trans} of ions when the nonlinear force causes a "dielectric explosion" to expel the plasma corona from the plasma interior which takes the recoil. We are starting from the nonlinear force of the following form (Eq. (16)) with the cut-off density n_{ec} ($n_e = n_{ec}$ where $\omega_p = \omega$ in Eq. (2)) and with the mass m_i, density n_i and velocity v_i of ions

$$f_{NL} = m_i n_i \frac{dv_i}{dt} = - \frac{E_v^2}{16\pi} \frac{n_e}{n_{ec}} \frac{\partial}{\partial x} \frac{1}{|\tilde{n}|}, \tag{36}$$

for the WBK approximation (Eq. (15)), but beyond this is valid for all cases of the Rayleigh profiles without collisions where the refractive index

$$|\tilde{n}(x)| = 1/(1 + \alpha\, x) \quad (\alpha \sim \text{constant}) \tag{37}$$

and the electron density is

$$n_c = n_{ec}(1 - 1/(1 + \alpha x)). \tag{37a}$$

The corresponding Airy case for linear density profiles has been discussed by Lindl and Kaw [87] beyond the WKB limitation.

Recalling that a velocity v_o gained by a body falling along a distance Δx with an acceleration $d\nu_i/dt$ is

$$v_o = \{2(dv_i/dt)\Delta x\}^{\frac{1}{2}}, \tag{38}$$

we find for the differential gain of the square of the velocity:

$$\delta v_o^2 = 2(dv_i/dt)\Delta x \tag{38a}$$

for varying acceleration. The increase of the kinetic energy of the ions due to the force (36) is then (using $n_e = Zn_i$, with ion charge Z, and avoiding the restriction $\nu << \omega$ of the collisionless case)

$$d(\tfrac{1}{2}m_i v_i^2) = m_i \frac{dv_i}{dt} dx = \left| \frac{E_v^2}{16\pi} \frac{Z}{n_{ec}} \frac{\partial}{\partial x}\left(\frac{\exp(-\bar{k}x)}{|\tilde{n}|}\right) \right| dx \tag{39}$$

Here k is the averaged absorption constant for the plasma between a point in the vacuum and the considered point x. The exact integral for the ion energy is then

$$\tfrac{1}{2}m_i v_i^2 = \frac{E_v^2}{16\pi} \frac{Z}{n_{ec}} \int_{x_1}^{x_2} \frac{\partial}{\partial x} \left(\frac{\exp(-\bar{k}x)}{|\tilde{n}|}\right) dx$$

$$= \frac{E_v^2}{16\pi} \frac{Z}{n_{ec}} \left(\frac{\exp(-\bar{k}x_2)}{|\tilde{n}(x_2)|} - \frac{\exp(-\bar{k}x_1)}{|\tilde{n}(x_1)|}\right) \tag{40}$$

The integration is performed between depth x_1, a point in the vacuum ($|\tilde{n}| = 1$; $k = 0$), and x_2, a point in the plasma corresponding to minimum refractive index $|\tilde{n}|$, assuming $k << 1$ in accordance with the numerical cases. [11,70] The maximum energy $\epsilon_{trans}^{i,abl}$ translational motion towards decreasing electron density which the ablated ions gain due to f_{NL} is then

$$\epsilon_{trs}^{i,abl} = \frac{E_v^2}{16\pi} \frac{Z}{n_{ec}} \left(\frac{1}{|\tilde{n}|_{max}} - 1\right), \tag{41}$$

The ion energy $\epsilon_{trans}^{i,cpr}$ for the compression of the plasma towards its interior (after selecting the appropriate sign by the same procedure as for equation (38a) is given by

$$\varepsilon_{trs}^{i,cpr} = \frac{E_v^{\ 2}}{16\pi} \frac{Z}{n_{ec}} \left(\frac{1}{|\tilde{n}|_{max}}\right) \tag{42}$$

where the limits of integration in equation (40) are now those for $|\tilde{n}|_{min}$ (with $\bar{k}x_2 << 1$) and for the interior of the super dense plasma with $\bar{k} >> 1/x$ and non-vanishing $|\tilde{n}|$. The condition $\bar{k} >> 1/x$ also holds for laser interaction with a collisionless plasma having a maximum density near n_{ec}. The acceleration of the plasma parallel to the laser beam corresponds to a (dynamical macroscopic) nonlinear absorption with $\bar{k} >> 1/x$.

The results (41) and (42) can be interpreted very easily by taking into account the fact that the average kinetic energy ϵ_{osc}^{e} of oscillation of the electrons in the laser field within the plasma is

$$\varepsilon_{osc}^{e} = E_v^{\ 2}\exp(-\bar{k}x)/16\pi n_{ec}|\tilde{n}|. \tag{43}$$

We assume again the $\bar{k}x_2 << 1$ holds for the maximum of $1/|\tilde{n}|$, but we include a damping mechanism for the non-WKB case, which prevents the too strong resonance-like increase of $1/|\tilde{n}|$ from reaching unrealistically high values, as pointed out. [89] This means that the maximum energy of the ablated ions

$$\varepsilon_{trs}^{i,abl} = Z\{(\varepsilon_{osc}^{e})_{max} - (\varepsilon_{osc}^{e})_{v}\} \tag{44}$$

is Z times the difference between the maximum oscillation energy of the electrons in the plasma and in the vacuum. In cases where the swelling of the oscillation energy over its vacuum value reaches a factor $1/|\tilde{n}| \gtrsim 100$, we can neglect the second term in equation (44). This correction, representing the radiation pressure in the vacuum, is then the difference of the ion energy for plasma compression to ablation from equations (42) and (43), namely,

$$\varepsilon_{trs}^{i,cpr} = Z(\varepsilon_{osc}^{e})_{max}. \tag{44a}$$

IV NUMERICAL RESULTS OF THE NONLINEAR FORCE

While the analytic calculations on the properties of the nonlinear force, as described in the preceding section, could describe some general behaviour, it was necessary to see the operation of the force in detail by numerical calculation of more or less special cases of laser-plasma interaction. An early numerical evaluation was described in the review (Ref. 18, Fig. 1) where a plasma of spatially constant temperature of 100 eV and with a density profile applicable for the WBK approximation was chosen, while a neodymium glass laser pulse of 10^{15} W/cm^2 was incident. The generated nonlinear force exceeded the thermokinetic force, absolutely and relatively, in wide regions. Because of the WBK-condition, the plasma corona (area between vacuum and cut-off density) was quite large, more than 100 vacuum wave lengths.

Another numerical example [69] (see Ref. 18, Fig. 5) used a plasma with a relatively thin corona of 2.5 vacuum wave length of a special density profile and 100 eV temperature. The laser radiation was then reflected (by more than 90%) generating a strong component of a standing wave (similar to the fully standing wave of the collisionless plasma treated analytically by Lindl and Kaw [87] and the plasma was driven to the nodes of the standing wave. Though the neodymium glass laser intensity was with 5×10^{14} W/cm^2 just near the threshold, a net acceleration force towards less density for a wide range could be calculated where the nonlinear force was larger than the thermokinetic force.

These numerical cases were of "static" kind, as they gave the forces only at a special time for a given plasma, though the optical constants, pressures, etc., were calculated exactly. It was an early desire to see the consequent operation of the nonlinear force during a longer period of the full gasdynamic development. It is the merit of Shearer, Kidder and Zink [68] and Shearer [68] to have shown such a case, though the first assumptions neglecting reflected radiation were quite crude, but not of such nature that the essential effects could have been reversed. It was remarkable, (see [68] Figures 6 to 8 in Ref. (18)), to see the initially monotonic density profile being kept in this shape at only thermokinetic laser-interaction, while the use of the nonlinear force at 10^{16} W/cm^2 neodymium glass laser intensity had changed the profile remarkably, producing a minimum of plasma density near the cut-off region. These results of Shearer, Kidder and Zink [68] have been reproduced recently by Brueckner [90], where the similar generation of a density minimum has been shown, and also experimentally (see next Section) [20,21].

For the application of the nonlinear force for high efficient transfer of optical energy into kinetic energy of plasma for plasma compression, the following numerical calculations were made, where a very generally developed one dimensional plasma code of E. Goldman with separate electron and ion fluids, fully developed optical constants including the nonlinear changes at very high laser intensities, and a general solution of the Maxwellian equations for the laser radiation for each time step was used. The mentioned calculation of the radiation for each time step neglected the retardation of the radiation which was questionable for times of 10^{-13} sec. We can report here that in recent calculations of the same kind, there is no essential change for times of 5×10^{-12} sec, for which the mentioned neglect cannot be of much influence on the physical results. The exact solution of the Maxwellian equations caused a high reflectivity of more than 50%, therefore a stronger motion of plasma to nodes of the standing waves and stronger Brillouin backscattering and dynamic absorption will occur than in the real cases, where the reflectivity is much less. As the nonlinear force effects for a nett acceleration of thick plasma blocks is better, when less reflection occurs, the following results are a pessimistic case with respect to this nonlinear acceleration, taking into account the low experimental reflectivity.

The equation of motion for the plasma in a one-dimensional geometry was given by the force density, distinguishing between electrons and ions,

$$f = -\frac{d}{dx} n_e kT_e - \frac{d}{dx} n_i kT_i + \frac{1}{8\pi}\frac{d}{dx}(E^2 + H^2) \quad (45)$$

where the electrons (e) and ions (i) of different temperatures T_e and T_i are coupled by a Poisson equation and by collisions. Besides the equation of continuity for both components, the energy equation contains internal and external thermodynamic energies, thermal conductivity, and the incident laser radiation absorbed by collisions or by nonlinear electrodynamic motion as a source term.

The calculation begins at a time $t = 0$ with a distribution of plasma density starting from solid state density and a temperature profile. Without any laser field there would be gasdynamic expansion and a conservation of the total energy. At the successive time steps, a time-dependent incident laser intensity is prescribed, for which at each time step an exact stationary solution of the Maxwell equations is calculated including the actual plasma density and its refractive index. The retardation of the waves, the swithcing-on mechanism, and the development of the reflected wave are neglected.

The motion of the plasma within the next step (with appropriately varying step size) is described generally according to Eq. (45a), with the first two terms of the gas-dynamic force and the last term of the nonlinear electrodynamic force included. The following two examples from extensive series of computer runs illustrate characteristic results. In both cases the initial temperature of the electrons and ions (100 eV) was spatially constant for the whole plasma. One may assume that this is produced by a pre-pulse from the laser. The laser pulse increases within 10^{-13} sec up to an intensity of 2×10^{16} W/cm^2, the increase is linear for 5×10^{-14} sec, and then follows smoothly a Gaussian profile. After reaching 2×10^{16} W/cm^2, the laser intensity remains constant.

The initial plasma density increased (for $x < 0$) quadratically above the cut-off density up to the solid state density of LiD. It then changed smoothly in Fig. 2 into a linear decrease up to the length $x = 50\mu$. The resulting electromagnetic momentum flux density $(E^2 + H^2)/8\pi$ is given in Fig. 3. Curve A is taken at an early time, when the laser intensity is 2×10^{14} W/cm^2. In a very thin plasma ($x \approx 50\mu$) the value of $E^2 + H^2$ is constant, as is well known, e.g., from Chen's [26] general derivation of the instabilities from laser-plasma interaction involving the nonlinear force. At about 20 microns we find an oscillation of $E^2 + H^2$, which increases in amplitude and wavelength between 0 and 8 microns. This behaviour is well known from the analytical work for the same linear density profile [87] (dealing, however, with a collisionless plasma). Near $x = 0$, the $E^2 + H^2$ value has increased due to the dielectric properties of the plasma.

The increase of the laser field E and of the wavelength λ over its vacuum values E_v and λ_o is given by a swelling factor $\theta = |\tilde{n}|^{-1} > 1$

$$E = \frac{E_V}{|\tilde{n}|} \; ; \; \lambda = \frac{\lambda_O}{|\tilde{n}|} \qquad (45)$$

where $\tilde{n}$ is identical with the optical (complex) refractive index of the plasma, provided the electromagnetic field can be described by the WKB approximation. In the case of curve A in Fig. 2, the plasma fits the WKB conditions at 1.3 micron quite well, and we find agreement between the swelling factor $|\tilde{n}|^{-1} = 6$, (taken from Fig. 2) and the value calculated from the actual refractive index.

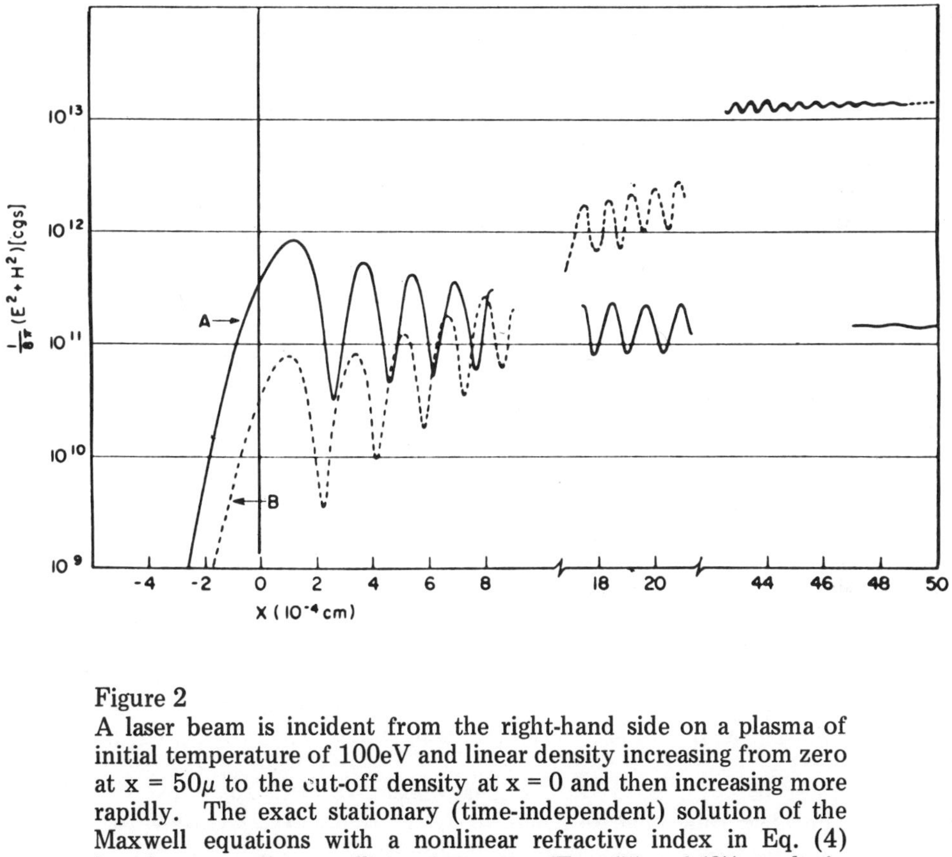

Figure 2
A laser beam is incident from the right-hand side on a plasma of initial temperature of 100eV and linear density increasing from zero at x = 50μ to the cut-off density at x = 0 and then increasing more rapidly. The exact stationary (time-independent) solution of the Maxwell equations with a nonlinear refractive index in Eq. (4) based on a nonlinear collision frequency (Eqs. (5) and (6)) results in an oscillation due to the standing wave and dielectric swelling of the amplitude (curve A). At a later time (2×10^{-13} sec) the laser intensity is 2×10^{16} W/cm^2 (curve B) where the relative swelling remains, but the intensity at x = 0 is attenuated by dynamic absorption.

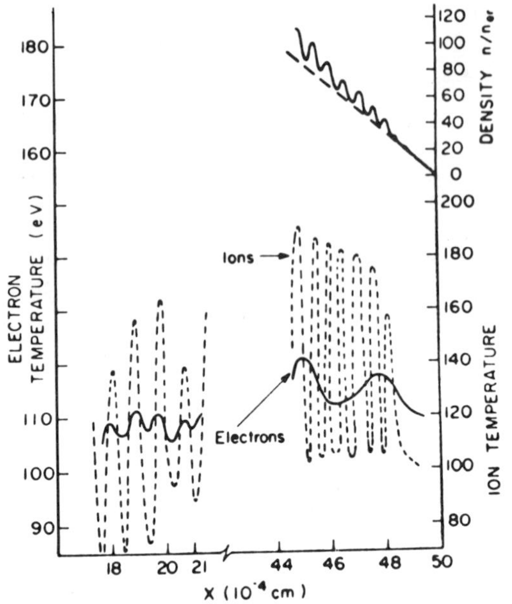

Figure 3
The initial density (dashed line) and the density along curve B of Fig. 2, where a ripple is created by the nonlinear force pushing plasma toward the nodes of the standing wave. The electron and ion temperatures are increased following the ripple by dynamic compression at conditions identical to curve B.

For later times, curve A moves in nearly parallel fashion to higher values. However, its upward shift is reduced near the cut-off density (x = 0). This decrease becomes so marked at $t = 2 \times 10^{-14}$ sec (and later even more pronounced) as shown by curve B that the intensity in the thin plasma ($x \approx 50\mu$) decreases by a factor of 10 up to $x \approx 20\mu$ and to a thousandth part of the initial intensity at $x \approx 0$. The reason is very simple: The standing wave pushes the plasma toward its nodes with ion velocities as high as 10^7 cm/sec even at $x \approx 48$. The gasdynamic velocities reached at that time are 10^3 cm/sec or less. The result is shown in Fig. 3.

At $x = 46\mu$ to 50μ, the initial linear density (upper part of Fig. 3) acquires a ripple. The slight total increase is due to net motion driven by the nonlinear force. The ion energy increases, oscillating up to 200 eV (from the initial value of 100 eV), out of phase with the ripple. There is a curious effect seen in the electrons, in which the temperature increase is only a fifth of that of the ions, as it should be for a compression similar to the early stages of a theta pinch; however, the periodicity is markedly different from that of the ions. At $x \approx 20\mu$ the periodicity of the temperature of the electrons and the ions is nearly the same, but the electron temperature is again less by a factor than the ion temperature.

The density ripple explains the strong decrease of laser intensity with depth at later time: It causes strong reflection of light and a transfer of the optical energy into the ions by collisionless ripple shocks. Because it is not connected with decay of photons into microscopic acoustic modes, one may call this process of dynamic ion absorption a "collisionless ion heating by nonlinear-force-induced macroscopic Brillouin decay". It is to be noted that depression of the laser intensity in interior regions becomes marked only after the density acquires pronounced maxima and minima (ripple).

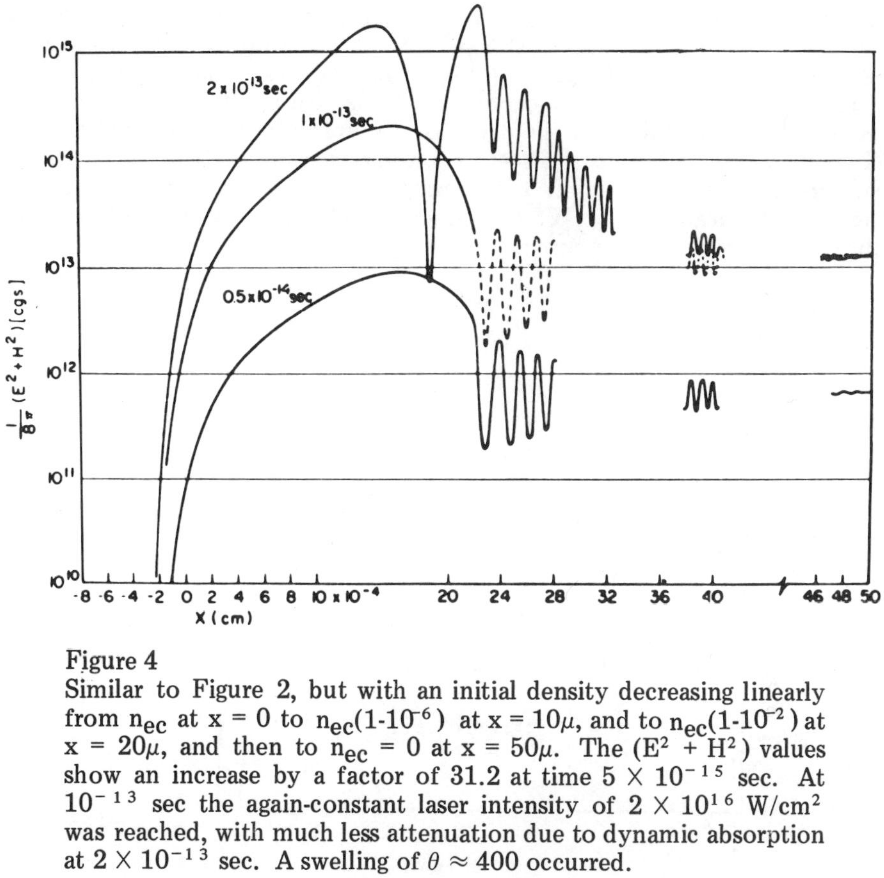

Figure 4
Similar to Figure 2, but with an initial density decreasing linearly from n_{ec} at $x = 0$ to $n_{ec}(1\text{-}10^{-6})$ at $x = 10\mu$, and to $n_{ec}(1\text{-}10^{-2})$ at $x = 20\mu$, and then to $n_{ec} = 0$ at $x = 50\mu$. The $(E^2 + H^2)$ values show an increase by a factor of 31.2 at time 5×10^{-15} sec. At 10^{-13} sec the again-constant laser intensity of 2×10^{16} W/cm^2 was reached, with much less attenuation due to dynamic absorption at 2×10^{-13} sec. A swelling of $\theta \approx 400$ occurred.

An obvious idea is to start with a modified density profile such that a change of the laser intensity I near cut-off results in a change of the actual nonlinear refractive index n, by virtue of the energy of the electron oscillation $\epsilon_{os}(I)$ determining the electron temperature. In Fig. 2 at 1.3 microns the absolute value of the refractive index is determined only by its real part (given by the actual density), while the imaginary part (determined by the temperature) is too small. Therefore, we calculated a case with the same overdense density profile and a density varying linearly first from $x = 0$ to 10 microns from cut-off to 10^{-6} less cut-off, then from there to $x = 20$ microns to 10^{-2} less cut-off, and last linearly at $x = 50$ microns to zero cut-off. Figure 4 shows the $(E^2 + H^2)/8\pi$ values for an initial time $(0.5 \times 10^{-14}$ sec) at a laser intensity of 10^{15} W/cm^2 and at subsequent times of 10^{-13} sec and

2×10^{-13} sec. The swelling factor at the beginning was

$$v = |\tilde{n}|^{-1} = 31.2 \qquad (45)$$

due to the dielectric properties of this density profile. At 10^{-13} sec, the little depression of the curve at $x = 20\mu$ indicates some tendency to decrease due to density rippling. The maxima and minima of $E^2 + H^2$, however, change locally very quickly, and at 2×10^{-13} sec, we find a swelling of $\theta = |n|^{-1} = 400$. The dynamic change of the density near $x = 20\mu$ is shown in Fig. 5. A process resembling tunneling of the electromagnetic field through the overdense plasma near x = 20 microns at 2×10^{-13} sec is very strong, by virtue of the depth of the overdense plasma ($\approx$ micron) being small compared with the actual effective wavelength of more than 20 microns. The generation of a density minimum is well pronounced, but it is of less magnitude than in the case of Shearer et. al. [68] or Brueckner [90], as the time of interaction is much shorter in the case of Figure 5.

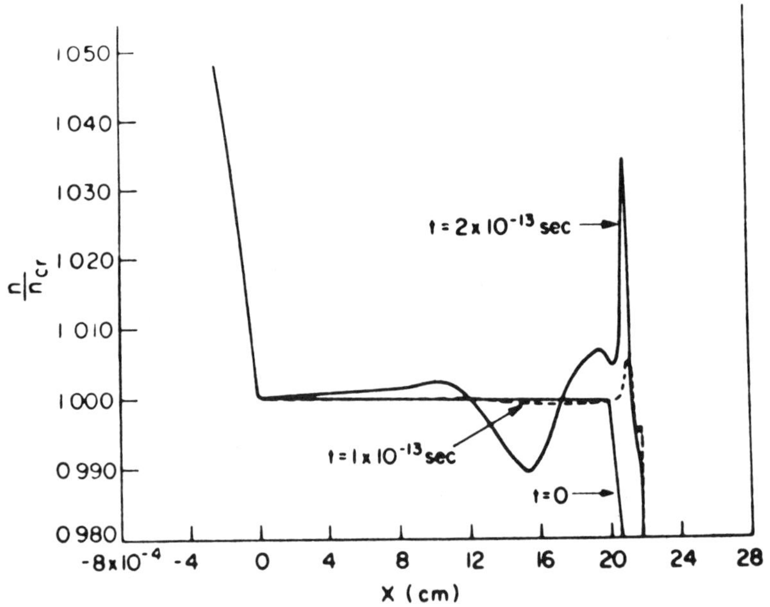

Figure 5
Initial and subsequent density profiles indicating inward motion of plasma toward the interior for x less than 14 microns.

The resulting velocity profile at 2×10^{-13} sec is shown in Fig. 6. The compression of plasma between x = 2 microns and x = 14 microns with velocities of 4×10^{7} cm/sec does not change much within subsequent time intervals, while the velocities at x = 50 microns, for example, are still changing. The increased ion temperature from collisionless processes at x = 40 microns is again much larger than the electron temperature, as could be expected from macroscopic dynamic ion absorption. A curious effect occurs in the interval from 20 to 28 microns, where the electrons are much hotter tan the ions and display an irregular oscillation. It can be assumed that a longitudinal acceleration of the electrons due to the nonlinear force occurs when the electrons have less interaction with the ions, comparable to hot electrons in solids at high electric fields. This phenomenon of dynamic electron absorption may be called "collisionless electron heating by nonlinear-force-induced macroscopic Raman decay".

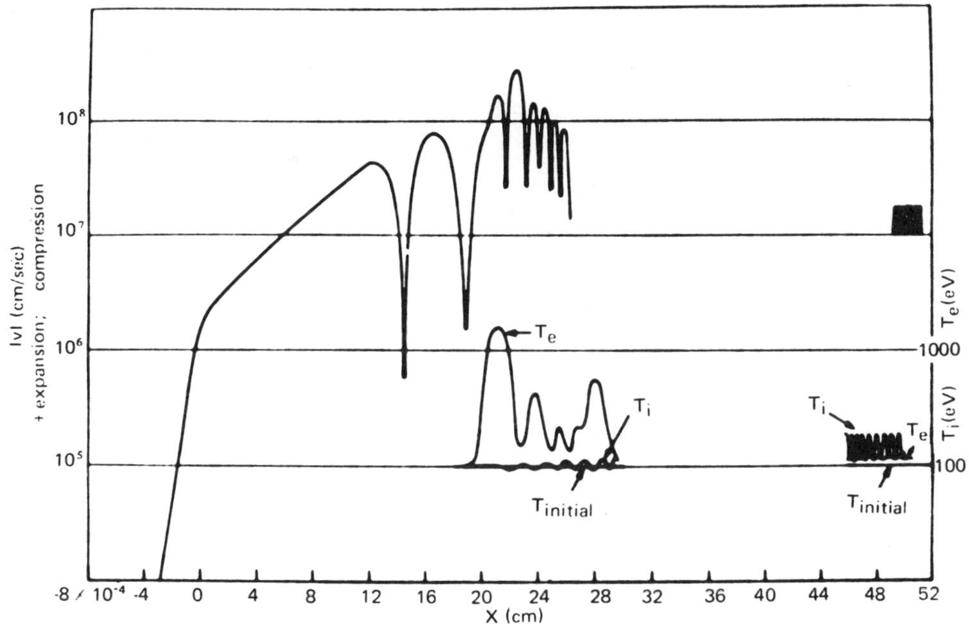

Figure 6
Dynamic heating of ions (at x = 48 microns) and of electrons (x = 20 microns) occurring at time $t = 2 \times 10^{-13}$ sec. The resulting plasma velocity v shows a compressing motion of the whole plasma between -4 and 14 microns with speeds up to 5×10^7 cm/sec, followed by expansion and alternating compression and expansion due to the standing-wave field.

The effects observed are highly sensitive to changes in the intensity I and the density profile. For an optimum case with $I = 4 \times 10^{16}$ W/cm^2 and the identical time and density profiles, the incident laser energy per cm^2 is 4.1 kJ up to 2×10^{-13} sec. The compression front between $x = -2\mu$ and $x = 14\mu$ absorbs 0.96 kJ in kinetic energy, the ablating plasma takes 0.68 kJ in kinetic energy of net motion and 2.2 kJ for dynamic heating. The remainder of 0.16 kJ goes into reflection and collisional heating of electrons.

The dynamic description of a plasma by this numerical model for neodymium-glass laser intensities exceeding 10^{16} W/cm^2 indicates that

(a) A strong dielectric increase of the laser field and the effective wavelength occurs, described by a swelling factor $|n|^{-1}$ of up to 400 (n is a parameter resulting from rigorous solution of the Maxwell equations and is identical with the complex refractive index in the WKB approximation);

(b) For further low-entropy compression, the nonlinear force accelerated thick (and not much heated = "cold") plasma fronts (15μ thickness) up to velocities of 6×10^7 cm/sec. The mechanical energy in this compression layer contains 23% of the incident laser radiation. (20% went into the outer plasma dynamically and 52% into reflection and some heating).

(c) A rippling of the density occurs in the whole plasma at densities below cut-off with nonlinear-force-generated velocities of 10^7 cm/sec or more. However, rippling can be suppressed by selective profiling of the plasma density and laser intensity with time;

(d) A preferential transfer of laser energy to heat ions by a collisionless compression (as in a theta pinch) occurs as a macroscopic Brillouin decay

(e) A preferential transfer of laser energy to electrons can occur at densities near cut-off (macroscopic Raman decay).

V EXPERIMENTAL VERIFICATION OF THE NONLINEAR FORCE

In the early stages of looking for nonlinear force mechanisms in laser produced plasmas, the most straight forward observation was the self-focusing of laser light in plasma at gas breakdown [22,91] as well as self-focusing in solid targets [92], where the agreement with the theory [23] was implicit in the presumed equilibrium of the nonlinear force with the gasdynamics at filamentation. The details of the dynamics of the self-focusing, were, e.g., demonstrated by Hohla [93] at irradiation of solid hydrogen, observing a development of the filament into the forward or in the backward direction which could be explained as energy deposition following the Basov-Krokhin [1] or Dawson [54] model [94] as justified by a quantitative agreement with the transmission times of the laser light [95]. These results indicated that, in the neodymium glass experiments with intensities of 10^9 to 10^{13} W/cm^2, self-focusing may be the reason for the complicated results including the initially mentioned superlinear fast ions, the superlinear momentum transfer, the high electron emission and the 10 keV - ions, being dependent on linearly by the ion charge [96] (see Eq. (44) while gasdynamics cannot result in such dependence because of internal plasma friction [66]).

The most detailed direct measurement of the nonlinear force is that by Wong and Stenzel [20] using microwaves. As described in Fig. 7, this experiment was very similar to that in the numerical cases of the preceding section. The interaction time was much less than the collision time, therefore, no remarkable heating could occur. The whole interaction was due to a dielectric swelling of $E^2 + H^2$ and the subsequent ablative explosion of the corona and the compression of the interior. The resonance-like swelling has been described in an early paper [18,89] and the whole mechanism of swelling and dynamic absorption of the radiation has been called in another connection "resonance absorption" and is at least in the case of Wong and Stenzel the mentioned dielectric swelling and dielectric explosion by the nonlinear force.

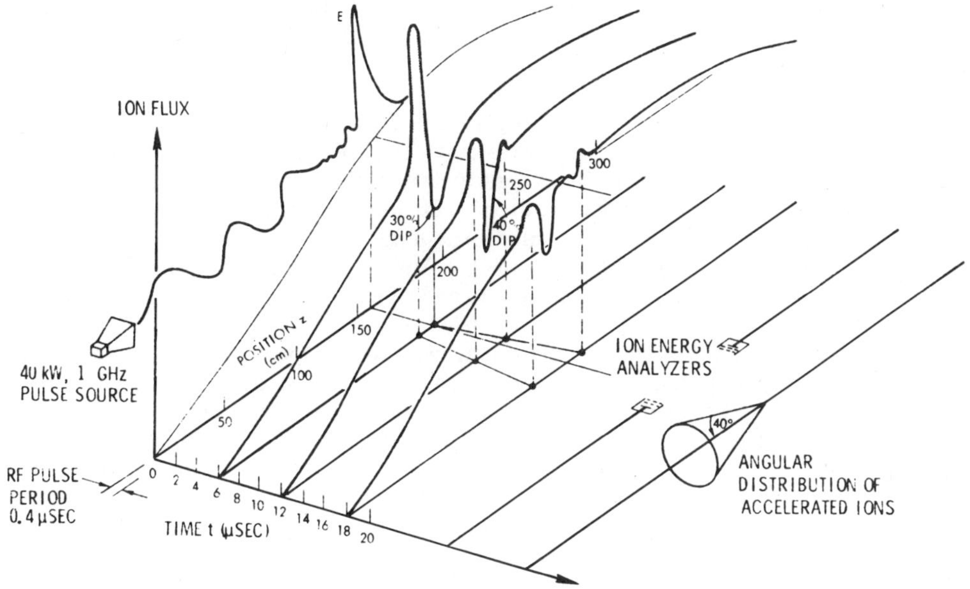

Figure 7
Experimental results of Wong and Stenzel [20], irradiating a 4 m long plasma with monotonically increasing density by a 1 μsec microwave pulse of an intensity beyond the corresponding limit of an equation [19]. The spatial measurement of E showed a swelling by a factor 700. The density profiles, measured at several time steps showed the generation of the density minimum near the critical density and the generation of a fast moving block of "cold" plasma. The initial plasma temperature of 2 eV was increased to less than 8 eV, while the moving blocks contained ions with 7 keV energy.

There is no uniform terminology in using the nonlinear force. This expression was used unlimitedly by some authors [26,66,67,72,73,74] while it is also called "nonlinear radiation force" [79,80], "v × B-force" [97], "electrostrictive force" [90], "ponderomotive force" [97], etc., which always describes some special property. It has been shown in Section II and in the preceding references that it is always the same phenomenon. The process where one highly swelled $E^2 + H^2$-value occurred, spread over some wide distance, and, driving the plasma as thick blocks, this all is the process known from the solitary waves (single strongwave mode). The aspect of a soliton is therefore relevant and the dynamics can be studied including all the knowledge of solitons, see e.g. Yu, Spatschek and Shukla [73].

The further experimental test of the action of the nonlinear force is the observation of blocks of 60keV-ions, taking the main part of the laser energy after spherical irradiation of pellets [19]. From their x-ray pinhole diagnostics it can be concluded that no self-focusing and fast-ion generation as at lower intensities can occur. Therefore, the pellets show the explosion of the whole corona as a block similar to the Wong-Stenzel experiment and similar to the numerical results of Lawrence et. al. [98].

In the case of self-focusing at Nd glass laser intensities exceeding 10^{15} W/cm^2, relativistic self-focusing [34-36] causes a fast shrinking of the beam to one wavelength diameter. The intensities are close to the relativistic threshold I_{rel} or above. This can be seen from x-ray emission of 100 keV or more (from the oscillating electrons) and the transfer of this oscillation energy of 2-times ionized ions of up to 6 MeV [30] or just protons [32,33] of 1.5 MeV. The acceleration of these ions then follows the mechanism described at the end of section III.

Finally, a more abstract experiment should be mentioned. Biskamp and Welter [99] have simulated a plasma by a large number of single electrons and ions in a computer code. An initially linearly increasing density is irradiated by a laser beam. At the critical density, or at one quarter of this value, the generation of a swelling of $E^2 + H^2$, the generation of a hole in the density and the motion of plasma away from this hole has been reproduced, as well as the generation of several reflected higher harmonics of the laser radiation. The advantage of this result is in an agreement of all the former results of the macroscopic theory with the microscopic simulation, which certainly represents a realistic description of the conversion of the laser energy into electron oscillation and the right build-up of the group velocity, which are questions that could not be ascertained by the macroscopic theory.

VI USE OF THE NONLINEAR FORCE FOR HIGHLY EFFICIENT COMPRESSION

For the concept of energy production from exothermal nuclear reactions in laser compressed plasmas, Nuckolls et. al. [37] and Brueckner [40] studied a low entropy (isentropic) compression caused by a thermal ablation of laser irradiated pellets. The ablation produces much entropy, and therefore, only low efficiencies are possible, e.g. a transfer of 5% of the laser energy into the compressed plasma [37, 40]. The question that these calculations may need a correction due to the delay in the thermalization processes, as pointed out recently [41], may cause even worse results.

In contrast to these results, the nonlinear force can also provide isentropy at the coupling, if nearly all optical energy is transfered into kinetic energy of fast moving thick and "cold" plasma shells. With this much more efficient compression it may be possible to work with clean nuclear reaction fuels, such as the hydrogen-boron reaction, which will be explained in this section.

1) The Nuclear Reactions.

The study of energy gain from the reactions of light nuclei goes back to the success of the acceleration techniques of Oliphant [100] from which one of the most important clean reactions used as advanced fuel was discovered [101].

$$H + {}^{11}B = 3\,{}^{4}He + 8.9\,MeV \tag{46a}$$

which preceded the discovery of the reactions of deuterium to form tritium T [102, 103]

$$2D \begin{array}{l} \nearrow \ {}^{3}He + n + 3.27\ MeV \\ \searrow \ T + p + 4.04\ MeV \end{array} \tag{46b}$$

and the reaction

$$D + T \rightarrow {}^{4}He + n + 17.8\,MeV. \tag{46c}$$

The last reaction is the basis of the very large world-wide efforts to reach conditions for a thermonuclear reactor. This reaction, however, is not considered as being a clean reaction because it produces neutrons which damage the reactor and the apparatus by thermalization of its energy; it further needs the breeding of the radioactive tritium T from ^{6}Li for which, in long terms, a shortage cannot be excluded [104].

The desire of the clean reaction (46a) or other reactions leading immediately to Helium-four, which can be released to the air, is therefore very high. The further advantage of this reaction is to avoid the neutrons and their strong damage of the reactor material (the estimated neutron-damage for D-T reactors may be a factor 10^4 higher than expected a few years ago [105], also the possibility of converting the nuclear reaction energy of the charged helium atoms (alphas) by electromagnetic means immediately into electrical energy avoids the serious environmental problems of heat pollution. The alphas may produce about 10 times the radioactivity [106] of the DT-reactor which can be reduced drastically by the electromagnetic decelerating conversion of the energy of the alphas into electricity too.

The difficulties of the clean reactions are their low reaction cross section compared to the DT-reaction. This is the reason why the immense efforts on thermonuclear energy production were based on the DT-reaction only. The success of the reactor concept with magnetic confinement (tokamak) has been quite remarkable during the last few years, however, this concept can be used for DT only, for DD or for $H^{11}B$ magnetic confinement as in the tokamak will produce more losses by cyclotron radiation [107] than nuclear energy can be gained. One way out may be a $\beta > 1$ magnetic confinement where the magnetic pressure works only in an outer layer and confines a field-free interior plasma [108]. The confinement densities, times and temperatures reached are however, on such a level that this concept may be only of academic interest at present.

The performance of nuclear reactions by gasdynamic laser compression is based on the irradiation of laser pulses of 10 nsec duration where a slow increase of the laser power requires 50% of the laser energy to be irradiated within the last 100 psec spherically symmetric to a pellet containing D and T. The coupling of the laser energy consists of a thermalization within an outer corona from where the energy is transferred by electrons to the gasdynamically generated compressed inner shell. This is heated by the electrons so that a further ablation of the outer part of the compressed shell drives its inner part to reach higher and higher compression. Computer simulations starting with pellets of solid state desity reach 10^4 times higher densities in the compressed core. Using laser pulses of 200 kJ energy result in DT plasma in overall gains in the order of 40 [37] or more, which seems to be realistic for the concept of a power reactor. The ideally spherical compression of pellets up to 100 times of its initial density (gas of 100 atmospheres or solids) has been measured as well as the generation of 10^7 to 10^8 fusion neutrons per incident laser pulse of up to 1 kJ energy [109]. The measurement of the ideal spherical compression holds as long as the neodymium glass laser intensity does not exceed 10^{14} - 10^{15} W/cm^2, one intensity which must not be exceeded for the reason that the collision times for the thermalization have to be less than the main pulse length of 100 psec, otherwise no thermalizing coupling can be expected [41]. The difficulties at 10^{14} W/cm^2 are also indicated experimentally: at these and higher intensities, the main part of the laser energy is transferred into very fast expanding plasma with ion energies of 50 keV and more, as first observed by Yamanaka [110] resulting in ion energies up to some MeV [30-33].

2) Inertial Confinement and Adiabatic Compression.

Laser compression plasmas follow the properties of inertial confinement for which the criteria for nuclear reaction gains have to be derived from the dynamic behaviour of such plasma. This is different from the Lawson criterion $n\tau > 10^{14}$ where n is the ion density in cm^{-3} and τ the confinement time (seconds) for DT and temperatures exceeding 7 keV are necessary, which criterion is specific for stationary confinement of plasma. A criterion for nuclear reaction gains for inertial confinement is known since 1964 [111] for DT which was extended [11] approximately for $H^{11}B$ and of which some details on exact calculations will be reported here [112]. The necessity for a different criterion for inertial confinement from Lawson's criterion was expressed also by Kidder [113] and conditions derived which agree with the author's earlier criterion [111].

Inertial confinement is the property of a hot plasma that it can undergo nuclear reactions without any external pressure but only by the fact that its adiabatic expansion is retarded by its inertia. One singular case is that of a plasma of the following initial conditions: spatially constant temperature, a Gaussian profile of the radial ion n_i density and a linear radial velocity profile. As was shown from the hydrodynamic equations by Jacquinot [114] et. al. and from the microscopic theory [115], the following behaviour of the plasma is of a self-similar type: it conserves its shape as a Gaussian density profile

$$n_i = n_o(t) \exp(-r^2/r_o(t)^2) \tag{47a}$$

and its linear velocity profile

$$v = W(t)r \tag{47b}$$

where the amplitude $n_o(t)$ is fixed by conservation of the number of ions N

$$N = \int_{-\infty}^{+\infty} n_i d\tau = \text{const} \tag{48}$$

and $r_o(t)$ and $W(t)$ is given by the adiabatic property of the dynamics, as has been derived [115] from the temporal behaviour of the temperature T(t)

$$T(t) = T_o (r_{oo}/r_o(t))^2. \tag{49}$$

W(t) depends on the initial conditions of the plasma. If neither energy is transferred to the plasma nor generated in the plasma, a negative initial value W corresponds to a compression until the kinetic energy of the plasma is transferred into thermal energy (Eq. (49)) when W = c and changes to positive values describing the expansion of the plasma. If the plasma without energy input or generation starts from different initial conditions with respect to density velocity or temperature profiles, compared to the ideal adiabatic case described, the maximum compression reached is less than that of the ideal case and the temperature is higher e.g. due to shock waves [116]. It is remarkable that numerical examples indicate the temporal development of the plasma towards the ideal adiabatic case [117]. If energy is transferred to the plasma, e.g. by laser radiation, and is generated by nuclear reactions or lost by emission of radiation, the dynamical behaviour of the plasma is very complicated and must be described by more or less general gasdynamic codes [66,67,118].

One self-similarity model used in the field of laser plasmas for the first time by Basov and Krokhin [1] and by Dawson [94] was based on a box-like density distribution which has an approximate equivalence to the ideal adiabatic case as could be demonstrated generally from the hydrodynamic equations [95]. All these considerations justify using a code based on the Dawson model to calculate a general relation for the nuclear fusion efficiency without the alpha particle or proton reheat in inertially confined plasmas, after heating the spherical plasma of an initial volume V_o and initial ion density n_o within a sufficiently short time (nanoseconds or less) by an input energy E_o to a temperature T_o. The reaction gain

$$a = \frac{\text{generated nuclear energy}}{\text{input energy } E_o} \tag{50}$$

showed maximum values at such selected E_o values for given V_o, where optimum temperatures T_o^{opt}

$$kT_o^{opt} = \begin{cases} 10.3 \text{ keV for Dt} \\ 32.8 \text{ keV for DD} \end{cases} \tag{51}$$

were reached. The relation between the initial radius R_o ($4\pi R_o/3 = V_o$) of the irradiated pellet and the initial energy E_o is given by the optimized temperature T_o^{opt} by

$$E_o = 4\pi k T_o^{opt} n_o R_o{}^3 \bar{Z} \tag{52}$$

where $\bar{Z}$ is the averaged charge of the ions. The general relation for G derived from the numerical treatment was [11, 111]

$$G = \left(\frac{E_o}{E_{BE}}\right)^{1/3} \left(\frac{n_o}{n_s}\right)^{2/3} \tag{53}$$

where n_s is the solid state density of D_2 or DT and E_{BE} is te break-even energy

$$E_{BE} = \begin{cases} 1.6 \text{ MJ} & \text{for DT} \\ 1.7 \times 10^{10}\text{J} & \text{for DD} \end{cases} \tag{54}$$

For the clean reaction of HB we have calculated [119] the gains based on cross sections given by Weaver [106] and collected by Titterton [120]. The result is

$$T_o^{opt} = 89 \text{ keV} \quad \text{for } H^{11}B \tag{55}$$

$$E_{BE} = 3.1 \times 10^{11}\text{J for } H^{11}B. \tag{56}$$

There is only a little algebraic reformulation of our gain formula (10) to arrive at the formulation [113] intended to substitute the Lawson criterion. We only have to substitute E_o from Eq. (52) into Eq. (53) to reach

$$G = \frac{(4\pi Z)^{1/3}}{E_{BE}^{1/3} n_s^{2/3}} n_o R_o. \tag{57}$$

Using the values of our numerical results we arrive at for DT

$$G = 1.3 \times 10^{-22} n_o R_o = 31.2 \rho_o R_o \tag{58}$$

where $[n_o] = \text{cm}^{-3}$; $[R_o] = \text{cm}$; $[\rho_o] = \text{g cm}^{-3}$. Kidder [113] arrived at $H = \rho_o R_o = 3\text{g/cm}^2$ to reach sufficient yields for gasdynamic compression of Nuckolls [37] which is from Eq. (58)

$$G = 93.6 \tag{59}$$

and agrees quite well with Nuckolls [37] postulate of needing G = 80 to 100, though the derivations of Kidder and in our case were completely different and independent. The advantage of the formulation (53) consists of the fact that it immediately shows the quadratic lowering of the input energy E_o if the initial density n_o is increased (by compression).

From our results we arrive at

$$G = 7.8 \times 10^{-24} n_o R_o = 2.34\ \rho_o R_o \tag{60}$$

for DD and

$$G = 3.51 \times 10^{-24} n_o R_o = 0.54\ \rho_o r_o \tag{61}$$

for HB.

The range of validity of Eq. (53), and therefore of Eqs. (58), (60) and (61) has been confirmed numerically [111] for $10^{-1}\,J < E_o < 10^{13}\,J$, $10^{-9}\,cm^{-3} < V_o < 10^{3}$ cm^{3} and a restriction has to be observed

$$G < G_o = \begin{cases} 300 \text{ for DT} \\ 30 \text{ for DD} \\ 30 \text{ for } H^{11}B \end{cases} \tag{62}$$

For higher gains, deviations from the formulas (53), (58), (60) and (61) are due to exhaustion of burning material. The gains G are then further and further below that of Eq. (53) depending on E_o. The question of losses of the ions of the fast tail of the Maxwellian distribution has been evaluated numerically [115] and was found not to be serious for the Dawson model and to be irrelevant for the ideal adiabatic case. The inclusion of the reheat due to helium nuclei or protons from the reactions does not lead to easily describable formulations. A further correction is necessary for the mean free path of the fast nuclei in plasma compared to the available formulations of the plasma theory, as could be shown from Bagges correction of the Bethe-Bloch-formula for plasmas [121].

3) Nonlinear-Force Laser Compression of Plasma.

The concept of compression of plasma by the nonlinear force of the interaction of these lasers became of interest, when it was possible to demonstrate numerically [70] that 23% of the energy of a laser pulse could be converted into the kinetic energy of a thick plasma layer, as described in Section IV. The aim of these calculations, in progress, is to arrive at a fast and "cold" imploding thick plasma shell which has a linear velocity profile on collapse to arrive at the conditions of the ideal adiabatic compression described in Section VI (2).

Here we are going to describe another tutorial example of a 50% transfer of laser energy into a thick layer of compressed plasma using the exceptional property of a Rayleigh density profile in dependence in the radius r

$$n_e = \begin{cases} 1, & R > R_o + do/2 \\ n_{ec}\{1-1/[1+\alpha(R_o + d_o/2-R)]^2\}, & R_o < R < R_o + d_o/2 \\ n_{ec}\{1-1/[1-\alpha(R_o - d_o/2-R)]^2\}, & R_o > R > R_o - d_o/2 \\ 1, & R_o - d_o/2 > R > 0 \end{cases} \tag{63}$$

where the cut-off density $n_{ec} = 10^{21}\,cm^{-3}$ for neodymium glass laser radiation and α was selected to be $10^4\,cm^{-1}$. The absorption due to collisions can be neglected. The averaged readius of the hollow sphere was chosen $r_o = 1$ cm. If radiation of an intensity $I = 2 \times 10^{16}$ W/cm² is incident, the outer half and the inner half are accelerated at each part with the same acceleration because the refractive has the profile (with a variable x) of the Rayleigh case

$$\tilde{n} = 1/(1 + \alpha x) \tag{64}$$

and the nonlinear force from Eq. (16) is

$$f_{NL} = \frac{E_v^2}{16\pi} \frac{\omega_p^2}{\omega^2} \alpha \tag{65}$$

with a constant acceleration

$$a = \frac{E_v^2}{16\pi} \frac{Z}{n_{ec}} \alpha \tag{66}$$

If the irradiated plasma is DT, the ions reach an averaged velocity of explosion (outer half of the hollow sphere) or implosion (inner half) corresponding to 10 keV ion energy after the time of 3.8×10^{-12} sec. During this time, the centre of the hollow sphere gets a minimum density of less than 3μ thickness. The time for Coulomb collisions of the electrons exceeds 80 psec, so that the plasma has no strong change of temperature and after the laser interaction one can describe the collapsing and compression separately. This example is of importance only for demonstrating the main properties of much more detailed numerical cases. One should note that the example described neglected the absorption of the kinetic energy transfer.

One aim for general codes is to generate an imploding Gaussian density profile with a linear velocity profile for implosion (see Section VI (2)) to reach ideal adiabatic compression, or at least a minimum of deviations from these conditions.

The other part of the calculation, the compression of the collapsed shell without laser radiation has been calculated by a detailed hydrodynamic spherical code by E. Goldman [122]. Starting from an initially box-like density and a constant velocity profile, one example of a DT plasma with an input energy of 200 Joules generated of nuclear energy. This, however, was about 30 times less than the ideal case without the shocks. A similar defect in the gains of 20 to 50 times less than the ideal gain resulted for cases with energy input of 100 J to 500 J. These calculations were always without the reheat by the reaction products. If the reheat was used, then gains were reached close to the ideal values or even higher by factors of up to 3.

It is therefore justified to calculate gains from Eqs. (11) to (14) for cases where the initially hollow spheres of DD, DT or $H^{11}B$ with centre densities close to n_{ec} are used of initial temperatures of few eV or less [123]. The result is that a total reaction gain of 80 can be reached if laser pulses with an energy of

$$E_{laser} = \begin{cases} 328 \text{ J for DT} \\ 131 \text{ kJ for DD} \\ 1.6 \times 10^6 \text{J for } H^{11}B \end{cases} \tag{67}$$

are used. Because the correction for reheat is more severe for the large pellets of $H^{11}B$ in the cases of Eq. (67), it is to be expected that the $H^{11}B$ case needs much less laser energy than that of Eq. (67) for same gains. The gain of 80 here is only of academic interest to arrive at a comparison with DT. For realistic cases one should work with a gain of 30 for HB, which would be reached by 84 kJ under the above assumptions.

The details of the cases of the nonlinear force laser compression must indeed be filled out by extensive work. The present state may not be more than simplified numerical results for the gasdynamic case, however, the difference is that in the nonlinear case each step described here is understandable and the whole concept convincing.

The comparison with the gasdynamic laser compression, as described in the Introduction, shows a nearly 1000 times less laser energy necessary for same reaction gains for DT. This can be understood very easily from Eq. (53), where the gain is given by the third root of the energy. The nonlinear forces were coupling 50% of the laser energy compared to 5% of gasdynamics, a factor of 10 which easily results in a factor 1000 less energy for the same gain. The energy E_{laser} cannot simply be compared, without taking into account that gasdynamic compression with some 100 kJ pulses have pulse lengths of τ = 100 psec, while the pulse in the example was of 3.8 psec length. For pulses of the 100 kJ region, the nonlinear force can work with pulses of 10 to 20 psec, so that the improvement given by the value of E_{laser}/τ is better only by a factor 100 to 200 in favour of the nonlinear force laser compression.

VII CONCLUSIONS

The derivation of the nonlinear force of the direct (nonthermalizing) laser-plasma interaction is based on several models (Landau-Lifshitz; two-fluid; quivering motion) which have to be complemented to arrive at the same result. While several general properties of the forces have been established as cases of oblique incidence, self-focusing, photon momentum, net transfer of electron oscillation into ion translation, analysis of parametric instabilities, etc., several questions are open, as the polarization dependence of electron acceleration in a low density discharge (Hollis), where the quivering model only arrives at a sufficient result.

The numerical results are characterized by the agreement in showing the predominance of the nonlinear force over the thermokinetic force according to the analytic theory. The fully dynamic calculation showed the generation of a density minimum (Shearer, Kidder and Zink, 1970) which was reproduced later by other codes. A further result was the low entropy (highly efficient) transfer of optical energy into kinetic energy of thick layers of plasma without a serious rise of the temperature.

Experimentally, the nonlinear force was evident a long time by self-focusing in plasma and complicated target interaction with (hidden) self-focusing. The microwave experiments of Wong and Stenzel gave an immediate confirmation of all theoretical expectations (density minima, strong swelling of $E^2 + H^2$, generation of fast, thick and cold plasma). The transfer of laser energy into fast ions at (non-self-focused) spherical pellet irradiation is further evidence of the nonlinear force as well as the measured MeV ions where the relativistic self-focusing effect (fast shrinking of the beam to a diameter of one wave length) has to be involved.

The consequence for laser compression of plasmas for exothermal nuclear reactions is the possibility of a high efficient coupling of the laser radiation. An example of 50% coupling (Rayleigh profile) arrives at the need of 1000 times less laser energy for some fusion gains compared with the usual gasdynamic compression. As the laser pulses have to be about 20 times shorter, the advantage are (approximately) 50 times smaller lasers for the same efficiency than in the gasdynamic scheme. This may permit the ideal reaction of hydrogen and boron (11).

REFERENCES

[1] N. G. Basov and O. N. Krokhin, Sov. Phys. JETP, 19, 123 (1964); [2] J. M. Dawson, Phys. Fluids 7, 981 (1964); [3] A. G. Engelhardt, Bull. Am. Phys. Soc. 9, 305 (1964); [4] H. Hora, Inst. Plasmaphys. Garching, Rept. 6/23 (1964); Laser Interaction and Related Plasma Phenomena, H. Schwarz and H. Hora, eds. (Plenum, New York (1971) Vol. 1, p.427; [5] W. I. Linlor, Appl. Phys. Letters, 3, 210 (1963); [6] N. R. Isenor, Appl. Phys. Letters 4, 152 (1964); [7] S. Namba, H. Schwarz, Proc. IEEE Symp. El. Ion and Laser Beam Tech., Berkeley, May 1967, p.861; [8] R. H. Honig, Appl. Phys. Letters, 3, 8 (1963); J. F. Ready, J. Appl. Phys. 36, 462 (1965); [9] G. Siller, K. Buchl and H. Hora, Laser Interaction and Related Plasma Phenomena, H. Schwarz and H. Hora, Eds. (Plenum, New York (1972) Vol.2, p.253; [10] S. A. Metz, Appl. Phys. Lett. 22, 211 (1973); H. Hora, Appl. Phys. Lett. 23, 39 (1973); [11] H. Hora, Laser Plasma and Nuclear Energy (Plenum, New York, 1975); [12] A. G. Engelhardt, T. V. George, H. Hora, J. L. Pack, Phys. Fluids 13, 212 (1970); C. Yamanaka, T. Yamanaka, T. Sasaki, Y. Yoshida and M. Waki, Phys. Rev. A6, 2335 (1972); [13] N. G. Basov, O. N. Krokhin and G. V. Sklizkov, IEEE J. QE-4, 988 (1968); [14] K. Buchl, K. Eidmann, P. Mulser, H. Salzmann and R. Sigel, Laser Interaction and Related Plasma Phenomena, H. Schwarz and H. Hora, Eds. (Plenum, New York, 1972) Vol. 2, p.503; [15] H. Hora, D. Pfirsch and A. Schluter, Z. Naturforsch. 22A, 278 (1967); [16] H. Hora, Phys. Fluids 12, 182 (1969); [17] H. Hora, Laser Interaction and Related Plasma Phenomena, H. Schwarz and H. Hora, Eds. (Plenum, New York, 1971) Vol.1, p.383; [18] H. Hora, ibid 1972, Vol. 2, p.341; [19] R. A. Grandey, P. Hammerling and F. J. Mayer, Bull. Am. Phys. Soc. 20, 682 (1975); see D. Slater and F. J. Mayer, Laser Interaction and Related Plasma Phenomena, H. Schwarz and H. Hora, Eds. (Plenum, New York, 1977) Vol. 4, p. 603 [20] A. Y. Wong and R. L. Stenzel, Phys. Rev. Letters 34, 727 (1975); [21] R. L. Stenzel, Phys. Fluids 19, 865 (1976); [22] V. V. Korobkin and A. J. Alcock, Phys. Rev. Letters 21, 1433 (1968); [23] H. Hora, Z. Phys. 226, 156 (1969); [24] A. J.

Palmer, Phys. Fluids 14, 2714 (1971); [25] J. W. Shearer and J. L. Eddleman, Phys. Fluids 16, 1753 (1974); [26] F. F Chen, Laser Interaction and Related Plasma Phenomena, H. Schwarz and H. Hora, Eds. (Plenum, New York, 1974) Vol. 3A, p.291; [27] M. S. Sodha, A. K. Ghatak and V. K. Tripathi, Progress in Optics, E. Wolf, Ed. (Acad. Press, New York, 1976) Vol. 13, p.171; [28] M. S. Sodha, D. P. Tewari, A. Kumar and V. K. Tripathi, Appl. Phys. 3, 141 (1974); 6, 119 (1975). [29] M. R. Siegrist, Opt. Comm. 16, 402 (1976); [30] M. R. Siegrist, B. Luther-Davies and J. L. Hughes, Opt. Comm. 18, 603 (1976); B. Luther-Davies and J. L. Hughes, Phys. Letters 58A, 399 (1976); [31] B. Luther-Davies, M. R. Siegrist, V. del Pizzo and J. L. Hughes, Laser Interaction and Related Plasma Phenomena, H. Schwarz and H. Hora, Eds. (Plenum, New York, 1977) Vol. 4, p. 639; [32] R. A. Haas, J. F. Holzrichter, H. G. Ahlstrom, E. Storm, and K. R. Manes, Opt. Comm. 18, 105 (1976); J. M. Boyle, Laser Interaction and Related Plasma Phenomena, H. Schwarz and H. Hora, Eds. (Plenum, New York, 1977) Vol. 4, p. 437; [33] Tai Ho Tan, et.al., Laser Interaction and Related Plasma Phenomena, H. Schwarz and H. Hora, Eds. (Plenum, New York, 1977) Vol. 4, p. 677; [34] H. Hora, Jour. Opt. Soc. Am. 65, 882 (1975). [35] H. Hora and E. L. Kane, Appl. Phys. 9, 221 (1977); [36] E. L. Kane and H. Hora, Laser Interaction and Related Plasma Phenomena, H. Schwarz and H. Hora, Eds. (Plenum, New York, 1977) Vol. 4, p. 913; Kvantovaja Elektronika (in print); [37] J. Nuckolls, L. Wood, A. Thiessen and G. Zimmermann, Nature 239, 139 (1972); [38] H. Hora, Ann. Physik, 22, 402 (1969); [39] S. Kaliski, Bull. W.A.T. (Poland) 5, No.2, 21 (1972); [40] K. A. Brueckner and S. Jorna, Rev. Mod. Phys. 46, 325 (1974); [41] H. Hora, Austral. J. Phys. 29, 375 (1976); [42] H. Hora, Atomkernenergie 24, 187 (1974); [43] H. Hora, Sov. J. Quantum Electronics 6, 154 (1976); [44] J. L. Hughes, Hemisphere 19, No.4, 28 (1975); [45] W. Weizel, Lehrbuch der Theoretischen Physik, (Springer, Berlin, 1949) Vol.1, p.350; [46] A. Schluter, Z. Naturforsch. 5A, 72 (1950); [47] H. Hora and H. Wilhelm, Nucl. Fusion 10, 111 (1960); [46] H. Hora, Opto-Electronics 2, 201 (1960); [47] S. Rand, Phys. Rev. 136B, 231 (1964); [48] T. P. Hughes and M. B. Nicholson-Florence, J. Phys. A2, 588 (1968); [49] L. D. Landau and E. M. Lifshitz, Electrodynamics of Continuous Media (Pergamon, Oxford, 1966) p.242; [50] H. Motz and C. J. H. Watson, Advances in Electronics and Electron Physics, D. B. Morton, Ed. (Acad. Press, New York, 1967) Vol. 23, p.153; [51] S. Spitzer, jr. Physics of Fully Ionized Gases (Wiley-Interscience, New York, 1956); [52] L. C. Steinhauer and H. G. Ahlstrom, Phys. Fluids 13, 1103 (1970); [53] H. Hora, Phys. Fluids 17, 939 (1974); [54] A. Schluter, Plasma Physics, 10, 471 (1968); [55] R. Klima and V. A. Petrzilka, Cz. J. Phys. B22, 896 (1972); [56] V. A. Petrzilka, Cz. J. Phys. B26, 151 (1976); [57] R. L. Dewar, Dept. Theor. Phys. - Austr. Nat. Univ., Rept. 1975; [58] N. G. Basov, Contribution to the European Meeting on Laser-Plasma Interaction, Paris, Nov. 1970; [59] M. G. Burt and Sir. R. Peierls, Proc. Roy. Soc. A333, 149 (1973); [60] Sir R. Peierls, Proc. Roy. Soc. A347, 475 (1976); [61] H. Hora, Phys. Fluids 17, 1042 (1974); [62] A. Ashkin, Phys. Rev. Letters 24, 156 (1970); [63] J. Gordon, Phys. Rev. A8, 14 (1973); [64] F. N. H. Robinson, Phys. Letters C16, 313 (1975); [65] W. Riedmuller (private communication); [66] P. Mulser, Z. Naturforsch. 25A, 282 (1970); [67] R. G. Rehm, Phys. Fluids 13, 921 (1970); A. F. Haught and D.H. Polk, Phys. Fluids 13, 2825 (1970); [68] J. W. Shearer, R. E. Kidder and J. W. Zink, Bull. Am. Phys. Soc. 15, 1483 (1970); J. W. Shearer, UCIO-15745 Rept. Dec. 1970; W. L. Kruer, Comments Plasma Physics 2, 139 (1976); [69] B. J. Green and P. Mulser, Phys. Letters 12A, 328 (1971); [70] H. Hora, Proc. Second Europ. Electro-Opt. Conf. ; H. A. Elion, Ed. (Mack Brooks, St. Alban,

1975) p.396; [71] K. A. Brueckner, Laser Interaction and Related Plasma Phenomena, H. Schwarz and H. Hora, Eds. (Plenum, New York, 1977) Vol. 4. p. 891; [72] E. B. Goldman, W. Leising, A. Brauer, M. Lubin, J. Appl. Phys. 45, 1158 (1975); [73] K. Boyer, Astronautics and Aeronautics 11, No.1, 28 (1973); M. Y. Yu, K. H. Spatschek and P. K. Shukla, Z. Naturforsch. 29A, 1736 (1974); [74] F. Winterberg, Z. Naturforsch. 30A, 976 (1975); [75] G. Beaudry and J. Martineau, Phys. Lett. 43A, 331 (1973); [76] M. Hugenschmidt, K. Vollrath and A. Hirth, Appl. Optics 11, 339 (1972); [77] M. Sodha and R. S. Mittal, Opto-Electronics 6, 167 (1974); [78] B. Bath and V. K. Tripathi, J. Appl. Phys. 46, 1141 (1975); [79] J. A. Stamper and D. A. Tidman, Phys. Fluids 16, 2024 (1975); [80] J. A. Stamper and S. E. Bodner, Phys. Rev. Letters 37, 435 (1976); [81] F. Schwirzke, Laser Interaction and Related Plasma Phenomena, H. Schwarz and H. Hora, Eds. (Plenum, New York, 1974) Vol. 3A, p.213; [82] M. G. Drouet, et.al., ibid, 1977, Vol. 4, p. 737; [83] M. J. Hollis, (submitted for publication); [84] H. Hora, (submitted for publication); [85] E. Yablontovich, Laser Interaction and Related Plasma Phenomena, H. Schwarz and H. Hora, Eds. (Plenum, New York, 1977) Vol. 4 p. 367; [86] H. J. Schwarz and R. Tabensky, Laser Interaction and Related Plasma Phenomena, H. Schwarz and H. Hora, Eds. (Plenum, New York, 1977) Vol. 4. p. 961; [87] J. Lindl and P. Kaw, Phys. Fluids 14, 2714 (1971), [88] M. E. Marhic, Phys. Fluids 18, 837 (1975); [89] H. Hora, Opto-Electr. 2, 201 (1970); [90] K. A. Brueckner, Proc. Nobel Conf. on Nonlinear Processes, July 1976; [91] M. H. Key, D. A. Preston and T. P. Donaldson, J. Phys. B3, L88 (1970); [92] W. Engelhardt, Appl. Phys. Letters 15, 219 (1969); [93] K. Hohla (unpublished, 1969); [94] J. M. Dawson, Laser Interaction and Related Plasma Phenomena, H. Schwarz and H. Hora, Eds. (Plenum, New York, 1971) Vol. 1, p.355; [95] H. Hora, ibid, p.365; [96] F. E. Irons and N. J. Peacock, J. Phys. B7, 1109 (1974); [97] C. E. Max and F. Perkins, Phys. Rev. Letters 27, 1342 (1971); [98] V. F. Lawrence and H. Hora, Laser Interaction and Related Plasma Phenomena, H. Schwarz and H. Hora, Eds. (Plenum, New York, 1977) Vol. 4 p. 877; [99] D. Biskamp and H. Welter, Phys. Rev. Letters, 34, 312 (1975); [100] M.L.E. Oliphant, Proc. Roy. Soc. A124, 228 (1929); [101] M.L.E. Oliphant and Lord Rutherford, Proc. Roy. Soc. A141, 259 (1933); [102] M.L.E. Oliphant, P. Harteck and Lord Rutherford, Proc. Roy. Soc. A144, 692 (1934); [103] P. Harteck, Laser Interaction and Related Plasma Phenomena, H. Schwarz and H. Hora, Eds. (Plenum, New York, 1971) Vol. 1, p.1; [104] W. Hafele, Atomwirtschaft 17, 384 (1972); [105] R. Behrisch, R. Gabler and J. Kalus J. Nucl. Mat. 53, 183 (1974); [106] T. A. Weaver, Lawrence Livermore Laboratory, Laser Report 9 (No.12), 1 (1973); R. W. B. Best, FOM rept. 75/061 (April 1976), Nucl. Instr. Meth. (to be published); [107] B. A. Trubnikov and V. S. Kudryatsov, UN Geneva Conf. 1958, 32, p.93; [108] A. Y. Wong, J. L. Dawson and F. F. Chen (private communication); [109] N. G. Basov, Y. S. Ivanov, O. N. Krokhin, Y. A. Mikhailov, G. V. Sklizkov and S. I. Fedotov, JETP Lett. 15, 417 (1972); [110] C. Yamanaka, T. Yamanaka, T. Sasaki, Y. Yoshida and M. Waki, Phys. Rev. A6, 2335 (1972); [111] H. Hora and D. Pfirsch, 6th Int. Quantum Electr. Conf. Kyoto, 1970, Digest, p.2-2; see [4]; [112] H. Hora and P. S. Ray, Bull. APS, 21, 73 (1976); [113] R. E. Kidder, Nucl. Fusion 14, 797 (1974); [114] J. Jacquinot, C. Leloup and F. Waelbroek, Rapp. C.E.A., No.12, 2617 (1964); [115] H. Hora and D. Pfirsch, Laser Interaction and Related Plasma Phenomena, H. Schwarz and H. Hora, Eds. (Plenum, New York, 1972) Vol. 2, p.515; [116] H. Hora and W. Lochte-Holtgreven, Atomkernenergie, 25, 53 (1975). [117] W. Fader, Phys. Fluids 11, 2200 (1968); [118] E. Goldman, Plasma Phys. 15, 289 (1973); [119] P. S. Ray and H. Hora,

Laser Interaction and Related Plasma Phenomena, H. Schwarz and H. Hora, Eds. (Plenum, New York, 1977) Vol. 4, p. 1081; [120] Sir Ernest Titterton, (private communication); [121] P. S. Ray and H. Hora, Nuclear Fusion 16, 3 (1976); [122] H. Hora, E. Goldman and M. Lubin, Univ. Rochester, LLE-Report 21 (1974); [123] H. Hora, Sov. J. Quantum Electr. 6, 154 (1976).

NUMERICAL CALCULATIONS OF LASER INTERACTION WITH PLASMAS INCLUDING MOMENTUM TRANSFER OF THE NONLINEAR FORCES*)**)

V.F. Lawrence and H. Hora

Dept. of Theoretical Physics, The University of New South Wales, Kensington, NSW Australia

Abstract

In order to study the action of the nonlinear force of electrodynamic laser-plasma interaction without thermalization, numerical calculations with a very general two-fluid numerical code were performed including nonlinear variation of the optical constants (dependence of the collision frequency on the laser intensity) and detailed description of the reflection of laser radiation based on Maxwell's equations. For laser pulses of less than 1 psec, the dynamics are characterized by the nonlinear force mainly. Using Rayleigh-like density profiles, the generation of instabilities as well as their suppression has been demonstrated. The energy transferred to two thick blocks of plasma (one moving against the laser light, the other with it under compensation of momentum) increases nearly quadratically on the laser intensity.

---

*) Presented at the Fourth International Workshop Conference on "Laser Interaction and Related Plasma Phenomena" at Rensselaer Polytechnic Institute, Troy, New York, 8-12 November, 1976.

**) Supported by the ARGC grant No. B75/15538.

I. INTRODUCTION

Plasma dynamics for the interaction of high intensity laser radiation with solid targets and the instantaneously generated plasma, can be determined by the nonlinear force which is the immediate electro-dynamic interaction, and by the gasdynamic pressure occuring after thermalization of the radiation and heating of the ions. The basic properties of these nonlinear forces have been developed over the last ten years[1-7] and the expansion of plasma and the subsequent compression has been demonstrated numerically[7-9] and experimentally[10,11], as well as indirectly by self focussing and generation of fast ions or fast plasma groups and ion separation by the charge number Z.[7]

The importance of the action of the nonlinear force, especially for short laser pulses (for neodymium glass lasers around 1 to 10 psec and for CO_2 lasers around .1 to 1 nsec) consists of the possibility of a non thermalizing transfer of optical energy into kinetic energy of plasma for compression, which has a minimum of entropy production and is therefore highly efficient. Followed by an isentropic compression, the same fusion reaction yields can be reached with about 1000 times less laser energy[12] than in the case of the gasdynamic laser compression scheme of Nucholls[13]. As the laser pulses have to be about 20 times less than in the Nucholls case the total advantage is that the required laser system could be smaller by a factor of 50, when using non linear force compression, based on a rough calculation. The advantage of less complications (instabilities) at short interaction times and the use of, experimentally verified[4], fast blocks of ions (fast ions[7]) as well as avoidance of the necessary delay due to thermalization[14] (all disadvantages in the Nucholls scheme) may be reasons enough to orient efforts towards the nonlinear force compressions. The use of high intensity CO_2 lasers[15] will necessarily require such a scheme, as the thermalization delay has been neglected unawares in numerical calculation of gasdynamic compression models, otherwise it would require[14] laser pulses of 10 nsecs or more.

The non linear force compression has not been studied numerically to the same extent as the gasdynamic compression. As the necessary laser intensities have only become available over the last two years a very intensive study is necessary now. This paper presents results of a very general one-dimensional code, where we used initial electron densities $n_e(x)$ of the Rayleigh type

$$n_e(x) = n_{ec}\left(1 - \frac{1}{(1+\alpha x)^2}\right) \tag{1}$$

where $x>0$ and α is a constant. Equation (1) corresponds to a refractive index $\tilde{n}$ of

$$\tilde{n} = \frac{1}{1+\alpha x} \tag{2}$$

for which the laser penetrates with a minimum of reflectivity[16].

II. THE ONE-DIMENSIONAL CODE

The calculations were based on a modified numerical code of which the very general gasdynamic part had been developed by E. Goldman[17] and the nonlinear forces have been included in the following way. All quantities depend on the one spatial coordinate α only and laser light is incident from +x. Electrons and ions are separately treated (index i and e) where fully ionised preferably DT-plasma or LiD plasma is presumed. The equations of continuity for the densities $n_{e,i}$ or velocity $\mathbf{v}_{e,i}$ is

$$\frac{\partial n_{e,i}}{\partial t} + \underline{\nabla}\cdot(n_{e,i}\,\underline{\mathbf{v}}_{e,i}) = 0 \tag{3}$$

and the equation of motion (force density) contains the temperatures $T_{e,i}$ ($k\sim$Boltzmann constant) and the electric and magnetic field strengths $\underline{E}$ and $\underline{H}$ for linearly polarized laser radiation describing the terms of the nonlinear force

$$m_i\left(\frac{d}{dt} v_i + v_i \frac{d}{dx} v_i\right) = -\frac{d}{dx} n_i k T_i \tag{4}$$

$$m_e\left(\frac{d}{dt} v_e + \mathbf{v}_e \frac{d}{dt} v_e\right) = -\frac{d}{dx} n_e k T_e + \frac{d}{dx}(E^2+H^2)(8\pi) \tag{5}$$

The equation of energy conservation

$$\tfrac{1}{2}\frac{\partial}{\partial}\, n_{e,i} m_{e,i} v_{e,i}{}^2 = -kT_{e,i}\frac{d}{dt} n_{e,i} - n_{e,i} k \frac{d}{dt} T_{e,i} + W_{e,i} \tag{6}$$

$T_{e,i}$ contains the change of the temperature both by the gasdynamic (adiabatic) motion as well as by thermal conductivity, using the collision frequency

$$\nu = \frac{\omega_p^{\,2}\pi^{3/2} m_e^{\,1/2} e^2 \ln\Lambda}{8\pi(2kT_e)^{3/2}} \qquad (7)$$

with the plasma frequency ω_p, coulomb logarithm $\ln\Lambda$, mass m_e and charge of electron e. The power density of thermal energy input from the laser field at each x is given by W_i, where the exact nonlinear optical constant was used

$$\tilde{n}^2 = 1 - \frac{\omega_p^{\,2}}{\omega^2+\nu^2}\left(1 + i\frac{\nu^x}{\omega}\right) \qquad (8)$$

where ν^x was identical with ν of Eq. (7), however including the energy ε_{osc} of the electrons from the quivering oscillation in the laser field[7] by substituting T_e by $T = T_e+\varepsilon_{osc}/k$, W_i is determined by the heating of the ions by the difference T_e-T_i including the delay given by the collision frequency ν. The coupling of the equations of the ions and electrons is due to Poisson's equation. The six equations (3)to (6) determine the six variables n_e, n_i, v_e, v_i, T_e, T_i in dependence on x and the time t. Further, initial condition (variables at t = 0) and boundary conditions (time dependent indent laser field) have to be chosen. The laser field is being calculated for each time step numerically as a solution of the Maxwellian equations.

III. INITIAL AND BOUNDARY CONDITION

In the following calculation, laser pulses were chosen with a $\sin^2\beta t$ amplitude such that after realising the first maximum the laser intensity is constant. β was chosen to reach this level at a time t=0.15psec. The initial conditions were chosen as

$$T_{i,e}(x;t = 0) = 100 \text{ eV} \qquad (9)$$

$$\mathbf{v}_{e,i}(x;t = 0) = 0 \qquad (10)$$

and $n_{e,i}(x;t = 0) = n_o(x)$ has been selected from the experience of the Rayleigh case[7,12,16] for neodymium glass laser radiation with $[n_o] = \text{cm}^{-3}$

$$n_o(x) = \begin{cases} 0; \quad x \geq 50\mu \\ 10^{21}\{1 - \frac{1}{(1-\alpha[x-50\mu])^2}\}; \quad 0 \leq x \leq 50\mu \\ 10^{21}\{1 - \frac{1}{(1+\alpha[x+50\mu])^2}\}; \quad -50\mu \leq x \leq 0 \\ 10^{18}; \quad x \leq -50\mu \end{cases} \tag{11}$$

to avoid total reflection[16], α has to fulfil the following restriction

$$x < \alpha_o = \frac{2\omega}{c} = 1.7 \times 10^5 \text{ cm}^{-1} \tag{12}$$

for neodymium glass laser radiation. It has to be noted[16] that the reflections at $x = \pm 50\mu$ and $x = 0$ are remarkably small in the collisionless case ($\nu^x = 0$) if $\alpha < 0.5\alpha_o$. The case of collisions results in modifications which will be reproduced by the programme automatically.

IV. GENERATION OF INSTABILITIES

From the numerous cases calculated, we describe the one where the generation of strong reflection has been generated in an early stage. The standing wave pushes the plasma towards the modes and the generated rippling of the density is parametrically increasing the reflection. The case of Fig. 1 we find $(E^2+H^2)/8\pi$ (corresponding fairly to the intensity) with one local maximum at the time where the laser reaches the maximum intensity. It is remarkable that this maximum is not at $x = 0$, where the initial density maximum is, as the absorption and the intensity dependence of the optical constants modify the laser field. The maximum corresponds to an intensity $I = I_{vac}/|\tilde{n}|$ related to its vacuum value I_{vac} by a swelling factor $\theta = 1/|\tilde{n}| \approx 7$ only the reached plasma velocity up to this time is shown in Fig. 2. and the electron density in Fig. 3. The velocity

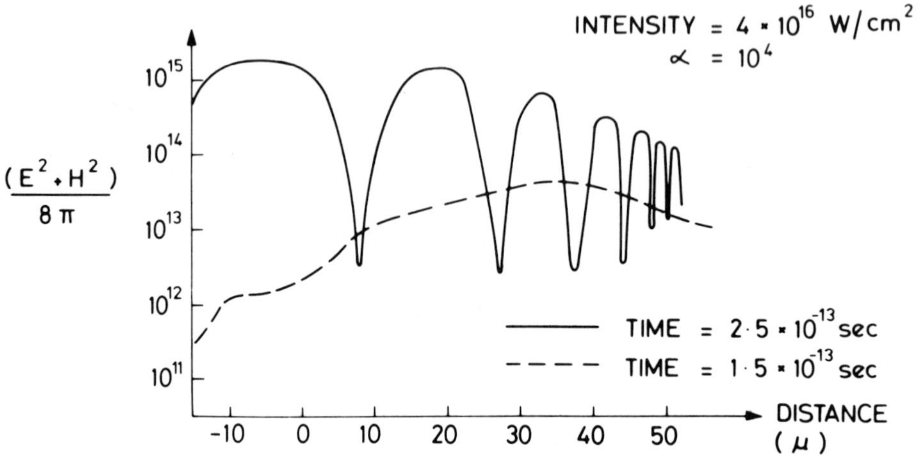

Fig. 1: Calculation of $(E^2+H^2)/8\pi$ for $\alpha=10^4$ at an intensity 4×10^{16}W/cm^2 at the times 1.5×10^{-13} sec and 2.5×10^{-13} sec.

profile at t = 0.15 psec is positive from x = 50μ down to 35μ which corresponds to an expansion (ablation) of this part of the plasma corona as a thick block with velocities up to 10^7 cm/sec, while the plasma below 35μ is moved as a whole block to the interior of the plasma (towards negative x). The density at that time (Fig. 3) shows very well a similar profile as the initial one, Eq. (11), especially the Rayleigh-like decay at 40 to 50μ. But it is remarkable, that the initial maximum at x = 0 has been moved to a value of 3.8μ which is due to some internal compression of the block moving towards -x, as the velocity profile is not constant. The generation of an instability can be seen at the time t = 0.45 psec, where the $(E^2+H^2)/8\pi$ (Fig. 1) is oscillating, corresponding to a standing wave. The velocity has then changed drastically into the oscillating profile of Fig. 2. The maxima and

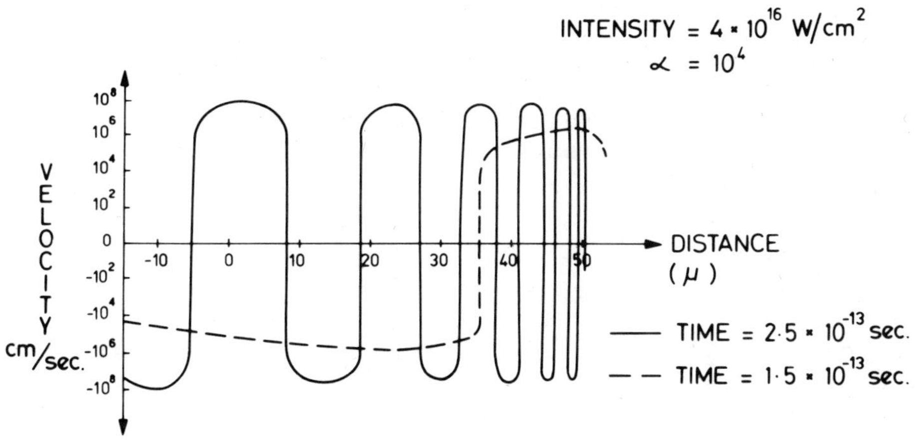

Fig. 2: Velocity profiles corresponding to the case of Fig. 1

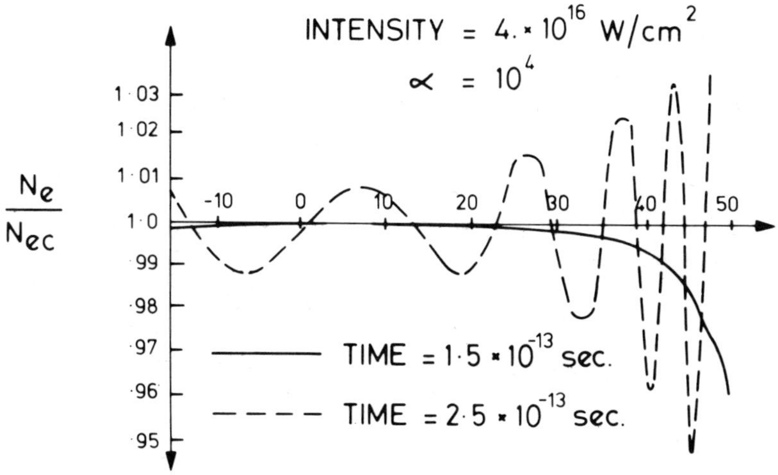

Fig. 3: Electron density (related to the cut off density) corresponding to the cases of Fig. 2.

minima correspond very close to the vacuum wave length $\lambda=\lambda_{vac}$ at x = 50μ, while a swelling $\lambda=\theta\lambda_{vac}$ of $\theta=9$ around x=0 can be seen. The density (Fig. 3) shows a ripple, easily understood by the motion due to Fig. 2. The following time steps show a very high reflectivity because of this Brillouin type dynamic instability[18].

V. ACCELERATION OF BLOCKS

Compared to the cases of Fig. 1 to 3, the situation is quite different, if an $\alpha=5\times10^4$ cm^{-1} and a maximum intensity of 10^{16}W/cm^2 is used (Fig. 4 and 5). It is

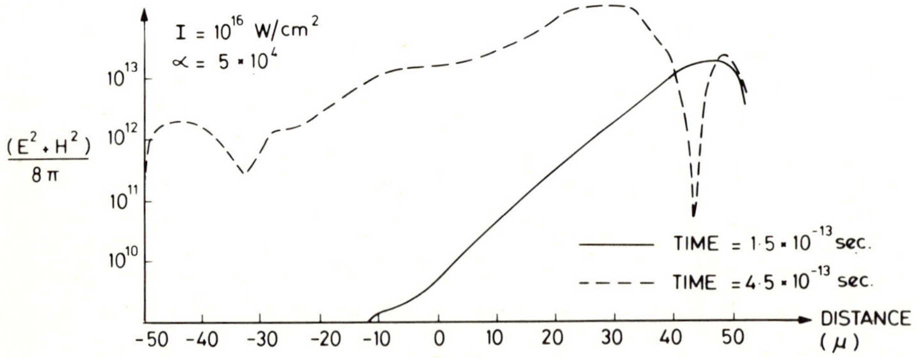

Fig. 4: Spatial profiles of $(E^2+H^2)/8\pi$ for a case with initial electron densities of Eq. (11) using $\alpha=5\times10^4$cm^{-1}, maximum laser intensity of 10^{16}W/cm^2 at the times 0.15 psec and 0.45 psec.

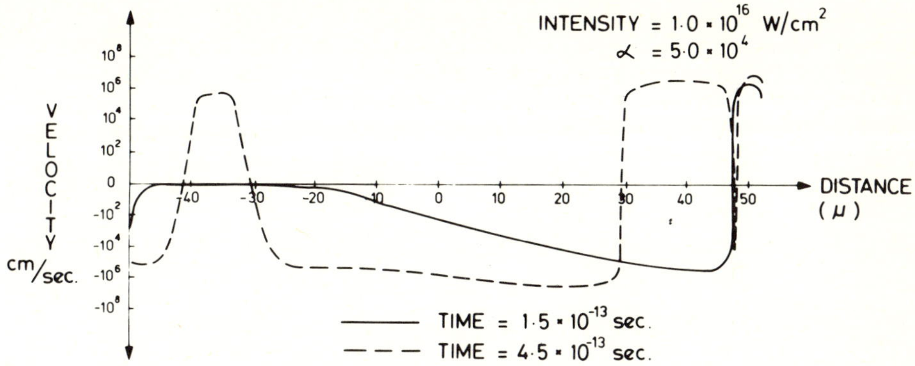

Fig. 5: Velocity profiles of the case of Fig. 4.

remarkable that at the early time 0.15 psec, the laser intensity has a maximum near 40μ (swelling 14) and then drops strongly towards $x = 0$. This is due to the fact that an $\alpha=5\times10^4$ cm^{-1} causes a density very close to the cut-off density, and the less intensity than the former case causes a less decrease of the collision frequency by the intensity dependent nonlinearity of ν^x in Eq.(8). Nevertheless, at $t = 0.45$ psec, the swelling of the intensity ($\theta=120$) at $x = 30\mu$ lets the light penetrate through the whole plasma without standing wave pattern (these did not appear even at later times) where the plasma was then (Fig. 5) moving in two blocks, from 30 to 50μ towards +x (ablation) and far less than 30μ towards -x (compression). The modification near -35μ is understandable from the minimum of $(E^2+H^2)/8\pi$ near $x = -35\mu$.

By a systematic search it was possible to find a series of cases where only a two block motion appeared for varying intensities and constant α with a separation of the blocks around the same depths x(±3μ).

Evaluation was made of the amount of kinetic energy E_k thathad been transferred by the nonlinear force which both the gasdynamic expansion and the thermokinetic acceleration had neglected. We have evaluated

$$E_k(I) = \frac{1}{x_2 - x_1} \int_{x_1}^{x_2} \frac{m_i n_i v_i^2}{2} \, d\xi \tag{13}$$

for dependence on the laser intensity (Fig. 6). The

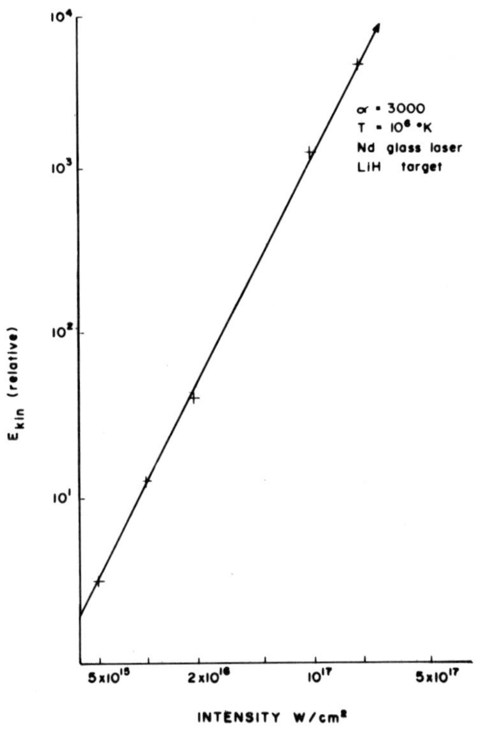

Fig. 6: Kinetic energy transferred to the plasma $E_k(I)$, Eq. (13) for $\alpha = 3\times10^3\,cm^{-1}$ up to a constant time t = 0.1 psec for independent temporal laser profiles.

result is that the transferred kinetic energy is increasing nearly quadratically against the neodymium glass laser intensity,

$$E_{kin} = \text{const} \times I^{1.8}$$
$$(5\times10^{15} < I < 10^{18}\ \text{W/cm}^2) \qquad (14)$$

expressing the nonlinear nature of the interaction. The energy transfer is a nonlinear, macroscopic, electrodynamic absorption process.

VI. CONCLUSION

The nonlinear force of laser plasma interaction showed various properties from numerical calculations for neodymium glass laser radiation of more than 10^{15} W/cm^2 intensity. The generation of strong reflection and that a standing wave could be generated from an initial Rayleigh density profile where nearly no reflection is expected for collisionless plasmas is unexpected. The reason for this behaviour is understandable from the intensity dependent thermal absorption process. Other cases show a nearly ideal transfer of the laser radiation into a block moving against the laser radiation and another in the same direction. The transferred energy was quadratically increasing against the laser light intensity. The momentum of each block was just compensating the momentum of the corresponding block moving in the opposite direction. Preliminary results for times of 5 psec were similar which confirms that the neglect of the retardation of the laser field in the numerical code is not important for the kinetic processes discussed here.

REFERENCES

1. H. Hora, D.Pfirsch and A. Schülter, Z. Naturforschung 22A, 278 (1967).

2. H. Hora, Phys. Fluids 12, 182 (1969); 17, 939 (1974); 17, 1042 (1974).

3. J. Lindl and P. Kaw, Phys. Fluids 14, 2714 (1971).

4. J.W. Shearer and J.L. Eddkeman, Phys. Fluids 16, 1753 (1974).

5. J.A. Stamper and S.E. Bodner, Phys. Rev. Letters 37, 435 (1970).

6. V.A. Klima and V.A. Petrzilka, Cz.J.Phys. B22, 896 (1972).

7. H. Hora, Laser Interaction and Related Plasma Phenomena, H. Schwarz and H. Hora eds. (Plenum, New York) 1971, Vol. 1 p. 383; 1972, Vol. 2, p. 341; 1977, Vol. 4, p. 841. Laser Plasmas and Nuclear Energy (Plenum, New York, 1975).

8. J.W. Shearer, R.E. Kidder and J.W. Zink, Bull.Am. Phys.Soc. 15, 1483 (1970).

9. K.A. Brueckner, Laser Interaction and Related Plasma Phenomena, H. Schwarz and H. Hora eds. (Plenum, New York) 1977, Vol. 4, p. 891.

10. M.E. Marhic, Phys. Fluids 18, 837 (1975).

11. A.Y. Wong and R.L. Stenzel, Phys. Rev. Letters 34, 727 (1975); R.L. Stenzel, Phys. Fluids 19, 865 (1976).

12. H. Hora, Sov. J. Quantum Elect. 6, 154 (1976).

13. J. Nucholls, Laser Interaction and Related Plasma Phenomena, H. Schwarz and H. Hora eds. (Plenum, New York) 1974, Vol. 3B, p. 395

14. H. Hora, Aust. J. Phys. 29, 375 (1976).

15. K. Boyer, Astonautics and Aeronautics, Vol. 11, No. 1, 28 (1973); G.H. McCall, D.V. Giovanielli and J.F. Kephart, Proc. 8th Conf. Laser and Nonlinear Optics, Tbilissi, May 1976, p. 287.

16. H. Hora, Jenaer Jahrbuch 1957, p. 131; H. Osterberg, Jour. Opt. Soc. Am. 48, 513 (1968).

17. E. Goldman, Plasma Physics 15, 289 (1973); H. Hora, E. Goldman and M. Lubin, Univ. Rochester -LLE Rept. No. 21 (1974).

18. F.F. Chen, Laser Interaction and Related Plasma Phenomena, H. Schwarz and H. Hora eds (Plenum, New York) 1974, Vol. 3A, p. 291.

WAVELENGTH-DEPENDENT PHENOMENA IN NONLINEAR LASER-PLASMA INTERACTIONS*

Keith A. Brueckner

University of California, San Diego

La Jolla, California 92093 U.S.A.

I. INTRODUCTION

The number and energy distribution of superthermal electrons produced in a laser-heated plasma can be quantitatively obtained directly from the experimental x-ray spectrum using only the assumption that the fast electrons lose energy by bremsstrahlung and electron-electron collisions. The result is independent of the spatial and temporal distribution of electron density and temperature. This analysis is given in Sec. II.

The effects of electrostrictive forces on a laser-heated plasma have been calculated for different laser wavelengths using a hydrodynamic code of standard form. Large density discontinuities and marked fluctuations in reflectivity are predicted. The analysis is given in Sec. III.

---

*Presented at the Fourth Workshop on "Laser Interaction and Related Plasma Phenomena" held at RPI, Troy, New York, November 8-12, 1976. Material was also presented at the Nobel Symposium, Gõteborg, Sweden, June 11-17, 1976 and will be published in the Proceedings, Nobel Symposium on Nonlinear Effects in Plasmas (Plenum Publishing Co., Ltd., London, 1976).

†Work supported in part by the National Science Foundation (Grant #PHY 75-02690).

Hydrodynamic calculations of shell implosions show a marked dependence on laser wavelengths for wavelengths less than one micron. The coupling to the dense imploded shell rises from 6% at one micron to 33% at 0.1 microns, increasing possible energy multiplication in a pellet from approximately 60 to 400. The details are given in Sec. IV.

II. ENERGY DEPOSITION IN LASER-HEATED PLASMAS

The x-ray spectra from laser-irradiated plasmas typically show an athermal component which cannot be attributed to the radiation from a plasma with a well defined temperature and hence Maxwellian electron distribution. The hard radiation is believed to be due to energetic electrons produced by resonant absorption, or by laser-plasma instabilities. The correlation between the observed x-ray spectrum and the athermal electron population is usually made by the use of computer models which incorporate various models of the absorption process. In this section we show that a simple direct correlation exists which allows essentially model-independent determination of the fast electron distribution and also direct determination of the energy deposited in fast electrons. We base our analysis on the simple and reasonable assumption that electrons produced in the absorption process, with initial energies well above the temperature of the bulk population, lose energy by radiation and by collisional exchange with the much colder thermal electrons, that collisions among the superthermal electrons can be ignored, and that the fast electron density is much less than the thermal electron density.

The radiation emitted into all solid angles by a single electron of energy ϵ per unit path length per unit radiated energy [1] is

$$\frac{d^2\epsilon_{rad}}{dx\,dh\nu} = \frac{8}{3}\frac{e^2}{\hbar c}\frac{e^4}{mc^2}\frac{\langle Z^2\rangle N_i g_{rad}}{\epsilon}, \qquad (2.1)$$

with $\langle Z^2\rangle$ the average value of Z^2, and the Gaunt factor given by

$$g_{rad} = \frac{x}{1-e^{-x}} \frac{x_o}{e^{x_o}-1} \ell n \frac{x+x_o}{x-x_o} , \quad x_o < x$$

$$= 0 , \quad x_o > x \tag{2.2}$$

$$x = \frac{2\pi z e^2}{\hbar v}$$

$$x_o = \frac{2\pi z e^2}{\hbar v_o}$$

$$\epsilon = \frac{1}{2} m v^2$$

$$\epsilon_o = \frac{1}{2} m v_o^2$$

$$h\nu = \epsilon_o - \epsilon$$

The multiplicative factors $x/(1-e^{-x})$ and $x_o/(\epsilon^{x_o}-1)$ correct the Born approximation logarithmic Gaunt factor. The logarithmic singularity of g_{rad} at $h\nu \to 0$ is removed by the correction for screening at $\Delta p_{min} = \hbar R_D^{-1}$.

The rate of energy loss by electron-electron collisions [2] is

$$\frac{d\epsilon}{dx} = -\frac{4\pi \langle Z \rangle N_i e^4}{\epsilon} \ell n \Lambda_{coll} , \tag{2.3}$$

with

$$\Lambda_{coll} \cong \frac{2}{\hbar k_D} (2m\epsilon)^{1/2} . \tag{2.4}$$

The total energy radiated by the electron of initial energy ϵ_o in stopping may be obtained by combining Eqs. (2.1) and (2.3) with the result

$$\frac{d\epsilon_{rad}}{dh\nu} = \frac{2}{3\pi}\frac{e^2}{\hbar c}\frac{\langle Z^2\rangle}{mc^2\langle Z\rangle}\int_{h\nu}^{\epsilon_o}\frac{g_{rad}\, d\epsilon}{\ell n\, \Lambda_{coll}} \quad . \tag{2.5}$$

The ratio of the two coulomb logarithms is slowly varying with ϵ and $h\nu$; we approximate the ratio by taking the average value for g_{rad} of two and by evaluating the collision logarithm at $\epsilon = 10\,\text{kev}$, $\theta = 1\,\text{kev}$, and $n_e = 10^{21}/\text{cm}^3$, where $\ell n\, \Lambda_{coll} = 7.85$. The result is

$$\frac{d\epsilon_{rad}}{dh\nu} = \lambda_Z \frac{\epsilon_o - h\nu}{mc^2}$$

$$\lambda_Z = 3.95 \times 10^{-4}\frac{\langle Z^2\rangle}{\langle Z\rangle} \quad . \tag{2.6}$$

The energy radiated by a distribution of electrons produced with initial energy ϵ_o is

$$\frac{dE_{rad}}{dh\nu} = \int_{h\nu}^{\infty} n(\epsilon_o)\frac{d\epsilon_{rad}}{dh\nu}\, d\epsilon_o \tag{2.7}$$

$$= \frac{\lambda_Z}{mc^2}\int_{h\nu}^{\infty} n(\epsilon_o)(\epsilon_o - h\nu)\, d\epsilon_o \quad . \tag{2.8}$$

From this result we find

$$\frac{d^2}{d(h\nu)^2}\frac{dE_{rad}}{dh\nu} = \frac{\lambda_Z}{mc^2}\, n(h\nu) \quad , \tag{2.9}$$

and, for the number and energy of electrons above the thermal spectrum,

$$N_{fast} = \int_{\epsilon_{min}}^{\infty} n(\epsilon_o)\, d\,\epsilon_o$$

$$= \frac{mc^2}{\lambda_Z}\left(\frac{d}{dh\nu}\frac{dE_{rad}}{dh\nu}\right)_{h\nu = \epsilon_{min}} \tag{2.10}$$

$$E_{fast} = \int_{\epsilon_{min}}^{\infty} n(\epsilon_o)\, \epsilon_o\, d\epsilon_o$$

$$= \frac{mc^2}{\lambda_Z}\left(\frac{dE_{rad}}{dh\nu}\right)_{h\nu = \epsilon_{min}}$$

$$= 2.28 \times 10^{-10} \frac{\langle Z \rangle}{\langle Z^2 \rangle}\left(\frac{dE_{rad}}{dh\nu}\right)_{h\nu = \epsilon_{min}} \text{ joules.} \tag{2.11}$$

The distribution of radiation emitted by the fast electrons at $h\nu = \epsilon_{min}$ may be readily estimated by extrapolation of the energetic spectrum [see Figs. 2.1 and 2.2].

The preceding calculation is valid only if the fast electrons slow down in a time considerably less than the time for the hydrodynamic expansion of the corona. If the slow-down is too long, the electrons can lose energy in work done in accelerating the plasma. The slow-down time depends on the density which is determined by the motion of the electron resulting from scattering. At the critical density for one-micron radiation, the slowing-down time as determined from Eq. (2.2) is $\tau = 1.26 \times 10^{-13}\, \epsilon_o(\text{kev})^{3/2}$ seconds or, for an energy of 20 kev, 12 picoseconds. The actual time is considerably less since the electron scattering causes diffusion into the dense plasma. This conclusion also holds in

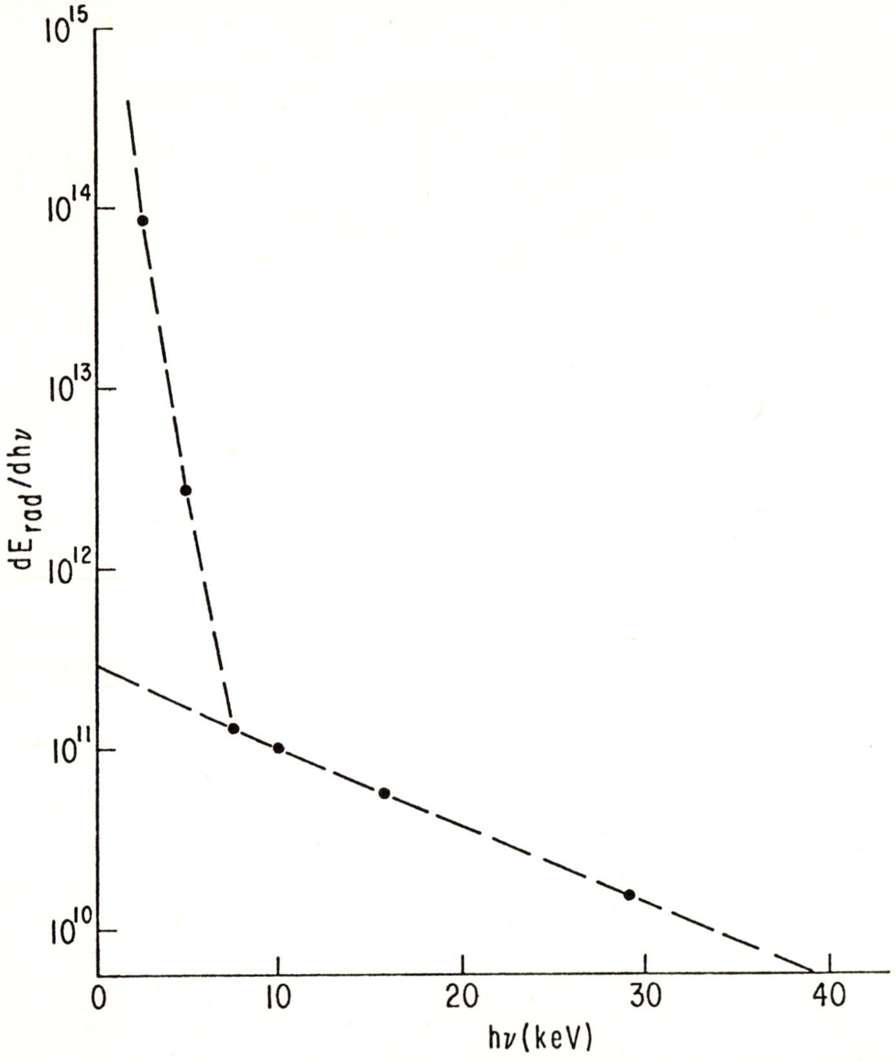

Fig. 2.1 X-ray spectrum for a 80 μ diameter glass microballoon target, with 30 joules incident in a 70 picosecond pulse. For uniform illumination, the power is 2×10^{15} w/cm^2. The dashed curves are best fits to the data and have been used in the estimates given in the text.

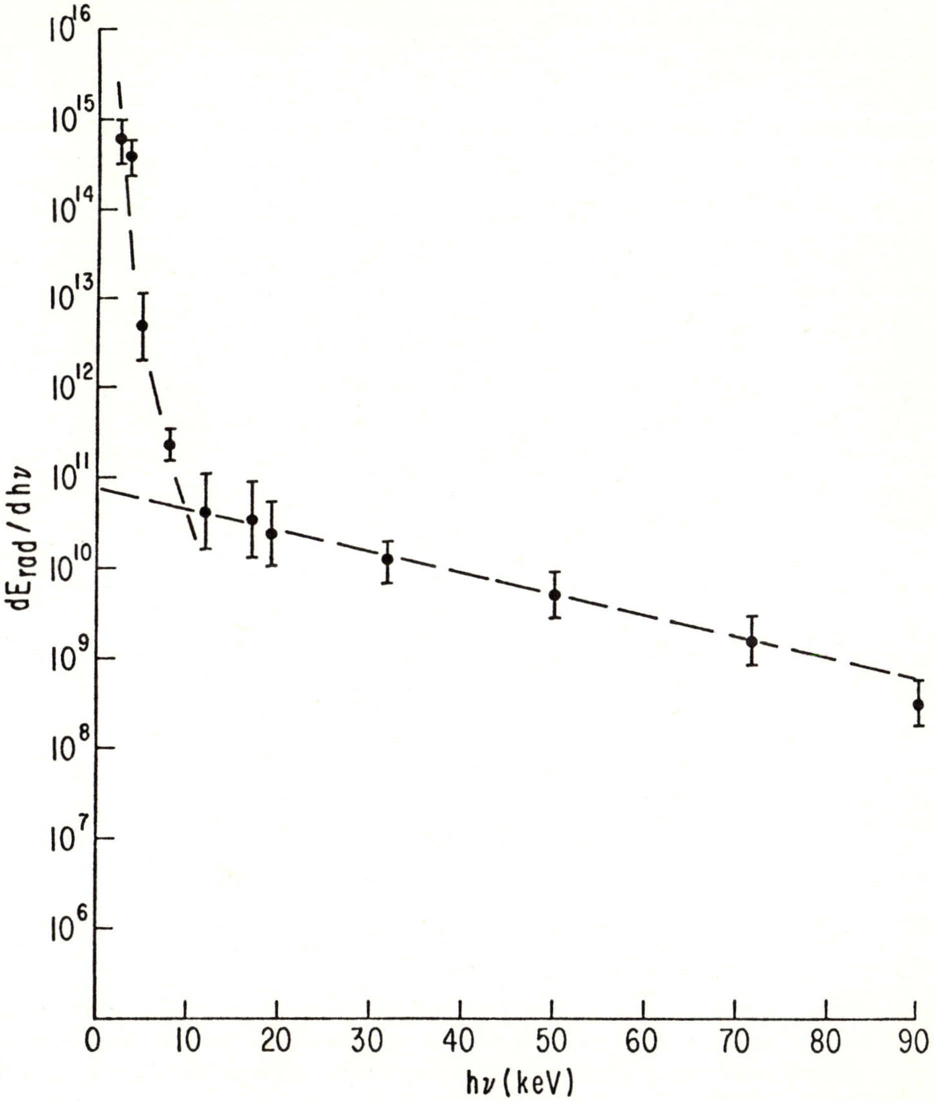

Fig. 2.2 X-ray spectrum from a parylene disk with 9.0 joules incident and a focal spot of 30μ diameter. The pulse length is approximately 130 picoseconds, giving a power of 10^{16} w/cm^2.

the presence of magnetic fields since scattering, drift resulting from the inhomogeneity of the fields, and motion along the fields will bring the electron into regions of high density where rapid energy loss occurs.

As examples, we consider two measurements reported by the Lawrence Livermore Laboratory [3, 4] shown in Figs. 2.1 and 2.2 and assume that ϵ_{min} in Eqs. (2.10) and (2.11) is equal to zero--which is reasonable since the thermal plasma temperature is low, about one kev. For the glass microballoon experiment, we find $\left(\text{for } \overline{Z} = 10,\ \overline{Z^2} = 108\right)$ $E_{fast} = 6.3$ joules and $\langle \epsilon \rangle = 10$ kev. The measured absorbed energy is about 20% of 30 joules incident, or 6.0 joules. For the parylene (C_2H_5) disk experiment, we find $\left(\text{for } \overline{Z} = 2.43,\ \overline{Z^2} = 11\right)$ $E_{fast} =$ 4.0 joules and $\langle \epsilon \rangle = 30$ kev. The measured absorption is 30-40% of 9.0 joules incident, or 2.7 - 3.6 joules, again in reasonable agreement with the spectral estimate. These results strongly imply that nearly all of the absorbed energy is in fast electrons, consistent with the previous interpretation [3, 4]. If the electrons are produced by resonant absorption, the average energy is [5]

$$\langle \epsilon \rangle = \left(\frac{m_e \omega_o \phi_{abs} L}{n_c} \right)^{1/2} , \tag{2.12}$$

with ϕ_{abs} the absorbed laser flux per unit area and L the electron density scale height evaluated at the critical density surface. For the glass microballoon, the absorbed flux is about 4×10^{14} watts/cm^2, giving $\langle \epsilon \rangle = 18$ kev $[L(\text{microns})]^{1/2}$. For the parylene disk, the absorbed flux is about 3×10^{15} w/cm^2, giving $\langle \epsilon \rangle = 49$ kev $[L(\text{microns})]^{1/2}$. These estimates are in approximate agreement with the x-ray spectrum only if the scale height is of the order of one micron, a surprising and somewhat implausible result. The average energy may, however, be reduced from Eq. (2.12) by the varying angles of incidence and polarization of the laser beam.

This analysis shows that the number of fast electrons produced in laser-plasma interaction can be directly determined from the x-ray spectrum. The production rate is approximately determined by the laser pulse length. The spatial origin is given by the illuminated area and may be reasonably assumed to be at the critical density surface. This provides a highly useful nearly model-independent input to computer codes and hence a semi-empirical procedure for determining other effects of the fast electron production such as shell and fuel preheat.

III. RADIATION-PRODUCED FORCES IN LASER-HEATED PLASMAS

The radiation pressure and ponderomotive forces [6] on a laser-heated plasma can produce substantial changes in the plasma density and expansion under conditions easily produced in present experiments. The radiation field interacts with the electrons through the average value of the $j \times B$ force, which gives

$$F_{es} = (1 - \epsilon) \nabla \frac{E^2}{8\pi} \tag{3.1}$$

$$\epsilon = 1 - \frac{\omega_{p\ell}^2}{\omega_L^2}$$

$$= 1 - \rho/\rho_c \quad ,$$

with ρ_c the critical density. Included in Eq. (3.1) is a force resulting from momentum deposition [7] by absorbed photons which results from the variation of the flux due to absorption. In the WKB approximation, neglecting absorption, the field is given by

$$\frac{E^2}{8\pi} = \frac{E_v^2}{4\pi} \frac{1}{\sqrt{\epsilon}} \quad , \tag{3.2}$$

including the factor of two arising from the average value of the incident and reflected flux.

The effect of the electrostrictive force on the plasma conditions is difficult to estimate analytically since the interaction is time-dependent and the plasma adjusts the density profile and the critical-density radius in response to the electrostrictive pressure. R. E. Kidder [8] estimated the density discontinuity and plasma density variation resulting from radiation pressure (omitting the electrostrictive forces) in a simplified model for one-dimensional isothermal expansion. His results are, however, only qualitatively similar to the numerical calculations of this paper. Density-profile steepening has also been observed in plasma simulation codes [9, 10] which, however, are incapable of describing the full hydrodynamic process or the time-dependent evolution of the profile over implosion times. The effects have usually been omitted in the large computer codes used for implosion calculations. The calculations reported here show that the principal effect may be easily obtained by a relatively simple model suitable for inclusion in standard codes. We have evaluated the effects of electrostriction by numerical solution of the hydrodynamic equations, using a hydrodynamic code [11] of standard form (Lagrangian with implicit heat conduction) simplified by the use of classical transport coefficients and by the assumption of a single temperature of electrons and ions.

The electric field was computed in under-dense zones according to Eq. (3.2), with $\epsilon \geq 0.2$. This limit roughly simulates the restriction [12] on the peak field given by solution of Maxwell's equations. The precise treatment of the variation of ϵ near $\rho = \rho_c$ is not important since the expanding plasma passes very quickly through the critical density. In the over-dense zones, the value of the WKB estimate of $E^2/8\pi$ averaged over the zone was used. The effect of the electrostrictive pressure on the plasma energetics was determined by evaluation of the overall plasma energy balance. The electrostrictive pressure was found to affect the energy balance by only a few percent. The absorbed laser energy was deposited in the first over-dense zone.

Typical results are given in Fig. 3.1 for one-micron radiation and in Fig. 3.2 for eight-micron radiation, for the conditions indicated in the figure captions. The detailed structure in the critical-density region could be obtained only if fine-zoning was used in the hydrodynamic calculations. In the one-micron case, the zoning in the initial glass shell in the region of the critical density surface shown in Fig. 3.1 was 35 Å at the initial glass density. For both cases, a marked density discontinuity forms which is convected outward at a velocity close to the natural velocity. The possibility of rapid convection of the critical density surface appears to have been omitted in plasma calculations [13] using numerical simulation. For the results shown in

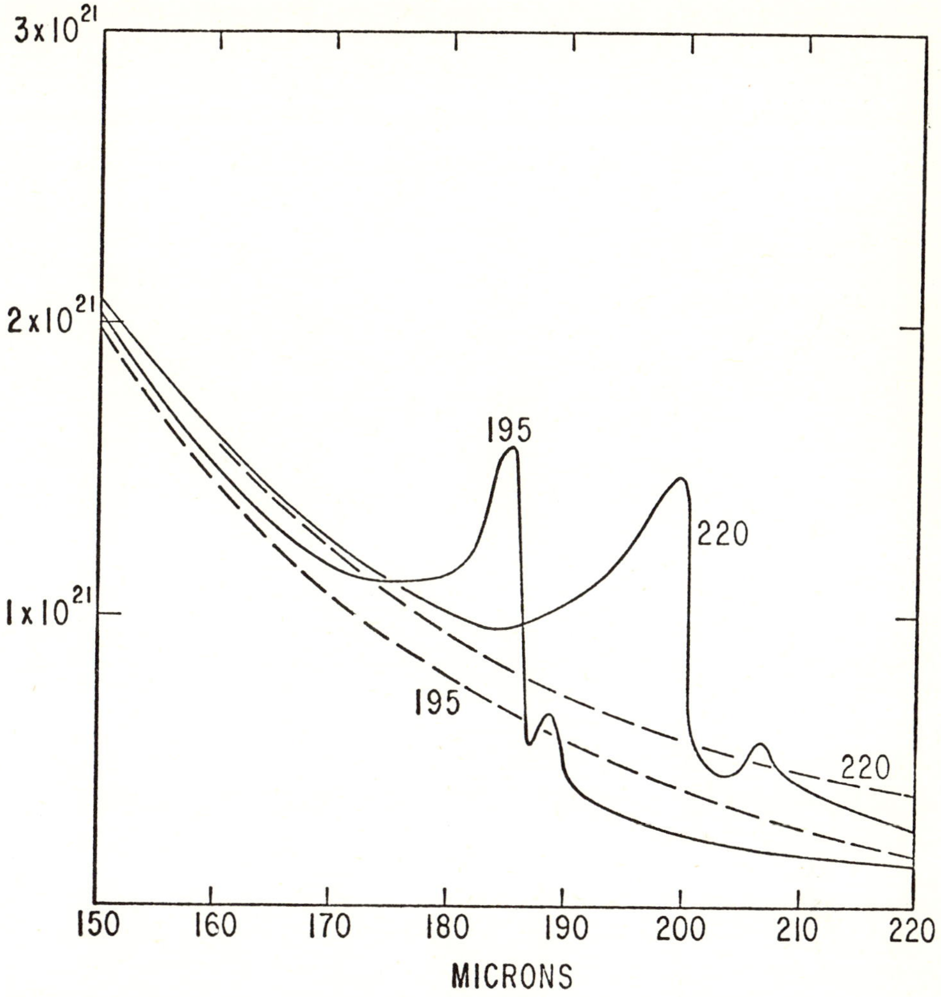

Fig. 3.1 Density (electrons/cm^3) profile in the vicinity of the critical density (10^{21}/cm^3) surface for one-micron radiation. The dashed curves give the density in the absence of the radiation pressure. The incident flux at the critical radius of 100 microns was 10^{16} watts/cm^2. Ten percent of the incident flux was assumed to be absorbed. The pellet configuration was a glass shell ($\overline{Z}$ = 10) with an initial wall thickness of 4 microns. Over the density range shown, the temperature was approximately constant at 3.3 kev. The material velocity was supersonic but subsonic with respect to the moving critical density interface. The curves are labelled with the time in picoseconds.

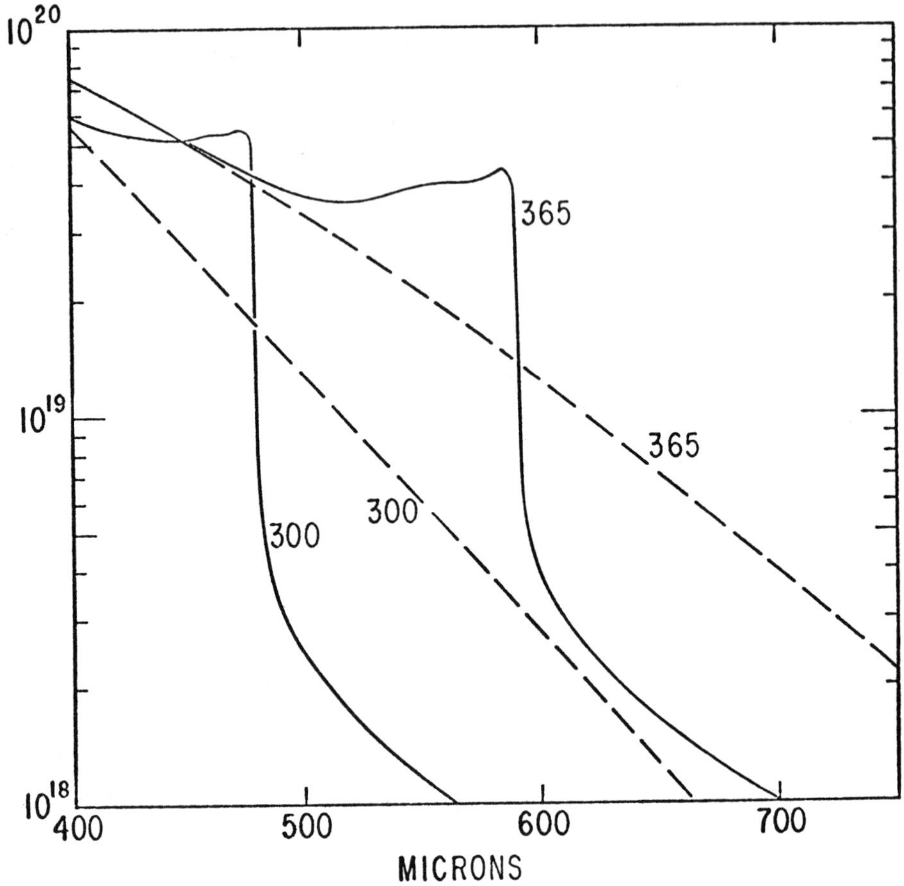

Fig. 3.2 The same as Fig. 3.1 for eight-micron radiation but with the absorption increased to 20%. The temperature in the plasma was approximately constant at 5.3 kev. The material velocity was again supersonic, but subsonic with respect to the moving critical density $\left(1.56 \times 10^{19}/\text{cm}^3\right)$ interface.

Figs. 3.1 and 3.2, the flow is supersonic into the critical density surface. The flux at the critical density surface as a result drops rapidly with time, particularly for the eight-micron case where the critical density surface moves to radii much larger than the critical radius.

As examples, for the one-micron case at 195 picoseconds, the critical density radius has moved outward to 186 microns and the incident flux has dropped to 2.9×10^{15} w/cm^2. The radiation pressure (assuming density much less than the critical density) is about 2.0 megabars while the pressure in the over-dense plasma is 8.3 megabars. For the eight-micron case at 365 picoseconds, the critical density radius has moved from 100 microns to 590 microns, reducing the laser flux to 2.87×10^{14} w/cm^2. The radiation pressure is 0.19 megabars while the pressure in the over-dense plasma is 0.36 megabars.

The fine structure near the critical density surface for the one-micron case persists at later times and is independent of the zoning if fine zoning is used. The effect may, however, be an artifact resulting from the numerical-difference scheme used. A striking phenomenon appears at late time in the one-micron case, as the density behind the density maximum near the critical density surface falls below critical density. This is shown in Fig. 3.3. As the critical density surface is convected outward into regions of decreasing electrostrictive pressure, the density maximum drops below critical density and the deposition point abruptly shifts inward to a much smaller radius. This effect, shown in this calculation which is spherically symmetric, may not persist in a 2-D calculation, particularly with non-uniform irradiation. Fluctuations in the critical density radius may, however, occur; the phenomenon would be conspicuous experimentally as a source of fluctuations in reflectivity.

The striking phenomena shown in this relatively simple model of the laser-plasma interaction are of particular interest in their effect on the processes of energy deposition and possible growth of unstable plasma and/or ion waves in the sub-critical density plasma. The electrostrictive pressure, which markedly increases the plasma density gradient and depresses the density of the sub-critical density plasma, can be expected to have a profound effect on the laser-plasma interaction [14].

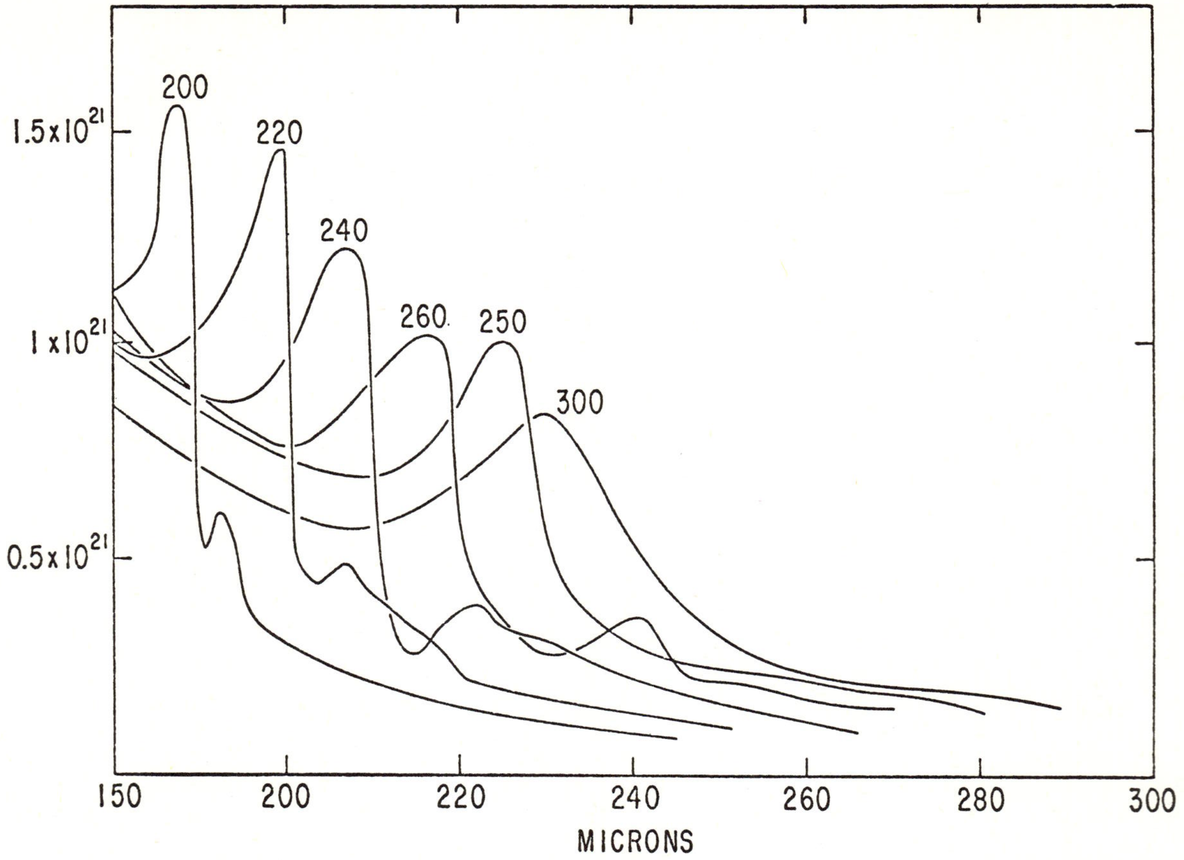

Fig. 3.3 Late-time variation of the density, for the same conditions as in Fig. 3.1. The laser deposition is in the outermost over-dense zone, allowing the density at smaller radii to become less than the critical density of $10^{21}/cm^3$. The deposition shifts abruptly inward at about 380 picoseconds to about 165 microns, as the radiation pressure at the larger radii becomes insufficient to keep the outer density maximum above the critical density.

IV. WAVELENGTH DEPENDENCE OF LASER COUPLING TO PELLET IMPLOSIONS

The coupling of laser energy into hydrodynamic motion in a laser-driven implosion is of critical importance in determining the possible fusion energy output per unit of absorbed laser energy. Analytic estimates [11] and numerical calculations [15] have indicated that the hydrodynamic coupling is in the range of 5 to 10% for laser wavelengths less than a few microns but may become considerably lower at larger wavelengths [16].

A recent suggestion by Dr. Raymond Grandey (private communication) that considerably stronger coupling can occur for shorter wavelengths has stimulated further study of the problem. A time-independent calculation by Kitomer et al. [17], ignoring acceleration and hence the dynamics of the implosion, suggested that the coupling is inversely proportional to the wavelength. These results are, however, inconclusive because of the limited model used.

We have calculated a series of implosions using a hydrodynamic code of standard form [11] (Lagrangian hydrodynamics with implicit heat condition) simplified by the assumption of a single electron and ion temperature and by the use of an ideal gas equation of state. Classical heat conduction by electrons without flux limiting was assumed. The calculation is inaccurate if the electron mean free paths are not small compared to the characteristic scale of the heated plasma or if the thermal flux exceeds the flux limit which is of the order of $n_e \epsilon_e v_e$. For electrons, the mean free path at the critical density surface is

$$\lambda_{mean} \cong \frac{10}{Z} E_{kev}^{2} \lambda_{\mu}^{2} \text{ microns} \quad , \qquad (4.1)$$

λ_{μ} = laser wavelength in microns.

Thus, for $Z = 10$, 2 kev electrons, and one micron wavelength the mean free path is four microns and a diffusion calculation is reasonable. For considerably larger wavelengths, the diffusion approximation fails and a more accurate model is necessary. Kidder and Zink [18] have shown in a simple model of the low-density corona that very marked corona-core decoupling (a factor of 50 compared with one micron) can occur at ten-micron wavelength. This effect is even more pronounced than the scaling we find.

The calculations were for a 100-micron-radius glass shell with wall thickness of four microns. The incident absorbed flux was 1.26 terawatts which was deposited in the first over-dense plasma zone. The effects of electrostrictive pressure were included but had a negligible effect on the hydrodynamic coupling.

The pressure profiles for several wavelengths from 0.1 to 1.0 microns are given in Fig. 4.1 at times selected to give the peak pressure at approximately the same radius. A significant variation of peak pressure with wavelength for wavelengths less than one micron is shown, which produces greater acceleration of the shell and hence increased coupling at short wavelength. The ratio of kinetic energy of the imploding shell material (at greater density than the initial glass density) to the absorbed energy at the times corresponding to the pressure profiles in Fig. 4.1 is given in Fig. 4.2. The ratio rises rapidly from 6.4% at 1.0 micron to 28.6% at 0.1 micron. Further analysis of the short wavelength implosions shows peak transfer efficiency to the dense shell of 33.4% at 0.1 microns and 19.4% at 0.175 microns.

The variation of coupling with wavelength arises from the deposition of the laser energy in increasingly dense material, steepening the temperature gradient in the ablation front and increasing the ablation rate. The temperature profile for the case given in Fig. 4.1 is shown in Fig. 4.3. The temperatures in the ablation front are also increased by the reduced corona temperatures and the corresponding decrease in the kinetic energy loss into the corona expansion as the laser wavelength decreases.

The possible increase in the efficiency of hydrodynamic coupling of the absorbed laser energy at short wavelengths can have a large effect on laser energy requirement for high pellet gain. The coupling of the incident laser energy to the pellet may also improve at short wavelength; we do not consider that effect in this paper.

Detailed implosion calculations [11] have shown that a practical limit in pellet gain, set by details of the hydrodynamics of compression and fuel burn-up, is about 1200 times the energy delivered to the compressed fuel core. This limits pellet gain

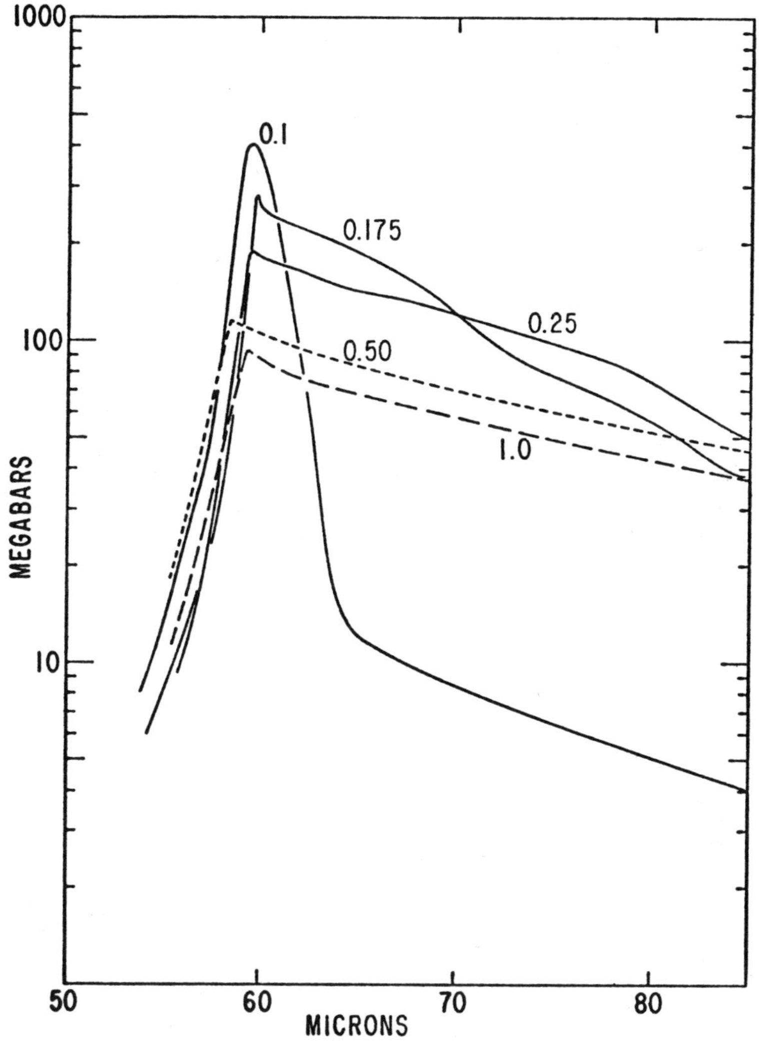

Fig. 4.1 Pressure as a function of radius. The curves are labelled by the wavelength in microns. The time in the implosion at which the peak pressure occurs at about 60-micron radius is 260, 250, 210, 180, and 130 picoseconds at 1, 0.5, 0.25, 0.175, and 0.1 micron wavelength.

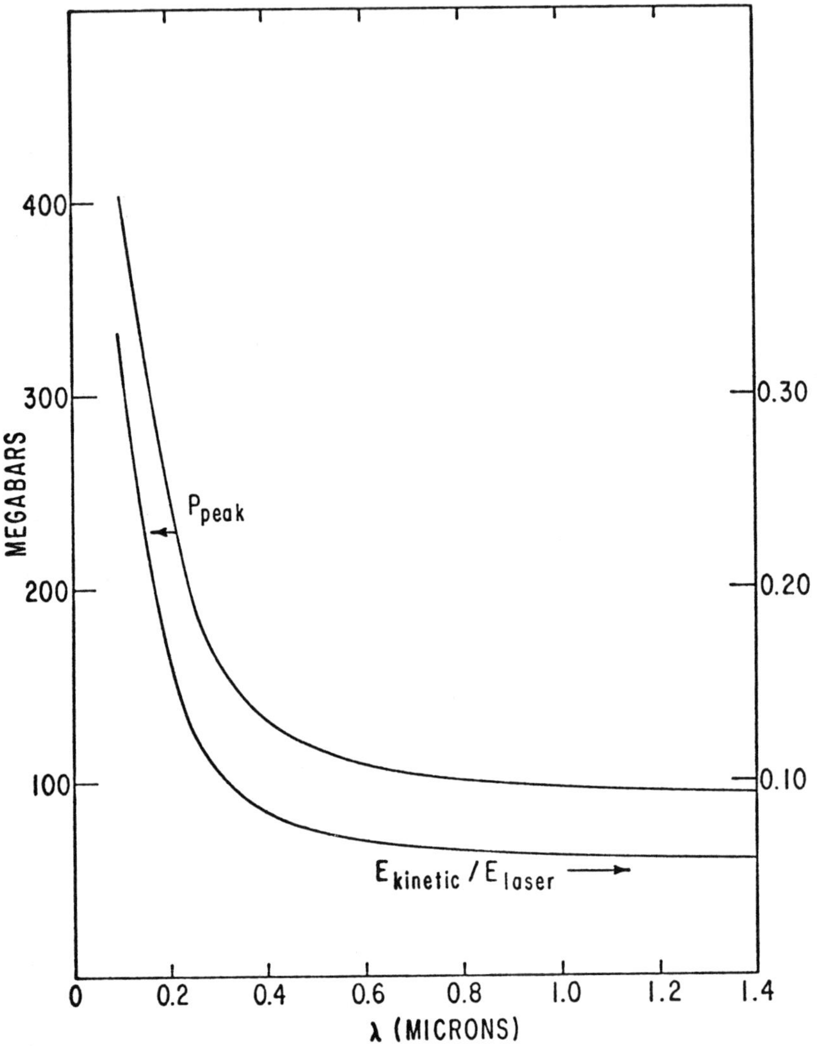

Fig. 4.2 Peak pressure from Fig. 4.1 and ratio of the kinetic energy of the dense shell (density greater than initial) to the absorbed laser energy.

(ratio of fusion energy output to absorbed laser energy) to about 60 for a coupling efficiency of 5% but gives a gain of 400 for the coupling efficiency of 33.4% we have computed at 0.1 microns. The practical usefulness of this high possible gain is limited by

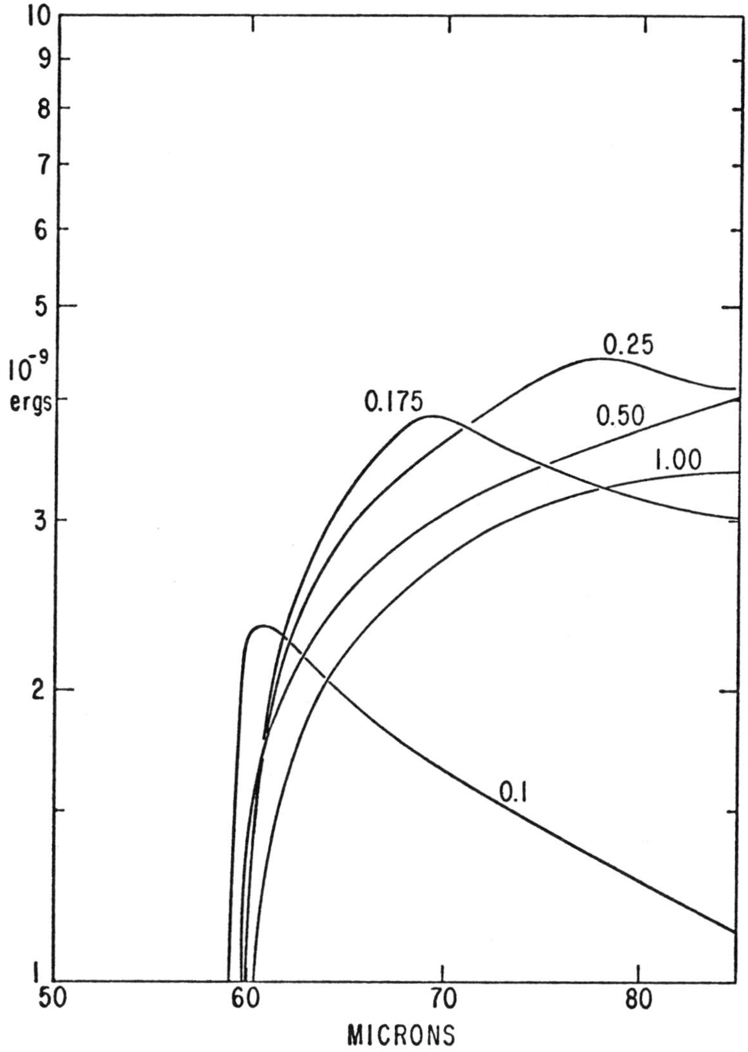

Fig. 4.3 Variation of temperature with radius for the evaluated wavelengths (in microns). The temperature is given in energy units (1.60×10^{-9} erg = 1 kev).

present laser technology. Our results emphasize the importance of development of lasers with wavelength considerably less than one micron.

REFERENCES

1. C. W. Allen, Astrophysical Quantities (University of London, The Athlone Press, 1955).

2. Lyman Spitzer, Jr., Physics of Fully Ionized Gases, Interscience Tracts on Physics and Astronomy, No. 3 (Interscience Publishers, Inc., New York, 1956).

3. H. G. Ahlstrom, UCRL Report 77094 Rev 1, Oct. 8, 1975; Bull. Am. Phys. Soc., Oct., 1975 (papers 3C9, 3C10, and 3C 11).

4. W. C. Mead, W. L. Kruer, J. D. Lindl, and H. D. Shay, Bull. Am. Phys. Soc., Oct., 1975 (paper 2C5).

5. J. P. Freidberg, R. W. Mitchell, R. L. Morse, and L. I. Rudsinski, Phys. Rev. Letters 28, 795 (1972).

6. See, for example, Landau and Lifshitz, Electrodynamics of Continuous Media (Pergamon Press, New York, 1960).

7. John A. Stamper, "Field Generation by a Generalized Ponderomotive Force," NRL Memorandum Report 3186, Naval Research Lab., Washington, D. C. (Dec., 1975).

8. R. E. Kidder, "Interaction of Intense Photon Beams with Plasmas (II)," Proc. Japan-U.S. Seminar, Laser Interaction with Matter, Sept. 24-29, 1972.

9. K. Estabrook, E. J. Valeo, and W. L. Kruer, Physics Letters 49A, 109 (1974).

10. K. Estabrook, E. J. Valeo, and W. L. Kruer, Phys. Fluids 18, 1151 (1975).

11. K. A. Brueckner and S. Jorna, Rev. Mod. Phys. 46, 325 (1974).

12. J. D. Lindl and P. K. Kaw, Phys. Fluids 14, 371 (1971).

13. K. Lee, D. Forslund, J. Kindel, and E. Lindman, Los Alamos Scientific Lab. Report LA-UR-75-2097 (submitted to J. Phys. Fluids).

14. Experiments at the Naval Research Lab., Washington, D.C. (Dr. Stephen E. Bodner, private communication) and at the University of Rochester (Professor Moshe Lubin, private communication) have suggested that marked density-gradient anomalies exist with one-micron radiation in the 10^{15} - 10^{16} watts/cm^2 range.

15. K. A. Brueckner, P. M. Campbell, and R. A. Grandey, Nucl. Fusion 15, 471 (1975).

16. R. E. Kidder, Nucl. Fusion 14, 797 (1974).

17. S. J. Kitomer, R. L. Morse, and B. S. Newberger, "Structure and Scaling Laws of Laser-Driven Ablative Implosions," Los Alamos Scientific Lab. Report #LA6079MS, to be published in Physics of Fluids.

18. R. E. Kidder and J. W. Zink, Nucl. Fusion 12, 325 (1972).

RELATIVISTIC SELF-FOCUSING*

E. L. Kane and H. Hora

Dept. of Theoretical Physics, The University of New South Wales

Kensington, NSW Australia

Abstract

Laser intensities of order and above 0.001 times the relativistic intensity I_{rel} coupled with ionized target electron number densities close to and above the laser cutoff density n_{ec} exhibit sudden relativistic self-focusing, with attendent beam converging lengths about equal to the diameter of the initial vacuum laser beam. The analytical model is a relativistic generalization, including mass and energy effects, of the plasma electron-ion collision process, combined with a local bending of the laser beam wave front resulting from influences on the plasma refractive index. Dependences on the plasma temperature, degree of ionization and instabilities, as given by an effective collision frequency, are studied. Self-focusing and dielectric swelling results in maximum oscillation energies in agreement with several high intensity experiments. The wavelength independence of electron energy, as shown in the recent experiments of G. H. McCall, as well as measured MeV ion emission from irradiated targets, can be explained by the model. It is concluded that Nd glass laser beams of 10^{14} Watts power will generate dense plasmas containing highly charged heavy ions with maximum energies approximately 20 GeV.

I INTRODUCTION

One of the self-focusing processes of laser beams in plasmas results from the relativistic mass and energy dependency of the refractive index at high laser intensities. It is obvious that such phenomena would be apparent near the relativistic threshold I_{rel} where the electron oscillation energy equals $m_o c^2$; however, it is surprising that this mechanism becomes significant at $I \geqslant 0.001\ I_{rel}$ (values exceeding 3.66 x 10^{15} W/cm^2 for an Nd laser and 3.66 x 10^{13} W/cm^2 for a CO_2 laser).[1] This phenomena is analagous to measurement of relativistic instabilities at nonrelativistic

*Presented at the Fourth International Workshop Conference on "Laser Interaction and Related Plasma Phenomena" at Rensselaer Polytechnic Institute, Troy, New York, 8-12 November, 1976.

energies.[2] The reaction time of this process is virtually instantaneous, of a scale significantly less than the self-focusing induced by nonlinear radiation forces on the plasma; comparatively long-time, high-inertial radial motion of the plasma initiates the optical self-focusing in this second process. In addition to this result it is significant that for plasma electron densities approximately equal to or above the critical cutoff density n_{ec} (10^{21} cm^{-3} for an Nd laser and 10^{19} cm^{-3} for a CO_2 laser) the self-focusing occurs within a distance of the order of the inital beam diameter in the relative vacuum outside of the plasma.[3,4] Parametric solutions of a simple geometric ray model describing the fast relativistic contraction is made for both Nd glass and CO_2 gas lasers, over a wide range of intensities, electron number densities and temperature, and ion charge numbers. The contraction mechanism can directly explain the observation of MeV ion generation[5,6] and the laser wavelength independence of this generation.[7] Coupling of the self-focusing with nonlinear radiation swelling results in ion energies of GeV levels; the results of Ref. 8 will be reviewed here and extended.

Self-focusing of laser beams in plasmas can also be caused by magnetic fields[9], by thermal effects[10], and by a nonlinear radiation force which expels plasma from the contracting beam. Calculated thresholds of this last process with associated beam diameters[11] correlate with measured values[12,13]; further confirmation has been made by several authors[14,15]. It should be emphasized that for short time scales and high intensities the relativistic self-focusing dominates over all the other plasma dynamic and thermal effects.

The form of the refractive index employed in this paper assumes that the electron-ion collision frequency plays a key role and that nonlinear intensity dependence, relativistic influences, and instability effects contained in the complex dielectric expression are all important. The neglect of collision dominated effects, as in Ref. 16, ignores the threshold appearance of the self-focusing process.

Although the most complete analysis of the problem involves direct solution of Maxwell's equations in a dielectric collisional medium, whereby the nonlinear radiation forces are also included in a time-dependent formalism, comparison of the model with a wave equation solution of Siegrist, et al[5] is favourable over a sufficiently high intensity region. The lower intensity results of Ref. 5 generally continue into the higher intensity calculations of this paper.

The electron oscillation energy within the beam focal region is determined from the dielectric swelling-net acceleration of plasma within that region. The conversion of electron oscillation energy into translational ion energy, where the electrons are relativistic but the ions are nonrelativistic, yields ion energies proportional to the laser intensity and ion charge number. Maximum ion energies are very high; as an example, an Nd glass laser of 10^{14} Watts power could energize nearly fully ionized heavy ions of 20 GeV energy. Ion beams of density up to 10^{20} cm^{-3}, which is more than 10^{10} times the density achieved with accelerator beams, may be produced from laser-plasma emission. This would result in particle pair production efficiencies 10^{10} times higher than that gained from an accelerator; in particular, proton pair production would be greatly facilitated[17].

II PLASMA REFRACTIVE INDEX

Relativistic self-focusing of a laser beam in a homogeneous plasma is derived from consideration of the dependence of the optical constants on the relativistic change of the mass of the electrons during their oscillation in a high intensity laser field. It is assumed that the square of the complex refractive index in a plasma is given by[18,19]

$$\tilde{n}^2 = 1 - \left(\frac{\omega_p^2}{\omega^2+\nu^2}\right)\left(1 - i\,\frac{\nu}{\omega}\right) \tag{1}$$

which has an absolute value of

$$|\tilde{n}|^2 = \left[\left(1 - \frac{\omega_p^2}{\omega^2+\nu^2}\right)^2 + \left(\frac{\nu}{\omega}\right)^2 \left(\frac{\omega_p^2}{\omega^2+\nu^2}\right)^2\right]^{\frac{1}{2}} \tag{2}$$

ω_p is the plasma frequency, ν the electron-ion collision frequency, and ω is the monochromatic laser frequency. For a collisionless plasma Eq. (1) reduces to

$$\tilde{n}^2 = 1 - \left(\frac{\omega_p}{\omega}\right)^2 \tag{3}$$

where the plasma frequency is given by the classical Langmuir value $(4\pi e^2 n_e/m_e)^{\frac{1}{2}}$, but with the electron mass possessing the relativistic velocity dependency

$$m_e = m_o/[1 - (v_e/c)^2]^{\frac{1}{2}} \tag{4}$$

with m_o the electron rest mass, v_e its oscillation velocity, and c the speed of light. Furthermore, the v_e/c velocity ratio is written in terms of the time average laser induced electric field E within the plasma by

$$[1-(v_e/c)^2]^{\frac{1}{2}} = [1 + \frac{e^2E^2}{m_o^2\omega^2c^2}]^{-\frac{1}{2}} \tag{5}$$

It is feasible that for a proper low intensity specification of long time ponderomotive effects, a kinetic theory obtained temperature dependency has to be included in the right hand side of Eq. (5). Insertion of Eqs. (4) and (5) into (3) yields

$$\tilde{n}^2 = 1 - \left(\frac{\omega_p^o}{\omega}\right)^2 \left(1 + \frac{e^2E^2}{m_o^2\omega^2c^2}\right)^{-\frac{1}{2}} \tag{6}$$

with $\omega_p^o = (4\pi e^2 n_e/m_o)^{\frac{1}{2}}$. Eq. (6) has been employed by Kaw and Dawson[20] in their collisionless study of the effects of the optical electric field on the refractive index and therefore on the relativistic self-focusing processes in the high energy plasma. As they pointed out, real values of a collisionless ñ are obtained for wave frequencies in the range

$$\omega_p^o \left(1 + \frac{e^2E^2}{m_o^2\omega^2c^2}\right)^{-\frac{1}{4}} < \omega < \omega_p^o$$

which is in contrast to propagation in a linear, non-relativistic plasma. It is also interesting to note that for very high irradiative electric fields, such that $e^2E^2/(m_o^2\omega^2c^2) >> 1$, Eq. (6) becomes

$$\tilde{n}^2 \simeq 1 - \left(\frac{\omega_p^o}{\omega}\right)^2 \left(\frac{m_o\omega c}{eE}\right) \tag{7}$$

retaining only the unity in the binomial expansion. Setting $E = E_o/\sqrt{2}$ where E_o is the peak amplitude of the electric field there results

$$\tilde{n}^2 = 1 - \sqrt{2}\left(\frac{\omega_p^o}{\omega}\right)^2 \left(\frac{m_o\omega c}{eE_o}\right) \tag{7a}$$

Now Max and Perkins,[16] in their study of the propagation of electromagnetic waves in an overdense collisionless relativistic plasma, have employed a refractive index of the form

$$\tilde{n}^2 = 1 - \frac{\pi}{2}\left(\frac{\omega_p^o}{\omega}\right)^2 \left(\frac{m_o\omega c}{eE_o}\right), \tag{8}$$

the difference of $[\sqrt{2}/(\pi/2)]^{1/2} = 0.95$ enters from their employment of an exact non-harmonic relativistic oscillation of the electron mass in the laser field while in our situation the time averaging is approximated by $\sqrt{2}$. It is seen, first of all, that the more general collisionless refractive index, Eq. (6), is not limited to exceedingly high field energies, values, as it turns out, that are way beyond the thresholds of relativistic influence. Secondly, the calculations presented here show that electron-ion collisional interactions are vital in the demonstration of relativistic processes.

The form of the collision frequency is that of the classical non-relativistic Lorentz gas expression in Spitzer[21], p. 138; where temperature is now, relativistically, also assumed to involve the electron kinetic energy and the electron mass possesses the aforementioned electric field dependency. It must be added that the high electron number densities created near the focal region may require that a more elaborate quantum mechanical dielectric and collisional model be employed; nevertheless, the approximations made in the ray focusing solution are probably of a larger magnitude than the neglect of quantum effects. The Spitzer formulation may be directly related to AC electrical conductivity and semi-classical inverse bremsstrahlung radiation absorption[22] so that its employment is, nevertheless, relatively general.

The electron-ion collision frequency ν is generalized from

$$\nu = \frac{\pi^{3/2} Z_I n_e e^4 \ln \Lambda}{2m_e^{\frac{1}{2}}(2KT)^{3/2}} \tag{9}$$

to

$$\nu = \frac{\pi^{3/2} Z_I n_e e^4 \ln \Lambda_R}{2m_o^{\frac{1}{2}}[1 + \frac{e^2E^2}{m_o^2\omega^2c^2}]^{\frac{1}{4}} \; [2(KT + \varepsilon_{kin})]^{3/2}} \tag{9a}$$

where Eqs. (4) and (5) have been employed. Z_I is the average ion charge number over the plasma mixture, KT is the thermal energy, Λ_R is the relativistic Coulomb factor (described below) and ϵ_{kin}, the electron kinetic energy, is given by[17]

$$\varepsilon_{kin} = m_o c^2 \left\{ \left(1 + \frac{e^2E^2}{m_o^2\omega^2c^2} \right)^{\frac{1}{2}} - 1 \right\} \tag{10}$$

where Eq. (5) has been employed in the relativistic particle energy expression. The Spitzer electron-electron collision correction factor γ_E, which may be included in the denominator of Eq. (9), is really only applicable for electron kinetic energies less than the thermal energy and $\omega << \nu$, conditions opposed to the high frequency heating processes studied here. Furthermore, even if it had applicability in this problem, the production of highly ionized plasmas yields γ_E values very close to one.

The non-relativistic expression of the Coulomb factor, given by

$$\Lambda = \frac{3(KT)^{3/2}}{2Z_I e^3 \; (\pi n_e)^{\frac{1}{2}}} \tag{11}$$

is equal to the ratio of the Debye length λ_D divided by the binary electron-ion collision impact parameter b_o where

$$\lambda_D = (KT/4\pi n_e e^2)^{\frac{1}{2}}$$

and

$$b_o = Z_I e^2/m_e v_e^2$$

The λ_D expression assumes that charged particle screening is entirely caused by the electrons; their relatively high thermal velocities induce more effectiveness than the comparatively sluggish ions. Strictly speaking, the m_e should be motion-dependent; however, since Λ is an average over all the electron velocities, the relation $m_e v_e^2$ =

3KT is employed. Λ_R is then assumed to be generalized to

$$\Lambda_R = \frac{3(KT + \varepsilon_{kin})^{3/2}}{2Z_I e^3 (\pi n_e)^{1/2}} \tag{12}$$

with ϵ_{kin} given by Eq. (10).

It is useful to write the $e^2E^2/m_o^2\omega^2c^2$ term as I/I_{rel} where I is the beam intensity within the plasma, $cE^2/8\pi$, and I_{rel} is a characteristic relativistic intensity defined by

$$I_{rel} \equiv (c/8\pi)(m_o^2c^2\omega^2/e^2) \tag{13}$$

$$= 3.66 \times 10^{18}\,\text{W/cm}^2 \text{ for Nd glass lasers}$$
$$3.66 \times 10^{16}\,\text{W/cm}^2 \text{ for } CO_2 \text{ gas lasers}$$

With this substitution Eqs. (9a) and (12) become

$$\left(\frac{\nu}{\omega}\right)^2 = \frac{C^* n_{ec} N^2 (\ln \Lambda_R)^2}{\left\{F\left[\frac{KT}{e^2} + \frac{\Gamma-1}{r_o}\right]\right\}^3 \Gamma} \tag{14}$$

and

$$\Lambda_R = \frac{D^*\left\{F\left[\frac{KT}{e^2} + \frac{\Gamma-1}{r_o}\right]\right\}^{3/2}}{(N\, n_{ec})^{1/2}} \tag{15}$$

with $C^* = \pi^2 Z_I^2/128$, $D^* = 3/(2\sqrt{\pi}Z_I)$, r_o (classical electron radius) = 2.81785 x 10^{-13} cm, n_{ec} is the non-relativistic cutoff density $\omega^2 m_o/4\pi e^2$ (for Nd glass 10^{21} cm^{-3}, for CO_2 gas 10^{19} cm^{-3}), $N = n_e/n_{ec}$ is the characteristic density ratio parameter, $\Gamma \equiv (1 + I/I_{rel})^{1/2}$, and F is a parametric instability quantity. F = 1 denotes the stable microscopic situation, while values less than unity represent an effective collectively-induced absorption that results in a certain level of instability. The resulting effective collision frequency is represented by $\nu_{eff} = F^{-3/2}\,\nu$; it also has to be included as an energy term multiplier in the Λ_R expression for reasons of consistency, although the logarithmic variation is of relatively minor influence.

As indicated previously, the square of the ratio of the relativistic plasma frequency ω_p to laser frequency ω is given by

$$\left(\frac{\omega_p}{\omega}\right)^2 = \left(\frac{\omega_p^o}{\omega}\right)^2 \left(1 + \frac{e^2E^2}{m_o^2\omega^2c^2}\right)^{-1/2}$$

$$= N/\Gamma \tag{16}$$

in terms of our newly defined variables. Insertion of this relation into Eq. (2) gives

$$|\tilde{n}| = \left\{\left[1 - \frac{N/\Gamma}{1+(\nu/\omega)^2}\right]^2 + \left(\frac{\nu}{\omega}\right)^2 \left[\frac{N/\Gamma}{1+(\nu/\omega)^2}\right]^2\right\}^{\frac{1}{4}} \tag{17}$$

with $(\nu/\omega)^2$ given by Eq. (14). The absolute value of the refractive index is therefore dependent on the local light intensity, electron number density and temperature, ionization number and, of course, laser frequency.

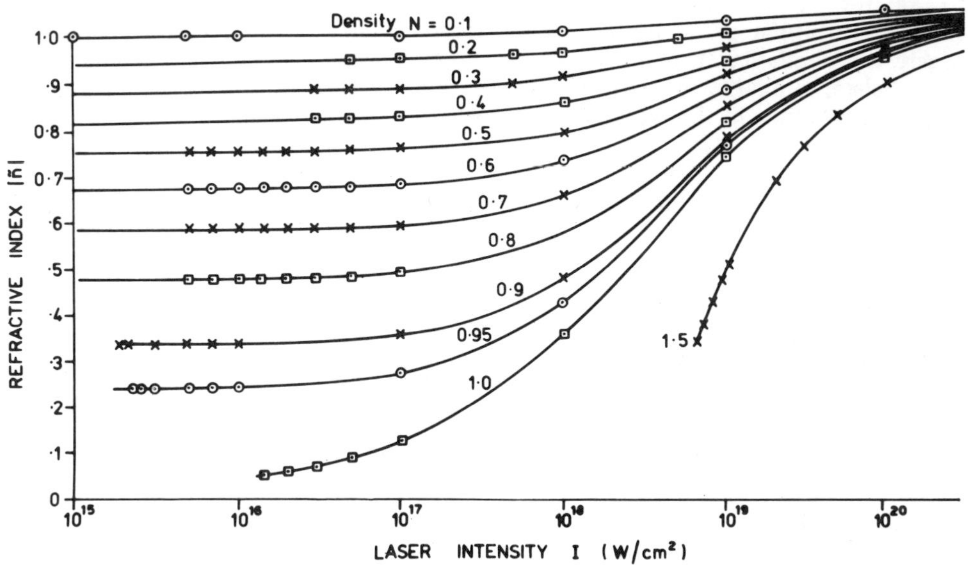

Fig.1
The absolute value of the refractive index ñ dependence on the vacuum neodymium glass laser intensity for various electron densities n_e contained in the parameter $N = n_e/n_{ec}$, with the cut-off density $n_{ec} = 10^{21}$ cm^{-3}. T = 10 eV, $Z_I = 1$ and F = 1.0.

Illustrated in Fig. 1 is |ñ| for various values of Nd laser intensity I and N with KT = 10 eV and $Z_I = 1$ (relatively cool hydrogen plasma). The symbols denote individual computer runs. The dependency on KT and Z_I, up to values of 10^3 eV and 10, respectively, is very slight except for the lowest values of intensity. More will be said about this later. Several interesting things in the figure should be pointed out: an increase in N, up to the cutoff value of unity, results in a steadily decreasing value of the refractive index; an intensity somewhat over 10^{16} W/cm^2 (Nd) gives an $|\tilde{n}| \approx .05$ value. Since within the plasma $I_p = I_v/|\tilde{n}|$, where I_v is the vacuum intensity value, this means that the local beam intensity is increased some twenty times upon plasma penetration. As the electron number density is increased to a supercritical value, N = 1.5, the minimum $|\tilde{n}|$ begins to increase. Cessation of the plots is due to a numerical threshold intensity I_s, above which the self-focusing solution becomes predominant (Section IV). It is to be noted that for the collisionless $|\tilde{n}|$,

$$|\tilde{n}|^2 = 1 - N(1 + I/I_{rel})^{-\frac{1}{2}} \tag{17a}$$

so that for the same value of N and a CO_2 laser intensity value 0.01 times the Nd value there results precisely the same refractive index plot. Collisions somewhat negate this similarity through the presence of the n_{ec} term in Eqs. (14) and (15); nevertheless, similarity is preserved down to intensities approximately within an order of magnitude of the threshold values for N values near unity.

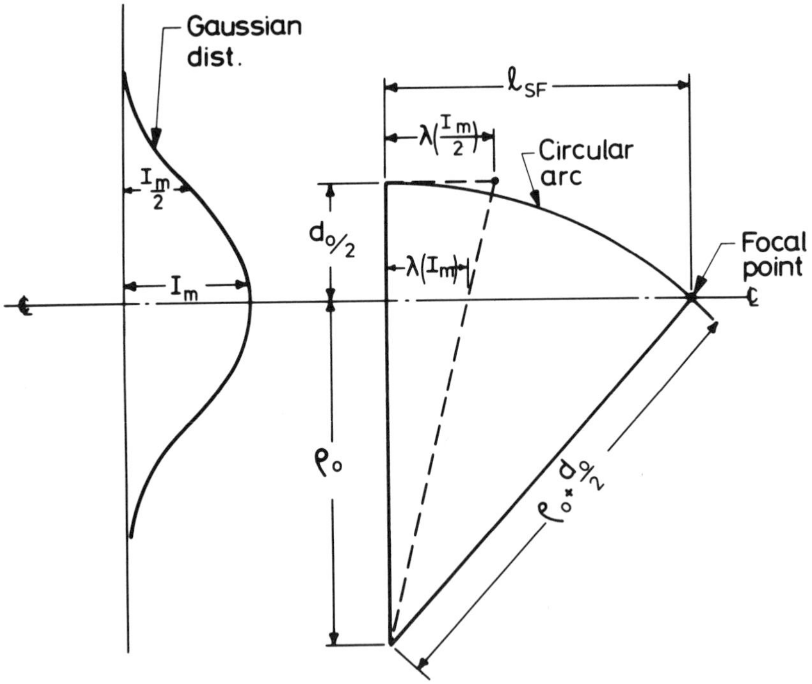

Fig. 2
Geometric construction for the evaluation of the laser beam self-focusing length l_{SF} assuming a Gaussian intensity profile with a distance d_o between the half-intensity maximum points. The initial plane wave front distorts since the effective wavelength λ is at a minimum at the highest intensity beam center.

III. GEOMETRIC RAY MODEL FOR RELATIVISTIC SELF-FOCUSING

Evaluation of the ratio of the self-focusing length l_{SF} to the distance between the half intensity maximum points on the initial Gaussian profile vacuum beam, denoted by d_o, is performed by geometric distortion of an initial plane wave assuming that a ray extending from the target surface to the focus traces out a circular arc. The geometry is illustrated in Fig. 2. Actually, the ray approaches the axis to within a distance of the order of the local wavelength in the plasma and then flares out to form a filamentary structure. In the construction ρ_o is an arbitrary "radial" distance which disappears in the final result. By proportional triangles

$$\frac{\lambda(I_{max})}{\rho_o} = \frac{\lambda(I_{max}/2)}{\rho_o + d_o/2} \tag{18}$$

Also, within the plasma volume, at any radial point of the irradiated region,

$$\lambda_I = \lambda_v / |\tilde{n}(I)| \quad , \tag{19}$$

with λ_v the vacuum wavelength. Due to the fact that $|\tilde{n}|$ increases with intensity (Fig. 1), the wave front propagates faster at a less intensity radial point than at its maximum intensity I_m and therefore self-focuses (ideally) without any time delay. Employment of Eq. (18) assumes that a relation holding immediately within the plasma surface can be applied throughout the self-focusing length. From Eqs. (18) and (19)

$$\frac{\lambda_v}{\rho_o |\tilde{n}(I_m)|} = \frac{\lambda_v}{(\rho_o + d_o/2)|\tilde{n}(I_m/2)|} \tag{20}$$

From the Pythagorean theorem (Fig. 2),

$$(\rho_o + d_o/2)^2 = l_{SF}^2 + \rho_o^2$$

or

$$\frac{l_{SF}}{d_o} = \tfrac{1}{2}\left[\frac{4}{d_o}(\rho_o + d_o/2) - 1\right]^{\frac{1}{2}} \tag{21}$$

Two relations may be derived from Eq. (20):

$$\frac{4\rho_o}{d_o} = \frac{2|\tilde{n}(I_m/2)|}{|\tilde{n}(I_m)| - |\tilde{n}(I_m/2)|} \tag{22a}$$

and

$$\rho_o + d_o/2 = \rho_o \frac{|\tilde{n}(I_m)|}{|\tilde{n}(I_m/2)|} \qquad (22b)$$

Insertion of Eqs. (22a) and (22b) into Eq. (21) gives

$$\frac{l_{SF}}{d_o} = \tfrac{1}{2} \frac{|\tilde{n}(I_m)| + |\tilde{n}(I_m/2)|}{|\tilde{n}(I_m)| - |\tilde{n}(I_m/2)|} \qquad (23)$$

which is independent of ρ_o. It is to be noted from Eq. (23) that the solution becomes singular and then imaginary for $|\tilde{n}(I_m/2)| \geqslant |\tilde{n}(I_m)|$; there is a threshold value of beam intensity (I_s) for each combination of parameters, below which physically correct values of l_{SF} may not be obtained. Equating the refractive indices at the intensity maximum and the half intensity maximum gives (dropping the m = max subscript)

$$\frac{2}{N}\{(1 + I_s/2I_{rel})^{-\frac{1}{2}} - (1 + I_s/I_{rel})^{-\frac{1}{2}}\}$$
$$+ \frac{[(\nu/\omega)^2_I + 1]}{1 + I_s/I_{rel}} - \frac{[(\nu/\omega)^2_{I/2} + 1]}{1 + I_s/2I_{rel}} = G(I_s) \qquad (24)$$

which is zero at a threshold intensity root. $(\nu/\omega)^2$ has been neglected compared to unity in Eq. (17) in the specification of this relation. The roots are determined by the Newton-Raphson technique. The proper numerical procedure is to first determine the threshold physical beam intensity I_s and then to gradually work upwards in intensity to determine solutions to Eq. (23).

IV. SELF-FOCUSING RESULTS

Illustrated in Fig. 3 is the l_{SF}/d_o ratio as a function of Nd glass and CO_2 gas laser intensity for N values of 0.10, 0.90, 0.95, 1.0, 1.5 and 3.0 (Nd glass) and 0.90, 0.95 and 1.0 (CO_2 gas), assuming a low temperature influence (KT = 10eV) and an ionization number $Z_I = 1$ (hydrogen plasma). The minimum intensity values for each parameter combination are illustrated by vertical shaded strips; calculations were made at initially narrow intensity increments so as to accurately determine the high gradient self-focusing length dependency. It is seen that as N exceeds unity the I_s limit is located further into the relativistic regime; in addition, the cusp-like distributions becomes more marked. It is obvious that extremely rapid self-focusing, involving lengths of the same order as the magnitude of the initial beam width between the half-irradiance maximum points, is generated by relativistic effects. The collisional dependence on n_{ec} removes the different laser plotting similarity at intensities near the thresholds.

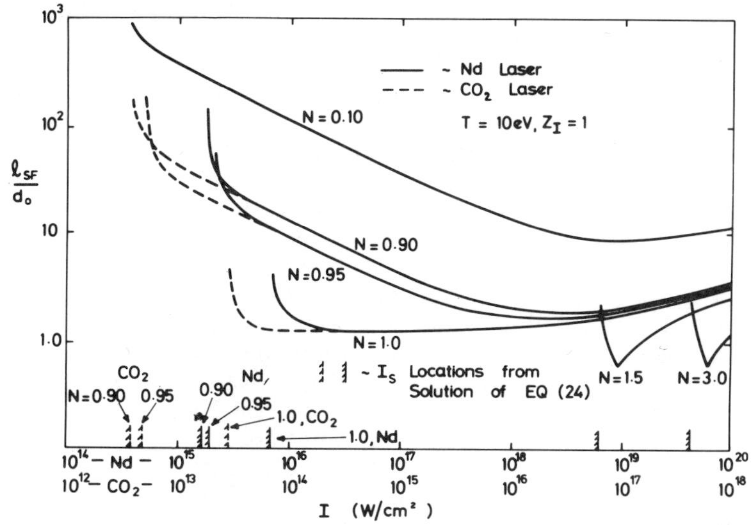

Fig. 3
Ratio of self-focusing length to beam width (distance between half-maximum intensity points) as a function of beam intensity for KT = 10 eV, $Z_I = 1$, and F = 1.0 for Nd glass and CO_2 gas laser radiation.

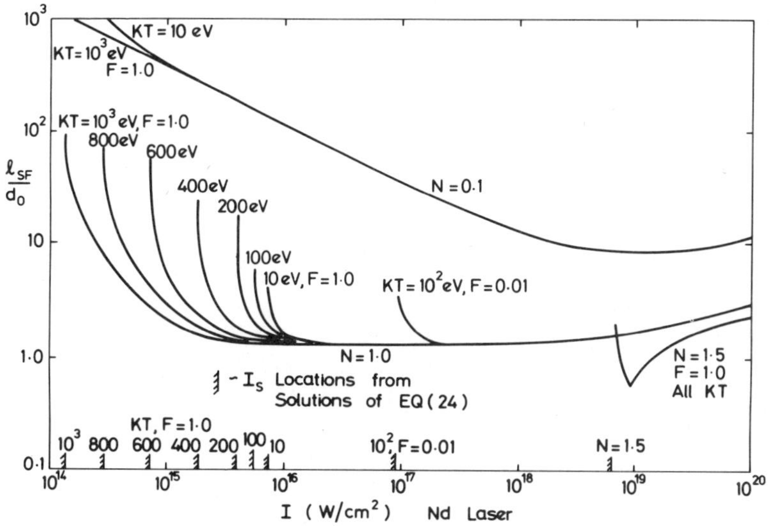

Fig. 4
Ratio of self-focusing length to beam width as a function of beam intensity for KT between 10 and 1000 eV, $Z_I = 1$, and F = 0.01, 1.0 for Nd glass laser radiation.

Fig. 4 shows, for an Nd laser and the three N values of 0.10, 1.0 and 1.5, the influence of plasma temperature on I_{SF}, which is most notable for $N \leqslant 1$. It is seen that an increase of plasma temperature to 1000 eV yields a two order of magnitude decrease in I_s with a corresponding change in the $I_{SF}(I)$ "low intensity" functional dependency. For N = 0.10 and KT = 10 eV an Nd laser I_s root was found at 1.23 x 10^{14} W/cm^2 while no root was determined for KT = 1000 eV; although the distributions continue off-scale it is seen that mutual divergence sets in at an intensity somewhat above 10^{15} W/cm^2. In contrast, the N = 1.5 distribution is the same irrespective of the temperature level. Employment of a parametric instability induced collision frequency is tested by letting F = 0.01 with KT = 100 eV, the I_s for the resulting $I_{SF}(I)$ distribution is about 15 times the F = 1.0, KT = 100 eV collisional case. The distribution blends smoothly into the F = 1.0 curves; nevertheless, collective oscillation instabilities would seem to delay the relativistic shrinkage until higher laser intensities are applied. It is of interest to plot the Nd laser threshold intensity I_s as a function of temperature near the N = 1 value (for fixed F and Z); Fig. 5 illustrates that for $N = 1 - \delta$, $\delta \ll 1$, temperature has a very marked effect on I_s and, of course, the entire near-threshold I_{SF} distribution. Conversely, for $N = 1 + \delta$, $\delta \ll 1$, temperature has essentially no effect, the electron number density being the all-important parameter.

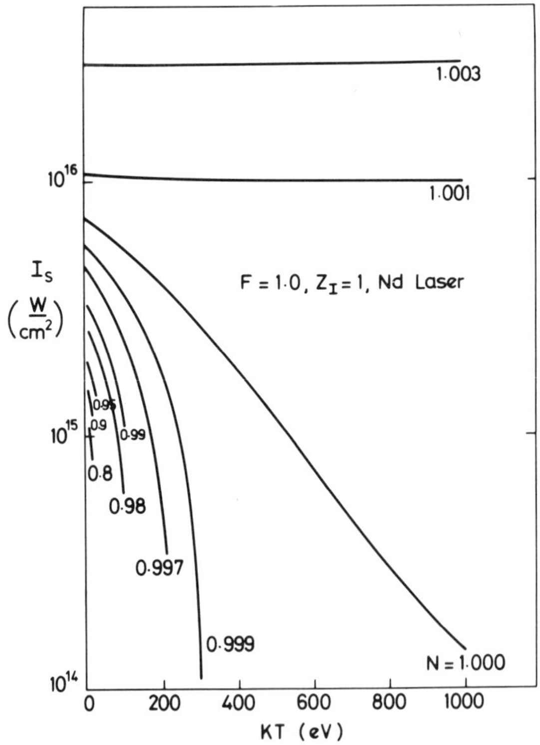

Fig. 5
Threshold intensity I_s as a function of temperature and electron number density for N values near unity (the critical cut-off density), assuming $Z_I = 1$ and Nd glass laser radiation.

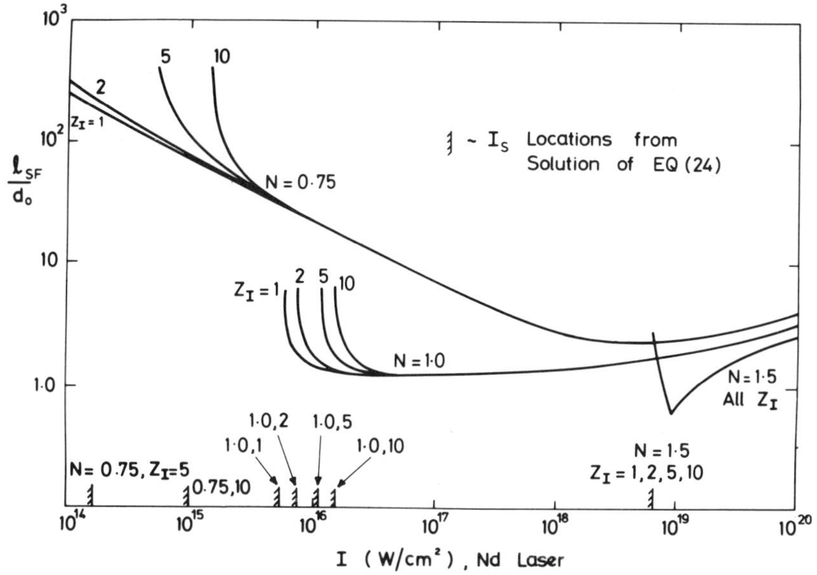

Fig. 6
Ratio of self-focusing length to beam width as a function of beam intensity for KT = 100 eV, F = 1.0 and Z_I = 1, 2, 5 and 10 for Nd glass laser radiation.

Fig. 6 shows, for an Nd laser and the three N values of 0.75, 1.0 and 1.5 (F = 1.0), the influence of ion charge number Z_I on $l_{SF}(I)$; analagous to temperature the dependency is the largest when $N < 1$. Z_I values of 1, 2, 5 and 10 are treated although laser irradiation of some metallic targets has produced ions with Z_I a good deal beyond 10 (see Section VI).

As indicated previously, I_s values calculated with $(\nu/\omega)^2$ neglected compared to one in Eq. 17 may be somewhat lower than what is physically obtainable with employment of the full $|\tilde{n}|$ relation. As an example, from Fig. 6, the Nd laser threshold I_s is determined to be 9.762 x 10^{14} W/cm² for N = 0.75, Z = 10 and KT = 100 eV, while l_{SF} is propertly determined only down to intensities around 1.5 x 10^{15} W/cm² for the same conditions. It seems that the computed value of $(\nu/\omega)^2$ at 1.5 x 10^{15} W/cm², 3.9 x 10^{-5}, which is around 10^7 times the $(\nu/\omega)^2$ value at I = 10^{18} W/cm², may not be sufficiently negligible compared to one so as to justify the approximation employed in the I_s determination. The collisional, geometric ray model, solution is indeed sensitive to small parameters within certain number density and intensity regions.

Functional dependency on the number density ratio decreases in the high relativistic regime while the influence of collisions in this region is negligible for values of N around unity or lower. This is essentially the region within which the collisionless refractive index relation of Max and Perkins,[16] Eq. (8), is most adequate. As $I/I_{rel} \to \infty$, from Eqs. (14) through (17),

$$|\tilde{n}| \simeq 1 - \frac{N}{2} \sqrt{\frac{I_{rel}}{I}}$$

and, from Eq. (23),

$$\left.\frac{l_{SF}}{d_o}\right|_{I/I_{rel} \to \infty} \simeq \frac{(I/I_{rel})^{\frac{1}{4}}}{N^{\frac{1}{2}}(\sqrt{2}-1)^{\frac{1}{2}}}$$

showing less than linear and therefore relatively weak influences of both intensity and number density on the self-focusing length. For N values around unity a small change in N produces a large change in both magnitude and shape of the $l_{SF}(I)$ distribution for $I < I_{rel}$; the distributions are mutually closer with $I > I_{rel}$.

V. COMPARISON TO WAVE EQUATION SOLUTIONS

Several groups[5,10,14,23,24] have, over the past few years, developed analytical or numerical solutions to the electric field wave equation

$$\nabla^2 \underset{\sim}{E} - \nabla(\nabla \cdot \underset{\sim}{E}) + \left(\frac{\omega}{c}\right)^2 (\tilde{n})^2 \underset{\sim}{E} = 0 \qquad (25)$$

which may be readily derived from Maxwell's electrodynamic equations and where $(\tilde{n})^2$ is given by Eq. (1). Most of the investigations have involved ponderomotive self-focusing with the neglect of collisions whereby Eqs. (3) or (6) may be employed for the refractive index and $(eE/m_o\omega c)^2 \ll 1$ (relatively low intensities) giving that

$$\tilde{n}^2 \simeq 1 - \left(\frac{\omega_p^o}{\omega}\right)^2 \left(1 - \frac{e^2E^2}{2m_o{}^2\omega^2c^2}\right) ,$$

or, approximately,

$$\tilde{n}^2 \simeq 1 - \left(\frac{\omega_p^o}{\omega}\right)^2 \exp\left\{-\frac{e^2E^2}{2m_o{}^2\omega^2c^2}\right\} \qquad (26)$$

This may be further written as, where $\beta \equiv e^2/2m_o\omega^2 KT$,

$$\tilde{n}^2 = 1 - \left(\frac{\omega_p^o}{\omega}\right)^2 \exp\{-(\beta E^2)(KT/m_o c^2)\}$$

This expression is similar to that derived from microscopic theory by Sodha, et al[25] and which is employed, for example, in Ref. 10. It is evident that for intensities approaching I_{rel} only the exact collisionless value, Eq. (6), may be employed. The majority of these solutions assume a gaussian intensity profile to solve for the geometric ray self-focusing, although a particle-in-cell form of numerical solution, whereby "photons" (defined corpuscles) are followed through a mesh field, has been performed[24]. Siegrist et al[5,14] have performed numerical-Gaussian profile solutions including both relativistic and ponderomotive influences; illustrated in Fig. 7 are some of their results together with calculations obtained by the model given in this paper. It is assumed that KT = 200 eV and Z_I = 38, consistent with results of experimental Nd glass laser irradiation of gold targets shown in Ref. 5. Correlation between the two models is only qualitative; nevertheless, it is reasonable to assume that extrapolation of the Ref. 5 result beyond the illustrated intensities would yield self-focusing lengths in good agreement with the geometric ray model calculation. The geometric ray model does, however, predict strongly collision-dependent threshold intensities at values somewhat below 10^{16} W/cm^2. The wave equation result are steep at lower intensities but possess no sudden thresholds. Nevertheless, both solution schemes predict similar values of minimum self-focusing lengths. Furthermore, even if a more exact solution is somewhat less singular than the geometric model the temperature, ionization number and instability trends illustrated in Figs. 4 and 6 should still be present.

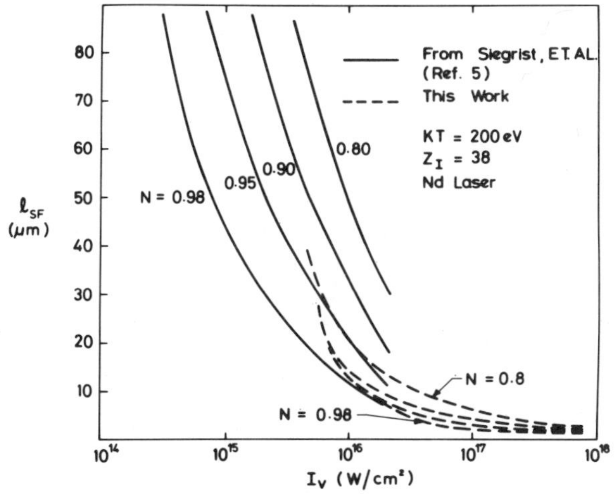

Fig. 7
Comparison of self-focusing length as a function of Nd laser intensity and electron number density to results obtained by Siegrist, Luther-Davies and Hughes (Ref. 5). KT = 200 eV and Z_I = 38.

VI. ENERGIES AND INTENSITIES WITHIN THE SELF-FOCUSING REGION

The automatic short range relativistic shrinking of intense laser beams in plasmas is a consistent explanation of the generation of electron oscillation energies exceeding 100 keV within the plasmas; the generation of multiply ionized ions of multi-GeV energy is then a consequence of nonlinear force acceleration[26]. Calculation is made in this section of the maximum intensities of the laser radiation near the focal region, where the dielectric swelling of the laser intensity near the critical density ($N \approx 1$) is included. It is shown that the swelling tends to oppose the attainment of the highest possible electron oscillation energy.

As shown in Fig. 1, the attainment of the highest laser intensities after shrinking from an initial value I_v down to a focus with a diameter near λ results in $|\tilde{n}| \approx 1$. This condition is most easily achieved in a plasma of low density ($N << 1$), but then only very long self-focusing lengths l_{SF} ($l_{SF}/d_o >> 1$) result. At densities near cut-off ($N \approx 1$), the effective wavelengths (denoted by λ_I in Eq. (19)) can swell up to much larger values than the vacuum value λ_v; therefore, the advantage of reaching very short shrinking lengths for the beam will decrease the maximum intensity near the focus. It is desirable to evaluate the optimum conditions for the shortest possible shrinking length and highest possible focal intensity and electron oscillation energy; a parallel consideration of both mechanisms is necessary.

The dielectric swelling of the laser intensity I_p in a plane wave, as discussed in a model of nonlinear force generation[15], results in an increase over the vacuum intensity I_v (with neglect of absorption) given by

$$I_p = I_v/|\tilde{n}| \tag{27}$$

For the situation of no plane waves but for a beam with a width d_o in vacuum, the shrinking causes a further geometric swelling due to self-focusing, resulting in an intensity (including the dielectric swelling of Eq. (27)) within the plasma related to the vacuum intensity of the beam and medium refractive index:

$$I = I_p\left(\frac{d_o}{\lambda_{eff}}\right)^2 = I_v\left(\frac{d_o}{\lambda_v}\right)^2 |\tilde{n}| \tag{28}$$

Now $|\tilde{n}|$ is a function of I, as may be seen from Eqs. (14) through (17); initial calculation of $|\tilde{n}|$ from specification of I_v (denoting it as I_m) gives a means by which a first order value of intensity near the beam focal point may be calculated (using Eq. (28)). This value is further inserted into Eqs. (14) through (17) and $|\tilde{n}|$ recalculated until convergence of $|(I/I_{rel})_{k+1} - (I/I_{rel})_k| \leqslant 10^{-6}$ is achieved, k being the iteration step.

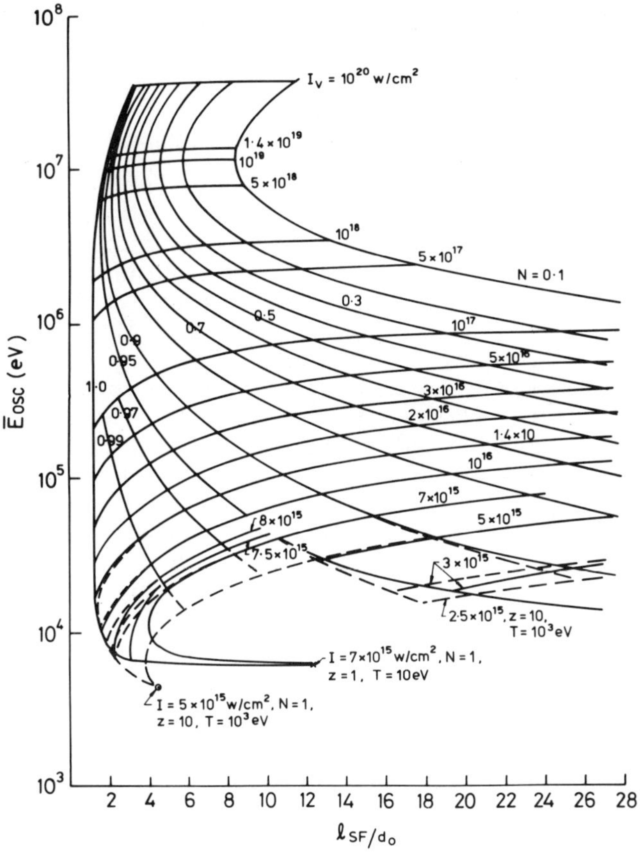

Fig. 8
Electron oscillation energy in electron volts as a function of self-focusing length, electron number density, and Nd laser intensity. KT = 10 eV, $Z_I = 1$ (solid lines) and KT = 10^3 eV, $Z_I = 10$ (dashed lines). First order non-iterative solution of laser focus intensity. Beam diameter is 30μm.

The average oscillation energy of the electrons, near the focus of the beam within the plasma, is obtained from Eqs. (10) and (13) as well as assuming that $\bar{\epsilon}_{osc} = \epsilon_{kin}/2$:

$$\bar{\varepsilon}_{osc} = \frac{m_o c^2}{2} \{(1 + I/I_{rel})^{\frac{1}{2}} - 1\} \quad (29)$$

Employment of Eq. (28) gives

$$\overline{\varepsilon}_{osc} = \frac{m_o c^2}{2} \{(1 + I_v|\tilde{n}| \; (d_o/\lambda_v)^2/I_{rel})^{\frac{1}{2}} - 1\} \tag{30}$$

Fig. 8 illustrates $\overline{\epsilon}_{osc}$ as a function of the dimensionless self-focusing length l_{SF}/d_o, for various values of the N number density ratio and the Nd vacuum laser intensity I_V. This plot is a somewhat conservative oscillation energy first iteration solution (I is not recalculated using Eq. 28) and $d_o|_{Nd}$ is assumed to be 30 μm, one-tenth the $d_o|_{CO_2}$ value of 300 μm. The $(d_o/\lambda_V)^2$ ratios are equal for the two types of lasers, so that only the aforementioned intensity translation of 10^2 W/cm^2 separates the different laser results at intensity values exceeding 10^{16} W/cm^2 (Nd) for N = 1.0 and exceeding lower intensities for decreasing values of N. As an example, for I_V = 10^{17} W/cm^2 (Nd), at any N number density, the self-focusing lengths and electron oscillation energies are entirely equivalent to the $I_V = 10^{15}$ W/cm^2 values with a CO_2 laser, where $(d_o/\lambda_V)^2$ for both lasers is 801.0. The solid lines refer to the conditions of KT = 10 eV, Z_I = 1 and F = 1.0, while the dashed lines characterise KT = 10^3 eV, Z_I = 10 and F = 1.0 results. It is seen that variations in temperature and ionization number produce negligible differences, except at the lower intensities and highest electron number densities. This result gives encouragement to the employment of this relatively simple model, uncoupled from detailed consideration of ionization chemistry and an energy balance. Another interesting result, as seen from the figure, is that the smallest self-focusing length does not correspond to the highest attained electron oscillation energy (and resulting ionkinetic energy) for any particular vacuum laser intensity. The cessation of the N = 1.0 curve (KT = 10 eV, Z_I = 1) for Nd laser I_V values below 7 x 10^{15} W/cm^2 is due to the threshold intensity numerical singularity, as discussed in Section III. The very rapid change in l_{SF}/d_o, resulting from a slight alteration in N with N = 1 - δ, $\delta \ll 1$, may be noted.

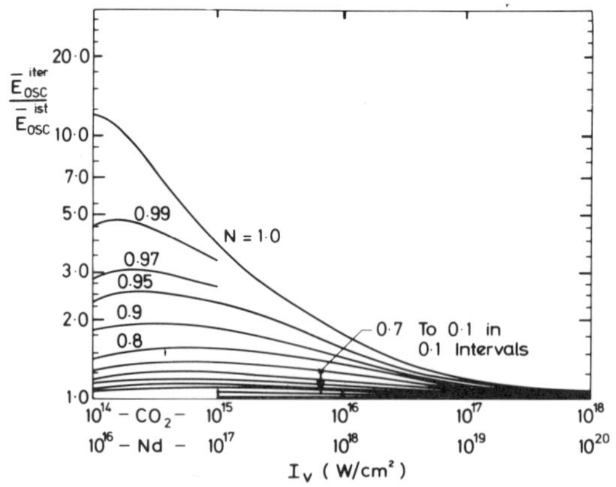

Fig. 9
Ratio of iterated to first order electron oscillation energies as a function of electron number density and both Nd and CO_2 vacuum laser intensity. KT = 10^3 eV and Z_I = 10.

Shown in Fig. 9 is the ratio of the iterated focal electron oscillation energy divided by the first iteration value, as a function of Nd glass and CO_2 gas vacuum laser intensity and electron number density. Values sufficiently beyond the collision dominated thresholds are plotted so that graph similarity between the laser types is assured; in addition, the (d_o/λ_v) ratio is equal for both lasers. Temperature and ionization number dependency is negligible. It is seen that the converged iterative value of oscillation energy exceeds the first iteration value by the greatest amount for N = 1.0 and the lowest plotted laser intensity; for lower values of N the ratio has maximums at increasingly higher intensities. Nevertheless, smaller values of N lead to a steadily decreasing oscillation energy ratio, irrespective of the intensity.

The initialization of relativistic electron self-focusing effects at intensities below 10^{-3} times I_{rel} presents the possibility that relativistic ion phenomena may enter at the very highest laser focus intensities. From Eqs. (4) and (5) we have that the electron mass ratio becomes, in terms of intensity,

$$m_e/m_o = (1 + I/I_{rel})^{\frac{1}{2}} \tag{31a}$$

Similarly, for the ion we have

$$m_I/m_I^o = \left\{1 + \frac{(Z_I e)^2 E^2}{(m_I^o)^2 \omega^2 c^2}\right\}^{\frac{1}{2}}$$

$$= [1 + (m_o Z_I/m_I^o)^2 (I/I_{rel})]^{\frac{1}{2}} \tag{31b}$$

Now m_o/m_I^o, the electron-to-ion rest mass ratio, is approximately equal to (m_o/Am_p), where A is the mass number (total number of nucleons) and m_p is the proton mass. Numerically, $m_o/m_I^o \approx 5.4466 \times 10^{-4}\ A^{-1}$. Plotted in the table are the electron and ionic mass ratios as a function of laser intensity. For the ionic masses singly ionized H ($A = 1$, $Z_I = 1$), highly ionized gold ($A = 197$, $Z_I = 38$), and highly ionized uranium ($A = 238$, $Z_I = 50$) are considered.

TABLE

(W/cm^2)					
I_{Nd}	10^{15}	10^{16}	10^{17}	10^{18}	10^{19}
I_{CO_2}	10^{13}	10^{14}	10^{15}	10^{16}	10^{17}
m_e/m_o	1.0001	1.0014	1.0136	1.1284	1.9319
m_I/m_I^o	{H; A = 1, Z_I = 1}				1.0000
m_I/m_I	{Au; A = 197, Z_I = 38}				1.0000
m_I/m_I	{U; A = 238, Z_I = 50}				1.0000

(W/cm)						
I_{Nd}	10^{20}	10^{21}	10^{22}	10^{23}	10^{24}	10^{25}
I_{CO}	10^{18}	10^{19}	10^{20}	10^{21}	10^{22}	10^{23}
m_e/m_o	5.3218	16.560	52.280	165.30	522.71	1652.9
m_I/m_I (H)	1.0000	1.0000	1.0004	1.0040	1.0397	1.3456
m_I/m_I (Au)	1.0000	1.0000	1.0000	1.0002	1.0015	1.0150
m_I/m_I (U)	1.0000	1.0000	1.0000	1.0002	1.0018	1.0177

It is seen that an increase in ion number decreases the threshold of relativistic ion phenomena; of course, increasing mass acts in the opposite direction. Employment of a 10^{18} W/cm^2 vacuum intensity Nd glass laser, giving a focal intensity of about 10^{21} W/cm^2 would, most probably, be below the onset of relativistic ionic phenomena, especially for heavier ions. Nevertheless, a 10^{19} - 10^{20} W/cm^2 Nd system would begin to enter the relativistic ion range and appropriate considerations of the ion dynamics would have to be made.

It is now possible to calculate, in terms of the relativistic electron oscillation energy at the focus, the kinetic energy of ions ejected from the plasma. The one-dimensional nonrelativistic momentum equation for the ions is

$$\frac{d}{dt}(m_I n_I v_I) = -\frac{E_v^2}{16\pi}\left(\frac{Z_I n_I}{n_{ec}}\right)\frac{\partial}{\partial z}\left(\frac{1}{|\tilde{n}|}\right) \quad (32)$$

where the right hand side comprises the nonlinear force for perpendicular incidence of radiation onto the plasma, within the WKB approximation[27], thermal pressure and viscous friction being neglected. Multiplying Eq. (32) by dZ gives for the left side

$$\frac{d}{dt}(m_I n_I v_I)\,dZ = \frac{dZ}{dt}\frac{d}{dZ}(m_I n_I v_I)\,dZ$$

$$= \frac{dZ}{dt}\, d(m_I n_I v_I) \simeq v_I d(m_I n_I v_I) \quad ,$$

considering that the mass flux is almost entirely ionic. With an assumed constant

n_I, uniform within the plasma region between the surface and the focus, we have that

$$m_I n_I d\left(\frac{v_I^2}{2}\right) = d\left(\frac{m_I n_I v_I^2}{2}\right) = -\frac{E_v^{\ 2}}{16\pi}\left(\frac{Z_I n_I}{n_{ec}}\right)\frac{d}{dZ}\left(\frac{1}{|\tilde{n}|}\right) dZ \ .$$

Integrating gives

$$\left.\frac{m_I n_I v_I^2}{2}\right]_{Z=0}^{Z=l_{SF}} = -\frac{E_v^2}{16\pi}\frac{Z_I n_I}{n_{ec}}\int_0^{l_{SF}}\frac{d}{dZ}\left(\frac{1}{|\tilde{n}|}\right) dZ$$

Assuming that ionic motion at the focus is zero and that $v_I = v_{I,max}$ at $Z = 0$ we have that

$$\frac{m_I v_{I,\ max}^2}{2} = \frac{E_v^2}{16\pi}\frac{Z_I}{n_{ec}}\left\{\frac{1}{|\tilde{n}|_{focus}} - 1\right\} \ ,$$

since at the plasma-vacuum interface $|\tilde{n}| = 1$. With $\epsilon_I^{trans} = m_I v_{I,\ max}^2/2$ being the ejected ion kinetic energy per particle there results

$$\varepsilon_I^{trans} = \frac{E_v^2}{16\pi}\frac{Z_I}{n_{ec}}\left\{\frac{1}{|\tilde{n}|_{focus}} - 1\right\} \qquad (33)$$

It is further postulated that the dielectric swelling in the focal region may be accommodated by writing Eq. (33) as

$$\varepsilon_I^{trans} = \frac{E_v^2}{16\pi}\frac{Z_I}{n_{ec}}\left\{\frac{(d_o/\lambda_{focus})^2}{|\tilde{n}|_{focus}} - 1\right\}$$

Since $\lambda_{focus} = \lambda_v/|\tilde{n}|_{focus}$ there results

$$\varepsilon_I^{trans} = \frac{E_v^2}{16\pi}\frac{Z_I}{n_{ec}}\left\{\left(\frac{d_o}{\lambda_v}\right)^2 |\tilde{n}|_{focus} - 1\right\} \qquad (34)$$

This quantity may be determined as a function of $\overline{\epsilon}_{osc}$ of the relativistic electrons by employment of Eqs. (10) and (28); we have that

$$E_v^2 = \frac{m_o^{\ 2}\omega^2 c^2}{e^2 |n|_{focus}}\left(\frac{\lambda_v}{d_o}\right)^2\left\{\left(\frac{2\overline{\varepsilon}_{osc}}{m_o c^2} + 1\right)^2 - 1\right\} \qquad (35)$$

Inserting this relation into Eq. (34) gives, since $n_{ec} \equiv \omega^2 m_o/4\pi e^2$,

$$\varepsilon_I^{trans} = \frac{m_o c^2 Z_I}{4} \left\{ 1 - \left(\frac{\lambda_v}{d_o}\right)^2 \frac{1}{|\tilde{n}|} \right\} \left\{ \left(\frac{2\bar{\varepsilon}_{osc}}{m_o c^2} + 1 \right)^2 - 1 \right\} \tag{36}$$

Eq. (36) may equivalently be written in terms of the laser intensity; employment of Eq. (30) gives

$$\varepsilon_I^{trans} = \frac{m_o c^2 Z_I}{4} \left\{ 1 - \left(\frac{\lambda_v}{d_o}\right)^2 \frac{1}{|\tilde{n}|} \right\} \left\{ \frac{I_v}{I_{rel}} \left(\frac{d_o}{\lambda_v}\right)^2 |\tilde{n}| \right\} \tag{37}$$

Therefore, ϵ_I^{trans} is directly proportional to intensity throughout the sub -- and superelativistic regimes. Fig. 10 illustrates the maximum ejected translational ion energy as a function of I_v (iterated solution at the focus), $N \equiv n_e/n_{ec}$, λ_v/d_o = 0.03533 (both lasers), and Z_I for $KT = 10^3$ eV. Heavy ion charge numbers of 50, 30, 20 and 10 over a range of Nd (CO_2) laser intensities between 7×10^{15} and 10^{20} (7×10^{13} and 10^{18}) W/cm² for $0.1 \leqslant N \leqslant 1.0$ are shown. The lines are drawn for N = 0.1, 0.3, 0.5, 0.7, 0.9 and 1.0; the N = 0.1 values being the maximum for a particular I_v and Z_I. The intensity values are sufficiently above the collisional-controlled threshold I_s values to enable both types of laser to be graphed by similar distributions (the n_{ec} difference has a negligible influence for the plotted conditions). It is seen that 100 GeV injected ion kinetic energies are exceeded with Z_I = 50, although as indicated above a relativistic ion solution may begin to have significance at the highest intensities. The ion energy dependence on electron number density is small for the higher laser intensities, as is obvious upon reflection on the Fig. 1 results; all the different curves yield nearly unity $|\tilde{n}|$ values for very high intensities.

The emission of Z_I = 38 gold ions in the experiments of Siegrist, et al[5], with the employment of a 2×10^{15} W/cm² Nd laser, may be analysed by employment of Eq. (37); assuming a 35 μ diameter vacuum laser beam gives that

$$\varepsilon_I^{trans} = 2.891 \times 10^6 \left(1 - \frac{0.000917}{|\tilde{n}|}\right) (|\tilde{n}|) \quad (\text{eV})$$

where Eq. (28) for the focal intensity has been used. With $|\tilde{n}| \approx 1$ there results ϵ_I^{trans} = 2.888 MeV with associated electron oscillation energies of 67.26 keV (from Eq. 30). These quantities are in agreement with the measured values of about 3 MeV.

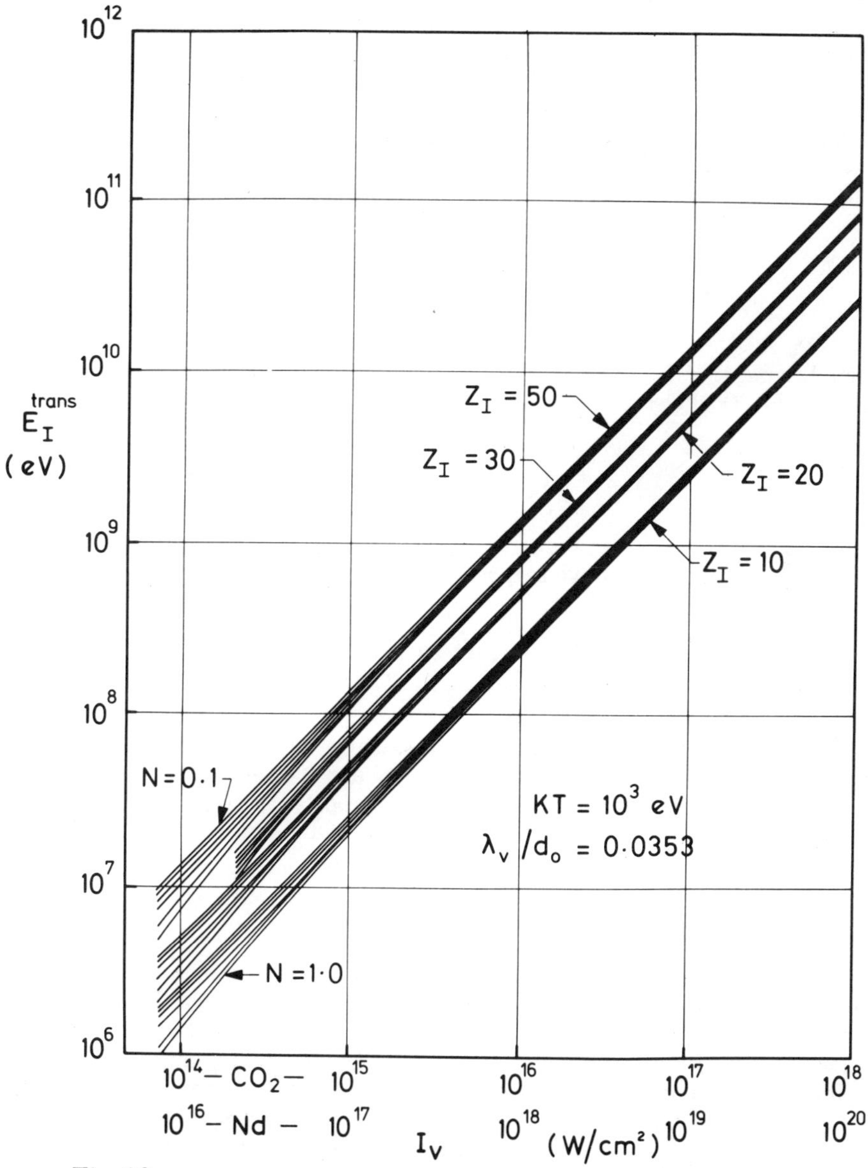

Fig. 10
Ejected translational ion energy, in eV, as a function of vacuum laser intensity, electron number density, and ion number for KT = 10^3 eV. I_V is based on 30μm for Nd laser and 300μm for CO_2 laser.

One phenomenon relating to beams (and not of plane waves with dielectric swelling) is relevant: McCall et al[7] have observed nearly the same multi-10^5 eV oscillation energies by measurement of X-rays and emitted electron energies, irrespective of whether CO_2 or neodymium glass laser radiation, at the same intensity (10^{15} W/cm^2), is employed. It is to be emphasized that the radiation wavelengths differ by a factor of ten (resulting in an equivalence of linear effects at one hundred times different intensities), yet the measured oscillation energies are nearly equivalent. In the case of these experiments a very direct explanation is possible by employing Eq. (29) in the limit of $I/I_{rel} << 1$ (the experimental vacuum intensity is still a great deal below the relativistic value); we have that

$$\bar{\varepsilon}_{osc} \simeq \frac{m_o c^2}{4} \frac{I}{I_{rel}} = \frac{e^2}{2m_o c^3 \pi} \lambda_v^2 I \tag{38}$$

with employment of Eq. (13) and $\omega = 2\pi c/\lambda_v$. In addition, the first and third equalities of Eq. (28) give

$$I = I_v \; |\tilde{n}| \, (d_o/\lambda_v)^2 \tag{39}$$

so that combination of the above two relations gives

$$\bar{\varepsilon}_{osc} = (e^2 \; |\tilde{n}| \; d_o^2/2\pi m_o c^3) I_v \tag{40}$$

or $\bar{\epsilon}_{osc} \propto |\tilde{n}| I_v$, indicating only a weak dependency on laser wave length through the refractive index. As Fig. 1 illustrates, for N density ratio $\leqslant 0.8$ and intensities $\leqslant 10^{17}$ W/cm^2, the refractive indices are nearly constant; an I shift of 10^2 W/cm^2 will result in the same $|\tilde{n}|$ results for the CO_2 and Nd lasers, so that $|\tilde{n}|$ should not have an effect on $\bar{\epsilon}_{osc}$ for the experimental conditions.

VII. CONCLUSIONS

A great deal of effort has been put into the attainment of a diffraction limited laser beam in vacuum. Conversely, high intensity light propagation in a medium automatically experiences self-focusing; indeed, relativistic effects in a plasma will produce a beam of one wavelength diameter within a penetration depth of the initial beam diameter, if the beam has one maximum and no further filamentation. This effect is not avoidable, two results of the process being the emission of highly energetic heavy ions and laser wavelength independence of oscillation energies (at equal intensities). Minimization of this phenomena, specifically for laser fusion applications, requires high spherical uniformity of pellet irradiation. Employment of thin plasma shells[28] is also helpful.

Evaluation of laser beam relativistic self-focusing lengths, electron oscillation energies, and induced ion kinetic energies is dependent upon the accurate specification of the plasma refractive index as well as the optical wave distortion process. As has been shown, an essential property of this type of wavefront bending self-focusing is the absorption described by various mechanisms, including instabilities. On the other hand, the refractive index expression (text after Eq. (8)) has a 9%

variation. This is simply due to the analytical difference between fully circular and linearly polarized waves. The plasma dielectric properties are treated in more detail in this paper, assuming that the circular ray profile is, at present, an adequate representation of the optical dynamics. It is shown that electron-ion collisional interactions impose definite threshold intensities for the initialization of relativistic self-focusing with the factors of electron temperature, ion charge number, parametric instabilities, and non-relativistic cut-off density presenting specific influences. Collisional-independent l_{SF}/d_o calculations utilizing a refractive index of the form of Eq. (17a) exhibit roughly symmetric and convex downward distributions with minimum values near I_{rel}. Dependence on temperature and ion charge number is significant for $N \leqslant 1$ values with temperature having somewhat more influence. For the most part, the minimum value of l_{SF}/d_o for a particular N is adequately determined in a KT = 10 eV, $Z_I = 1$ calculation; the threshold intensity I_s for a physically correct relativistic self-focusing solution is nevertheless strongly dependent on KT and Z_I.

Combination of self-focusing with dielectric swelling of the electric field near the cut-off density somewhat diminishes the effective focal intensities at subrelativistic conditions. Nonlinear radiation force acceleration of plasma from superintense focus areas produces translational ion energies that exceed the product of ion number times electron oscillation energy. It is therefore feasible to expect that highly charged 20 GeV ions will be produced with the application of laser powers of 10^{14} Watts.

ACKNOWLEDGEMENT

Discussions with Drs. J. L. Hughes and B. Luther-Davies are gratefully acknowledged. This work was supported in part by the Australian Research Grant Committee.

REFERENCES

1. H. Hora, J. Opt. Soc. Am. 65, 882 (1975).

2. N. L. Tsintsatse and E. G. Tsikarishvili, Astrophys. and Space Science 39, 191 (1976).

3. E. L. Kane and H. Hora, submitted to Aust. J. Phys.

4. E. L. Kane and H. Hora, Conference on Lasers and Nonlinear Optics, Tbilissi, May, 1976; R. Castillo, et al., Nucl Inst. and Methods (in print).

5. M. R. Siegrist, B. Luther-Davies and J. L. Hughes, Opt. Comm. 18, 603 (1976).

6. A. W. Ehler, J. Appl. Phys. 46, 2464 (1975).

7. G. H. McCall, Conference on Lasers and Nonlinear Optics, Tbilissi, May, 1976.

8. H. Hora and E. L. Kane, to be published in Appl. Phys. (1977).

9. S. A. Akhmanov, D. P. Krindach, A. P. Sukhorkokov and R. V. Khokhlov, JETP Lett. 6, 38 (1967).

10. M. S. Sodha, A. K. Ghatak and V. K. Tripathi, Progress in Optics XIII, Ed. E. Wolf, p. 171, (1976).

11. H. Hora, Z. Phys. 226, 156 (1969).

12. V. V. Korobkin and A. J. Alcock, Phys. Rev. Lett. 21, 1433, (1968).

13. W. M. Manheimer and E. Ott, Phys. Fluids 17, 1413, (1974); G. V. Ostrovskiya and A. N. Zaidel, Soviet Physics Uspekhi, 16, 834 (1974); J. A. Stamper and D. A. Tidman, Phys. Fluids 16, 2024 (1973); D. A. Tidman, Phys. Rev. Lett. 32, 1179 (1974); B. Bhat and V. K. Tripathi, J. Appl. Physics 46, 1141 (1975).

14. M. R. Siegrist, Opt. Comm. 16, 402 (1976).

15. J. W. Shearer and J. L. Eddleman, Phys. Fluids 16, 1753 (1973); F. F. Chen in Laser Interaction and Related Plasma Phenomena, H. Schwarz and H. Hora, eds. (Plenum, New York, 1974) Vol. 3A, p.291.

16. C. Max and F. Perkins, Phys. Rev. Lett. 27, 1342 (1971).

17. H. Hora, Laser Interaction and Related Plasma Phenomena, H. Schwarz and H. Hora, eds. (Plenum, New York, 1974) Vol. 3B, p. 819.

18. H. Hora, Laser Plasmas and Nuclear Energy, (Plenum, New York, 1975).

19. R. E. Kidder in Proc. of International School of Physics "Enrico Fermi", Physics of High Energy Density, P. Caldirola and H. Knoepfel, eds. (Academic Press,

1971) Vol. XLVIII, p. 306.

20. P. Kaw and J. Dawson, Phys. Fluids 13, 472 (1970).

21. L. Spitzer, jr., Physics of Fully Ionized Gases, (Interscience, New York, 1962).

22. J. Dawson and C. Oberman, Phys. Fluids 5, 517 (1962); T. W. Johnston and J. Dawson, Phys. Fluids 16, 722 (1973).

23. S. A. Akhmanov, A. P. Sukhorukov and R. V. Khokhlov, JETP, 23, 1025 (1966).

24. C. J. Elliott and D. B. Henderson, J. Appl. Phys., 46, 354 (1975).

25. M. S. Sodha, D. P. Tewari, A. Kumar and V. K. Tripathi, J. Phys. D: Appl. Phys., 7, 345 (1974).

26. H. Hora, Phys. Fluids, 12, 182 (1969).

27. H. Hora, Aust. J. Phys. 29, 375 (1976).

28. H. Hora, Sov. J. Quantum Electronics, 6, 154 (1976).

SELF-FOCUSING OF ELECTROMAGNETIC BEAMS IN PLASMAS: SOME RECENT INVESTIGATIONS

M. S. Sodha* and V. K. Tripathi**

*Drexel University, Philadelphia, Pa. 19104

**University of Maryland, College Park, Maryland 20742

ABSTRACT

Using the earlier expressions for nonlinear dielectric constant ($\underline{\underline{\varepsilon}}=\underline{\underline{\varepsilon}}_0+\underline{\underline{\Phi}}(EE^*)$) of isotropic and magnetoactive plasmas, we have investigated the self focusing of gaussian electromagnetic beams in short ($t<<\tau_\varepsilon$), and long ($t>\tau_\varepsilon$) time scales; τ_ε is the energy relaxation time of electrons with heavy particles and $\underline{\underline{\varepsilon}}_0=\underline{\underline{\varepsilon}}(EE^*=0)$. The wave equation has been solved in the paraxial ray and WKB approximations by expanding $\underline{\underline{\varepsilon}}$ as $\underline{\underline{\varepsilon}}=\underline{\underline{\varepsilon}}_a(EE^*|r=0)+\underline{\underline{\gamma}}r^2$. Unlike previous analyses[1] no restriction has been put on the relative magnitudes of $\underline{\underline{\Phi}}$ and $\underline{\underline{\varepsilon}}_0$. Consequently, the treatment is applicable beyond critical electron density. In a magnetoplasma, the self focusing of extraordinary mode has been studied and the treatment is applicable for all values of plasma frequency. Saturating effects of nonlinearity have been discussed in detail. Special attention to self focusing of whistlers and interpretation of Stenzel's experimental results has been given. We have also studied the filamentation of a plane uniform whistler and the growth rate of filamentation instability is seen to be a saturating function of the intensity of the beam.

1. INTRODUCTION

The earlier analyses[1-4] of the phenomenon of self focusing are not applicable around and beyond critical electron density ($\omega\gtrless\omega_p$) as they assume $\phi(EE^*)<<\varepsilon_0$ where $\varepsilon_0(=1-\omega_p^2/\omega^2)$ is the linear

*Presented at the Fourth Workshop on "Laser Interaction and Related Plasma Phenomena" held at RPI, Troy, New York, November 8-12, 1976.

part of the dielectric constant and ϕ is the change in the dielectric constant on account of nonlinearity. In the case of self focusing of ordinary and extraordinary modes in a magnetoplasma the analyses are restricted to the case when $\varepsilon_{\pm}(EE^{*})-\varepsilon_{\pm}(EE^{*}=0)<<\varepsilon_{\pm}(EE^{*}=0)(\varepsilon_{\pm}=\varepsilon_{xx}\mp i\varepsilon_{xy}$ and ε_{xx} and ε_{xy} are the components of the dielectric tensor). However, the experiments acquire interest and significance for considerably high values of plasma frequency and power density when the above mentioned restrictions are not valid.

In the present paper we have investigated the self focusing of electromagnetic beams in isotropic and magnetoactive plasmas when nonlinear part of dielectric constant is arbitrarily large as compared to linear part. The dielectric constant has been expanded as $\varepsilon=\varepsilon_a+\gamma r^2$ in the paraxial ray approximation and the wave equation has been solved in the WKB approximation, i.e., when ε_a is a slowly varying function of z $((1/\varepsilon_a)(\partial\varepsilon_a/\partial z)<(\varepsilon_a^{1/2}/\lambda_0))$; λ_0 being the free space wavelength). The conditions for uniform and oscillatory waveguide propagation have been derived. Special attention to the self focusing of whistlers and a quantitative explanation of some of the experimental results of Stenzel[5] has been given.

The nonlinearity in the dielectric constant of the plasma arises on account of redistribution of electron density in the transverse direction on account of (i) the ponderomotive force and (ii) nonuniform heating of electrons by the beam. The ponderomotive nonlinearity is set up in a shorter time and predominates in the time domain $t<<\tau_\varepsilon$ in a collisional plasma and throughout in a collisionless plasma where τ_ε is the energy relaxation time of the beam. For $t>\tau_\varepsilon$ the nonlinearity is mainly on account of nonuniform heating of electrons.

Using the earlier expressions for nonlinear dielectric constant as given by Sodha, Ghatak and Tripathi,[1] in Section 2 we have solved the wave equation for the propagation of an initially gaussian beam in an isotropic plasma. A beam width parameter f has been introduced and various aspects of self focusing have been investigated. It is seen that the self focusing of an electromagnetic beam for a given set of parameters occurs only when its power lies between two critical values P_{cr1} and P_{cr2}. For $P<P_{cr1}$ the beam suffers monotonic divergence; for $P_{cr1}<P<P_{cr2}$ the beam propagates in a self made oscillatory waveguide, with the intensity on the axis being always greater than the initial value, while for $P>P_{cr2}$ the propagation of the beam takes place in an oscillatory waveguide with the intensity on the axis being always less than the initial value. In the case of an inhomogeneous plasma (electron density increasing monotonically in the direction of propagation) a beam of high power is seen to penetrate much beyond the region of critical electron density. In Section 3 the treatment is extended to self focusing of extraordinary mode in a magnetoplasma; special attention has been given to self focusing of whistlers. On the short time ($t<\tau_\varepsilon$) scale in collisional plasmas and always in collisionless plasmas the self focusing of whistlers is possible for all frequencies whereas, on long time scale for collisional plasmas ($t>\tau_\varepsilon$) whistlers can be self focused only when $\omega>\omega_c/2$; ω and ω_c are the wave and electron cyclotron frequencies respectively. However, on both time scales the power P of whistler for self focusing must be in the range $P_{cr1}<P<P_{cr2}$. In Section 4, we have given a quantitative interpretation of experimental results of Stenzel for self focusing of whistlers. In Section 5, the growth of a small perturbation in the intensity distribution of a plane uniform whistler has been analyzed. A detailed discussion of results is given in Section 6.

2. SELF FOCUSING IN AN ISOTROPIC PLASMA: ARBITRARY NONLINEARITY

We consider the propagation of a gaussian electromagnetic beam, having time dependence $\exp(i\omega t)$ in a plasma along z direction. The intensity distribution of the beam at z=0 is given by

$$EE^*\big|_{z=0} = E_0^2 \exp(-r^2/r_0^2) \,, \tag{1}$$

where r refers to a cylindrical coordinate system. Considering slowly converging/diverging beams and following Akhmanov et al.[6] (consistent with the present analysis), the intensity distribution (as shown later in this section) for z>0 may be written as

$$EE^* = \frac{E_0^2}{f^2} \frac{\varepsilon_a^{1/2}(z=0)}{\varepsilon_a^{1/2}} \exp(-r^2/r_0^2 f^2) \tag{2}$$

where ε_a is the value of dielectric constant on the axis of the beam and f is the beam width parameter. In the presence of the beam the electron density is redistributed on account of (I) ponderomotive force or (II) nonuniform collisional heating of electrons. On short time scale ($t<<\tau_\varepsilon$) in collisional plasmas and always in collisionless plasmas (I) predominates and on long time scale ($t>\tau_\varepsilon$) (II) is dominant in collisional plasmas. Following Sodha, Ghatak and Tripathi[1] the nonlinear dielectric constant of the plasma may be written as

$$\varepsilon = \varepsilon_0 + \Phi(EE^*) \tag{3}$$

where

$$\varepsilon_0 = 1 - \omega_p^2/\omega^2 \,,$$

$$\Phi = \frac{\omega_p^2}{\omega^2} (1-\exp(-\alpha_1 EE^*))$$

for ponderomotive nonlinearity and

$$\Phi = \frac{\omega_p^2}{\omega^2}\left(1 - \frac{1}{1+\alpha_2 EE^*}\right)$$

for heating nonlinearity, where

$$\alpha_1 = \frac{e^2}{8m\omega^2 k_B T_0} \quad , \quad \alpha_2 = \frac{e^2}{6\delta m\omega^2 k_B T_0} \quad ,$$

$$\omega_p = (4\pi n_0 e^2/m)^{1/2} \quad ,$$

$-e$, m and n_0 are the electronic charge, mass and equilibrium electron density, δ is the mean fraction of excess electron energy transferred in a collision with a heavy particle (for perfectly elastic collisions with scatterers of mass M, $\delta=2m/M$ and for inelastic collisions $\delta>2m/M$), T_0 is the equilibrium plasma temperature and k_B is the Boltzmann's constant. In the paraxial ray approximation, using Eqs. (2) and (3) ε may be expanded as

$$\varepsilon = \varepsilon_a + \gamma r^2 \quad ; \quad \gamma r^2 << \varepsilon_a \; , \tag{4}$$

where

$$\varepsilon_a = \varepsilon_0 + \Phi(EE^*|_{r=0}) \; ,$$

$$\gamma = -\Phi' \frac{E_0^2 \, \varepsilon_a^{1/2}(0)}{r_0^2 \, \varepsilon_a^{1/2} f^4}$$

and $\Phi' = \partial\phi/\partial EE^*|_{r=0}$.

The scalar wave equation governing the propagation of the beam may be written as

$$\frac{\partial^2 E}{\partial z^2} + \frac{1}{r}\frac{\partial}{\partial r}\left(r \frac{\partial}{\partial r} E\right) + \frac{\omega^2}{c^2}(\varepsilon_a + \gamma r^2)E = 0 \; . \tag{5}$$

In writing Eq. (5) we have neglected a term $\underline{\nabla}(\underline{E}\cdot\underline{\nabla}\ell n\varepsilon)$ which is justified as long as $\omega_p^2/\omega^2 \ell n\varepsilon_a/\varepsilon_a << 1$. However, in the case of a two dimensional beam if we choose $\underline{E} \parallel$ y axis and $\partial/\partial y=0$, this term is identically zero. In the WKB approximation, E may be expressed as

$$E = A \frac{\varepsilon_a^{1/4}(z=0)}{\varepsilon_a^{1/4}} \exp i\left(\omega t - \frac{\omega}{c}\int_0^z \varepsilon_a^{1/2} dz\right) \; . \tag{6}$$

Then Eq. (5) gives

$$-2i\,\frac{\omega}{c}\,\varepsilon_a^{1/2}\,\frac{\partial A}{\partial z} + \frac{1}{r}\frac{\partial}{\partial r}\,r\,\frac{\partial}{\partial r}\,A + \frac{\omega^2}{c^2}\,\gamma r^2 A = 0\ . \qquad (7)$$

On separating the real and imaginary parts of A as

$$A = A_0 e^{-is}$$

we obtain from the above equation

$$\frac{\omega}{c}\,\varepsilon_a^{1/2}\,\frac{\partial A_0^2}{\partial z} + \left(\frac{\partial^2 s}{\partial r^2} + \frac{1}{r}\frac{\partial s}{\partial r}\right)A_0^2 + \frac{\partial s}{\partial r}\frac{\partial A_0^2}{\partial r} = 0 \qquad (8)$$

and

$$2\varepsilon_a^{1/2}\,\frac{\partial s}{\partial z} + \frac{c}{\omega}\left(\frac{\partial s}{\partial r}\right)^2 + \frac{\omega}{c}\,\Phi'\,\frac{E_0^2 r^2}{r_0^2 f^4} = \frac{c}{\omega A_0}\left(\frac{\partial^2 A_0}{\partial r^2} + \frac{1}{r}\frac{\partial A_0}{\partial r}\right)\ . \qquad (9)$$

Following Akhmanov, Sukhorukov and Khokhlov[6] and Sodha, Ghatak and Tripathi[1] the solution to these equations satisfying the boundary conditions expressed by Eq. (1) may be written as

$$S = \beta\,\frac{r^2}{2} + \phi_1(z)\ ,$$

$$A_0^2 = \frac{E_0^2}{f^2}\,\exp(-r^2/r_0^2 f^2)\ , \qquad (10)$$

and $\beta = \frac{\omega}{c}\,\varepsilon_a^{1/2}\,\frac{1}{f}\frac{df}{dz}$,

where f is the beam width parameter, governed by

$$\varepsilon_a\,\frac{d^2 f}{dz^2} + \frac{1}{2}\frac{d\varepsilon_a}{dz}\frac{df}{dz} = \frac{c^2}{\omega^2 r_0^4 f^3} - \Phi'\left(\frac{E_0^2\varepsilon_a^{1/2}(0)}{\varepsilon_a^{1/2} f^2}\right)\cdot\frac{E_0^2\varepsilon_a^{1/2}(0)}{\varepsilon_a^{1/2} r_0^2 f^3}\ . \qquad (11)$$

Equation (2) follows from Eqs. (6) and (10). The initial conditions on f are

$$f(z=0)=1 \quad \text{and} \quad df/dz\big|_{z=0} = 0$$

corresponding to an initially plane wavefront. Eq. (11) may be rewritten as follows for homogeneous plasmas

$$\left(\frac{df}{dz}\right)^2 = \frac{2}{\varepsilon_a}\left[\frac{c^2}{2\omega^2 r_0^4}\left(1-\frac{1}{f^2}\right) - \frac{1}{r_0^2}\int_1^f \frac{E_0^2\varepsilon_a^{1/2}(0)}{\varepsilon_a^{1/2}\,f^3}\,\Phi'\left(\frac{E_0^2\varepsilon_a^{1/2}(0)}{\varepsilon_a^{1/2}\,f^2}\right)df\right]. \tag{11A}$$

The first term on RHS of Eq. (11) corresponds to diffraction divergence and the second term corresponds to convergence due to nonlinearity. When these terms balance each other, the solution of Eq. (11) is f=1 for all values of z, i.e., the beam propagates without convergence or divergence. This is referred to as uniform waveguide propagation or self trapping of the beam. The condition for self trapping is therefore

$$\Phi'(E_{0cr}^2)E_{0cr}^2 = \frac{c^2}{\omega^2 r_0^2}\ ; \tag{12}$$

whence

$$\alpha_1 E_{0cr}^2 \exp(-\alpha_1 E_{0cr}^2) = \frac{c^2}{r_0^2\omega_p^2} \tag{12A}$$

for ponderomotive nonlinearity and

$$\alpha_2 E_{0cr}^2/(1+\alpha_2 E_{0cr}^2)^2 = \frac{c^2}{r_0^2\omega_p^2} \tag{12B}$$

for heating type nonlinearity. Eqs. (12A) and (12B) are the same as Eqs. (4.18) and (4.26) of reference 1 and the variation of E_{0cr}^2 with $(r_0\omega_p/c)$ is displayed in Figs. 4.1 and 4.5 of the reference. Eqs. (12A) and (12B) have roots E_{0cr1} and E_{0cr2} ($E_{0cr1}<E_{0cr2}$); E_{0cr1} increases with decreasing $r_0\omega_p/c$ whereas E_{0cr2} decreases with decreasing $r_0\omega_p/c$. For some value of $r_0\omega_p/c=(r_0\omega_p/c)_{min}$ the two roots are coincident; $(r_0\omega_p/c)_{min}=e^{1/2}$ for $t<<\tau_\varepsilon$ and $(r_0\omega_p/c)_{min}=2$ for $t>\tau_\varepsilon$. For lower values of ω_p ($r_0\omega_p/c<(r_0\omega_p/c)_{min}$) Eqs. (12A) and (12B) do not have real roots hence the self trapping can not occur. The beam can be self focused only when $E_{0cr1}<E_0<E_{0cr2}$ and the range ($E_{0cr2}-E_{0cr1}$) increases rapidly with increasing $r_0\omega_p/c$. Thus increasing the value of $r_0\omega_p/c$ is advantageous in two ways: (i) the critical power for onset of self focusing is reduced and (ii) the beam can be focused over a wider range of powers; the power of the beam is related to E_0 through the expression

$$P = \frac{c}{8\pi}\int_0^\infty 2\pi r dr\ \varepsilon^{1/2}\exp(-r^2/r_0^2)E_0^2\ ; \tag{13}$$

$\varepsilon=\varepsilon_a$ to a good approximation.

In the range $P_{cr1}<P<P_{cr2}$ the second term on RHS of Eq. (11) is larger than the first term and hence the beam is focused. The variation of beam width parameter with distance of propagation is displayed in Fig. 1 (for $P_{cr1}<P<P_{cr2}$); f is seen to be an oscillatory function of z. This may be explained as follows. For a given value of $r_0\omega_p/c$, when $E_0>E_{0cr1}$ the second term on RHS of Eq. (11) is greater than the first term at z=0 and d^2f/dz^2 is negative. Consequently f decreases with z. At some value of z, where $(E_0/f)\cdot(\varepsilon_a^{1/4}(0)/\varepsilon_a^{1/4})=E_{0cr2}$, the two terms on RHS cancel each other; for larger values of z, RHS becomes positive. However, the beam still continues to converge due to the curvature it has already gained though $\partial f/\partial z$ becomes less and less negative. At $z=z_f$, df/dz=0 and $f=f_{min}$. Beyond this point f increases with z, attains a maximum f=1 at $z=2z_f$ and then repeats its behavior. Thus the beam propagates in self made oscillatory waveguide. For a given value of E_0 the effect of increasing $r_0\omega_p/c$ is to increase the value of E_{0cr2} and hence to decrease the value of $(E_0/f_{min})(\varepsilon_a^{1/4}(0)/\varepsilon_{amin}^{1/4})$ (which is always greater than E_{0cr2}), i.e., f_{min} decreases with increasing $r_0\omega_p/c$.

For $P<P_{cr1}$ the diffraction effects predominate over nonlinear effects and the beam suffers divergence. For $P>P_{cr2}$ the diffraction effects at z=0 dominate over the nonlinear effects and the beam diverges. However, as the beam propagates f increases and the nonlinear effects become more and more important. Beyond a certain value of z, d^2f/dz^2 becomes -ve and df/dz starts decreasing. At $z=z_0$, df/dz becomes zero and f attains a maximum $f=f_{max}$. For larger values of z, f decreases and attains a minimum f=1 at $z=2z_0$. Thus an oscillatory waveguide is formed with r_0 and r_0f_{max} as minor and major radii. This may be referred to as oscillatory waveguide propagation with higher effective radius.

Eq. (11) is valid in an inhomogeneous plasma also. We have solved this equation for wave propagation in a plasma with equilibrium plasma frequency varying as

$$\omega_p^2=\omega_{p0}^2(1+qz) \quad , \quad (a>0)$$

and the results are displayed in Fig. 2. A beam of low power could not penetrate beyond $\omega=\omega_p$. However, the high power gaussian beam penetrates much deeper into the plasma.

3. SELF FOCUSING IN A MAGNETOPLASMA

In this section we consider the propagation of an extraordinary mode (E_x+iE_y) in a magnetoplasma along the direction of static magnetic field $\underline{B}_0$. The intensity distribution of the beam at z=0 is gaussian,

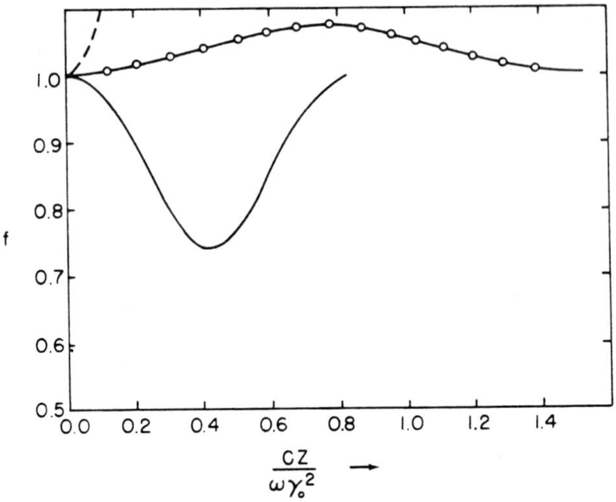

Fig. 1 Variation of beam width parameter with distance of propagation on short time scale for $\omega_p^2/\omega^2=0.9$, $r_0^2\omega_p^2/c^2=50$, $\alpha_1E_0^2=4$ (——— $E_{0cr1}<E_0<E_{0cr2}$), $\alpha_1E_0^2=6$ (—o—o— $E_0>E_{0cr2}$), $\alpha E_0^2=0.002$ (--- $E_0<E_{0cr1}$).

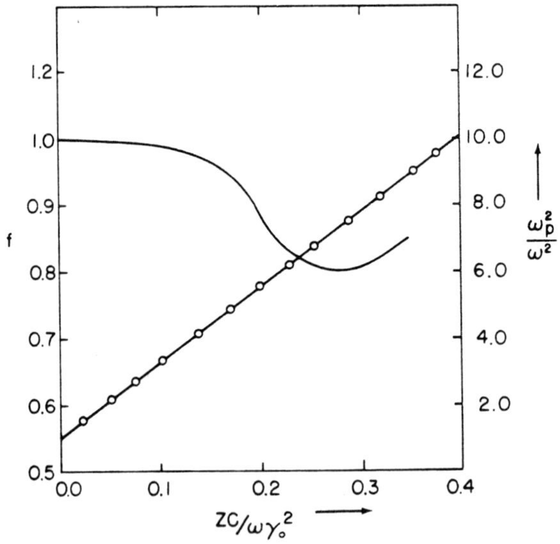

Fig. 2 Variation of beam width parameter (——) and axial plasma frequency (—o—o—) with distance of propagation for $r_0^2\omega^2/c^2=50$, $\alpha_1E_0^2=5.6$.

$$A_1A_1^*\Big|_{z=0} = E_0^2 \exp(-r^2/r_0^2) , \tag{14}$$

where $A_1=E_x+iE_y$. For $z>0$, $A_1A_1^*$ may be written as

$$A_1A_1^* = \frac{E_0^2\varepsilon_{a+}^{1/2}(0)}{\varepsilon_{a+}^{1/2} f^2} \exp(-r^2/r_0^2f^2) , \tag{15}$$

where ε_{a+} is the value of $\varepsilon_{xx}-i\varepsilon_{xy}$ on the axis of the beam and f is the beam width parameter. In the presence of the beam the components of nonlinear dielectric tensor of the plasma are given by,[1]

$$\varepsilon_{\pm}=\varepsilon_{xx}\mp i\varepsilon_{xy}=\varepsilon_{0\pm}+\Phi_{\pm}(A_1A_1^*) ,$$

$$\varepsilon_{zz}=\varepsilon_0+\Phi(A_1A_1^*) , \quad \varepsilon_{yy}=\varepsilon_{xx} , \quad \varepsilon_{yx}=-\varepsilon_{xy} ,$$

$$\varepsilon_{xz}=\varepsilon_{zx}=\varepsilon_{zy}=\varepsilon_{yz}=0 ,$$

$$\varepsilon_{0\pm} = 1 - \frac{\omega_p^2}{\omega(\omega\mp\omega_c)} , \quad \varepsilon_0 = 1 - \frac{\omega_p^2}{\omega^2} ,$$

$$\Phi_{\pm} = \frac{\omega_p^2}{\omega(\omega\mp\omega_c)}\psi , \quad \Phi = \frac{\omega_p^2}{\omega^2}\psi \tag{16}$$

$$\psi=1-\exp(-\alpha_1A_1A_1^*) \tag{16A}$$

for ponderomotive nonlinearity and

$$\psi=1-1/(1+\alpha_2A_1A_1^*) \tag{16B}$$

heating type nonlinearity where

$$\alpha_1 = \frac{e^2(1-\omega_c/2\omega)}{16mk_BT_0\omega^2(1-\omega_c/\omega)^2} , \tag{16C}$$

$$\alpha_2=e^2/12mk_BT_0\omega^2(1-\omega_c/\omega)^2\delta , \tag{16D}$$

ω_c is the electron cyclotron frequency and other symbols are the same as defined earlier. It must be noted here that α_1 is positive for $\omega>\omega_c/2$ and negative for $\omega<\omega_c/2$.

Employing the paraxial ray approximation $\varepsilon_{\pm}$ and ε_{zz} may be expanded as

$$\varepsilon_{\pm}=\varepsilon_{a\pm}+\gamma_{\pm}r^{2} \tag{17}$$

and

$$\varepsilon_{zz}=\varepsilon_{a}+\gamma r^{2} \tag{18}$$

where

$$\varepsilon_{a\pm}=\varepsilon_{0\pm}+\Phi_{\pm}(A_1A_1^{*}\big|_{r=0}) \ ,$$

$$\varepsilon_{a}=\varepsilon_{0}+\Phi(A_1A_1^{*}\big|_{r=0}) \ ,$$

$$\gamma_{\pm}=-\Phi'_{\pm}\,\frac{E_0^2\varepsilon_{a+}^{1/2}(0)}{r_0^2\,\varepsilon_{a+}^{1/2}f^4} \ ,$$

$$\gamma=-\Phi'\,\frac{E_0^2\,\varepsilon_{f+}^{1/2}(0)}{r_0^2\,\varepsilon_{a+}^{1/2}\,f^4} \ ,$$

$$\Phi'_{\pm}=\partial\Phi_{\pm}/\partial A_1A_1^{*}\big|_{r=0} \ , \qquad \Phi'=\partial\Phi/\partial A_1A_1^{*}\big|_{r=0} \ .$$

On assuming $\gamma_{+}r_0^2 \ll \varepsilon_{f+}$ and $\gamma r_0^2 \ll \varepsilon_f$ the wave equation governing the propagation of extraordinary mode can be written as (cf. Eq. 4.36a of reference 1),

$$\frac{\partial^2 A_1}{\partial z^2} + \frac{1}{2}\left(1+\frac{\varepsilon_{a+}}{\varepsilon_a}\right)\left(\frac{\partial^2}{\partial r^2}+\frac{1}{r}\frac{\partial}{\partial r}\right)A_1 + \frac{\omega^2}{c^2}\left(\varepsilon_{a+}+\gamma_{+}r^2\right)A_1=0 \ . \tag{19}$$

Following the treatment of earlier section the solution of Eq. (19) can be obtained in the WKB and paraxial ray approximations. The result is

$$A_1 = A_0\,\frac{\varepsilon_{a+}^{1/4}(0)}{\varepsilon_{a+}^{1/4}}\,\exp i\left(\omega t - \frac{\omega}{c}\int_0^z \varepsilon_{a+}^{1/2}dz - s\right) \tag{20}$$

$$S = \beta\,\frac{r^2}{2} + \phi_1 \ ,$$

$$\beta = \frac{\omega}{c}\,\varepsilon_{a+}^{1/2}\,\frac{2}{\left(1+\frac{\varepsilon_{a+}}{\varepsilon_a}\right)}\,\frac{1}{f}\frac{df}{dz} \ ,$$

and

$$A_0^2 = \frac{E_0^2}{f^2(z)} \exp(-r^2/r_0^2 f^2) \; , \tag{21}$$

where f is governed by

$$\varepsilon_{a+} \frac{d^2 f}{dz^2} + \frac{1}{2} \frac{d}{dz} \varepsilon_{a+} \frac{df}{dz} = \frac{1}{4} \left(1 + \frac{\varepsilon_{a+}}{\varepsilon_a}\right)^2 \frac{c^2}{\omega^2 r_0^4 f^3}$$

$$- \Phi_+'(E_0^2 \varepsilon_{a+}^{1/2}(0)/\varepsilon_{a+}^{1/2} f^2) \frac{E_0^2 \varepsilon_{a+}^{1/2}(0)}{2\varepsilon_{a+}^{1/2} f^3 r_0^2} \left(1 + \frac{\varepsilon_{a+}}{\varepsilon_a}\right) . \tag{22}$$

The first term on RHS corresponds to diffraction effects and the second is on account of nonlinear effects. The self focusing of the beam occurs when the second term is negative and has a magnitude larger than the diffraction divergence term. Thus the frequency range in which the self focusing of the beam is possible (i.e., the second term is negative) corresponds to

$$\left(1 + \frac{\varepsilon_{a+}}{\varepsilon_a}\right) \alpha_1 \frac{\omega_p^2}{\omega(\omega-\omega_c)} > 0 \tag{23A}$$

for $t \ll \tau_\varepsilon$ (ponderomotive nonlinearity) and

$$\left(1 + \frac{\varepsilon_{a+}}{\varepsilon_a}\right) \alpha_2 \frac{\omega_p^2}{\omega(\omega-\omega_c)} > 0 \tag{23B}$$

for $t > \tau_\varepsilon$ (heating type nonlinearity). Eq. (23A) may be satisfied at all frequencies but (23B) can be satisfied only for $\omega > \omega_c/2$, i.e., the waves of frequency $\omega < \omega_c/2$ cannot be at all focused in the plasma (regardless of power) by the heating type nonlinearity. The value of E_0 (i.e., E_{0cr}) for uniform waveguide propagation of the wave can be obtained by equating RHS of Eq. (22) to zero, i.e.,

$$\alpha_1 E_{0cr}^2 = \frac{c^2}{2\omega^2 r_0^2} \left(1 - \frac{\omega_c}{\omega}\right) \frac{1}{P_1} \left[1 + \frac{1 - \frac{1}{(1-\omega_c/\omega)} P_1}{1-P_1}\right] \tag{24A}$$

for ponderomotive nonlinearity and

$$\frac{\alpha_2 E_{0cr}^2}{1+\alpha_2 E_{0cr}^2} = \frac{c^2}{2\omega^2 r_0^2} \left(1 - \frac{\omega_c}{\omega}\right) \frac{1}{P_2} \left[1 + \frac{1 - \frac{1}{1-\omega_c/\omega} P_2}{1-P_2}\right] \tag{24B}$$

for heating type nonlinearity where,

$$P_1 = \frac{\omega_p^2}{\omega^2} e^{-\alpha_1 E_{0cr}^2} \quad \text{and} \quad P_2 = \frac{\omega_p^2}{\omega^2} \frac{1}{1+\alpha_2 E_{0cr}^2} .$$

Eqs. (24A) and (24B) again have two roots E_{0cr1} and E_{0cr2}. The variation of E_{0cr} with ω_p^2/ω^2 is displayed in Fig. 3 for a typical set of parameters. In this case again the self focusing of the beam takes place only when $E_{0cr1}<E_0<E_{0cr2}$. For E_0 in the above mentioned range the beam propagates in an oscillatory waveguide as in the case of isotropic plasma. The variation of f with z is similar as in the earlier case and we have not shown its behavior explicitly for the present case.

In the case of whistlers, viz., $\omega_p^2 \gg \omega(\omega_c-\omega)$, $\omega<\omega_c$, conditions (23A) and (23B) are very sensitive to the value of E_0. The upper limit of E_{0cr} (viz. E_{0cr2}) for self focusing of whistlers $\omega_c/2<\omega<\omega_c$ turns out to be, given by

$$\frac{\omega_p^2}{\omega^2} e^{-\alpha_1 E_{0cr2}^2} = 1 \qquad (25A)$$

on the short time scale (ponderomotive nonlinearity) and

$$\frac{\omega_p^2}{\omega^2} \frac{1}{1+\alpha_2 E_{0cr2}^2} = 1 \qquad (25B)$$

on the long time scale (heating type nonlinearity). In both cases E_{0cr2} increases with ω_p; for $E_{0cr}>E_{0cr2}$ the whistler diverges as it propagates but later as the axial wave amplitude decreases the nonlinear convergence effects become more important than the diffraction effects and the beam as a competition of these two processes propagates in an oscillatory waveguide of larger dimension. In the range $\omega<\omega_c/2$, E_{0cr2} tends to infinity on short time scale (i.e., ponderomotive nonlinearity) and a beam with $E_{0cr}>E_{0cr1}$ tends to focus to a point. On the long time scale a beam of $\omega<\omega_c/2$ can not at all be focused irrespective of power.

The lower limit on E_{0cr} (viz. E_{0cr1}) for self focusing of whistlers can be easily obtained by realizing that $\alpha_1 E_{0cr1}^2$ and $\alpha_2 E_{0cr1}^2$ are usually less than 1 and then Eqs. (24A) and (24B) simplify to give

$$E_{0cr1}^2 \simeq \left(\frac{4c(\omega_c-\omega)}{\omega_p r_0} \right)^2 m \frac{k_B T_0}{e^2} ,$$

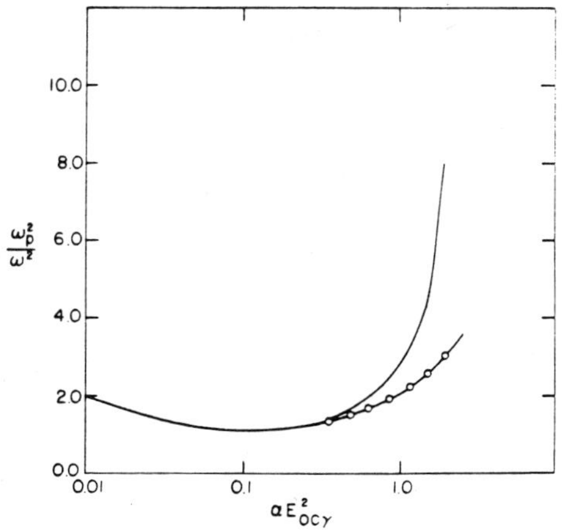

Fig. 3 Variation of ω_p^2/ω^2 with αE^2_{ocr} for uniform wave-guide propagation when $\omega_c/\omega=1.2$, $r_0^2\omega^2/c^2=50$. The continuous curve refers to $t<<\tau_\varepsilon$, $\alpha=\alpha_1$ and the —0—0—0— curve refers to $t>\tau_\varepsilon$, $\alpha=\alpha_2$.

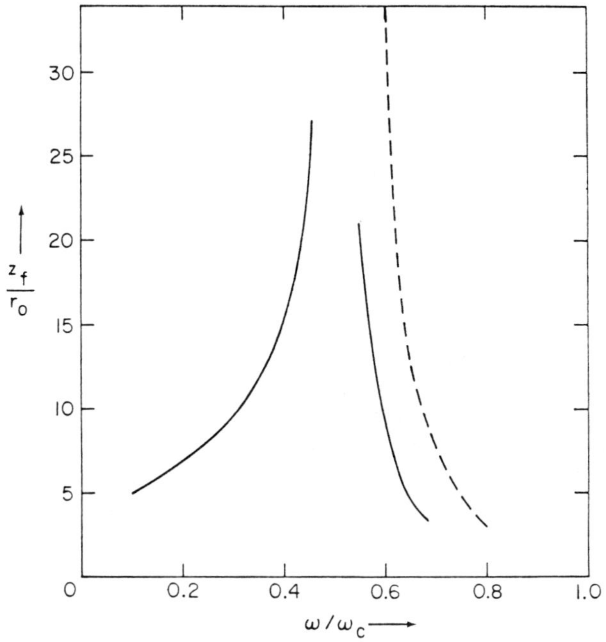

Fig. 4 Variation of self focusing length with the frequency of whistler on short time scale for $r_0\omega_p/c = 5$ and $e^2E_0^2/16mk_BT_0\omega_c^2=0.01$ (---), 0.05 (—).

$$P_{cr1} = \frac{c^3(\omega_c-\omega)^{3/2}}{\omega_p\,\omega^{1/2}}\,\frac{mk_BT_0}{e^2} \tag{26A}$$

for $t \ll \tau_\varepsilon$ and

$$E^2_{0cr1} \simeq 3\left(1-\frac{\omega_c}{2\omega}\right)\left(\frac{2c(\omega_c-\omega)}{r_0\,\omega_p}\right)^2\frac{\delta mk_BT_0}{e^2}$$

$$P_{cr1} = \frac{3}{8}\,\frac{c^3(2\omega-\omega_c)(\omega_c-\omega)^{3/2}}{\omega_p\,\omega^{3/2}}\,\frac{\delta mk_BT_0}{e^2} \tag{26B}$$

for $t > \tau_\varepsilon$. It must be noted from Eq. (26B) that E_{0cr1} could be real only when $\omega > \omega_c/2$, i.e., heating type nonlinearity can cause self focusing only when $\omega > \omega_c/2$. For $P > P_{cr1}$ (i) the beam with $\omega > \omega_c/2$ propagates in an oscillatory waveguide with a period of $2z_f$ and (ii) the beam with $\omega < \omega_c/2$ tends to focus to a point at a distance z_f. The behavior of z_f with ω_c/ω is displayed in Figs. 4 and 5.

4. INTERPRETATION OF STENZEL'S RESULTS

In Stenzel's experiment[5] the specifications of plasmas are,

plasma diameter $\sim$50 cm

$N_0 = 5\times10^{11}\,\mathrm{cm}^{-3}$, $T_0 = 6\times10^{3}\,{}^\circ\mathrm{K}$,

$B_0 = 100$ Gauss , Krypton gas pressure$=2\times10^{-4}$Torr ,

$\tau_\varepsilon \simeq 10^{-4}$sec , Power of the antenna $P_A = 0$–100 Watts.

From power of antenna, the power contained in the gaussian beam can be estimated as follows. The gaussian beam of size r_0 at a distance R ($\simeq$20 cm) from the antenna may be considered to have an effective area πr_0^2. On assuming the gain of the antenna in the direction of gaussian beam as G (which is of the order of 1 in the experiment), the power contained in the beam can be written as

$$P \simeq (\pi r_0^2/4\pi R^2)\,GP_A \;.$$

I) Experimentally a gaussian whistler ($r_0 \simeq 2$ cm, $\omega \simeq 0.73\ \omega_c$) obtained from a 12 watt antenna propagates in the steady state without any convergence or divergence. Theoretically if we choose $\delta \sim 10^{-3}$ the critical power for self focusing (i.e., for the beam to propagate without convergence or divergence) on account of heating type nonlinearity comes out to be $P_{cr} \simeq 0.05$ watts. This corresponds to $GP_A \simeq 15$ Watts and compares well with the experimental value.

II) The expressions for local electron temperature T_e and electron density N_e are given by Eqs. (5.11) and (5.21) respectively of reference 6. On taking $r_0/c_s \ll \tau_\varepsilon$ and assuming local value of wave amplitude to be quasistatic in time (which is valid up to many wavelengths from the antenna for above parameters), these expressions simplify to

$$T_e = T_0(1+2\alpha_2 A_1 A_1^*(1-e^{-t/\tau_\varepsilon}))$$

and

$$N_e = 2N_0/(T_e+T_0) \ .$$

For $\omega/\omega_c=0.72$, $P_A=80$ Watts and $\omega/2\pi=150$ MHz the radial variation of electron temperature as given by this equation is in agreement with the experimental curve (Fig. 8a, ref. 1); see Fig. 6. For other frequencies also the agreement is good. However, in the near vicinity of antenna the agreement is poor because the approximation of local heating does not remain valid.

III) From Eq. (21) the radius of curvature of the wavefront $(=(\omega/c)(\varepsilon_{a+}^{1/2}/\beta))$ of whistler (as viewed in the +z direction) in the frequency range $\omega>\omega_c/2$ is negative if f is an increasing function of z, i.e., a diverging beam (as is the case for weak power) has a converging wavefront. This conclusion is in agreement with the experiment.[5]

IV) At power of the antenna exceeding 20 watts, $\alpha_2 E_0^2 > 0.3$ the present analysis predicts an oscillatory waveguide propagation, however, experimentally, the beam appears to propagate in a uniform waveguide. Probably the resolution is insufficient to distinguish between the two types.

5. FILAMENTATION INSTABILITY OF WHISTLERS

We have seen in earlier sections that the nonlinearity introduced by a whistler of sufficiently high power has a tendency to concentrate energy around intensity maximum. A natural extrapolation of this result would suggest that a small intensity fluctuation over a uniform beam should grow in course of its propagation, i.e., the beam in course of its propagation in a plasma should break into filaments. This phenomenon is known as filamentation instability. The earlier analyses[7,8] of filamentation instability are applicable only to high frequency waves in isotropic plasmas. Here we study the filamentation instability of a plane uniform whistler in a magnetoplasma. The time scale of intensity fluctuations is usually much smaller than the energy relaxation time τ_ε and hence only ponderomotive nonlinearity is considered; the analysis can be easily extended to slow fluctuations where heating type nonlinearity predominates. For the sake of mathematical simplicity we consider the frequency range $\omega \ll \omega_c < \omega_p$. This range is very important from the point of view of ionospheric whistlers and helicons in solid state plasmas.

We consider the propagation of a plane uniform whistler in a magnetoplasma in the direction of static magnetic field. The circularly polarized electric vector of the wave may be expressed as

$$E_{0x}+iE_{0y}=A_{10}\exp i(\omega t-k_0 z) \ ,$$

$$E_{0x}-iE_{0y}=0 \ , \quad E_{0z}=0 \ , \tag{27}$$

where $k_0=(\omega/c)(\varepsilon_{0+}^{1/2})$; the nonlinear part of the dielectric tensor in the absence of any perturbation is zero because the ponderomotive force due to a uniform beam is zero. Over this uniform beam we superimpose a perturbation $\underline{E}_1(x,z)\exp i(\omega t-k_0 z)$ where $E_1(x,z)$ is not necessarily a slowly varying function of space variables. The total electric vector of the whistler may now be written as

$$E_x+iE_y=(A_{10}+A_1(x,z))\exp i(\omega t-k_0 z)$$

$$E_x-iE_y=A_2(x,z)\exp i(\omega t-k_0 z)$$

and

$$E_z=E_{z1}(x,z)\exp i(\omega t-k_0 z) \ , \tag{28}$$

where A_1, A_2 and E_{z1} are first order quantities. Following Sodha, Ghatak and Tripathi[6] (2.6.1) the ponderomotive force on electrons may now be obtained as

$$\underline{F}_p=-2\alpha_1 k_B T_0 \underline{\nabla}(A_1+A_1^*)A_{10} \ . \tag{29}$$

In writing Eq. (29) the products of perturbation quantities (i.e., $A_2A_2^*, E_{z1}E_{z1}^*$) have been neglected. This ponderomotive force gives rise to redistribution of electrons and the nonlinear dielectric tensor of the plasma is given by Eq. (16) with $A_1A_1^*$ replaced by $(A_1+A_1^*)A_{10}$.

To obtain the spatial growth of perturbation we solve the linearized wave equation

$$\nabla^2\underline{E}_1-\underline{\nabla}(\underline{\nabla}\cdot\underline{E}_1) = -\frac{\omega^2}{c^2}(\underline{\underline{\varepsilon}}_L\cdot\underline{E}_1+(\underline{\underline{\varepsilon}}-\underline{\underline{\varepsilon}}_L)\cdot\underline{E}_0) \tag{30}$$

where $\underline{\underline{\varepsilon}}_L$ is the dielectric tensor in the absence of perturbation. At this stage we introduce $\partial/\partial y=0$. Using the Maxwell's equation

$$\underline{\nabla}\cdot\underline{\underline{\varepsilon}}\cdot\underline{E}=0$$

and employing the notations of Eq. (28) we obtain,

$$2ik_0\frac{\partial A_1}{\partial z}+\frac{\partial^2 A_1}{\partial z^2}+\frac{1}{2}\frac{\partial^2}{\partial x^2}(A_1-A_2)=-k_0^2(\psi_{2+}/\varepsilon_{0+})\cdot A_{10}^2(A_1+A_1^*)-A_{10}^2\frac{\psi_{2+}}{\varepsilon_0}\cdot\frac{\partial^2}{\partial x^2}(A_1+A_1^*) \quad (30A)$$

and

$$2ik_0\frac{\partial A_2}{\partial z}+\frac{\partial^2 A_2}{\partial z^2}-\frac{1}{2}\frac{\partial^2}{\partial x^2}(A_1-A_2)=2k_0^2A_2-\frac{\psi_{2+}A_{10}^2}{\varepsilon_0}\cdot\frac{\partial^2}{\partial x^2}(A_1+A_1^*) \quad (30B)$$

where $\psi_{2+}\simeq-(\omega_p^2/\omega\omega_c)\alpha_1$ is a positive quantity. The last term on RHS of coupled Eqs. (30A) and (30B) is ω/ω_c times smaller than the other nonlinear term and hence may be dropped. On expressing A_1 and A_2 as $A_{1,2}=A_{1,2r}+iA_{1,2i}$ and separating real and imaginary parts of Eqs. (30A) and (30B) we obtain

$$\frac{\partial^2 A_{1r}}{\partial z^2}+2k_0\frac{\partial}{\partial z}A_{1i}+\frac{1}{2}\frac{\partial^2}{\partial x^2}(A_{1r}-A_{2r})=2\alpha_1A_{10}^2k_0^2A_{1r}\ , \quad (31A)$$

$$\frac{\partial^2 A_{2r}}{\partial z^2}+2k_0\frac{\partial}{\partial z}A_{2i}-\frac{1}{2}\frac{\partial^2}{\partial x^2}(A_{1r}-A_{2r})=\frac{2\omega^2}{c^2}\varepsilon_{0+}A_{2r}\ , \quad (31B)$$

$$\frac{\partial^2 A_{1i}}{\partial z^2}-2k_0\frac{\partial}{\partial z}A_{1r}+\frac{1}{2}\frac{\partial^2}{\partial x^2}(A_{1i}-A_{2i})=0\ , \quad (31C)$$

and

$$\frac{\partial^2 A_{2i}}{\partial z^2}-2k_0\frac{\partial}{\partial z}A_{2r}-\frac{1}{2}\frac{\partial^2}{\partial x^2}(A_{1i}-A_{2i})=\frac{2\omega^2}{c^2}\varepsilon_{0+}A_{2i}\ . \quad (31D)$$

Taking the (x,z) variations in the above equations as $\exp-ik_0(k_xx+k_zz)$ one may obtain the following dispersion relation

$$(-k_z^2-\frac{1}{2}k_x^2-2\alpha_1A_{10}^2)\Big[-(k_z^2+\frac{1}{2}k_x^2)\{(2+k_z^2+\frac{1}{2}k_x^2)^2-4k_z^2\}$$

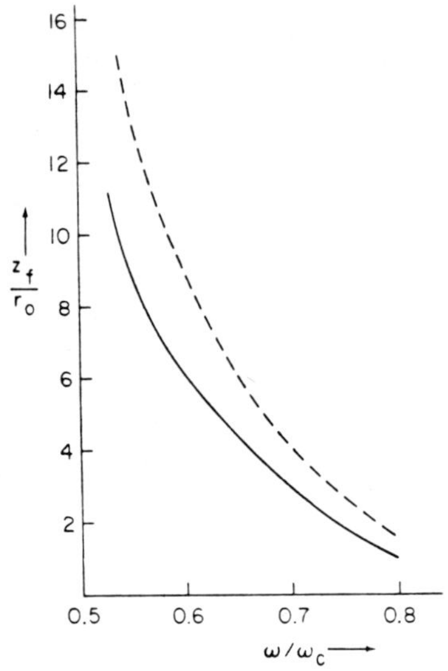

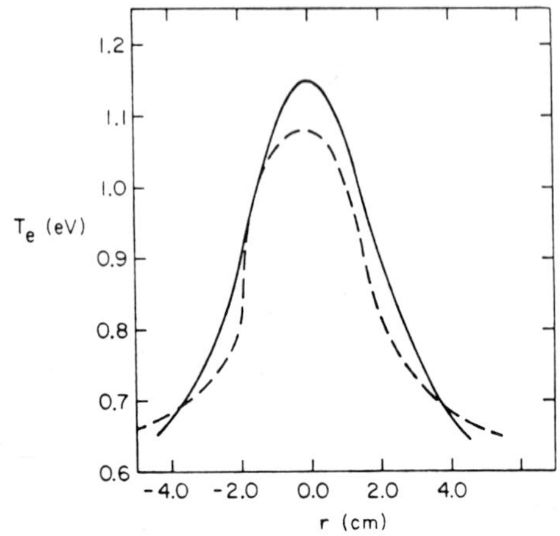

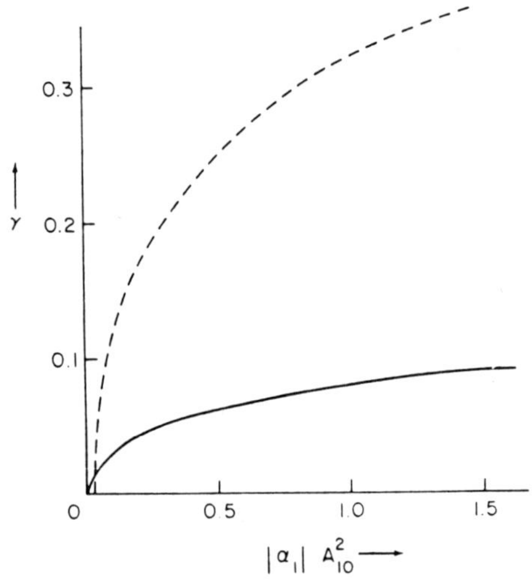

Fig. 5 Variation of self focusing length with the frequency of whistler on long time scale for $r_0\omega_p/c = 5$ and $e^2E_0^2/12mk_BT_0\omega_c^2\delta=0.01$ (---), 0.03 (——).

Fig. 6 Radial profile of electron temperature for r_0=2.5 cm, $\omega=9.4\times10^8 sec^{-1}$, ω/ω_c=0.72, $t_a\delta\nu$=16, P=0.25 Watts (P_A=80 Watts); (---) experimental curve (Fig. 8a, ref. 1), (——) theoretical curve.

Fig. 7 Variation of growth rate γ (=+ik_z) of perturbation with intensity of whistler for k_x=0.2 (——) and k_x=0.8 (---), γ is in units of k_0.

$$+ \frac{k_x^4}{4}(2+k_z^2+k_x^2/2)\Bigg] - 4k_z^2\left[\frac{k_x^4}{4} - 4k_z^2 + \left(2 + k_z^2 + \frac{k_x^2}{2}\right)^2\right] + \frac{k_x^4}{4}$$

$$\cdot \left[\frac{k_x^4}{4} - 4k_z^2 - \left(k_z^2 + \frac{1}{2}k_x^2\right)(2+k_z^2+k_x^2/2)\right] = 0 \qquad (32)$$

A general solution of this equation has to be obtained on computer. However, for $k_x^2 \gg k_z^2$ (i.e., for small scale fluctuation), Eq. (32) may be solved to obtain

$$k_z^2 = + \frac{2\alpha_1 A_{10}^2\left(2k_x^2 + \frac{k_x^4}{2}\right) + \frac{1}{4}k_x^4}{-2\alpha_1 A_{10}^2\left(\frac{k_x^4}{2} + 2k_x^2 + 4\right) + 16 + 4k_x^2 - 2k_x^4} . \qquad (33)$$

It must be remembered that k_x and k_z are dimensionless quantities as they are expressed in units of k_0. Also α_1 is negative as given by Eq. (16D). Eq. (33) shows that for high values of $|\alpha_1|A_{10}^2$, k_z^2 may become negative, i.e., the perturbation would grow as it propagates in the z direction. It must be mentioned here that the perturbation is not circularly polarized. The polarization of the perturbation has to be determined from Eqs. (31A)-(31D) after substituting the x and z dependences. The threshold for the growth of perturbation can be obtained from Eq. (33) as

$$A_{10th}^2 = -\frac{k_x^4}{\alpha_1(16k_x^2+4k_x^4)} .$$

To have an explicit nature of variation of growth rate γ_z $(=ik_z)$ with the intensity of the beam we have solved Eq. (32) numerically and the results are displayed in Fig. 7.

6. DISCUSSION

The self focusing of a gaussian beam is caused by the self induced inhomogeneity in the nonlinear dielectric constant of the plasma. The dielectric constant is a saturating function of power density of the beam. Consequently, both at very low and very high power densities the inhomogeneity in dielectric constant across the cross section of the beam is too small to overcome the diffraction effects and the beam suffers divergence. The self focusing of the beam occurs when its power P is in the range $P_{cr1}<P<P_{cr2}$. In the case of an extraordinary mode in a magnetoplasma a similar behavior is observed when $\omega>\omega_c/2$. In the range $\omega<\omega_c/2$ the nonlinear dielectric constant of the plasma on short time scale $(t<\tau_\varepsilon)$ is an exponentially increasing function of power density and hence P_{cr2} tends to infinity. A beam of power $P>P_{cr1}$ should focus to a point. However, the approximations made above for the validity of the wave equation break down near the focal point. On the long time scale the beam with $\omega<\omega_c/2$ can not at all be focused irrespective of its power. The theoretical results obtained for long time scale explain some of the experimental results of Stenzel.[5]

In the case of an inhomogeneous dense plasma a high power Gaussian beam, after redistribution of carriers, penetrates far beyond the critical density region in the paraxial approximation.

A plane uniform whistler of high intensity is seen to be unstable for small scale fluctuations, i.e., it must break up into filaments in course of its propagation. The growth rate increases with decreasing scale length of perturbation and is seen to be a saturating function of intensity of the beam.

Loter et al.[9] have recently reported experimental observations of nonstationary self focusing of a ring shaped laser in a magnetoplasma. The details are however, not sufficient to make a comparison with theory.

REFERENCES

1. M. S. Sodha, A. K. Ghatak, and V. K. Tripathi, Progress in Optics 13, 169 (1976).
2. V. V. Demchenko and A. M. Hussein, Physica 65, 396 (1973).
3. K. B. Dysthe, Phys. Lett. 27A, 59 (1968).
4. A. G. Litvak, Zh. Eksper. Teor. Fiz. (USSR) 57, 629 (1969).
5. R. L. Stenzel, Phys. Fluids 19, 865 (1976); 19, 857 (1976).
6. S. A. Akhmanov, A. P. Sukhorukov, and R. V. Khokhlov, Sov. Phys. Usphekhi 10, 609 (1968).
7. P. Kaw, G. Schmidt, and T. Wilcox, Phys. Fluids 16, 1522 (1973).
8. M. S. Sodha, A. Kumar, V. K. Tripathi, and P. Kaw, Optoelectronics 5, 509 (1973).
9. N. G. Loter, D. R. Cohn, W. Halverson, and B. Lax, J. Appl. Phys. 46, 3302 (1975).

*Work supported by National Science Foundation.
**Work supported by Center for Theoretical Physics, University of Maryland, College Park, Maryland 20742.

KINETIC ENERGY OF LASER ACCELERATED CHARGED PARTICLES IN A PLASMA AND THE POSSIBILITY OF PAIR PRODUCTION[+*]

Helmut Schwarz** and Romualdo Tabensky‡

Universidade de Brasília, Departamento de Física

70,000 Brasília, D.F., Brazil

ABSTRACT

An exact solution of the general equation of motion of charged particles in a medium of refractive index n subjected to irradiation of electromagnetic waves at any intensity and frequency is given. No approximations are necessary even for the relativistic case. Linearly polarized as well as circularly polarized waves are considered. Self-induced fields due to the generated charged particle currents have very little influence on the solutions, especially at higher intensities, where electron-positron pair production can be expected.

INTRODUCTION

Soon after the advent of lasers it had been suggested that electron-positron pair production should be possible in the almost perfectly coherent and monochromatic electromagnetic field of laser pulses.

+ Work partially supported by Brazilian Research Council (CNPq).

* Presented at the 4th Workshop on *Laser Interaction and Related Plasma Phenomena* held at Rensselaer Polytechnic Institute, Troy, N. Y., November 8-12, 1976 and at the 2nd US-Japan Seminar on *Laser Interaction with Matter*, University of Rochester, N. Y., November 2-5, 1976; to be published in part in "Optical and Quantum Electronics"(London 1977).

**Also with Rensselaer Polytechnic Institute, Troy, N.Y. 12181 USA.

‡ Present address: Joseph Henry Laboratory, University of Princeton, Princeton, New Jersey 08540 USA.

There are mainly two possibilities of such pair production:

a) by the collision of two[1] or more[2] photon beams in vacuum

or b) by, for example, scattering of an electron in the Coulomb potential of a nucleus[3], while the electron is being accelerated within the high intensity electromagnetic field.

Relativistic motion is certainly required. Exact solutions will be presented for the movement of the electrons in a dispersive medium, a plasma, of known refractive index n. Most previous solutions of the equations of motion of electrons in a periodically changing field were given in vacuum; for reference to such solutions see J. H. Eberly's review[4].

CONDITIONS FOR PAIR PRODUCTION

Vacuum Pair Production

Possibility a), vacuum pair production by multiphoton absorption, has originally been worked out by H. R. Reiss[5,6] and followed up by A. I. Nikishov and V. I. Ritus[7], F. V. Bunkin and I. I. Tugov[8] a. o.[2,9], but has, as far as we know, never been observed due to the very high laser intensities required. However, H. R. Reiss's suggestion[6] to experiment with two photon beams of energies differing by nine orders of magnitude - one 1.17eV (Nd doped high intensity glass laser), the other 18GeV (bremsstrahlung beam from a linear accelerator) - brings vacuum pair production closer to experimental reality. Also, A. A. Varfolomeyev's[2] idea of intersecting a third light beam under certain spatial configuration of the three fields makes the cross section for pair production depend on the third power of the light intensity.

Trident Pair Production

Nevertheless, possibility b), the so-called trident pair production, as first pointed out by V. P. Yakovlev[10], appears to be closer to experimental realization than possibility a) since laser intensities several orders of magnitude lower[11,12] should yield appreciable pair production rates with beams in the optical range. Hereby, it is sufficient that the laser field does not change much during the scattering process. This imposes conditions on the density of the plasma in which the electron has to be accelerated to a maximum kinetic energy T; the "scattering duration" t_s should be much shorter than one light pulse ($t_s \ll 1/\omega$; ω being the frequency of the laser light). For kinetic energies $T > m_o c^2$ (m_o - electron rest mass, c - velocity of light in vacuum) t_s is mainly determined by the plasma density N through the Debye length[13] $h = (\varepsilon_o k T_p N^{-1} e^{-2})^{1/2} \simeq 69.1\sqrt{T_p/N}$ (h in m, if T_p temperature of plasma is entered in K and N in m^{-3}; $\varepsilon_o = 8.85\times10^{-12}$ F m^{-1} is the permittivity of free space; $k = 1.38\times10^{-23}$ J/K is Boltzmann's constant; $e = 1.60\times10^{-19}$ C is the elementary charge). It would be approximately $t_s = h/c$. Applying a Neodymium doped glass laser ($\omega = 1.78\times10^{15}$ rad/s), the density of the plasma suitable for trident pair production should be $N \gg 10^{17}T_p$ or for $T_p = 10^6$K, $N \gg 10^{23}$ m^{-3}, a value three orders of magnitude below the classical cut-off density of the plasma for that radiation frequency.

The probability for trident pair production has been calculated by H. J. Bhabha[3] and T. Erber[14]. J. W. Shearer et al[11] have refined their calculations; for pair production due to scattering in the Coulomb field of a nucleus of proton number Z they obtained a cross section σ_T in cm^2:

$$\sigma_T \simeq 4.1\times10^{-33}\, Z^2 \left(\frac{T}{m_o c^2} - 2 \right)^{3.6} \tag{1}$$

Electron-positron pair creation with the trident process can be expected for kinetic energies $T > 2m_oc^2 \simeq 1$ MeV.

DETERMINATION OF KINETIC ENERGY

Losses

Inverse Faraday effect. A. D. Steiger and C. H. Wood[15] have given a solution of the electron motion in a circularly polarized field considering also the self-induced magnetic field due to the circling electrons; they found that the superimposed self-induced field has little influence on the solutions for higher intensities. The influence of such fields should be even less in a linearly polarized field due to the 8-shape trajectories.

Radiation. Also, radiation losses should be higher in the circularly polarized case. Radiation of an electron per unit time moving with a velocity $\vec{v}$ and acceleration $\dot{\vec{v}}$ is given by[16]:

$$\frac{dE_{rad}}{dt} = -\frac{e^2}{6\pi\varepsilon_o c^3}\, \frac{\dot{\vec{v}}^2 - (\dot{\vec{v}}\times\vec{v}/c)^2}{\left(1 - \frac{v^2}{c^2}\right)^3} \tag{2}$$

($c=3X10^8 ms^{-1}$ - speed of light in vacuum.) For a circularly polarized electromagnetic field of period $\tau = 2\pi/\omega$ the acceleration is simply $\dot{v} = \omega v$ and is always perpendicular to the velocity. The radiation loss ε_{rad} per oscillation is then given by:

$$\varepsilon_{rad} = - \frac{e^2\omega}{3\varepsilon_o c} \gamma^2 (\gamma^2 -1) \tag{3}$$

where
$$\gamma = (1 - \frac{v^2}{c^2})^{-1/2} \tag{4}$$

The ratio between energy loss due to radiation and kinetic energy T is:

$$\varepsilon_{rad} /T = \frac{e^2\omega}{3\varepsilon_o m_o c^3} \gamma^2 (\gamma + 1) \tag{5}$$

For circularly polarized light originating from a neodymium-doped glass laser ($\omega = 1.78X10^{15}$ rad s^{-1}) in vacuum this ratio is plotted against the light intensity I in Fig. 1, whereby $\gamma=\sqrt{1+\alpha^2}$ with $\alpha^2 = \frac{e^2 I}{\varepsilon_o \omega^2 m_o^2 c^3}$ as given later. One can see from Fig. 1 that within the range of light intensities needed for pair production the radiation losses are less than 1%; only at an intensity $I \simeq 10^{23}$ W/cm^2 all energy would be lost by radiation.

Refractive Index

In this paper we give an exact analytical solution for the kinetic energy of the electron movement at all intensities for all refractive indices $n \leqq 1$. It turns out that at a given light intensity I, kinetic energies of electrons accelerated in a plasma

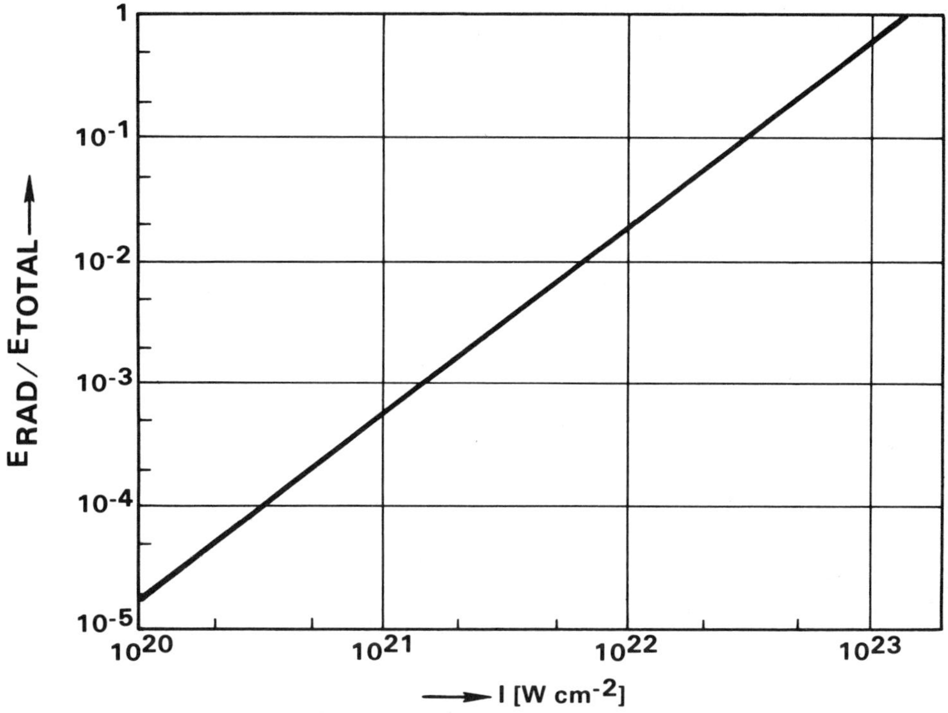

Fig. 1: Radiation energy loss from electron accelerated within a circularly polarized laser field. (ω=1.78x10^{15}rad s^{-1})

of refractive index n<1 could be between n^{-1} and $\sqrt{n}^{-1}$ times larger than those in vacuum. Dense plasmas can yield a refractive index as low as 10^{-5}. The refractive index n depends[17] on the intensity of the electromagnetic field, the plasma density and temperature as well as on the light frequency. Its value for circularly[18-21] polarized light differs from that for a linearly[20,21] polarized field. The dispersion relation (n^2) had been derived analytically for the linearly polarized case either for large[20-22] α or for relatively small, but still weakly relativistic[23,24] α; however, numerical solutions for an arbitrary α have been derived by A. Ferrari et al[25].

In our solutions we assume that the refractive index is known. For circularly polarized fields[18-21] in a collisionless plasma of density N we have

$$n = [1 - \frac{\omega_p^2}{\omega^2} (1 + \alpha^2)^{-1/2}]^{1/2} \tag{6}$$

for the linearly polarized case

$$n \simeq (1 - \frac{\pi}{2} \frac{\omega_p^2}{\omega^2} \alpha^{-1})^{1/2} \quad , \text{ if}^{20} \quad \alpha >> 1 \tag{7}$$

and

$$n \simeq \{ 1 - \frac{\omega_p^2}{\omega^2} [1 - \frac{\alpha^2}{2} (\frac{3}{4} - \frac{\omega^2 - \omega_p^2}{4\omega^2 - \omega_p^2})]\}^{1/2} \tag{8}$$

in weakly relativistic electron movements[23]; hereby ω_p is the plasma frequency as determined by:

$$\omega_p = (\frac{Ne^2}{m_o \varepsilon_o})^{1/2} \tag{9}$$

For more general cases see references 17, 25 and 26.

Exact Solutions

We shall consider the exact motion of an electron in a linearly polarized plane wave and in a circularly polarized wave within a dispersive plasma with refractive index n in order to calculate the kinetic energy T of the electron being:

$$T = m_o c^2 (\gamma - 1) \tag{10}$$

The exact general solutions of the classical relativistic electron movement in vacuum (n=1) had been given for both cases already as early as 1947-1949 by N. D. Sengupta[27] and A. H. Taub[28] and also as a problem by L. D. Landau and E. M. Lifshitz[29]. From their solutions it is easy to obtain the relativistic momentum p which leads directly to $\gamma = E/m_o c^2$(E - total relativistic energy); in the vacuum case we arrive at:

$$\gamma = \sqrt{1 + a^2\alpha^2} \tag{11}$$

whereby $a^2 = 1$ for circularly polarized light and

$$a^2 = \left(1 + \frac{9}{16}\alpha^2\right)\left(1 + \frac{1}{2}\alpha^2\right)^{-1} \tag{12}$$

for the maximum kinetic energy in linearly polarized light. For very large α where the kinetic energy becomes $T = m_o c^2 a\alpha$, it turns out that the factor a for the linearly polarized case in vacuum becomes $\frac{3}{4}\sqrt{2}$, which means that the maximum kinetic energy in a linearly polarized field is about 6% greater than that obtained in a circularly polarized field. This factor has been omitted by H. Hora[30,31] and is smaller than the value which M. B. Nicholson-Florence[32] extracted from the work of C. E. Max and F. Perkins[20].

Solutions of the equation of motion in a plasma (n<1) have also been attempted -- besides by those already mentioned -- by W. Lünow[33] and C. S. Lai[34].

It has been shown[35] that in order to obtain the kinetic energy of a charged particle in a circularly polarized field it is not necessary to solve the equation of motion. One simply enters with the expression of momentum p from

$$\omega p = eE_o \tag{13}$$

(E_o - being the peak electric field) into the relativistic energy equation:

$$T = m_o c^2 \left[\sqrt{1 + \left(\frac{p}{m_o c}\right)^2} - 1\right] \tag{14}$$

and arrives at:

$$T = m_o c^2 \left(\sqrt{1+\alpha^2} - 1\right) \tag{15}$$

with

$$\alpha = \frac{eE_o}{m_o c\omega} \tag{16}$$

α^2 can be written in terms of the field intensity I considering the relationship

$$E_o^2 = \frac{I}{n\varepsilon_o\ c} \tag{17}$$

and will then yield with Eqs. 11 and 12 the correct kinetic energy, but only for the vacuum case (n=1) and for the circularly polarized case in any medium having a refractive index $n<1$; a=1. But one cannot proceed this way -- as this had been done before[30,31] -- for the linearly polarized case in a medium of refractive index $n<1$ due to the contribution of the longitudinal component originating from the $\vec{v} \times \vec{B}$-term. This cannot be neglected in the relativistic case, since it contains as factors one velocity component divided by the phase velocity c/n. The omission of this term results incidentally in the correct value of the kinetic energy for the circularly polarized case even for refractive indices $n<1$ but not for the linearly polarized case as assumed[30]. Exact solutions for the linearly polarized cases are more complicated.

We consider the linearly polarized plane wave as propagating in the x-direction which means that the plane wave is given by:

$$\vec{E} = \vec{j}\ E_o \cos[\omega(t - \frac{n}{c}x)] \tag{18}$$

$$\vec{B} = \vec{k}\ \frac{n}{c}E_o \cos[\omega(t - \frac{n}{c}x)] \tag{19}$$

where $\vec{E}$ denotes the electric field and $\vec{B}$ the magnetic flux density within the medium of refractive index n. $\vec{i}$, $\vec{j}$, and $\vec{k}$ are the Cartesian unit vectors along the x-, y-, and z-axes.

A circularly polarized wave can be represented by the following two equations:

$$\vec{E} = E_o \{ \vec{j} \cos[\omega(t - \frac{n}{c} x)] + \vec{k}\lambda \sin[\omega(t - \frac{n}{c}x)]\} \tag{20}$$

$$\vec{B} = \frac{n}{c} E_o \{ -\vec{j}\lambda \sin[\omega(t - \frac{n}{c}x)] + \vec{k} \cos[\omega(t - \frac{n}{c} x)]\} \tag{21}$$

whereby $\lambda = +1$ for left-circular helice
and $\lambda = -1$ for right-circular helice.

The equation of motion for the electron is given by the Lorentz equation:

$$\frac{d\vec{p}}{dt} = e(\vec{E} + \vec{v}\times\vec{B}) \tag{22}$$

whereby $\vec{p} = m_o\vec{v}\gamma$ is the relativistic momentum.

Vacuum case (n=1). Solutions of the linearly polarized case, as, for example given by L. D. Landau and E. M. Lifshitz[29], show that the trajectories of the electrons describe an eight. The maximum longitudinal amplitude x_{max} can become as large as 17.7% of the maximum lateral amplitude y_{max}. The ratio between the two is given by:

$$y_{max}/x_{max} = 4\sqrt{2}\ \sqrt{1 + 2/\alpha^2} \tag{23}$$

The trajectory for the case $\alpha = 0.2023$, corresponding to a laser intensity $I = 10^{17}$ W/cm^2 (see Eqs. 16 and 17) is depicted in Fig. 2; x- and y-axes are given in units of the maximum longitudinal amplitude ($x_{max} = 8.44$ Å).

The exact value for γ in the linearly polarized case is:

$$\gamma = [1 + \frac{1}{4} \alpha^2(1 + 2 \sin^2\phi_v)](1 + \frac{1}{2}\alpha^2)^{-1/2} \tag{24}$$

whereby ϕ_v is the phase of the wave in vacuum (n=1):

$$\phi_v = \omega(t - \frac{x}{c}) \tag{25}$$

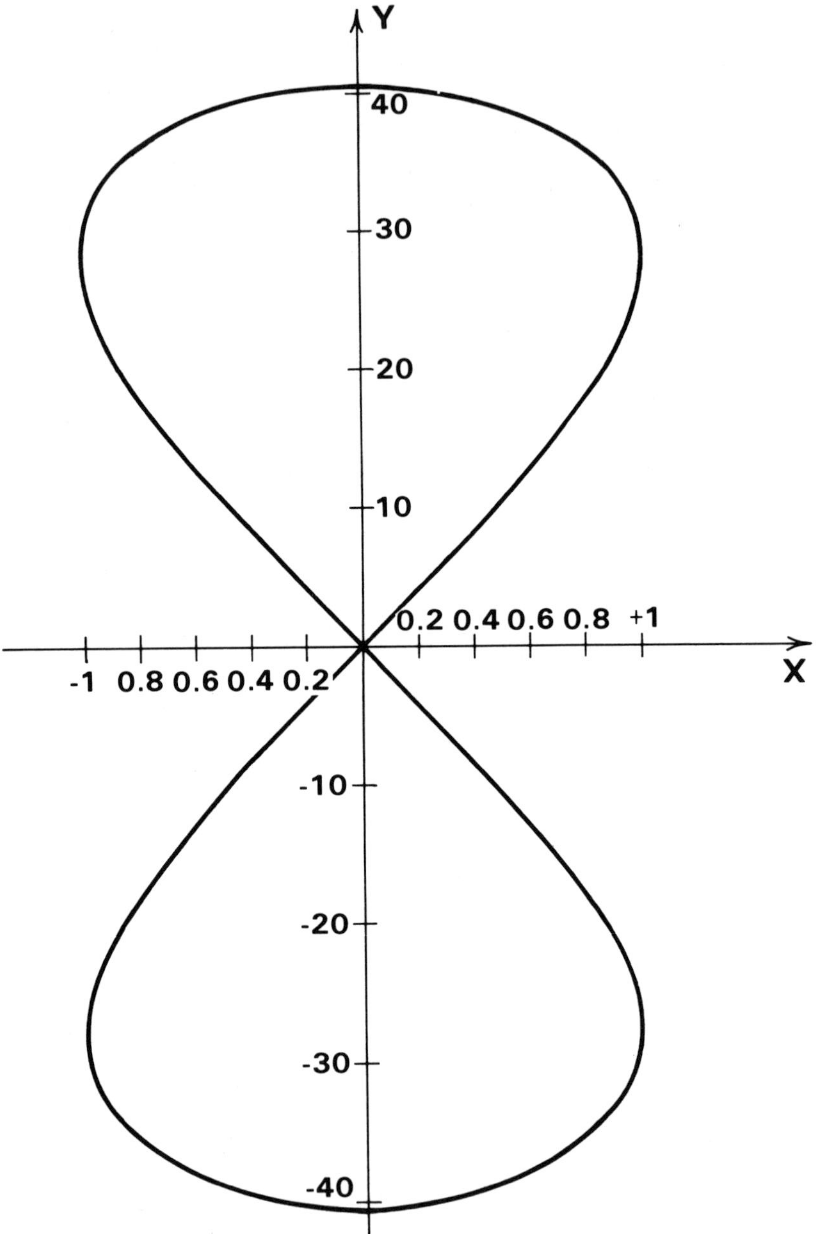

Fig. 2: Trajectory of electron accelerated within a linearly polarized laser field in vacuum for α = 0.2023, corresponding to I = 10^{17} W/cm^2; (x_{max} = 8.44 Å) (ω = 1.78x10^{15} rad/s).

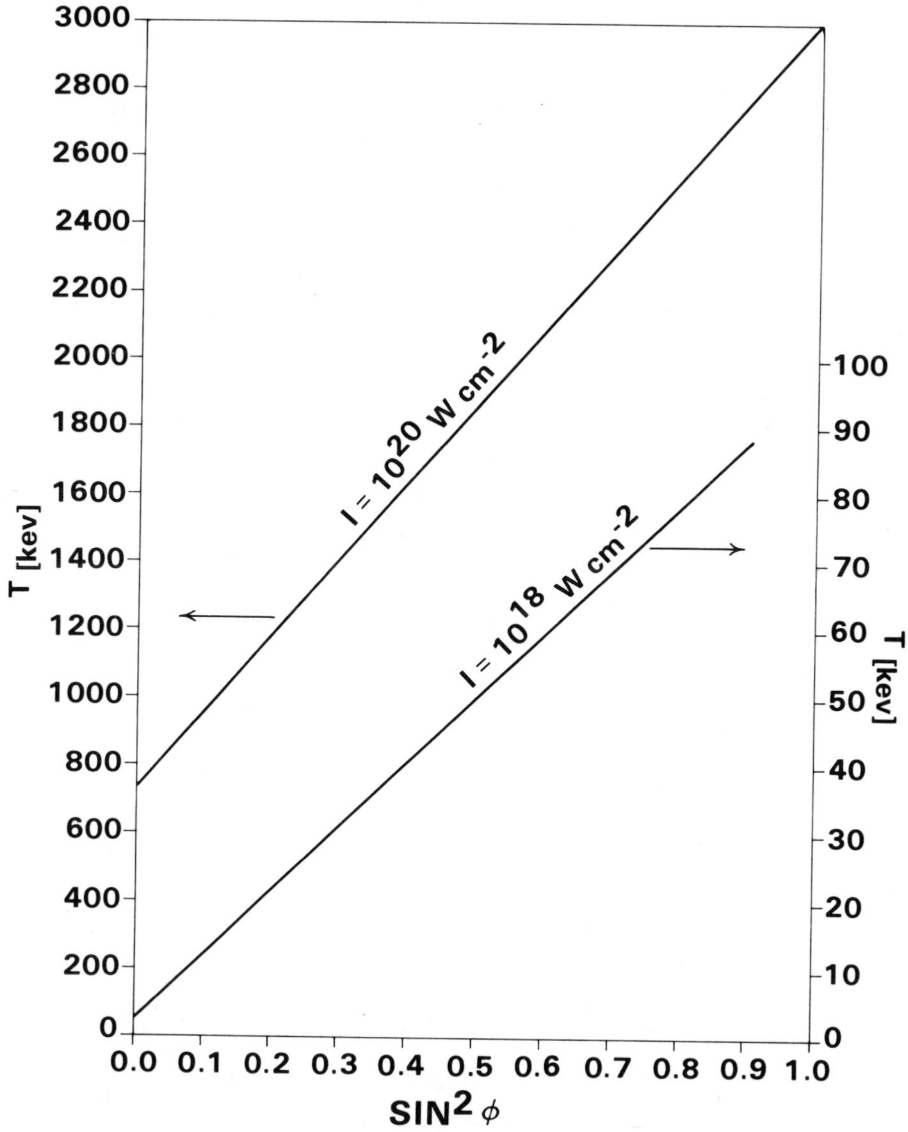

Fig. 3: Kinetic energy of an electron accelerated in a linearly polarized laser field as a function of phase angle ϕ_v for two different light intensities, I = 10^{18} W/cm^2 and 10^{20} W/cm^2 (ω = 1.78x10^{15} rad/s).

Figure 3 depicts two curves showing the dependence of the kinetic energy T on this phase angle ϕ_V for two different light intensities $I = 10^{18}W/cm^2$ and $10^{20}W/cm^2$ originating from an Nd doped glass laser ($\omega = 1.78x10^{15}$rad/s).

The maximum ($\phi_V = \pi/2$) and minimum ($\phi_V = 0$) kinetic energies of electrons accelerated in a linearly polarized laser field (ω = $1.78x10^{15}$ rad/s) in vacuum are plotted on Fig. 4 as a function of light intensities I [W/cm^2] as well as the corresponding curve in a circularly polarized field.

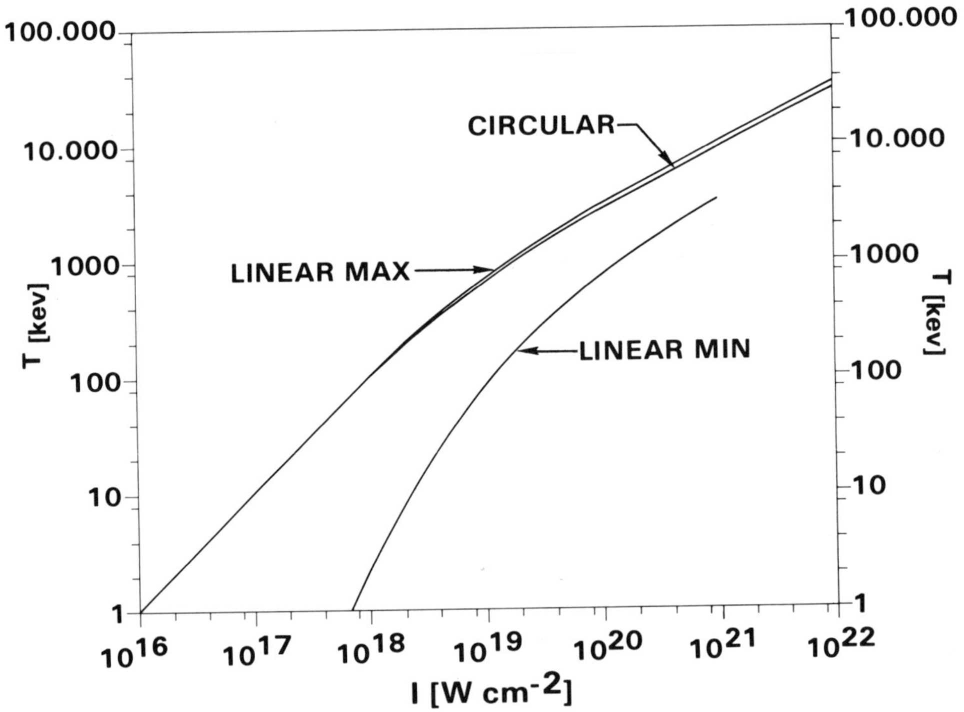

Fig. 4: Maximum ($\phi_V = \pi/2$) and minimum ($\phi_V = 0$) kinetic energies of electrons accelerated in a linearly polarized laser field ($\omega = 1.78x10^{15}$ rad/s) as well as in a circularly polarized field in vacuum (n = 1) as a function of light intensity I [W/cm^2].

Non-vacuum case ($n < 1$).
Linearly polarized. In order to solve Eq. 22 it is customary and frequently useful to write the associated energy equation:

$$\frac{d}{dt}(m_o c^2 \gamma) = e\vec{v}\cdot\vec{E} \tag{26}$$

Combining Eqs. 18, 19, 22, and 26, we obtain in component form the following four equations:

$$\frac{d}{dt}(\dot{z}\gamma) = 0 \tag{27}$$

$$\frac{d}{dt}(\dot{x}\gamma) = \frac{eE_o n}{m_o c}\,\dot{y}\cos[\omega(t - \frac{n}{c}x)] \tag{28}$$

$$\frac{d}{dt}(\dot{y}\gamma) = \frac{eE_o}{m_o}(1 - \frac{n}{c}\dot{x})\cos[\omega(t - \frac{n}{c}x)] \tag{29}$$

and

$$\frac{d\gamma}{dt} = \frac{eE_o}{m_o c^2}\,\dot{y}\cos[\omega(t - \frac{n}{c}x)] \tag{30}$$

Dots indicate as usually time differentiations. Equation 30 is the same as Eq. 26 divided by the constant rest energy and performing on the right hand side the scalar product of $\vec{v}$ with Eq. 18. Very useful information is obtained from comparing Eqs. 28 and 30, namely:

$$\frac{d}{dt}[\gamma(1 - \frac{\dot{x}}{nc})] = 0 \tag{31}$$

or

$$\gamma(1 - \frac{\dot{x}}{nc}) = \gamma_o = \text{constant} \tag{32}$$

We now use the phase of the wave ϕ as a new variable instead of time, i.e.

$$\phi = \omega(t - \frac{n}{c}x) \tag{33}$$

This substitution allows immediate integration of Eq. 29, because by elimination of the factor $(1 - n\dot{x}/c = \dot{\phi}/\omega)$ Eq. 29 can be written now as:

$$\frac{d}{d\phi}(\gamma\dot{y}) = \frac{eE_o}{m_o\omega}\cos\phi \tag{29a}$$

and we arrive at:

$$\gamma\dot{y} = \frac{eE_o}{m_o\omega}(\beta + \sin\phi) \tag{34}$$

where β is an integration constant.

We find it convenient to define a new dependent variable η:

$$\eta = (n\omega\gamma_o)^{-1}\gamma\dot{x} \tag{35}$$

For the integration of Eq. 28 we also change the time variable to ϕ and use the variable η in the following way:

$$\omega\left(1 - \frac{n}{c}\dot{x}\right)\frac{d}{d\phi}(\gamma\dot{x}) = \frac{eE_o n}{m_o c\gamma}\gamma\dot{y}\cos\phi$$

or

$$\frac{d\eta}{d\phi} = \alpha^2 c\,\frac{\beta + \sin\phi}{\omega\gamma_o}\,\frac{\cos\phi}{\gamma\left(1 - \frac{n}{c}\dot{x}\right)} \tag{36}$$

Now we express γ in terms of Eqs. 32 and 35:

$$\gamma = \gamma_o\left(1 + \frac{\omega}{c}\eta\right) \tag{37}$$

and

$$\dot{x} = \frac{n\omega\eta}{1 + \frac{\omega}{c}\eta} \tag{38}$$

so that Eq. 36 can be re-written as:

$$\frac{d\eta}{d\phi} = \frac{\alpha^2 c}{\gamma_o^2\omega}\,\frac{(\beta + \sin\phi)\cos\phi}{1 + \frac{\omega}{c}(1 - n^2)\eta} \tag{39}$$

which is now trivial to integrate. However, this is not necessary, because using Eqs. 27, 34, and 37 we can express $\dot{y}$ and $\dot{z}$ in terms of η by:

$$\dot{y} = \frac{\alpha c}{\gamma_o}\,\frac{\beta + \sin\phi}{1 + \frac{\omega}{c}\eta} \tag{40}$$

and

$$\dot{z} = \frac{\delta}{\gamma_o\left(1 + \frac{\omega}{c}\eta\right)} \tag{41}$$

whereby δ is just another integration constant arising from Eq. 27. It is now easy to see that since $\dot{x}$, $\dot{y}$, and $\dot{z}$ are expressed in terms of the variable η, we obtain from Eq. 32, once Eqs. 38, 40, and 41 are used, a single algebraic equation for η, which is nothing but the solution of Eq. 39.

$$\frac{\omega^2}{c^2}(1 - n^2)\eta^2 + 2\frac{\omega}{c}\eta + 1 - \frac{1 + \delta^2/c^2}{\gamma_o^2} - \frac{\alpha^2}{\gamma_o^2}(\beta + \sin\phi)^2 = 0 \quad (42)$$

It is now easy to work out a scheme for solving the problem completely. From Eqs. 38, 40, 41, and 33 we obtain:

$$\frac{dx}{d\phi} = \frac{n\eta}{1 + \frac{\omega}{c}(1 - n^2)\eta} \quad (43)$$

$$\frac{dy}{d\phi} = \frac{c\alpha}{\omega\gamma_o}\;\frac{\beta + \sin\phi}{1 + \frac{\omega}{c}(1 - n^2)\eta} \quad (44)$$

$$\frac{dz}{d\phi} = \frac{\delta(\gamma_o\omega)^{-1}}{1 + \frac{\omega}{c}(1 - n^2)\eta} \quad (45)$$

Therefore, for every solution $\eta = \eta(\phi)$ of the quadratic equation 42 we obtain the trajectories $x = x(\phi)$, $y = y(\phi)$, and $z = z(\phi)$ by simple quadratures of Eqs. 43, 44, and 45. The solutions of physical interest are the stationary ones which correspond to the periodic solutions. Therefore, we have to find those integration constants β, γ_o, and δ that yield periodic trajectories. From physical reasons or simply by considering Eq. 41 we see that $\delta = 0$, which means that the electrons move in the x-y plane. The mathematical periodicity requirements are:

$$\int_o^{2\pi}\frac{dx}{d\phi}\,d\phi = 0 \qquad \text{and} \qquad \int_o^{2\pi}\frac{dy}{d\phi}\,d\phi = 0 \qquad (46a,b)$$

These conditions are obviously necessary, but it can be shown that they are also sufficient. We, therefore, have two equations for two unknowns, β and γ_o. In order to impose the periodicity conditions it is useful to re-write Eq. 42 with $\delta = 0$ in the following form:

$$1 + \frac{\omega}{c}(1 - n^2)\eta = \frac{\alpha}{\gamma_o}(1 - n^2)^{1/2}[\,\varepsilon^2 + (\beta + \sin\phi)^2]^{1/2} \quad (47)$$

where

$$\varepsilon^2 = \frac{n^2\gamma_o^2 + (1 - n^2)}{\alpha^2(1 - n^2)} > 0 \quad (48)$$

Simple symmetry arguments are enough to realize that β has to be zero, so that Eq. 46b can be fulfilled for all values of γ_o :

$$\int_0^{2\pi} \frac{c\alpha}{\omega\gamma_o} \frac{(\beta + \sin\phi)\ d\phi}{1 + \frac{\omega}{c}(1 - n^2)\ \eta} = 0 \tag{49}$$

In order to determine γ_o we have to solve Eq. 46a whose integrand is identical with Eq.43:

$$\int_0^{2\pi} \frac{n\eta}{1 + \frac{\omega}{c}(1 - n^2)\ \eta}\ d\phi = 0 \tag{50}$$

This is done by substituting the denominator of Eq. 50 by Eq. 47 and eliminating η in the nominator of Eq. 50. Equation 50 thus becomes:

$$\int_0^{2\pi} \frac{d\phi}{\sqrt{\varepsilon^2 + \sin^2\phi}} = \frac{2\pi\alpha\ \sqrt{1 - n^2}}{\gamma_o} \tag{51}$$

By using Ref. 36, formula 259.00 we arrive at a complete elliptic integral of the first kind for Eq. 51:

$$\kappa\ K(\kappa) = \frac{\pi}{2} \frac{\alpha\sqrt{1 - n^2}}{\gamma_o} \tag{52}$$

where $\kappa^2 = (1 + \varepsilon^2)^{-1}$. With the help of Eq. 47 we can cast Eq. 52 into the more suitable form:

$$K(\kappa) = \frac{\pi}{2} \frac{n}{\sqrt{1 - \nu^2\kappa^2}} \tag{53}$$

where

$$\nu^2 = 1 + \frac{1}{\alpha^2} \tag{54}$$

It is easy to show that there is always a solution for Eq. 53.

We are mainly interested in the maximum kinetic energy transmitted by the laser field to the electrons. This we obtain most easily from Eq. 37, after it has been realized from Eq. 47 that $\sin\phi = 1$ corresponds to a maximum of η and consequently to a maximum of the kinetic energy T_{max} :

$$T_{max} = m_o c^2 \left[\frac{\alpha}{\kappa\sqrt{1 - n^2}} \left(1 - n\sqrt{1 - \nu^2\kappa^2} \right) - 1 \right] \tag{55}$$

Therefore, the maximum kinetic energy of electrons accelerated by a linearly polarized laser field within a plasma of refractive index n is determined by a solution κ of Eq. 53, n and the laser intensity I. ν^2 and α can be calculated from Eqs. 16, 17 and 54.

On Fig. 5 are depicted curves at five different intensity levels (10^{14} W/cm^2 up to 10^{22} W/cm^2) of linearly polarized electromagnetic waves of radial frequency $\omega = 1.777 \times 10^{15}$ rad/s (neodymium doped glass laser); they are calculated from Eq. 55 and give the maximum kinetic energy (ordinate) of electrons accelerated by the laser field within a plasma as a function of its refractive index n (abscissa). The intersections of the curves with the right side ordinate scale give the values of the kinetic energy in vacuum as this does Fig. 4 also. The factors g_m at the left side are the factors by which the maximum kinetic energy has increased after the vacuum has been filled with a plasma of refractive index $n = 10^{-4}$ for the corresponding intensity. One can notice that this factor lies between n^{-1} and $n^{-1/2}$; this means that at given lower intensities the kinetic energy is almost proportional to the reciprocal value of n whereas at given higher intensities it is almost proportional to the reciprocal value of the square root of n.

Circularly polarized. After entering with Eqs. 20 and 21 into Eq. 22, the circularly polarized case for a plasma with refractive index $n < 1$ could be treated analogously to the linearly polarized case. One can also express $\dot{x}$, $\dot{y}$, and $\dot{z}$ in terms of the same variable η as defined in Eq. 35 and one arrives at the same equations as Eqs. 38 and 40, except that Eq. 41 reads now:

$$\dot{z} = \frac{\alpha}{\gamma_o} \lambda \frac{\delta_o - \cos\phi}{1 + \frac{c}{\omega}\eta} \tag{56}$$

Also, one obtains a quadratic equation in η similar to Eq. 42 having now the form:

$$\frac{\omega^2}{c^2}(1 - n^2)\eta^2 + 2\frac{\omega}{c}\eta + 1 - \frac{1}{\gamma_o^2} - \frac{\alpha^2}{\gamma_o^2}[(\beta + \sin\phi)^2 + (\delta_o - \cos\phi)^2] = 0 \tag{57}$$

The periodicity requirements:

$$\int_0^{2\pi} \frac{dx}{d\phi} d\phi = 0; \quad \int_0^{2\pi} \frac{dy}{d\phi} d\phi = 0; \quad \text{and} \quad \int_0^{2\pi} \frac{dz}{d\phi} d\phi = 0 \tag{58a,b,c}$$

will yield the integration constants, after having obtained the equations for $dx/d\phi$, $dy/d\phi$, and $dz/d\phi$, which again are identical

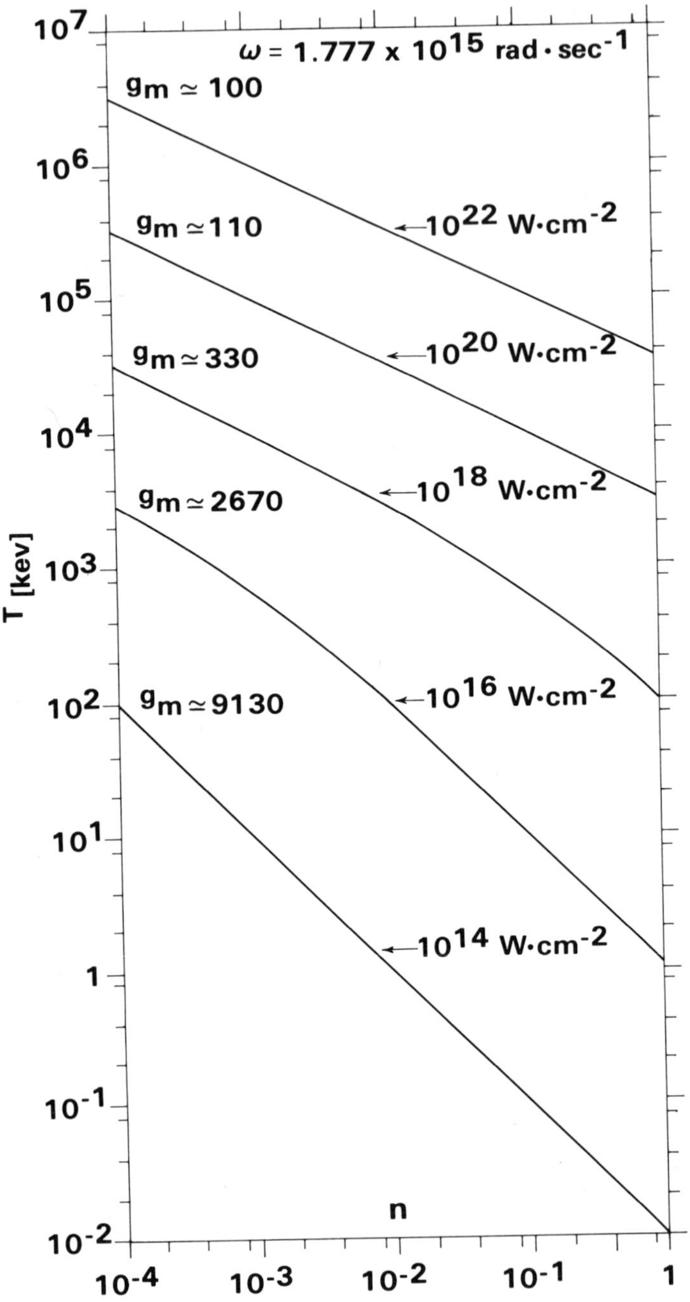

Fig. 5: Kinetic energy of electrons accelerated in a laser field of radial frequency $\omega=1.777\times10^{15}$rad/s in a plasma as a function of refractive index n at different laser intensities.

with Eqs. 43 and 44, except for Eq. 45 which should be replaced by:

$$\frac{dz}{d\phi} = \lambda \frac{c\alpha(\gamma_o\omega)^{-1} (\delta_o - \cos\phi)}{1 + \frac{\omega}{c} (1 - n^2) \eta} \tag{59}$$

For the same reason as in the linearly polarized case, it follows immediately from the integration conditions of Eqs. 58 b and c applied to Eqs. 44 and 59 that $\beta = \delta_o = 0$. For physical reasons (no component in the x-direction) and from Eqs. 43 and 58a follows that η has to be zero at all times. This then simplifies Eq. 57 to:

$$\gamma_o = \sqrt{1 + \alpha^2} \tag{60}$$

which according to its value taken from Eq. 37 is for $\eta = 0$ the same as γ to be used for the determination of the kinetic energy:

$$T = m_o c^2 \left(\sqrt{1 + \alpha^2} - 1 \right) \tag{61}$$

which shows that in the circularly polarized case the refractive index n enters only in α of Eq. 16 through using the relationship of Eq. 17.

REFERENCES

1. G. Breit and J. A. Wheeler, Phys. Rev. 46, 1087 (1934).
2. A. A. Varfolomeyev, Sov. Phys.-JETP 23, 681 (1966).
3. H. J. Bhabha, Proc. Roy. Soc. (London) A 152, 559 (1935).
4. J. H. Eberly, in Progress in Optics Vol. VII, Editor: E. Wolf (North-Holland Publishing Co., Amsterdam 1969) pp. 361.
5. H. R. Reiss, J. Math. Phys. 3, 59 (1962).
6. H. R. Reiss, Phys. Rev. Lett. 26, 1072 (1971).
7. A. I. Nikishov and V. I. Ritus, Sov. Phys.-JETP 19, 529 and 1191 (1964).
8. F. V. Bunkin and I. I. Tugov, Sov. Phys.-Dokl. 14, 678 (1970).
9. E. Brezin and C. Itzykson, Phys. Rev. D 2, 1191 (1970).
10. V. P. Yakovlev, Sov. Phys.-JETP 22, 223 (1966).
11. J. W. Shearer, J. Garrison, J. Wong, and J. E. Swain, in Laser Interaction and Related Plasma Phenomena, Editors: H. Schwarz and H. Hora (Plenum Press, New York 1974) Vol. 3B, pp. 803.
12. F. V. Bunkin and A. E. Kazakov, Sov. Phys.-Dokl. 15, 758 (1971).
13. All formulae in this paper are written in such a way that the SI system of units (Système Internationale) can be used which is identical with the rationalized MKSA system.
14. T. Erber, Rev. Mod. Phys. 38, 626 (1966).
15. A. D. Steiger and C. H. Wood, Phys. Rev. A 5, 1467 (1972).
16. see for example: A. N. Matveyev, Principles of Electrodynamics.

Translation Editor: L. F. Landovitz (Reinhold Publishing Corp., New York 1966) p. 358.

17. see for example: R. E. Kidder in Physics of High Density. Editors: P. Caldirola and H. Knoepfel (Academic Press 1971). Course Number 48.
18. A. I. Akhiezer and R. V. Polovin, Sov. Phys.-JETP 3, 696 (1956).
19. P. Kaw and J. Dawson, Phys. Fluids 13, 472 (1972).
20. C. E. Max and F. Perkins, Phys. Rev. Lett. 27, 1342 (1971).
21. C. E. Max, Phys. Fluids 16, 1277 and 1480 (1973).
22. M. Dobrowolny, A. Ferrari, and G. Bosia, Plasma Phys. 18, 441 (1976).
23. F. W. Sluijter and D. Montgomery, Phys. Fluids 8, 551 (1965).
24. N. L. Tsintsadze, Sov. Phys.-JETP 32, 684 (1971).
25. A. Ferrari, S. Massaglia, and M. Dobrowolny, Phys. Lett. 55A, 227 (1975).
26. E. L. Kane and H. Hora, these Proceedings p. 913.
27. N. D. Sengupta, Calcutta Math. Soc. Bull. 39, 147 (1947); and 41, 187 (1949).
28. A. H. Taub, Phys. Rev. 73, 786 (1948).
29. L. D. Landau and E. M. Lifshitz, The Classical Theory of Fields (Addison-Wesley Publishing Co., Inc., Reading, Mass. 1951) [p. 118 (problems) in 3rd Edition of 1971].
30. H. Hora, Opto-electronics 5, 491 (1973).
31. H. Hora, in Laser Interaction and Related Plasma Phenomena. Editors: H. Schwarz and H. Hora (Plenum Press, New York 1974) Vol. 3B, pp. 819.
32. M. B. Nicholson-Florence, these Proceedings p. 981.
33. W. Lünow, Plasma Phys. 10, 879 and 973 (1968).
34. C. S. Lai, Phys. Rev. Lett. 36, 966 (1976).
35. R. M. Lichtenstein and H. Schwarz, Bull. Am. Phys. Soc. 19, 676 (1974).
36. P. F. Byrd and M. D. Friedman, Handbook of Elliptic Integrals for Engineers and Scientists. 2nd Edition (Springer Verlag, Berlin, Heidelberg, New York 1971).

RELATIVISTIC OSCILLATIONS OF ELECTRONS IN LASER FIELDS AND TRANSFER OF ENERGY AND MOMENTUM*

M. B. Nicholson-Florence

Dept. of Theoretical Physics, The University of New South Wales

Kensington, NSW Australia

Abstract

Solutions of the relativistic equation of motion of an electron in an intense radiation field are examined for both linear and circularly polarised cases. The effect of the switching-on process on the longitudinal drift is studied and the results of Marburger & Tooper and Lai are applied to estimate the momentum and energy transfer in a plasma.

Non-relativistic theory of the motion of an electron in an electromagnetic field gives the velocity of the electron in terms of the electric field amplitude E_o and the angular frequency ω:

$$u(t) = (e/m\omega)\, E_o \sin(\omega t) = u_o \sin \omega t$$

A parameter of velocity amplitude is clearly u_o/c with c the vacuum velocity of light.

$$\mu = u_o/c = e\, E_o/m\omega c$$

At light intensities above 10^{18} W/cm^2 the directed velocity approaches the velocity of light c and the parameter exceeds unity.

---

* Presented at the Fourth International Workshop Conference on "Laser Interaction and Related Plasma Phenomena" at Rensselaer Polytechnic Institute, Troy, New York, 8-12 November, 1976.

This parameter μ can be written in many forms:

$$\mu = u_o/c = (e\ E_o/m\omega^2)/(c/\omega)$$

$$= k\ x_o = x_o/\lambda\!\!\!^{-}$$

$$= e\ A/mc^2$$

$$\mu^2 = e^2 I/m^2\ \omega^2\ c^3$$

$$= r_o \lambda\!\!\!^{-2} I/mc^3$$

in terms of the vector potential A, intensity I, electron radius r_o and wavelength λ. The parameter measures the ratio of electron oscillation amplitude to radiation wavelength and μ^2 measures the amount of electromagnetic energy in contact with an electron compared to its rest energy mc^2. In Nd laser light at 1.06μ the electric field for which $\mu = 1$ is:

$$E_r = m\omega c/e \simeq 3 \times 10^{10}\ \mathrm{V/cm}$$

$$\simeq 10^{18}\ \mathrm{W/cm^2}$$

The relativistic equation of motion for an electron in a monochromatic radiation field may be written:

$$\frac{d}{dt}\left\{\frac{mv}{(1-v^2/c^2)^{\frac{1}{2}}}\right\} = e(E + (v/c) \times B)$$

Sengupta (1949) and Vachaspati (1963) showed that these equations may be reduced to quadratures and integrated to give the electron orbits. Another method is to use the Hamilton-Jacobi equation (Landau & Lifshitz (1951), Brown & Kibble (1964), Eberly & Sleeper (1968) and Sarachik & Schappert (1970)). From the equation for Hamilton's principal function S and a choice of boundary conditions and vector potential, the electron motion can be expressed in terms of its proper time τ,

$$\tau = t - \underline{n}\cdot\underline{r} \qquad \text{where}$$

$$\underline{n} = \underline{k}/\omega \qquad \text{and } \underline{k} \text{ is the propagation vector.}$$

Eberly & Sleeper used a vector potential:

$$\underline{A}(r,t) = a\ \underline{\varepsilon}\ \exp(-\gamma|\tau|)\cos\omega\tau$$

to represent a wave packet with zero displacement and velocity for the electron at time $t = -\infty$. For $\tau > 0$ we find

$$\underline{\varepsilon}\cdot\underline{r}(\tau) = -\frac{ea}{n}\left[\frac{2\gamma}{\omega^2+\gamma^2} - \frac{e^{-\gamma\tau}}{\omega^2+\gamma^2} S(\omega\tau)\right]$$

$$\underline{n}\cdot\underline{r}(\tau) = \tfrac{1}{2}\frac{e^2a^2}{m^2}\left[\frac{1}{2\gamma} - \frac{1}{4\gamma}e^{-2\gamma\tau} + \frac{1}{\omega^2+\gamma^2}(\tfrac{1}{2}\gamma - \tfrac{1}{4}e^{-2\gamma\tau}S(2\omega\tau))\right]$$

The resulting electron orbit shows a displacement along the propagation direction n, a much smaller displacement along the vector potential direction ϵ which disappears in the limit of a monochromatic wave. There is an oscillation of frequency ω along ϵ (here the same as the electric vector) and one of frequency 2ω along n the propagation direction. The displacement perpendicular to ϵ and n is always zero. Thus for a plane polarised wave the electron motion lies in the form of a figure of eight in the E and k plane with no motion along the H direction. (see figure 1, Sarachik & Schappert (1970)).

The energy and momentum are derived from the time and space derivatives of the principal function and give for the electron mass:

$$E^2 - p^2 = m^2 + e^2A^2(\tau)$$

The mass increase is given by $\Delta m^2 = e^2 \langle A^2 \rangle$

$$= \tfrac{1}{2}\, e^2 \cdot a^2$$

For a circularly polarised wave there is a purely transverse motion, a rotation of the electron around the n axis again with mass increase and with a displacement along this axis due to the switching-on process.

These were solutions for a single electron in an intense radiation field first produced in studies of Compton scattering in an intense field. Solutions for the motion of a plasma in an intense field were first found by Akhiezer & Polovin (1956). They started from the Maxwell equations and the Lorentz force equation and derived differential equations for the dimensionless momentum $\rho = p/mc$ in terms of the proper time τ and the phase velocity $\beta = v_p/c$ valid for homogeneous plasma.

These equations were used by Kaw & Dawson (1970) to evaluate an effective plasma frequency changed from the non-relativistic value

$$\omega_p^2 = 4\pi ne^2/m$$

by virtue of the mass change of the electron. This allows penetration of the otherwise overdense plasma by an intense laser beam. As with Akhiezer & Polovin and others to follow, they consider a cold plasma, which is reasonable for a very large directed velocity, and fixed ions which is valid for intensities up to $10^{22}\,W/cm^2$ at 1μ. They show that only circularly polarised waves are pure transverse at high intensities giving fields of the form:

$$E = \mu E_r \sin \omega\tau$$

$$n^2 = 1 - (\omega_p{}^2/\omega^2)(1 + \mu^2)^{-\frac{1}{2}}$$

$$\omega_{p_{eff}} = \omega_p/(1 + \mu^2)^{\frac{1}{4}} < \omega_p$$

At large enough field ($\mu >> 1$) we find for circular polarisation (Max and Perkins (1971)):

$$\omega_{p_{eff}}{}^2 = (4\pi nec\omega/E) = \omega_p{}^2 (\frac{mc^2}{\varepsilon_e})$$

where $\epsilon_e = e\,c\,E\,/\,\omega$ the high energy oscillation energy given by Hora (1973). For linear polarised waves one must average over the electric field E. If E_o be the amplitude of the E vector taking $E^2 = \frac{1}{2} E_o{}^2$ gives:

$$\omega_{p_{eff}}{}^2 = \sqrt{2}\ \omega_p{}^2 (\frac{mc^2}{\varepsilon_e})$$

differing only by 5% from an exact treatment by Max and Perkins (1971) where

$$\omega_{p_{eff}}{}^2 = \frac{\pi}{2}\ \omega_p{}^2\ (\frac{mc^2}{\varepsilon_e}) \quad .$$

Max & Perkins found analytic solutions for a linear polarised waves using these same equations and demonstrated that while the energy flux of circularly polarised waves may be regarded as purely electromagnetic with for phase velocity β_c flux S

$$S = cE^2/4\pi\beta$$

a linear polarised wave has contributions from both particles and fields:

$$S = (cE^2/12\pi\beta)\cdot(1 + \frac{3}{2} < \rho_z\beta/\rho_{x_{max}} >)$$

where ρ_z, ρ_x are the electron momenta along and perpendicular to the propagation direction. The effective plasma frequency for $\mu > 1$ and a mildly inhomogeneous plasma of scale length L was found to be

$$\omega_{p_{eff}} \simeq \omega_p (k_v \; L \; u^2)^{-\frac{1}{4}}$$

Kidder (1971) produced an iterative solution to the Lorentz force equation for linear polarisation which demonstrated the longitudinal oscillation and drift due to the v X B force and the correction to the transverse part due to the change in phase experienced by the electron after the longitudinal drift. The method shows that the effect of the steady drift cancels and the transverse part is unaffected if the drift is reduced to zero by an opposing electrostatic field.

Steiger & Woods (1972) considered circularly polarised waves and calculated the dipole moment generated by the rotating electrons in an inverse Faraday effect. The effective plasma frequency is again modified, not only by the factor due to the change in mass, namely $\gamma = (1 + \mu^2)^{\frac{1}{2}}$ but also by an additional cyclotron frequency. Thus the refractive index is:

$$n^2 = 1 - \omega_p{}^2/\gamma\omega(\omega + \Omega)$$

$$= 1 - (\omega_p{}^2/\omega^2) \cdot (R/x_o)$$

$$\text{where } \Omega = - e \, B_{ind}/\gamma mc$$

and the induced magnetic field is B_{ind}, the orbit radius R.

At these intensities considered, the induced B_{ind} reaches 10 MG and can be expressed as:

$$B_{ind} = E \cdot \tfrac{1}{2} (\omega_p{}^2/\omega^2) \cdot (\mu^2/(1 + \mu^2))$$

comparing with the transverse field B

$$B = E \; (1 - \omega_p{}^2/\omega^2 (1 + \mu^2)^{\frac{1}{2}})^{\frac{1}{2}}$$

At large μ, B_{ind} tends to a value independent of intensity E/μ and may exceed B near a cutoff since $B = E \, n_{eff}$. Otherwise B_{ind} is most significant near $\mu = 1$ but is less important than the intensity and density. Steiger & Woods found that the electron kinetic energy is a strong function of laser intensity but almost independent of plasma density. Shearer, Garrison et.al. found a power law approximation for intensities close to relativistic threshold:

$$\varepsilon_e/mc^2 = \left(\frac{\lambda^2 I}{9 \times 10^{18}}\right)^{0.657}$$

for λ in microns and I in watts/cm^2.

Marburger & Tooper (1975) examined the equations of Akhiezer & Polovin for the case of total reflection and found standing wave solutions for overdense plasmas in a steady state with fixed ions and zero temperature. For a completely standing wave with separate space and time factors for the field variables in an inhomogeneous plasma they found that the waves must be circularly polarised and have zero longitudinal momentum. Felber & Marburger (1976) extended this work to mixed standing and travelling waves as also did Lai (1976) in a generalisation of Marburger's method. In terms of Lai's variables, the method is to try a solution for the vector potential in the form:

$$A = (e/mc^2)R(\zeta)\left\{-\underline{i}\ \sin\ (\omega t - \phi(\zeta)) + \underline{j}\ \cos(\omega t - \phi(\zeta))\right\}$$

where $\zeta = \omega z/c$ and R and ϕ are functions to be evaluated from the differential equations produced by inserting this expression for A into the original equations. Expressing the ion and electron densities in terms of the ratio of plasma to radiation frequencies $a_{i,e} = \omega^2{}_{p_{i,e}}/\omega^2$ a Lagrangian is found for R and ϕ and the motion expressed as the one-dimensional motion of a particle of mass $(1 + R^2)^{-1}$ in terms of a 'time' variable ζ. A constant of motion while neglecting radiation reaction is, using primes to indicate $d/d\zeta$

$$S = R^2\ \phi'$$

which is actually the dimensionless Poynting vector. S equals zero for complete standing waves. There is an effective potential

$$V_{eff} = \tfrac{1}{2}\frac{S^2}{R^2} - \tfrac{1}{2}R^2 - \alpha_i(1 + R^2)^{\frac{1}{2}}$$

which gives the maxima and minima for the range of permitted R. There is a Hamiltonian for the motion, constant for given ion density. The value of R = 0 is attained only by standing waves or when the plasma is not overdense.

The equations reveal that when the value of the parameter R exceeds a given amplitude R_m, the electron density must vanish giving regions that are depleted of electrons while still maintaining overall charge neutrality within each quarter wavelength. These depletions may be identified with the striations described by Hora (1974) and Lindl and Kaw (1971).

The condition for depletion may be expressed as:

$$d/d\zeta\ (E_z/E_r) < -\alpha_i$$

The parameter R may be expressed explicitly in terms of ζ using Jacobi elliptic functions. R has been evaluated by Marburger and his co-workers and by Lai. It is in fact equal to the parameter μ already introduced. The analysis gives for the electron velocity for circular polarisation:

$$R = \gamma v/c = eA/mc^2$$
$$v^2/c^2 = R^2/\ (1 + R^2)$$
$$\gamma = (1 - \frac{v^2}{c^2})^{-\frac{1}{2}} = (1 + R^2)^{\frac{1}{2}}$$

and for linear polarisation:

$$\gamma \frac{v}{c} = - R\underline{i} + \tfrac{1}{2}R^2\underline{k}$$
$$v^2_{max}/c^2 = \frac{R^2 + \frac{1}{4}R^4}{1 + R^2 + \frac{1}{4}R^4}$$
$$\gamma^2_{max} = 1 + \tfrac{1}{2}R^2$$

The drift velocity is at its peak

$$v_d/c = \underline{n}\ \tfrac{1}{2}R^2/(1 + \tfrac{1}{2}R^2)$$

Hora (1973) examined the Lorentz force equations and from the non-vanishing of a radical in the equations identified a parameter Z with maximum value $Z^* = (1 + e^2E^2/m^2c^2\omega^2)^{-1}$ which we can identify with $Z^* = (1 + R^2)^{-1}$ From this parameter, Hora deduced the kinetic energy of the oscillating electron in the form:

$$\varepsilon_e/mc^2 = (1 + e^2E^2/m^2\omega^2c^2)^{-\frac{1}{2}} - 1$$

leading to the usual non-relativistic $\epsilon_e = \frac{1}{2}mu^2$ and the relativistic limit of eEc/ω independent of mass. This is again $\gamma = (1 + R^2)^{\frac{1}{2}}$ tending to R at high R. The correction factor of $\pi/2\sqrt{2}$ for linear polarised waves mentioned above due to Max & Perkins (1971) must be applied to Hora's result.

The analytical derivation of this correction for the whole range of intensities was given by Schwarz and Tabensky in these Proceedings (pp. 961-980).

In the steady state the vectors E and v are exactly 90° apart. One can estimate the effect of radiation damping from the equations for a circular pendulum (Zeldovich (1975)) and find the angle ϕ by which v would advance thus:

$$\phi \simeq \frac{4\pi}{3}\left(\frac{r_o}{\lambda}\right)(1 + R^2)^{3/2}$$

and r_o/λ for 1μ is about 10^{-9}.

During the switching on and off process however the velocity v will deviate from the steady state orientation. One can estimate, approximately, the component of v perpendicular to the steady state direction and parallel to B_1 in the figure by crudely inserting a factor q(t) into the expression for the vector potential. Doing this one obtains a term $v_1/c = \omega R\, q/\omega_p^2$ along E which gives a force v X B along Oz of $\omega^2 R^2 q\dot{q}\, cm/\omega_p^2$. This leads to a velocity $v_z = \frac{1}{2}\omega^2\, S\, q^2\, c/\omega_p^2$. For a complete standing wave $S = 0 = v_z$ but for travelling wave the electron gains speed while q is positive and is brought to rest when the wave is passed. The displacement is proportional to the time of passage, T, depends on the pulse shape given by q(t) and is of magnitude $\frac{1}{2}R^2$ cT.

In Lai and Marburger's model the lack of any momentum along Oz is effected by the time independent field $E_z = E_r\, d/d\zeta\, (1 + R^2)^{1/2}$. The field E_z balances the v X B force present in the steady state due to the B_2 component of B of amplitude E_rR'.

REFERENCES

A. I. Akhiezer & R. V. Polovin — Sov. Phys. JETP 3 (5), 696, (1956).

L. S. Brown & T. W. B. Kibble — Phys. Rev. 133A, 705 (1964).

J. H. Eberly — Prog. in Optics VII, 361 (1969).

J. H. Eberly & A. Sleeper — Phys. Rev. 176 (5), 1570, (1968).

F. S. Felber & J. H. Marburger — Phys. Rev. Lett. 36 (20), 1176 (1976).

H. Hora — Phys. Fluid 12, (1), 182 (1969).
Phys. Fluid 17 (5), 939, (1974).
Laser Interaction & Related Plasma Phenomena, Vol. 3, 819, (1973).

P. Kaw & J. M. Dawson — Phys. Fluid 13 (2), 472, (1970).

T. W. B. Kibble — Phys. Rev. 138B, (3), 740, (1964).
Phys. Lett. 20, (6), 627, (1966).
Phys. Rev. Lett. 16, (23), 1054, (1966).
Phys. Rev. 150, (4), 1060, (1966).

R. E. Kidder — UCRL Report 71775 (1971).

R. H. Kohler — Phys. Rev. 151 (1), 165, (1966).

C. S. Lai — Phys. Rev. Lett. 36 (16), 966, (1976).

L. D. Landau & E. M. Lifshitz — Classical Theory of Fields, Pergamon, Oxford (1951).

J. H. Marburger & R. F. Tooper — Phys. Rev. Lett. 35 (15), 1001, (1975).

C. Max & F. Perkins — Phys. Rev. Lett. 27 (20), 1342, (1971).

J. J. Sanderson — Phys. Lett. 18 (2), 114, (1965).

E. S. Sarachik & G. T. Schappert — Phys. Rev. D1 (10), 2738, (1970).

Sengupta — Bull. Math. Soc. (Calcutta) 41, 187, (1949).

J. W. Shearer, J. Garrison, et.al. — Phys. Rev. A8, (3), 1582, (1973).

A. D. Steiger & C. H. Woods — Phys. Rev. A5, (3), 1467, (1972).

Vachaspati — Phys. Rev. 128, (2), 664, (1962).

Ya. B. Zeldovitch — Sov. Phys. Uspekhi 18, (2), 79, (1975).

LASER ACCELERATORS

W.J. Willis

CERN

Geneva, Switzerland

The large electrical fields present in laser beams have long prompted speculations on their use for accelerating particles. A perspective on the specifications of a competitive accelerator for high-energy physics may be provided by reflecting that the existing proton synchrotron at Fermilab delivers 10^{13} particles at $1/2 \times 10^{12}$ eV, or $\sim 10^6$ J, in one pulse. A useful laser accelerator would have to accelerate a bunch of at least 10^{10} particles, with a gradient of more than 10^9 eV/m. Then the power going into the beam is greater than $1/2 \times 10^9$ W. Since the acceleration efficiency is likely to be small, we see that very powerful lasers will be required, independent of the details of the acceleration mechanism.

A number of laser acceleration mechanisms have been suggested. Lawson[1)] has recently reviewed these, most of which have problems in principle or in practice. One experimental demonstration of laser acceleration has been carried out by Pantell and collaborators[2)] using helium gas to produce phase matching between the light wave and the particles. An attempt to extrapolate that experiment to an interesting accelerator[3)] resulted in a design shown in Fig. 1. Although this device should work in principle, there are several serious problems:

1) It seems very doubtful that the required mechanical tolerances can be maintained.

2) As in any linear accelerator, there is a choice of accelerating phase, with radial focusing in one and longitudinal focusing in the other. The radial defocusing is too strong to overcome with lenses, ruling out the latter case. The resulting requirements on longitudinal bunching at injection are severe.

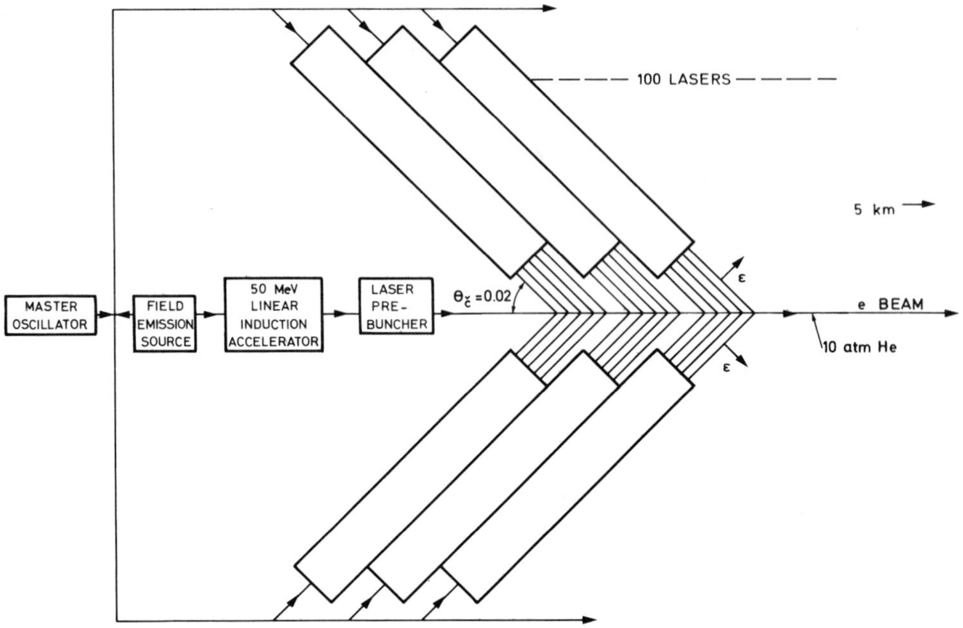

Fig. 1 Diagram of the laser linear accelerator described in Ref. 3. The phase matching is achieved by the index of refraction. The light is incident at the Čerenkov angle, and the acceleration mechanism may be described as an inverse Čerenkov effect. All the lasers must be locked in phase.

3) The accelerator is very inefficient. Even with an optimistic beam current, less than 0.1% of the light goes into the beam.

4) The rate of energy gain is only marginally interesting. It is limited by the breakdown threshold in the helium.

It seems unlikely that this straightforward approach will achieve a practical design.

The author has suggested another approach, based on the laser-plasma interaction[4], which may overcome some of these problems. The laser is used to produce a concentration of electrons by means of the ponderomotive force. The electric field due to the electron bunch then accelerates the positive ions. This can be recognized as a type of "collective field" accelerator.

The origin of the collective field can be understood by considering the quivering motion of a single electron in a strong electromagnetic field. A calculation by Reiss and Eberly[5] shows that the passage of a wave packet leaves an electron with a longitudinal displacement ξ, where

$$\xi = C\tau c$$

where 2τ is the duration of the wave packet, c the velocity of light, and

$$C = \frac{I r_0 \lambda\!\!\!^{-2}}{mc^3} ,$$

where I is the intensity of the light, r_0 is the classical electron radius, $2\pi\lambda\!\!\!^{-}$ the wavelength of the light, and m is the mass of the electron. Note that mc^3/r_0 is a power, 8.72×10^9 W, so the C = 1 implies an intensity of $\sim 10^{10}$ W/$\lambda\!\!\!^{-2}$. If the light wave covers an area A, the energy w is

$$w = IA\tau$$

and we can write the energy required to displace the electron a given distance

$$\frac{w}{\xi} = 11.5 \frac{A}{\lambda^2} \text{ J/cm.}$$

The effect of a light wave introduced into a dilute plasma can be seen in Fig. 2, which shows snapshots of electron positions at successive times. Each electron tends to be carried with the wave packet for a time, then dropped behind. This leads to an accumulation of excess negative charge in the wave packet, balanced by a depletion of electrons at the place where the wave packet enters the plasma. If the plasma is sufficiently dilute that we may ignore the Coulomb force on the electrons, the *net* charge has a density of

$$q_- = Cq_0 .$$

Our purpose is to use the Coulomb force due to the bunch of electrons to accelerate positive ions. The excessive charge should be as large as possible, and the case of interest is a plasma sufficiently dense so that the ponderomotive force producing the charge excess is balanced by the Coulomb force. In this case, the longitudinal field at the end of the wave packet will be that given by the ponderomotive force, which may be written[6)]

$$E_L = 4\pi \frac{e}{m} \frac{1}{\omega^2} \nabla \frac{\langle E^2 \rangle}{8\pi}$$

$$\simeq \frac{e\lambda\!\!\!^{-}}{mc^2} \langle E^2 \rangle \frac{\lambda\!\!\!^{-}}{\tau c}$$

if we approximate the pulse shape as triangular. It is interesting to compare the longitudinal field with the peak transverse field in the light wave, E_0, as a measure of the efficiency of the acceleration process $E_L/E_0 = \sqrt{C}(\lambda\!\!\!^{-}/\tau c)$. The two requirements are strong intensity, so that $\sqrt{C} \sim 1$, and a short pulse. Briefly, the acceleration is driven by the rate of change of the power.

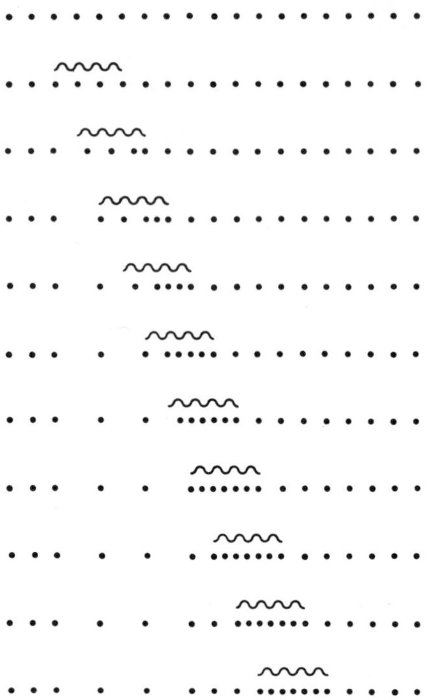

Fig. 2 Effects of a short laser beam on an electron beam of uniform density. Each line of dots represents the charge density at successive times. The position of the light pulse is indicated above each line. The only mechanism for interaction is assumed to be the free motion of the electrons under the influence of the light waves, and space-charge forces are neglected. In this case a modulation of increased electron density runs along with the light pulse, and a modulation of decreasing density is left stationary in the electron rest frame. In practice, other interactions and space-charge forces lead to a rarification which follows the light pulse. In the laboratory frame, the rarification follows the increased density modulation over the depth of field of the laser focus in any case.

Since at best $\tau c \gg \lambda\!\!\!^{-}$, a strongly focused light pulse will tend to produce a charge distribution which is long compared to its transverse width. There arises the problem of the charge bunch blowing up radially. This problem can be avoided by forming a ring image with a diameter approximately equal to τc. The ponderomotive force due to the distribution of power in the ring image generates a radial field

$$E_R \simeq \frac{e\lambda\!\!\!^{-}}{mc^2} E_0^2 \frac{\lambda\!\!\!^{-}}{4r} ,$$

where r is the ring image radius, or

$$\frac{E_R}{E_0} = \sqrt{C}\,\frac{\lambdabar}{4r} \ .$$

For a reasonable optical system, $r \gtrsim 10\ \lambda$. If $4r \simeq \tau c$, the forces on the electrons is the bunch will be isotropic, and result in a stable distribution. A qualitative illustration of the distribution of the light power and the electric fields is shown in Fig. 3.

The positive ions are subject to an accelerating force if they are on the trailing side of the light pulse. There is a radially focusing force due to the negative charge concentration on the axis. For longitudinal stability, the ions must experience an accelerating force increasing with the distance from the front of the bunch. This condition is satisfied from the centre of the bunch to a position near the half-power point. The phase stable region for a finite acceleration is somewhat smaller, as shown qualitatively by the region enclosed by dotted lines in Fig. 3.

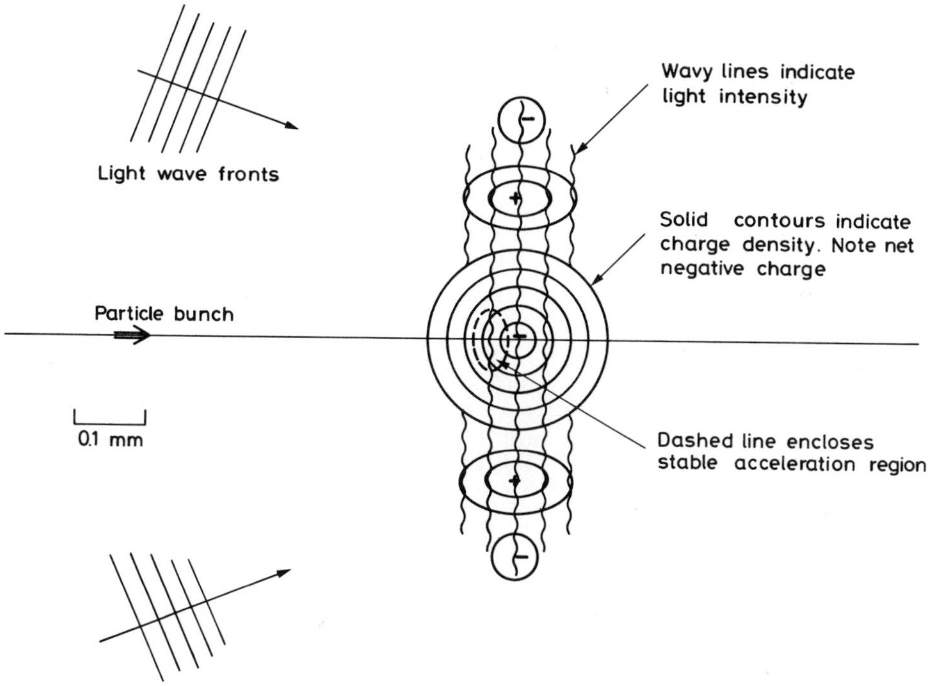

Fig. 3 Schematic diagram of the light wave packet intensity and the net charge density. There is an over-all net negative charge as explained in the text.

Although the electrons in the bunch are not relativistic, for $C \lesssim 1$, the charge excess moves with velocity $\simeq c$. If relativistic particles (with $\beta \gtrsim 0.99$) are being accelerated, the phase error introduced by the difference between the particle velocity and the wave packet velocity does not become appreciable over the depth of focus of one laser image. The time of the laser pulse in successive stages of the accelerator may be ajusted to give over-all phase matching.

A numerical example can be given for $\lambda = 10$ µm and $C = 1$. Then $E = 2.8 \times 10^{11}$ V/m. If $r = 0.1$ mm and $\tau c = 1.3$ psec, $E_L = E_R \simeq$ $\simeq 10^9$ V/m, while the power is $\sim 3 \times 10^{12}$ W, and the energy in the pulse is ~ 6 J. The line density of the electrons in the bunch is $\simeq 10^{11}$ electrons/cm. The number of positive ions allowed in the stable region is at least 10^{10}, a number which does not disturb the field distribution in a major way. The effective depth of field has been computed according to the calculation of Linfoot and Wolf[7)] for an aperture with the central half blocked, for an f-ratio of 1/70, giving a result of 8 cm. The energy given to the ions as they pass through the depth of focus of the image is $\sim 1/20$ J. Thus the efficiency of the conversion of the light energy into particle energy is $\sim 10^{-2}$.

This efficiency is not much lower than that for conversion of RF power into particle energy in a conventional linac, but RF power may be generated with higher efficiency than ultra-short pulse laser light, of course. On the other hand, the voltage gradient is larger by $\sim 10^3$.

In comparison with the phase-matched laser linac described in the first part of the paper, this collective accelerator has several advantages. The alignment tolerances are set by the size of the focused image, rather than the wavelength of the light, and it is not necessary to maintain any phase relationship between the lasers driving successive stages. The accelerated particles are both longitudinally and radially stable, unlike those in the other device. Also, since this device is not limited by breakdown in the accelerating region, and given an acceleration proportional to the power rather than the electric field, there is no advantage to stretching it out over a longer distance, and it can be made more compact, and can give increased gradients if more powerful lasers are built.

The theory of operation outlined here is clearly very crude. Only linear effects have been described. The plasma density required for the example chosen is small compared with the critical plasma density, and therefore the parametric instabilities in the plasma may not be strongly excited, but a careful examination should be made to see which co-operative effects might affect the operation of the device.

REFERENCES

1) J.L. Lawson, Laser accelerators?, Report RL-75-043, Rutherford Laboratory (1975).

2) M.A. Piestrup, G.B. Rothbart, R.N. Fleming and R.H. Pantell, J. Appl. Phys. 46, 132 (1975).

3) W.J. Willis, Novel mechanisms for particle acceleration, *in* Proc. 1975 Isabelle Summer Study, BNL 20550 (Brookhaven National Laboratory, Upton, N.Y., 1975).

4) W.J. Willis, CERN 75-9 (1975).

5) H. Reiss and J. Eberly, Phys. Rev. 151, 1058 (1966); also E. McMillan, Phys. Rev. 79, 498 (1950); and J. Eberly, Progress in optics (Ed. E. Wolf) (North-Holland Publ. Co., Amsterdam, 1969), Vol. 7.

6) H. Hora, Laser interactions and related plasma phenomena, Proc. 3rd Workshop on Laser Interaction and Related Plasma Phenomena, Rensselaer Polytechnic Institute, Troy, N.A., 1973 (Eds. H.J. Schwarz and H. Hora) (Plenum Press, N.Y., 1974), p. 383.

7) E.H. Linfoot and E. Wolf, Proc. Phys. Soc. B 66, 145 (1953).

*This paper has been included in these Proceedings, since it was scheduled for presentation at the 4th Workshop on "Laser Interaction and Related Plasma Phenomena" held at Rensselaer Polytechnic Institute, 8-12 November 1976; however, the author could not attend.

RECOIL AND GRAVITATIONAL EFFECTS OF ELECTRONS IN PLASMAS WITHIN VERY INTENSE LASER RADIATION[a,b]

H. Hora and M. M. Novak

Dept. of Theoretical Physics, The University of New South Wales, Kensington, N.S.W. Australia

Abstract

The relativistic oscillation energy ε_{osc} of electrons in laser fields of an intensity I which follows a square root law is understandable for black body radiation of very high temperature T, since $\varepsilon_{osc} \sim T$.

The subrelativistic behaviour $\varepsilon_{osc} \sim T^3$ is a kind of Debye T^3- law with a Debye energy of mc^2. In analogy to Einstein's law of recoil for the emitters and absorbers with kT/2 energy per degree of freedom, we find the same for ε_{osc}, if the reciprocal fine structure constant were 142.17. The correction due to neglect of self damping is in the right direction. The relativistic change of the mass of the electrons is compared with the energy density of the laser field. Near the cut-off density, a generation of a gravitational wave of the laser frequency should be possible.

a) Presented at the Fourth International Workshop Conference on "Laser Interaction and Related Plasma Phenomena" at Rensselaer Polytechnic Institute, Troy, New York, 8-12 November, 1976.
b) Work supported by ARGC Grant No. B75 15538.

I. INTRODUCTION

The properties of the oscillations of plasma electrons in very high intensity laser fields have been discussed by several authors during this conference. This contribution is devoted to the following two questions:

1) If the oscillation energy of the electrons in electromagnetic radiation can be considered in analogy to the recoil energy of the molecules in Einstein's derivation of the black body radiation[1], what are the consequences for the energy per degree of freedom?

2) Will the inertial mass generated during the relativistic oscillations of the electrons in a laser field give rise to gravitational waves of the laser frequency?

Both questions are devoted to the correct description of the oscillation energy of the plasma electrons in a high intensity electromagnetic field. The details of such an oscillation are still being developed, with the efforts directed mainly towards finding the exact relativistic description of the optical constants in plasma, the self-damping by emission of radiation and the inclusion of the polarization of the vacuum[2]. We shall discuss the mentioned properties of electrons in oscillation on the basis of presently known parameters and shall, more hypothetically, compare them with well known factors, as the fine structure constant or with general expectations on the laser produced gravitational waves. This is just a preliminary treatment to a later, more reasonable discussion which should improve our understanding of relativistic optical constants, self-damping and vacuum polarization.

II. RECOIL ENERGY IN BLACK BODY RADIATION

In the discussion of pair-production in very high intensity laser fields, using the Bhabha trident process, some authors assumed that the oscillation energy ε_{osc} of electrons in a laser field with an electric field amplitude E and an intensity $I=CE^2/(8\pi)$ will follow the subrelativistic relation

$$\varepsilon_{osc} = \frac{e^2E^2}{2m\,\omega} \sim I \qquad (1)$$

even for such high laser intensities, as $I>I_{rel}$, where I_{rel} produces $\varepsilon_{osc} = mc^2$ (e ~ electron charge, m ~ electron rest mass, laser frequency ω)[3], while in other calculations[4-6]

$$\varepsilon_{osc} \sim \sqrt{I} \qquad (I>>I_{rel}=\frac{3m^2\omega^2c^3}{8\pi e^2}). \qquad (2)$$

This relation has been discussed by other authors[7,8] and can be considered generally accepted.

We shall discuss here a property of black body radiation which makes evident that only a relation in Eq. (2) is in agreement with the radiation laws. The oscillation energy, without radiation damping, can be described by

$$\varepsilon_{osc} = m_o c^2[(1+A\frac{e^2E^2}{m^2\omega^2c^2|\tilde{n}|})^{\frac{1}{2}} - 1] \qquad (3)$$

where $\tilde{n}$ is the complex refractive index, the factor A(I) is 1 for circularly polarized light, and for linearly polarized light it has the following form

$$A = \begin{cases} 1 \text{ for } I<<I_{rel} \\ 1 \text{ to } A(\infty) \text{ for } I\geq I_{rel} \end{cases} \qquad (4)$$

where $A(\infty)$ is $\pi(2)^{-3/2}$ following from Max et al[8] and it is $3(2)^{-3/2}$ according to a general analytic evaluation of Schwarz and Tabensky[9] with a detailed expression of A(I) being a monotonically increasing function of the intensity. From (3) and (4), Eq. (1) is reproduced for $I<<I_{rel}$. For $I>>I_{rel}$ we find from Eq. (3)

$$\varepsilon_{osc}{}^2 = \frac{2^5\pi e^2cA}{\omega^2} I \qquad (5)$$

This can be used for a comparison with the Stefan-Boltzmann law if the electrons are in black body radiation at very high temperature T. Thus, since

$$I = \sigma T^4 \qquad (6)$$

and Wien's displacement law describes the maximum frequency ω_{max} of the spectral distribution

$$\omega_{max} = \alpha T \qquad (7)$$

where α and σ are constants, substitution of (6) and (7) into (5) gives

$$\varepsilon_{osc} \sim T \qquad (8)$$

as it has to be. If we had used the nonrelativistic relation (1), $\varepsilon_{osc} \sim T^2$ would have followed, which is impossible. This is a very transparent result, to understand why a square root law $\varepsilon_{osc} \sim \sqrt{I}$ must hold for very high (relativistic) intensities. It is remarkable that for free electrons in black body radiation a kind of Debye-T^3-law is valid, where however the Debye energy for the electrons is mc^2.

The relation (8) will now be compared with Einstein's result[1] that each emitting and absorbing particle in black body radiation has a recoil energy of

$$\varepsilon_{recoil} = \tfrac{1}{2} kT \qquad (9)$$

per degree of freedom, if emission and absorption of radiation are fully directed (needle radiation). From relation (8) we arrive at the average electron oscillation energy in a (relativistic) black body radiation

$$\varepsilon_{osc}{}^2 = \frac{A 2^5 \pi e^2 c}{\omega^2} \iint I d\omega d\Omega \qquad (10)$$

where $d\Omega$ describes the integration over all directions and

$$I d\omega = \frac{\hbar\ \omega^3}{2^2 \pi^3 c^2} \ \frac{d\omega}{\exp(+\hbar\omega/kT)-1} \qquad (11)$$

is Planck's radiation law with Planck's constant $h=2\pi\hbar$. Integration of Eq. (10) leads to

$$\varepsilon_{osc}{}^2 = A\frac{2^7\pi}{3}\ \alpha(\tfrac{1}{2}kT)^2 \qquad (12)$$

where $\alpha=e^2/(hc)$ is the fine structure constant. We see that Einstein's recoil energy (9) is equal to the energy of the oscillating electron in black body radiation at very high T, if

$$1/\alpha = A\frac{2^7\pi}{3} = 142.17 \qquad (13)$$

This value is a little larger than 137.03 and may indicated that we have neglected the self-damping in our derivation. It may also be possible that vacuum polarization is involved too, while the usual plasma effect of optical constants can be neglected in this case, as very low electron densities can be assumed.

III. GRAVITATIONAL EFFECT IN THE LASER FOCUS IN PLASMAS

During the relativistic oscillation of electrons in a laser field of frequency ω, the electrons gain mass as their speed approaches c. However, the appearance of this mass for an outside detector has to take into account the fact that the electrons in the radiation field form together a system with certain energy. The increase in the electron mass can be looked at as being due to the transfer of the mass from the radiation field. This is of prime importance in investigating the possibility of generating gravitational waves. Thus it is important to know the ratio of the generated mass to the mass of the radiation field. We shall see that at low plasma densities the electron mass increase is very small. On the other hand, the region of higher electron densities, where the change in the optical constants has been taken into account, will produce a relativistic electron mass increase in excess of the radiation mass. This conclusion is based on the presently known relativistic optical constants, although their values may still need a correction. If there were a theorem that the relativistic oscillation mass is due to exchange with the radiation mass only, then this comparison should show a way to derive the corrections; otherwise, the method of detecting the generated mass of the electrons is to be found employing the most sensitive gravitational detectors[10] and bearing in mind that the frequency of such a gravitational wave would be equal to the laser frequency ω.

Using the momentum of photons p_ψ in plasma[11,12] of an electron density n_e and with the nonrelativistic cut-off density n_{ec}^{NR}

$$p_\psi = \frac{h\omega}{c\tilde{n}} \left(1 - \frac{\omega_p^2}{2\omega^2}\right) = \frac{h\omega}{2\tilde{n}c} (1 + \tilde{n}^2) \qquad (14)$$

where we use a collisionless plasma

$$\tilde{n}^2 = 1 - \frac{\omega_p^{\ 2}}{\omega^2} = 1 - \frac{n_e}{n_{ec}^{NR}} \left(1 + A\frac{I}{I_{rel}\tilde{n}}\right)^{-\frac{1}{2}} ;$$

$$\omega_p^{\ 2} = \frac{4\pi e^2 n_e}{m} , \qquad (15)$$

we find the density of the laser radiation in the plasma is

$$\varepsilon_\psi = n_\psi \frac{h\omega}{2\tilde{n}} (1 + \tilde{n}^2) = \frac{I}{2c\tilde{n}} (1 + \tilde{n}^2) . \qquad (16)$$

The time averaged relativistic gain of mass by an electron at the oscillation is

$$m_{osc} = m\left(1 + A\frac{I}{I_{rel}\tilde{n}}\right)^{\frac{1}{2}} - m \qquad (17)$$

leading to a mass density

$$M_{osc} = m_{osc} n_e = n_e m\left[\left(1 + A\frac{I}{I_{rel}\tilde{n}}\right)^{\frac{1}{2}} - 1\right] \qquad (18)$$

$$M_{osc} = n_e m\sqrt{A\frac{I}{I_{rel}\tilde{n}}} \quad \text{at } I >> I_{rel} \qquad (19)$$

The corresponding energy density (oscillating mass energy density) is for $I>>I_{rel}$

$$\varepsilon^m_{osc} = M_{osc} c^2 = c^2 n^{NR}_{ne} (1-\tilde{n}^2) m \; AI/(I_{rel}\tilde{n}) \qquad (20)$$

Comparison with the energy density Eq. (16), of the laser radiation results in

$$\frac{\varepsilon^m_{osc}}{\varepsilon_\psi} = \frac{4A}{3} \frac{1-\tilde{n}^2}{1+\tilde{n}^2} \qquad (21)$$

It is obvious, if the electron density is very small ($n \approx 1$), the ratio of Eq. (21) is zero. But it is remarkable, that at $n \approx 0$ (electron density near the relativistic cut-off density), we find a ratio of a value 1.33A, where $A \geq 1$. Therefore under these conditions, the generation of a gravitational wave would be possible in the focus of a high density plasma with relativistic intensities.

The result that a gravitational effect is possible near the cut-off density is not in discrepancy with the postulate so that the energy density of the vacuum has to be conserved. It has to be taken into account that according to the group velocity and the collective effects, an additional emission of energy (gravitational energy) should be possible causing a certain degree of absorption. The same happens if a collisionless plasma with a Coulomb collision frequency zero causes a loss of the energy of the transmitted laser beam and therefore absorption, e.g. by Thomson scattering.

REFERENCES

1. A.Einstein, Physikalische Zeitschrift, 18, 121 (1917).

2. W. Heisenberg and E. Euler, Z. Physik, 98, 714 (1936).

3. F.V. Bunkin and A.E. Kazakov, Sov. Phys. Doklady, 15, 758 (1971).

4. H. Hora, Laser Interaction and Related Plasma Phenomena, H. Schwarz and H. Hora eds (Plenum, New York 1974, Vol 3B, p. 819.

5. J.W. Shearer, J. Garrison, J. Wong and J.E. Swain, ibid. p. 803.

6. M.B. Nicholson-Florence, Diss. Essex 1972.

7. P. Kaw and J. Dawson, Phys. Fluids 13, 472 (1970).

8. C. Max and F. Perkins, Phys. Rev. Letters 27, 1342, (1971).

9. H. Schwarz and R. Tabensky, Laser Interaction and Related Plasma Phenomena, H. Schwarz and H. Hora eds. (Plenum, New York), 1977, Vol. 4, p. 961.

10. J. Weber, Nuovo Cimento, 4B, 197 (1971).

11. R. Klima and V.A. Petrzilka, Czech.J. Phys. B22, 896 (1972).

12. H. Hora, Phys. Fluids 17, 1042 (1974).

BASIC REQUIREMENTS OF LASER FUSION POWER PRODUCTION AND LASER FUSION-FISSION PLUTONIUM BREEDING*

Ray E. Kidder

University of California, Lawrence Livermore Laboratory

Livermore, California 94550

ABSTRACT

Required values of thermonuclear energy-gain, inertial confinement, laser-pulse energy and power for both electric power production by laser-fusion and for plutonium breeding by laser fusion-fission are obtained from simple considerations, and are tabulated for purposes of comparison.

It is found that breeder requirements are very significantly less demanding, suggesting that plutonium breeding is a much more feasible near-term goal of laser fusion.

I. INTRODUCTION

A central problem of laser-driven pellet fusion is achievement of the very large thermonuclear energy gain necessary to compensate for the low efficiency of short-pulse lasers and of the fuel-compression process. To achieve the required energy gain with the least amount of laser-pulse energy, a small amount of thermonuclear fuel must be highly compressed; its inertial confinement, measured by its ρR product, must attain a large, specified value. This minimum energy and required ρR product can be estimated by considering a simple model of the ignition and burn of the compressed fuel. (1,2)

---

*Presented at the Fourth Workshop on "Laser Interaction and Related Plasma Phenomena" held at RPI, Troy, New York, November 8-12, 1976.

It is found that isentropic fuel compression at near-minimum entropy is needed to avoid excessive laser-pulse energy requirements.

An elementary treatment of the hydrodynamics of isentropic compression shows that the ρR product that can be achieved is proportional to the square root of the peak hydrodynamic power expended.(3) A more detailed consideration of the pellet ablation process shows that this square-root relation also applies, approximately, to the peak laser power absorbed by the pellet, the coefficient of proportionality being an increasing function of the pellet aspect ratio (inner radius divided by wall thickness). (4) Thus there is a motivation to employ thin hollow shells, rather than the solid DT spheres originally proposed, (5) to reduce the peak laser-pulse power required to drive the implosion. The reduction in peak power that can be achieved by this means, however, is limited by fluid instability of the Rayleigh-Taylor type.

We shall apply the relatively simple relations between thermonuclear energy gain, laser-pulse energy, ρR-product, and laser-pulse power to evaluate the basic requirements of laser fusion electric power production and of laser fusion-fission plutonium breeding. We find that the breeder requirements are very significantly less demanding than those of electric power production.

II. REQUIRED PELLET GAIN

The basic energy-balance equation of a power plant based on laser-driven pellet fusion is

$$\varepsilon_{He} f_e \varepsilon_{eL} f_a G_P = 1 \qquad (2.1)$$

where ε_{He} is the conversion efficiency of heat into electricity, f_e is the fraction of the total electric power generated that is supplied to the laser, ε_{eL} is the efficiency of the laser in converting its electrical input to light output, f_a is the fraction of the light output that is ultimately absorbed by the target pellet, and G_P is the pellet gain; i.e., the ratio of thermonuclear energy produced to the light energy absorbed by the pellet. A typical set of values for these factors might be: $\varepsilon_{He} = 0.4$, $f_e = 0.25$, $\varepsilon_{eL} = 0.10$, $f_a = 0.5$, $G_P = 200$. (6)

It is useful to write the pellet gain, G_P, as

$$G_P = \varepsilon_{LF} G_F \quad , \qquad (2.2)$$

where ε_{LF} is the fraction of absorbed laser light that has been converted into internal energy of the highly compressed core of thermonuclear fuel (DT) at the instant of fuel ignition. A typical value for ε_{LF} is 0.05, (6) so that a fuel gain, G_F, of 4000 would be required to achieve a pellet gain of 200. The fuel gain is defined as the ratio of the thermonuclear energy yield to the internal energy imparted to the compressed core of DT.

III. REQUIRED INERTIAL CONFINEMENT

The inertial confinement necessary to achieve the large pellet gains required for laser fusion can be evaluated by means of a simple model. (1) The model used is that of a uniformly compressed pellet of DT ignited by a small, hot, central "spark." The spark is specified to be large enough and hot enough to ignite the remainder of the compressed fuel via an out-going thermonuclear detonation.

The fuel gain G_F is shown in Figs. 1 and 2 as a function of the inertial confinement parameter H_F denoting the density-radius product ρR of the compressed fuel for a sequence of values of the internal energy W_F supplied to the compressed core ranging from 50 J to 50 kJ. If ε_{LF} equals 0.05, and f_a equals 0.5, this corresponds to laser-pulse energies, W_L (= $W_F/f_a\varepsilon_{LF}$) ranging from 2 kJ to 2 MJ.

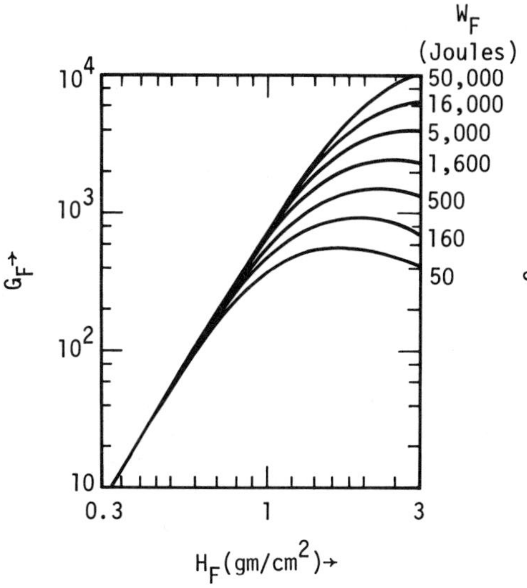

Figure 1

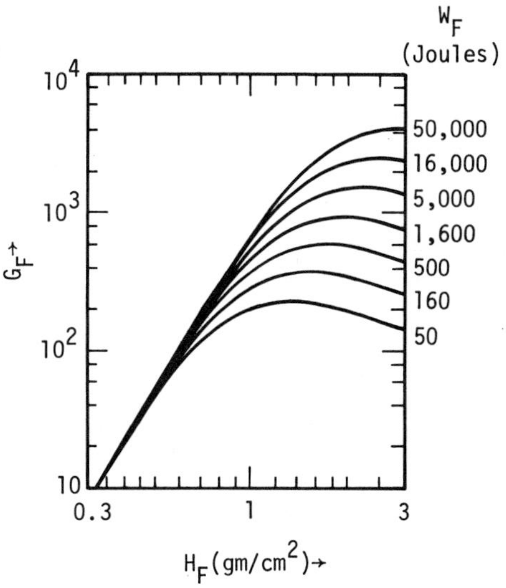

Figure 2

Figure 1. Fuel-gain G_F versus inertial confinement parameter H_F for fuel-energies W_F of 50, 160, 500, 1600, 5000, 16000, and 50000 joules, assuming minimum-entropy compression (α = 1).
Figure 2. Same as Fig. 1 but with α = 2. (Twice as much work must be done to achieve the same compression η.)

Figure 1 applies to the most favorable case, that of minimum-entropy compression ($\alpha = 1$). Figure 2 applies when twice as much work ($\alpha = 2$) must be done to achieve the same compression. Inspection of Figs. 1 and 2 shows that H_F must amount to 2-3 g/cm^2 if fuel gains of several thousand are to be realized.

IV. REQUIRED LASER-PULSE ENERGY

Inspection of Figs. 1 and 2 also shows that laser-pulse energies of 200 kJ ($\alpha = 1$) to 1.6 MJ ($\alpha = 2$) are necessary if a pellet gain as large as 200 is to be achieved, again assuming the efficien ε_{LF} to be 5% and f_a to be 50%.

The scaling of the least laser-pulse energy W_L^* required to achieve a desired pellet gain G_P, in terms of the principal factors describing the implosion and burn of the fuel, is found to be (1)

$$W_L^* \propto \left(G_P/q_{DT}\right)^{12/5} H_B^{9/5} \left(A_S/\varepsilon_{LF}\right)^{2/5} / \varepsilon_C^3 f_a \quad , \tag{4.1}$$

where

$$H_B = 8M_i/\sigma \geq 7 \text{ g/cm}^2 \tag{4.2}$$

$$\sigma = \left\langle \sigma_{DT} v \right\rangle / c_s \tag{4.3}$$

$$A_S = T_S(\text{eV})[H_S(\text{g/cm}^2)]^3/540 \geq 1 \quad . \tag{4.4}$$

Here, M_i is the average ion mass, c_s is the isothermal speed of sound in DT, $\langle \sigma_{DT} v \rangle$ is the Maxwell-averaged DT cross section, and q_{DT} is the energy of the DT fusion reaction. T_S and H_S are the temperature and ρR of the central ignition spark, respectively.

The factor ε_C is the <u>compression efficiency</u> defined as: the least possible work required to achieve a given compression (i.e., isentropically at minimum entropy), divided by the (absorbed) laser-pulse energy actually needed to achieve it. It is proportional to ε_{LF}/α, i.e., (1)

$$\varepsilon_C \cong (7/8)(\varepsilon_{LF}/\alpha) \quad . \tag{4.5}$$

Compression efficiencies ε_C predicted by detailed implosion calculations are low, typically 1-2%. (Our choice of $\varepsilon_{LF} = 0.05$ and $\alpha = 2$ implies that $\varepsilon_C \sim 2\%$.)

We note that the laser-pulse energy W_L^* is a strong function of all factors appearing in Eq. (4.1) except the size A_S of the ignition spark.

The value of the confinement parameter H_F^* that minimizes the required laser-pulse energy scales as

$$H_F^* \propto \left\{H_B^{3/4} A_S G_P / q_{DT} \varepsilon_{LF}\right\}^{4/15} \tag{4.6}$$

It is a far less sensitive function of the implosion parameters than is the laser-pulse energy itself.

V. REQUIRED LASER-PULSE POWER

A. Solid DT-Sphere Target

We demonstrate as follows that the mechanical power P_M required to homogeneously and isentropically compress an ideal ($\gamma = 5/3$) gas sphere is proportional to the square of the ρR product achieved by the compression:

$$P_M = 4\pi R^2 p u = \underbrace{(4\pi/\gamma)(c_a^3/\rho)}_{\text{const.}} \; \underbrace{M}_{\sqrt{3}} (\rho R)^2 \tag{5.1}$$

$$\propto \alpha^{3/2} (\rho R)^2 \quad . \tag{5.2}$$

$M (= u/c_a)$ is the Mach number of the compression, and c_a is the adiabatic sound speed. The quantity c_a^3/ρ is constant along an isentrope and is proportional to $\alpha^{3/2}$.

The Mach number M is determined by our desire to minimize the laser-pulse energy required to drive the implosion, a result that is accomplished by carrying out the compression as rapidly as possible without the formation of shocks. This requires that the Mach number be equal to $\sqrt{3}$ for either homogeneous isentropic compression (HIC: compression-rate uniform throughout pellet volume (7,8)) or centered isentropic compression (CIC: characteristics arrive simultaneously at the pellet's center (8,9)).

The time-dependence of the mechanical power P_M required to carry out either HIC or CIC is given (at late times) by

$$P_M \propto \left(1 - t/t_c\right)^{-2} \quad , \quad \left(1 - t/t_c \ll 1\right) \quad , \tag{5.3}$$

where t_c is the asymptotic time of "total collapse" (= R_o/c_{so} in the case of CIC).

A more detailed treatment that takes into account the nature of the pellet-ablation process (3) leads to expressions for the peak laser power P_L, and for the time-dependence of P_L, that are closely similar to those given for the mechanical power P_M by Eqs. (5.2) and (5.3). They are

$$P_L \propto \alpha^{7/6}(\rho R)^{11/6}/f_a \quad , \tag{5.4}$$

and

$$P_L \propto \left(1 - t/t_c\right)^{-11/6}/f_a \quad . \tag{5.5}$$

Combining the results of Eqs. (5.4) and (4.6), and identifying H^*_F with ρR, we obtain the following scaling relation for laser-pulse peak power P^*_L,

$$P^*_L \propto \alpha^{7/6}\left\{H_B^{3/4} A_S G_P/q_{DT}\varepsilon_{LF}\right\}^{1/2}/f_a \quad . \tag{5.6}$$

We note that the required laser-pulse peak power is a considerably less-sensitive function of the implosion and burn parameters than is the laser-pulse energy.

B. Hollow DT-Shell Target

In the case of hollow shell targets, the coefficient of proportionality relating P_L and $H_F^{11/6}$ depends on the aspect ratio (ζ_u) of the unablated portion of the hollow shell; i.e., the aspect ratio of the payload of the ablation-driven rocket. A semianalytical result, obtained by a best-fit to detailed computer runs of imploded shells with a wide range of aspect ratios (including solid spheres), is (4)

$$P_L(\mathrm{TW}) = (160/\ f_a)\left[H_F(\mathrm{g/cm^2})\right]^{11/6} \quad , \tag{5.7}$$

where

$$\Omega = \sqrt{1 + 3\zeta_u} \quad . \tag{5.8}$$

A solid sphere is characterized by $\Omega = 1$. The Ω has, therefore, the physical significance of the factor by which the peak laser power can be reduced (by imploding hollow shells with aspect ratio $\zeta_u (\neq 0)$ instead of solid pellets).

We see that there is considerable advantage in imploding hollow shells instead of solid pellets. Not only is the peak power requirement reduced (by the factor Ω), but the intensity of the absorbed light will be lower, leading to reduced excitation of nonlinear optical effects and plasma instabilities, and improved corona-core coupling. However, we expect that the minimum feasible shell thickness will be limited by the growth of fluid instabilities during implosion.

Anticipating the result of the next section that fluid instabilities may limit the minimum shell thickness to aspect ratios $\zeta_u \sim 30$ and reduction factors $\Omega \sim 10$, and using the values $H_F = 2.7\ g/cm^2$ and $f_a = 0.5$, we obtain from Eq. (5.7) the required peak laser power

$$P_L \sim \begin{cases} 2000\ \text{TW (solid DT-sphere)} , \\ \\ 200\ \text{TW (hollow DT-shell)} . \end{cases} \tag{5.9}$$

VI. REQUIRED SHELL THICKNESS

The amplitude of a surface perturbation of a homogeneously and isentropically imploded shell increases, due to Rayleigh-Taylor instability growth, by the factor (4)

$$f_\ell = \frac{\cosh(\sqrt{\ell}\chi) - \beta^{2\ell}\cos(\sqrt{\ell}\chi)}{1 - \beta^{2\ell}} \tag{6.1}$$

where

$$\chi = \cosh^{-1}\eta^{1/3} \quad , \quad \beta = R_{10}/R_{20} \quad , \tag{6.2}$$

η is the compression of the shell, and $\beta(< 1)$ is the ratio of the inner and outer shell radii. The surface perturbation is described by the spherical harmonic $Y_{\ell m}(\theta,\phi)$. Equation (6.2) reduces to the following simple form for large values of η and ℓ,

$$f_\ell \cong \frac{1}{2}(8\eta)^{\sqrt{\ell}/3} \qquad (\beta^{2\ell} << 1, \eta >> 1) \quad . \tag{6.3}$$

Application of these results to the question of shell break-up leads to the following limit on the shell aspect ratio ζ_u

$$\zeta_u < \left[3\ \ell n(5\ \varepsilon_0)/\ell n\ \eta\right]^2 - 1 \quad , \tag{6.4}$$

and on the peak power reduction factor Ω

$$\Omega < 3\sqrt{3}\ |\ell n(5\ \varepsilon_0)/\ell n\ \eta| \quad , \qquad (6.5)$$

where ε_0 is the initial fractional surface perturbation amplitude $\delta R_{20}/R_{20}$. If $\eta = 10^4$, and $\varepsilon_0 = 2 \times 10^{-5}$, for example, then ζ_u must be less than 8 to avoid shell break-up, and Ω can therefore not exceed 5. (Since η and ε_0 enter Eqs. (6.4) and (6.5) only logarithmically, these results are relatively insensitive to the values of η and ε_0 chosen.)

In obtaining these results we have not taken into account the effects of ablation, viscosity, heat diffusion, or finite density gradient at the fluid surface, each of which will tend to decrease the growth rate of the shorter wavelength modes. This omission appears to be crudely justified by two-dimensional computer simulations of isentropically imploded shells that include the above-named effects. (10) Such simulations exhibit growth rates amounting to 50 to 100% of the Rayleigh-Taylor values for wavelengths considered here ($\ell \lesssim 10$).

If we assume that the actual growth rate is 50% as large as we have calculated neglecting these effects, then we conclude that the initial aspect ratio of the unablated or payload portion of the shell cannot exceed about thirty, which implies a ten-fold reduction in peak laser power as compared with the peak laser power required to implode solid spheres to the same density.

VII. REQUIRED LASER APERTURE

The useful aperture-averaged output light intensity that a laser is capable of radiating is limited by optical damage to laser materials and/or beam-quality degradation due to nonlinear propagation effects. If, for example, we consider this limiting intensity to be 3-6 GW/cm^2 as is presently the case for Nd-doped silicate glass, (11) then a 20 cm diameter output aperture is capable of radiating 1-2 terawatts of laser power. It follows that the number N_L of 20 cm diameter laser beams that are required to provide the total peak laser power listed in Eq. (5.9) for pellet compression are

$$N_L \sim \begin{cases} \text{1000-2000 beams} & \text{(solid DT-sphere)} \quad , \\ \\ \text{100-200 beams} & \text{(hollow DT-shell)} \quad . \end{cases} \qquad (7.1)$$

These numbers clearly indicate the need for optical material capable of transmitting greater light intensity without deleterious effects.

VIII. SIGNIFICANCE OF FUSION-FISSION

The pulse-energy W_L^* and pulse-power P_L^* of Eqs. (4.1) and (5.6), respectively, can be written as

$$W_L^* \propto (A_S/\varepsilon_{LF})^{2/5}/\phi^{12/5}\varepsilon_C^3 f_a \quad , \tag{8.1}$$

$$P_L^* \propto \alpha^{7/6}(A_S/\varepsilon_{LF}\phi)^{1/2}/\Omega f_a \quad , \tag{8.2}$$

where

$$\phi = q_{DT}\varepsilon_{He} f_e \varepsilon_{eL} f_a / H_B^{3/4} \quad , \tag{8.3}$$

by eliminating G_P and generalizing P_L^* to the case of hollow shells (divide by Ω).

We note that a significant reduction in required laser-pulse power and a major reduction in laser-pulse energy could be made if the energy q_{DT} released per DT reaction could be substantially increased.

The energy q_{DT} can be increased approximately ten-fold (12) by adding natural or depleted uranium to the lithium blanket that is needed anyway to breed replacement tritium. The nuclear reactions and energetics, with and without a fissionable blanket, are compared below.

<u>PURE FUSION vs FUSION-FISSION</u>

(22 MeV vs 224 MeV + 2Pu)

PURE FUSION: $H^2 + Li^6 = 2He^4 + \underline{22\ \text{MeV}}$

FUSION-FISSION: $H^2 + Li^6 + 3U^{238} = 2He^4 + \underline{224\ \text{MeV}}$
(t = ∞)

$+ \underline{2Pu^{239}(400\ \text{MeV})} + \{Zr^{90} + Nd^{145} + 12e\} + n$
typical fission products

In addition to the ten-fold increase in energy produced directly, at least two fissile plutonium atoms can be created per DT-reaction with the potential for providing 400 MeV when burned in a fission reactor. This suggests a fissile-breeder mode of operation with zero net power output. In this mode of operation q_{DT} is increased ten-fold, and f_e is increased four-fold (from 0.25 to 1). We shall take advantage of the four-fold increase in f_e to reduce our assumed laser efficiency to 2.5%. The factor ϕ appearing in Eqs. (8.1) and (8.2) is then increased 10-fold, assuming that other quantities remain unchanged. The result is a 3-fold reduction in required peak laser power (to 60 TW) and a 250-fold reduction in required laser-pulse energy (to 6 kJ).

Since it may be more important to reduce pulse-power than to minimize pulse-energy, it may be desirable to operate at points in Fig. 2 other than the locus of curve-maxima. For example, a fuel-gain of 400 is also obtained with W_F = 500 J and H_F = 1 gm/cm^2, corresponding to a pulse-energy of 20 kJ and a pulse-power of 30 TW.

The requirements of pure-fusion electric power production and of fusion-fission plutonium breeding are summarized in Table I below.

TABLE I. Requirements[a] for pure-fusion electric power production and fusion-fission plutonium breeding.

	Power Production	Plutonium Breeding
Laser-pulse power P_L (TW)	200	60 (30)
Laser-pulse energy W_L (kJ)	1,600	6 (20)
Laser efficiency ε_{eL}	10%	2.5%
ρR (g/cm^2)	2.7	1.5 (1.0)
Pellet gain G_P	200	20

[a]Assuming 30:1 aspect ratio hollow DT-shell target (ζ_u = 30), compressive work equals twice minimum entropy value (α = 2), 50% of laser output absorbed by pellet (f_a = 0.5).

IX. SUMMARY AND CONCLUSIONS

We conclude that a 200 terawatt, 1.6 megajoule, 10% efficient (ε_{eL} = 0.1) laser may be required for practical (ε_{He} = 0.4, f_e = 0.25) laser-fusion power production. In arriving at this conclusion we have assumed that half the laser output is ultimately absorbed (f_a = 0.5) by a hollow DT-shell target having a 30:1 (unablated) aspect-ratio (ζ_u = 30), and that the fuel-compression is accomplished within a factor of 2 of minimum entropy conditions (α = 2) with an efficiency of 2% (ε_C = 0.02). The uncertainty of our estimate of required pulse-energy (1.6 MJ) is considered to be much greater than that of our estimate of required pulse-power (200 TW), since it depends much more sensitively on implosion parameters whose values are uncertain.

We have also shown that these very onerous projected laser requirements are importantly relaxed in the case of a zero net power ($f_e = 1$) plutonium breeder. Here it appears that a 2.5% efficient laser providing 30 TW, 20 kJ pulses may suffice, requirements that are clearly far easier to meet.

It also should be noted that nothing has been said about the wavelength-dependnece of these laser requirements. The values of the important parameters f_a, ε_C and α all depend on the details of the light-plasma interaction and therefore are surely wavelength-dependent. There is, however, little agreement at present about this important dependence either theoretically or experimentally. The values we have employed are those that generally describe the results of one-dimensional, three-temperature, optimized computer runs with 0.25-1 micron laser light. Since these runs do not include hot-electron preheat, corona-core decoupling, anomalous electron transport, hot ion ejection, spontaneous magnetic fields, nonuniform illumination, etc., our values are almost certainly optimistic, perhaps grossly so.

We believe that the much-less-demanding requirements of the fusion-fission breeder, together with economic considerations (why use valuable 14 MeV neutrons merely to produce heat?), indicate that breeder development should be considered as the interim goal of laser fusion. (It tends to be forgotten that the possibility of plutonium breeding was an important motivation of the early CTR program, and the principal reason that the CTR program was originally classified.)

ACKNOWLEDGMENT

This work was performed under the auspices of the U. S. Energy Research and Development Administration, contract No. W-7405-Eng-48.

1. R. E. Kidder, Nuclear Fusion 16, 405 (1976).

2. K. A. Brueckner and S. Jorna, Rev. Mod. Phys. 46, 325 (1974).

3. R. E. Kidder, Nuclear Fusion 14, 797 (1974).

4. R. E. Kidder, Nuclear Fusion 16, 3 (1976).

5. J. Nuckolls *et al.*, Nature 239, 139 (1972).

6. L. R. Hafstad *et al.*, *Final Report of the Special Laser-Fusion Advisory Panel*, ERDA Report No. 28, January 31, 1975. (Note: Higher values for f_e(= 1/3) and f_a(= 1) are postulated in this reference, leading to a required pellet-gain G_P of only 75 instead of 200.)

7. R. E. Kidder, Nuclear Fusion 14, 53 (1974).

8. D.E.T.F. Ashby, Nuclear Fusion 16, 231 (1976).

9. C. Ferro-Fontan *et al.*, Phys. Lett. 55A, 35 (1975).

10. J. D. Lindl and W. C. Mead, Phys. Rev. Lett. 34, 1273 (1975).

11. K. A. Brueckner *et al.*, *Assessment of Laser-Driven Fusion*, Final Report, KAB & Associates, Inc., La Jolla, California, March 31, 1976.

12. J. A. Maniscalco, *Fusion-Fission Hybrid Concepts for Laser Induced Fusion*, UCRL-76763, Lawrence Livermore Laboratory, May 1, 1975.

THERMONUCLEAR REACTION WAVE IN HIGH-DENSITY PLASMA*

Katsunobu Nishihara

Institute of Laser Engineering, Osaka University, Osaka, Japan

Kazuhiro Nozaki, Toshiya Taniuti

Department of Physics, Nagoya University, Nagoya, Japan

Naruyoshi Asano

Department of Engineering Mathematics, Utsunomiya University, Utsunomiya, Japan

I. INTRODUCTION

Thermonuclear reaction waves in plasmas have recently attracted considerable attentions in connection with the laser-induced thermonuclear fusion. In a highly compressed laser plasma, the burning process may develop as follows[1] : The inner region of a D-T sphere is compressed adabatically by a factor 10^3 to 10^4 above its solid density, and the center of the compressed fuel is brought to an ignition temperature, while the rest of the fuel remains relatively cold. Then, the enrgy produced in the ignited and burning central region of the fuel can heat the surrounding cold fuel sufficiently for ignition to occur. This process leads to the formation of a self-driven reaction wave which can propagate radially outwards igniting the cold fuel.

To our knowledge, there are three typical models of the thermonuclear reaction waves. One is a shock heating model which was investigated for a low density gas (15^{15} cm^{-5}) by Fuller and Gross[2] based on a Zel'dovich-Neumann-Doring model. Namely, the shock is sufficiently strong so that it heats the gas to a thermonuclear temperature. Because of the low density assumption, this

*Presented at the Fourth Workshop on "Laser Interaction and Related Phenomena" held at RPI, Troy, New York, November 8-12, 1976.

model is not directly applicable to the laser-plasma. The history of the development to the self-driven reaction wave in a solid density plasma was studied by Chu[3]. His numerical results show that the structure and mechanism of the reaction wave are dominated by electron thermal conduction. These two studies did not, however, take into account the effect of nonlocal deposition of α-particle energy. The propagation of the wave due to the conduction of high energy α-particles was discussed by Bruckner and Jorna[1] They have shown by means of energy conservation that a burning front advances supersonically if a temperature is higher than 15 KeV. However, their simple model does not include the effect of finite deposition time of the α-particle energy to a plasma. Namely, they assumed that the total energy of the α-particles produced by the reaction is instantaneously deposited to the fuel and so the effect of α-particle heating is overestimated. Therefore, according to their results, the propagation velocity of the burning front becomes greater than the α-particle velocity, when a temperature is higher than 30 KeV, which is apparently inconsistent with their assumption of the α-particle conduction.

The aim of the present study is thus to investigate how the characteristics of the thermonuclear reaction wave in the high-density plasma depend on such processes as the shock heating, the electron thermal conduction and the α-particle conduction. In the following four sections, we will present the ignition condition and development of a weak reaction wave which was analytically investigated[4]. The propagation and structure of a strong one, which was numerically studied by two of the authors (K.N.)[5], are given in the sections VI, VII and VIII.

II. WEAK REACTION WAVE

In order to analyze the hydrodynamic motion of the plasma which is heated by the α-particles produced by D-T reaction, we consier a weakly reacting sound wave in a quasi-equilibrium high temperature plasma. The quasi-equilibrium state is realized when the deposition of the α-particle energy is balanced with the bremsstrahlung loss. This condition is satisfied at about 4 KeV temperature when the D-T plasma is transparent to the radiation, which is assumed throughout. At this temperature, the deposition of neutron energy is not important since the neutron range is much longer than the α-particle range.[2] The D-T fuel depletion is negligible because the depletion time t_R (a reation time) is much longer than a characteristic time of the weakly reacting sound wave, which is estimated as $t_c \sim t_R T_o/E_\alpha$, where T_o/E_α is the ratio of plasma temperature to the α-particle energy and so $T_o/E_\alpha << 1$. The self-collision times t_{ii} and t_{ee} of ions and electrons are much smaller than t_c but the relaxation time between electrons and ions is not exceedingly smaller than t_c. Hence electrons and ions are described by a system of two-temperature fluid equations. The nonlocal heating by high energy α-particles is taken into account by using an appropriate diffusion equation.

Under the schemes described above, we derive two simple equations governing the hydrodynamical properties of a weakly reacting wave and discuss some characteristic phenomena such as the ignition condition and a steady-state weak reaction wave. It is shown that the initial disturbance with the wavelength longer than a several fold of the α-particle range will grow and under the periodic boundary condition it will eventually reach a steady-state sawtooth wave.

III. DERIVATION OF SIMPLE GOVERNING EQUATIONS

As was mentioned in the preceeding section, the basic equations are the two-temperature Navier-Stokes equations, and one-dimensional geometry is assumed for mathematical simplicity.

$$\frac{\partial n}{\partial t} + \frac{\partial nv}{\partial x} = 0, \tag{1}$$

$$\frac{\partial v}{\partial t} + v\frac{\partial v}{\partial x} = -\frac{1}{m_i n}\frac{\partial}{\partial x}n(T_i+T_e) + \frac{1}{m_i n}\frac{\partial}{\partial x}(\mu_i+\mu_e)\frac{\partial v}{\partial x} \tag{2}$$

$$\frac{\partial T_i}{\partial t} + v\frac{\partial T_i}{\partial x} = -\frac{2}{3}T_i\frac{\partial v}{\partial x} + \frac{2}{3}\frac{1}{n}\frac{\partial}{\partial x}(K_i\frac{\partial T_i}{\partial x}) + \frac{2\mu_i}{3n}(\frac{\partial v}{\partial x})^2 + \frac{T_e-T_i}{t_{ei}} + \frac{2}{3}E_\alpha \omega_{\alpha i}\frac{n_\alpha}{n}, \tag{3}$$

$$\frac{\partial T_e}{\partial t} + v\frac{\partial T_e}{\partial x} = -\frac{2}{3}T_e\frac{\partial v}{\partial x} + \frac{2}{3}\frac{1}{n}\frac{\partial}{\partial x}(K_e\frac{\partial T_e}{\partial x}) + \frac{2\mu_e}{3n}(\frac{\partial v}{\partial x})^2 + \frac{T_i-T_e}{t_{ei}} + \frac{2}{3}E_\alpha \omega_{\alpha e}\frac{n_\alpha}{n} - \frac{2}{3}P_B. \tag{4}$$

The subscripts i and e denote ions and electrons respectively. The thermal conductivities $K_{i,e}$ and the viscosity coefficients $\mu_{i,e}$ have their classical values. P_B is the bremsstrahlung loss per unit time. E_α is the α-particle energy and is equal to 3.5 MeV, and $\omega_{\alpha i,e}$ are the slowing down frequencies of α particle due to the collisions with ions and electrons respectively. Since

$\omega_{\alpha e} \gg \omega_{\alpha i}$ at 4 keV, $\omega_{\alpha} = \omega_{\alpha i} + \omega_{\alpha e} \simeq \omega_{\alpha e} \simeq 5\times10^{-13} n/T_e^{3/2}$ sec^{-1} (T_e in keV). n_{α} is the number density of high energy α particles slowed down mainly by collisions with electrons. For those particles we may assume the following diffusion equation.[6]

$$\frac{\partial n_{\alpha}}{\partial t} + \frac{\partial}{\partial x} n_{\alpha} v = \frac{\partial}{\partial x} D_{\alpha} \frac{\partial n_{\alpha}}{\partial x} + S - \omega_{\alpha} n_{\alpha} , \tag{5}$$

where the diffusion coefficient is approximated as

$$D_{\alpha} = \frac{1}{3} v_{\alpha}^2/\omega_{\alpha}, \qquad v_{\alpha} = (E_{\alpha}/2m_{\alpha})^{1/2} .$$

S is the reaction rate per unit volume and is given by

$$S = n/t_R = \frac{1}{4} n^2 \langle\sigma v\rangle ,$$

$$\langle\sigma v\rangle = \frac{3.68\times10^{-12}}{T_i^{3/2}} \exp(-19.94/T_i^{1/3}) \ \sec^{-1} cm^3 \quad (T_i \text{ in keV}).$$

In the system of eqs. (1)∿(5), a homogeneous equilibrium state is determined by the following equations.

$$\frac{T_{e0} - T_{i0}}{t_{ei0}} + \frac{2}{3} \omega_{\alpha i0} \frac{n_{\alpha 0}}{n_0} E_{\alpha} = 0, \tag{6}$$

$$\frac{T_{i0} - T_{e0}}{t_{ei0}} + \frac{2}{3} \omega_{\alpha e0} \frac{n_{\alpha 0}}{n_0} E_{\alpha} - \frac{2}{3} P_{B0} = 0, \tag{7}$$

$$S_0 = n_0/t_{R0} = \omega_{\alpha 0} n_{\alpha 0} , \tag{8}$$

where the subscript 0 indicates equilibrium values. From eqs. (6), (7) and (8), we obtain the relations

$$E_{\alpha}/t_{R0} = P_{B0}, \qquad n_{\alpha 0}/n_0 = (t_{R0}\omega_{\alpha 0})^{-1}, \tag{9}$$

and the former relation gives $T_{io} \sim T_{eo} \sim 4$ keV $\equiv T_o$ and the latter gives $n_{\alpha o}/n_o \sim 2{,}4\times10^{-5}$ and so the reaction is weak.

In order to estimate the magnitude of each term in eqs. (1) ∿ (5), we normalize the variables as

$$t \to t/t_{ii}, \quad x \to x/v_S t_{ii}, \quad v \to v/v_S,$$

$$n \to n/n_0, \quad T_{i,e} \to T_{i,e}/T_0,$$

where v_s is the sound speed at 4 keV. Then, noting eqs. (9) one sees that the energy deposition terms in eqs. (3) and (4) and the bremsstrahlung loss are $O\{(E_\alpha/T_0)(t_{ii}/t_{RO})\} \sim 10^{-4}$ and the other terms in eqs. (1) $\sim$ (4) are $O(1)$ or $O(\sqrt{m_i/m_e})$ at most or $O(\sqrt{m_e/m_i})$ at least. On the other hand, in eq. (5), the diffusion term is $O\{D_{\alpha 0}(n_{\alpha 0}/n_0)\} \sim 10^4 \times O(n_{\alpha 0}/n_0)$ while the others are $O(n_{\alpha 0}/n_0)$. We now introduce a smallness parameter ε so that ε^2 represents weakness of the reaction, that is, $\varepsilon^2 \sim O\{(E_\alpha/T_0)(t_{ii}/t_{RO})\}$, then from the numerical estimations given so far we find that $\varepsilon^2 \sim 10^{-4} \sim O(D_\alpha^{-1}) \ll 1$. Consequently the α-particle range λ_α becomes $O(\varepsilon^{-1})$ because $O(\lambda^2) \sim O(D_\alpha) \sim O(\varepsilon^{-2})$. Following the conventional procedure of the reductive perturbation method,[7] we stretch the independent variables as

$$\xi = \varepsilon(x - v_s t), \quad \tau = \varepsilon^2 t,$$

and expand the dependent variables around the equilibrium state as

$$n = n_0 + \varepsilon n^{(1)} + \varepsilon^2 n^{(2)} + \cdots\cdots ,$$

$$v = \varepsilon v^{(1)} + \varepsilon^2 v^{(2)} + \cdots\cdots ,$$

$$T_{i,e} = T_0 + \varepsilon T_{i,e}^{(1)} + \varepsilon^2 T_{i,e}^{(2)} + \cdots\cdots ,$$

$$n_\alpha = n_{\alpha 0} + \varepsilon n^{(1)} + \varepsilon^2 n_\alpha^{(2)} + \cdots\cdots ,$$

Here it may be noted that the slow time scale is associated with the long reaction time and the slow spatial scale is given by the long range of the α-particles. The lowest order terms in eqs. (1) $\sim$ (4) give the ordinary linear relations of the sound wave.

$$n^{(1)} = \frac{3}{2}\frac{n_0}{T_0} T_i^{(1)}, \quad v^{(1)} = \frac{3}{2}\frac{v_s}{T_0} T_i^{(1)}, \quad T_e^{(1)} = T_i^{(1)} = T^{(1)}.$$

The next order terms in eqs. (1) $\sim$ (4) give the modified Burgers equation.

$$\frac{\partial T^{(1)}}{\partial \tau} + aT^{(1)}\frac{\partial T^{(1)}}{\partial \xi} - b\frac{\partial^2 T^{(1)}}{\partial \xi^2} - qn_\alpha^{(1)} + fT^{(1)} = 0, \qquad (11)$$

where

$$a = 2v_s/T_0, \qquad b \simeq K_{e0}/15n_0, \qquad q = \frac{E_\alpha \omega_{\alpha 0}}{15\varepsilon^2}\frac{n_{\alpha 0}}{n_0},$$

$$f = \frac{1}{15\varepsilon^2}\{\frac{3}{2}\frac{P_{B0}}{T_0} + (\frac{\partial P_B}{\partial T})_0 + \frac{3}{2}\frac{E_\alpha}{T_0}\frac{n_{\alpha 0}}{n_0}\omega_{\alpha 0}\}.$$

In the expression b, contributions from other terms such as the viscosity and etc. can be neglected as small, hence the electron thermal conductivity is responsible for the dissipation. The coefficient q represents the heating by α particles while f represents the loss due to the bremsstrahlung and the loss of the energy gain per plasma particle due to the increase of the plasma density. The slowing down time of α-particle increases with temperature so that the α-particle heating seems to be supressed as the temperature increase. However it is shown that this increase of slowing down time is exactly cancelled by its decrease due to the increasing density following the temperature rise. On the other hand noting that the diffusion coefficient D_α is $O(\varepsilon^{-2})$, we find that the lowest order terms in eq. (5) give

$$\lambda_\alpha^2 \frac{\partial^2 n_\alpha^{(1)}}{\partial \xi^2} = n_\alpha^{(1)} - \gamma T^{(1)}, \tag{12}$$

where

$$\lambda_\alpha^2 = \frac{\varepsilon^2 D_{\alpha 0}}{\omega_{\alpha 0}}, \qquad \gamma = \{3\frac{n_0}{T_0}\frac{1}{t_{R0}} + n_0\frac{\partial}{\partial T}(\frac{1}{t_R})_0\}\frac{n_0}{n_{\alpha 0}}/\omega_{\alpha 0}.$$

The diffusion, the deposition and the production of α particles are balanced one another in eq. (12). Now we have two simple equations (11) and (12) which govern the weak reaction wave.

IV. IGNITION CONDITION

The linearized version of eqs. (11) and (12) yields a growth rate for wavenumber k,

$$\Omega(k) = \frac{c}{1 + k^2\lambda_\alpha^2} - bk^2 - f, \tag{13}$$

where

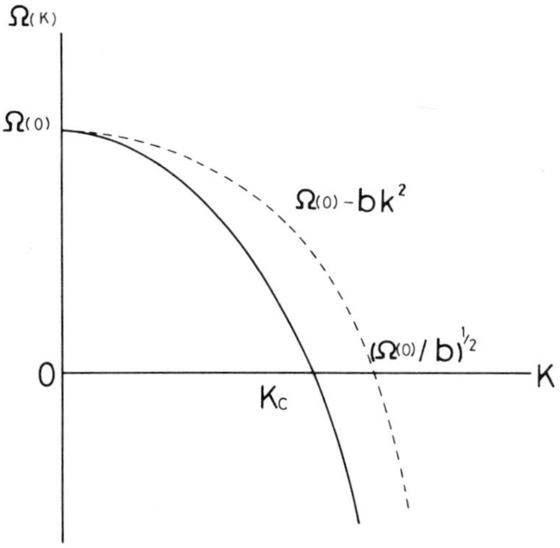

Fig. 1. Growthrate $\Omega(k)$ versus wavenumber k.

$$c = q\gamma = \frac{E_\alpha}{15\varepsilon^2}\left\{\frac{3}{T_0 t_{R0}} + \frac{\partial}{\partial T}\left(\frac{1}{t_R}\right)_0\right\},$$

$$\Omega(0) = c - f = \frac{1}{15\varepsilon^2}\frac{\partial}{\partial T}\left(\frac{E_\alpha}{t_R} - P_B\right)_0,$$

where $\Omega(0)$ is positive at 4 keV. The first term on the right hand side of eq. (13) represents the energy gain by the nuclear reaction and the factor $k^2\lambda_\alpha^2$ in the denominator exhibits that the energy supply is reduced by the α-particles escaping from a burning region. $\Omega(k)$ is plotted as a function of k in Fig. 1. The ignition condition is given by $\Omega(k) > 0$, which is satisfied when k is less than a critical wavenumber k_c, that is,

$$k^2 < k_c^2 = [-(b+f\lambda_\alpha^2) + \{(f\lambda_\alpha^2+b)^2 + 4b\lambda_\alpha^2(c-f)\}^{1/2}]/2b\lambda_\alpha^2.$$

The corresponding critical wavelength $\lambda_c = 2\pi/k_c$ becomes nearly equal to $12\lambda_{ii}/\varepsilon \sim 5\lambda_{\alpha o}$ at 4 keV.

V. DEVELOPMENT AND STEADY-STATE STRUCTURE OF REACTION WAVE

If $k^2 << k_c^2$, then $k^2\lambda_\alpha^2 << 1$, and the reaction wave grows.

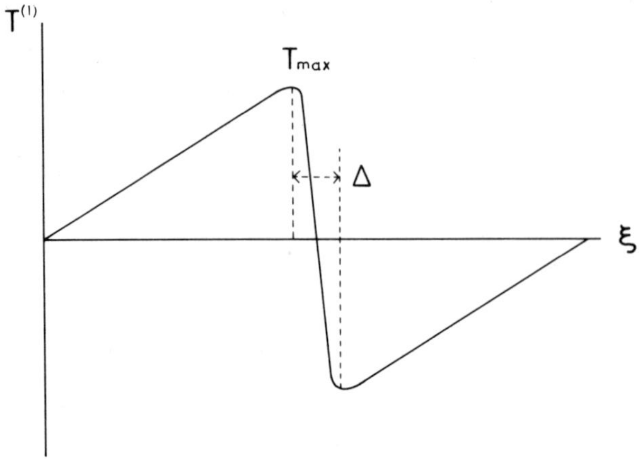

Fig. 2. Sawtooth solution of eq. (14) for wavenumber small compared with the critical wavenumber $k << k_c$.

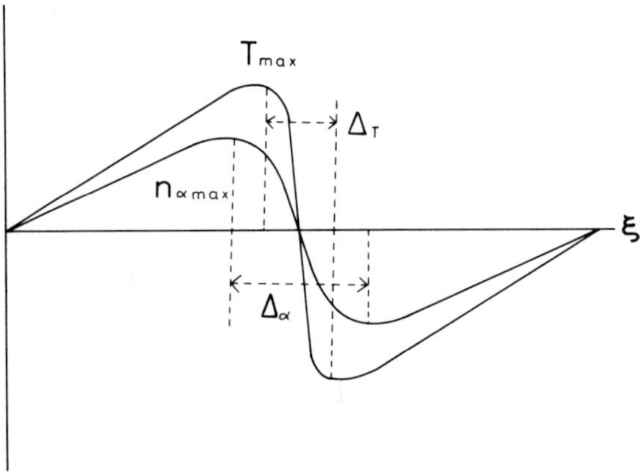

Fig. 3. Sawtooth solution of eqs. (11) and (12) for wavenumbers small compared with the critical wavenumber $k < k_c$.

Table I. Shock strength T_{max} and shock widths Δ_T associated with the effect of finite α-particle range against various wavenumber k or wavelengths λ for $k_c = 0.52$, $\lambda_c = 12.0$ and $\lambda_\alpha = 2.6$. $\Omega(0)\lambda/2a$ and $2b/T_{max}$ are shock strength and shock widths respectively without the effect of finite α-particle range.

k	λ	T_{max}	$\Omega(0)\lambda/2a$	Δ_T	b/aT_{max}	n_{max}	Δ_α
0.50	12.5	2.3	10.8	4.5(±0.25)	1.0	10	5.5(±0.25)
0.36	17.6	9.8	15.2	3.1(±0.22)	0.2	46	6.2(±0.25)
0.22	28.8	22.1	24.8	2.9(±0.48)	0.1	118	7.7(±0.25)
0.20	31.8	25.1	27.4	3.2(±0.53)	0.09	135	7.4(±0.53)
0.18	34.8	28.1	30.0	3.5(±0.58)	0.08	155	8.1(±0.58)

The growth rate is approximately equal to $\Omega(0)-bk^2$ and eqs. (11) and (12) are combined to give a single equation

$$\frac{\partial T^{(1)}}{\partial \tau} + aT^{(1)}\frac{\partial T^{(1)}}{\partial \xi} - b\frac{\partial^2 T^{(1)}}{\partial \xi^2} - \Omega(0)T^{(1)} = 0, \qquad (14)$$

where $n_\alpha^{(1)} \simeq \gamma T^{(1)}$ has been used. The equation of this type, which is the modified Burgers equation with instability, was once investigated by E. Ott et al.[8] in order to discuss the ion-acoustic instability. By means of numerical computations, they found that under the periodic boundary condition a small initial disturbance of long wavelength grows and eventually goes over to a steady-state sawtooth wave which is shown in Fig. 2. From eq. (14) it is easily deduced that the shock strength $T_{max} \sim \lambda\Omega(0)/2a$, in which λ is the period of the system and the shock width $\Delta \sim b/a\ T_{max}$, and so the size of the steady-state sawtooth wave is determined by the period of the system. In our case, the early stage of the development of the small initial disturbance of long wavelength is governed by eq. (14). The disturbance grows and becomes steep in front. When the width of the front layer becomes as short as λ_α, eq. (14) is no longer valid in the front layer and the effect of finite α-particle range must be taken into account. Steady-state structures of the reaction wave obtained by numerical calculations of eqs. (11) and (12) are illustrated in Fig. 3 and Table I. It is found that T_{max} is smaller than $\lambda\Omega(0)/2a$ and so the effect of finite α-particle range reduces the shock strength especially for $k \leq k_c$, while Δ_T is nearly equal to $2\lambda_\alpha$ and so the effect of finite α-particle range plays a dominant role to determine the shock width even for $k \ll k_c$.

VI. STRONG REACTION WAVE

As was shown in the previous section, if an initial slightly heated region has a large size in comparison with the critical wavelength λ_c, the initial disturbance will grow and develop to a strong reaction wave. The development, propagation and structure of the strong reaction wave were numerically studied, being assumed the spherical symmetry. We don not have precise knowledge about the final compression stage and the central ignition region in a laser-compressed plasma. However, so long as the characteristics of the completely developed reaction wave is concerned, detail inspections for the initial and boundary conditions may not be necessary. Thus, we initially set $T_e=T_i=1$ keV everywhere except in the central ignition region. In order to avoid complexity arising from the outer boundary conditions, we consider a homogeneously compressed plasma of which density is 5×10^{26} cm^{-3}. The compressed plasma is also assumed to be quiescent and the

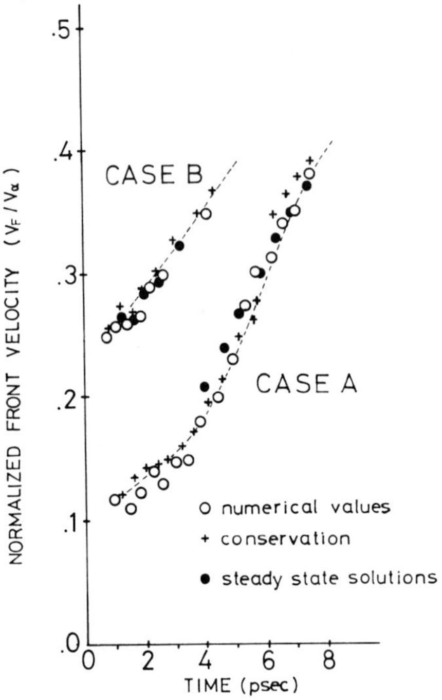

Fig. 4. The propagations velocities of the wave normalized by the α-particle velicity for two different initial conditions. Three different marks indicate the values obtained from the numerical results, calculated by means of the conservation relation $v_F=\{p_1\rho_1/\rho_0(\rho_1-\rho_0)\}^{1/2}$ and the steady state solution of the pre-heating zone.

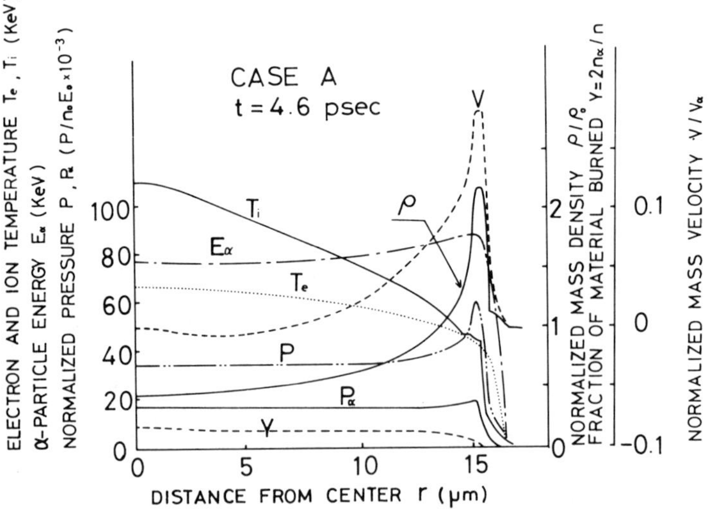

Fig. 5. Profiles of the various physical variables at 4.6 psec in Case A.

initial input energies in the central ignition region are

Case A; $E_{imp} \simeq 200$ J for $r \leq 0.7$ μm,
Case B; $E_{imp} \simeq 2$ kJ for $r \leq 1.3$ μm.

VII. THERMONUCLEAR REACTION WAVE AS AN ACCELERATING OVERCOMPRESSED DETONATION WAVE

The numerical results are presented in this and the following sections. The propagation velocity of the buring front normarized by the α-particle velocity is plotted as a function of time in Fig. 4. After some ignition period (about 1 psec), the front accelerates due to the increase of the plasma temperature. A typical wave profile is shown in Fig. 5. The maximum compression rate is about 2.2 which is larger than $1.6 = (\gamma+1)/\gamma$ ($\gamma=5/3$), (and much larger than 1.3 which was obtained by Chu[3]). This suggests that the strong reaction wave is an overcompressed detonation wave, which can be also confirmed by the following results. As the physical variables change slowly with time after the ignition of the wave is completed, the adiabatic approximation can be available. From the mass and momentum conservation relations across the thin wave front, we have

$$\frac{(v_F - v_1)^2}{c_1^2} = 1 + \frac{\eta_1 \zeta_1}{\gamma(1-\eta_1)} \frac{c_0^2}{c_1^2} \{(\gamma+1) - (\gamma-\gamma_\alpha)\frac{\zeta_\alpha}{\zeta_1}\}(\eta_1 - \eta_c), \quad (15)$$

$$\eta_c = \frac{\gamma - (\gamma-\gamma_\alpha)\zeta_\alpha/\zeta_1}{\gamma + 1 - (\gamma-\gamma_\alpha)\zeta_\alpha/\zeta_1}, \quad (16)$$

where v_F and c are the propagaion velocity of the wave and the sound speed ; $c_1^2 = \{\gamma p_1 + (\gamma_\alpha - \gamma)p_\alpha\}/\rho_1$ and $c_0^2 = \gamma p_0/\rho_0$, ζ and η are the normalized pressure and specific volume ; $\zeta_1 = p_1/p_0$, $\zeta_\alpha = p_\alpha/p_0$ and $\eta_1 = \rho_0/\rho_1$, γ_α is the effective adiabatic constant for the α-particle, given by $\gamma_\alpha = 1 + p_\alpha/n_\alpha E_\alpha$, subscripts 0 and 1 indicate the values in the front of the wave and those behind it, respectively. The Chapman-Jouguet condition is met when $\eta_1 = \eta_c$ and $(v_F - v_1)/c_1 = 1$. If $\eta_1 < \eta_c$, the reaction wave is an overcompressed detonation wave, and the propagation velocity with respect to the fluid behind the wave is subsonic, i.e., $(v_F - v_1) < c_1$. η_1, η_c and $(v_F - v_1)/c_1$ are plotted as a function of time in Fig. 6, where we have used ζ, η and v obtained from the numerical results. Now, we are convinced that the thermonuclear reaction wave driven by the laser implosion is an accelerating overcompressed detonation wave rather than the undercompressed one shown by Chu[3] and Bruckner[1]. This implies that the shock heating contributes a great deal to the propagation mechanism as well as the electron thermal conduction

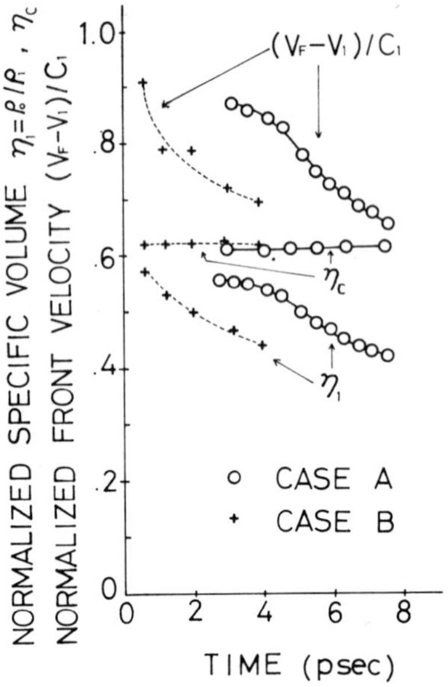

Fig. 6. η_1 is the normalized specific volume in the shock region (ρ_0/ρ_1), and η_c is that for the Chapman-Jouguet detonation wave defined by eq. (16). $(v_F-v_1)/c_1$ is the normalized propagation velocity with respect to the shock compressed fluid, calculated from eq. (15).

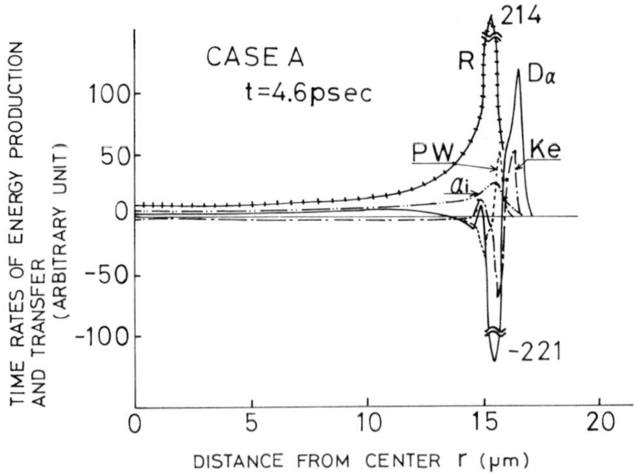

Fig. 7. The time variations of the energy of the plasma species due to the various mechanism of the energy production and transfer, R, D_α, α_i, K_e and PW refer, respectively, to the contributions due to the α-particle production by the thermonuclear reaction, the α-particle diffusion, the α-particle heating to the ions, the electron thermal conduction, and the electron pressure work which is nearly equal to the ion pressure work.

and the α-particle conduction (see also Fig. 7). Since the deposition time of the α-particle enrgy to the plasma becomes longer as the electron temperature increases, the α-particle heating is relatively suppressed and the shock heating becomes effective.

Figure 7 shows the time variation of the energy of the plasma species due to the various mechanism of the energy production and transfer, such as the reaction, the α-particle heating, the pressure work, the electron thermal conduction and the α-particle diffusion. As was expected from the previous discussion, the energy increase of the plasma due to the pressure work (PW) is as large as that due to the α-particle heating (α_i) in the thin compressed region. It is also found that the high energy α-particles are mainly produced in the thin front compressed region (R) and diffuse ahead (D_α). Since the main reaction zone is not separated from the shock region, the Zel'dovich-Neumann-Doring models is not valid for this class of the reaction waves.

VIII. FRONT STRUCTURE OF THE REACTION WAVE

In order to understand in detail the internal pysics, more precise profiles of the time rate of increase of the energies due to the different energy production and transfer processes are present in Figs. 8 and 9. It is found that their exist two different stages.

At the first stage (Fig. 8), the temperature is relatively low ($T_e \leq 15$ keV) and the reaction is weak so that the α-particle energy ($E_\alpha = (n_\alpha/n_0)$x3.5 MeV) is less than the electron and ion energy. Because of the relatively low temperature, the α-particles deposite their energy mainly to the electrons. The ions are heated up by the shock and the electron-ion relaxation, which is so effective that T_i is nearly equal to T_e. The energy transfer by the electron thermal conduction is not remarkable. This stage of the reaction wave is thus quite similar to the weak reaction wave dicussed before.

At the second stage (Fig. 9), the temperature increases higher than 30 keV, and the reaction becomes stronger so that the α-particle energy is greater than the electron and ion energies. This stage can be characterized by the distinct appearance of a preheating zone in the front of the shock, which consists of both the large electron and α-particle conductions. In the preheating zone the energy transfer by the electron thermal conduction exceeds the α-particle heating to the electrons. The ions are heated by the shock and the α-particles. In this stage, the electronion relaxation is not effective. It should be noted that

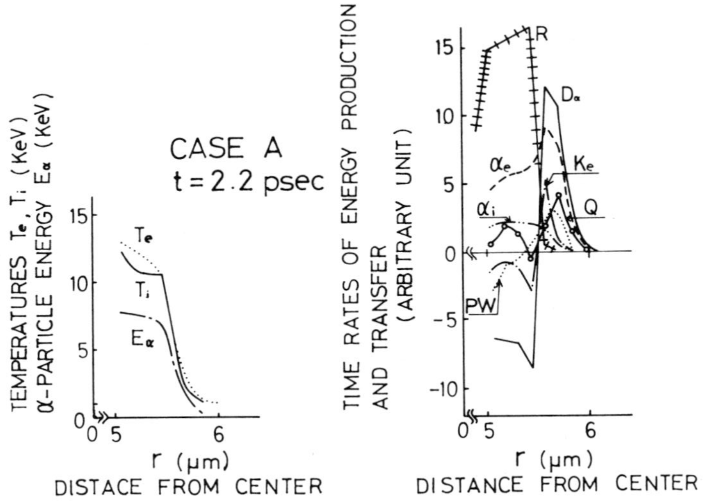

Fig. 8. Profiles of the energies of the plasma species and different heating and energy transfer processes at 2.2 psec in the Case A (the first stage). R, D_α, K_e, α_e, α_i, Q and PW refer, respectively, to contributions due to the α-particle production by the reaction, the α-particle conduction, the electron thermal conduction, the α-particle heating to the electrons, the α-particle heating to the ions, the ion heating due to the electron-ion relaxation and the electron pressure work which is nearly equal to the ion pressure work.

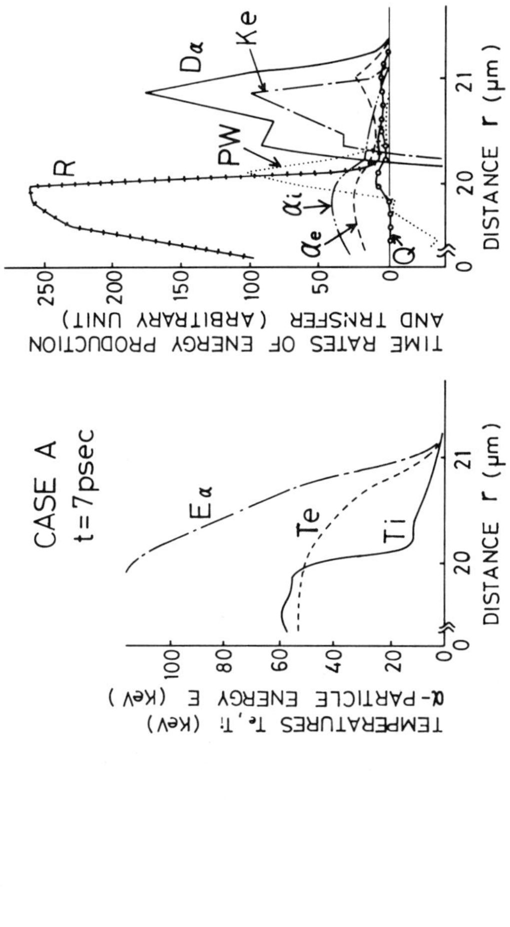

Fig. 9. Profiles of the energies of the plasma species and different heating and energy transfer processes at 7 psec in the Case A (the second stage). The notations are the same as in Fig. 8.

in Case B (the large initial energy in the ignition region) the first stage does not clearly exist but only the second stage.

The structure of the pre-heating zone is approximately calculated as follows. As the pre-heating zone is relatively thin, a plane geometry can be assumed. The fluid motion and the reaction can be ignored in the zone. Thus, in the steady propagation, the energy equations for the plasma species can be written as

$$- v_F \frac{dE_\alpha}{dx} = \frac{d}{dx}\left(D_\alpha \frac{dE_\alpha}{dx}\right) - \omega_\alpha E_\alpha, \qquad (17)$$

$$- v_F \frac{dT_e}{dx} = \frac{2}{3n_0} K_{e0}(\bar{T}_{e0}) \frac{d^2 T_e}{dx^2} + \frac{2}{3}\omega_{\alpha e} \frac{n_\alpha}{n_0} E_\alpha + \frac{T_i - T_e}{t_{ei}}, \qquad (18)$$

$$- v_F \frac{dT_i}{dx} = \frac{2}{3}\omega_{\alpha i} \frac{n_\alpha}{n_0} E_\alpha + \frac{T_e - T_i}{t_{ei}}, \qquad (19)$$

where $\bar{T}_{e0}$ is an average temperature in the pre-heating zone and $E_\alpha = n_\alpha \times 3.5$ MeV. The boundary conditions are chosen as

$$E_\alpha = 0, \quad T_e = T_i = 1 \text{ keV} \quad \text{at } x \to \infty$$

$$E_\alpha = E_{\alpha 0}, \quad T_e = T_{e0}, \quad T_i = T_{i0} \quad \text{at } x = 0.$$

The solution of eq. (17) is

$$E_\alpha = E_{\alpha 0} \exp(-\kappa_1 \xi), \quad \xi = x/\lambda_\alpha, \quad \beta = v_F/v_\alpha,$$

$$\kappa_1 = \frac{2+3\beta+(16-12\beta+9\beta^2)^{1/2}}{2(1-2\beta)} \qquad (20)$$

where $\beta < 1/2$. When $\beta \geq 1/2$, $E_\alpha = 0$, that is, the α-particle can not plunge into the pre-heating zone (in this case, the electron thermal conduction is dominant). The solutions of eqs. (18) and (19) are

$$T_e = T_{e1}\exp(-\kappa_1 \xi) + T_{e2}\exp(-\kappa_2 \xi) + 1, \qquad (21)$$

$$T_i = T_{i1}\exp(-\kappa_1 \xi) + T_{i2}\exp(-\kappa_2 \xi) + 1, \qquad (22)$$

where

$$T_{e1} = \frac{E_{\alpha 0}(1+b\alpha_{e0}\kappa_1\beta)b}{(b\kappa_1\beta+1)^2-\bar{K}_{e0}b^2\kappa_1^2(b\kappa_1\beta+1)-1}, \tag{23}$$

$$T_{i1} = \frac{T_{e1}+b\alpha_{i0}E_{\alpha 0}}{1+b\kappa_1\beta}, \qquad T_{i2} = \frac{T_{e2}}{1+b\kappa_2\beta}, \tag{24}$$

$$T_{e1}+T_{e2} = T_{e0}-1, \qquad T_{i1}+T_{i2} = T_{i0}-1,$$

$$\kappa_2 = \frac{\beta^2-\bar{K}_{e0}+\{(\bar{K}_{e0}-\beta^2)^2+8\bar{K}_{e0}\beta^2\}^{1/2}}{2\bar{K}_{e0}b\beta}, \tag{25}$$

$$\alpha_{e0} = 32/(32+\bar{T}_{e0}), \qquad \alpha_{i0} = 1-\alpha_{e0},$$

$$b = \omega_\alpha(\bar{T}_{e0})t_{ei}(\bar{T}_{e0}), \qquad \bar{K}_{e0} = K_{e0}/t_{ei}.$$

κ_1 is the term due to the α-particle diffusion, while κ_2 essentially due to the electron thermal conduction.

In the first stage, $\kappa_2 \gg \kappa_1$ and $\beta^2 \gg \bar{K}_{e0}$ because of the relatively low temperature. The effect of the electron thermal conduction is then ignored and the length of the pre-heating zone can be given by κ_1^{-1}. The normalized electron-ion relaxation time b is nearly equal to the normalized time $(\kappa_1\beta)^{-1}$ during which the α-particle heats the plasma (mainly the electrons) in the pre-heating zone. Then eqs. (21)$\sim$(25) yield

$$T_{e1} = E\alpha_0(1+b\alpha_{e0}\kappa_1\beta)/\kappa_1\beta(2+b\kappa_1\beta), \qquad \kappa_2 = \beta/\bar{K}_{e0}b,$$

$$T_{i1} = T_{e1}/(1+b\kappa_1\beta), \qquad T_{e2} = T_{i2} = 0,$$

where $\bar{K}_{e0}b/\beta$ $(=\kappa_2^{-1})$ is an ordinary electron thermal conduction range.

In the second stage, $\kappa_2 \geq \kappa_1$ and $b \gg (\kappa_1\beta)^{-1}$ because of the relatively high temperature. The contribution of the electron-ion relaxation is then negligible and the electron thermal conduction plays an important role as well as the α-particle conduction in the pre-heating zone. Then eqs. (21)$\sim$(25) yield

$$\kappa_2 = \beta/\bar{K}_{e0}b, \qquad T_{e1} = \alpha_{e0}E_{\alpha 0}/\kappa_1\beta(1-\kappa_1/\kappa_2),$$

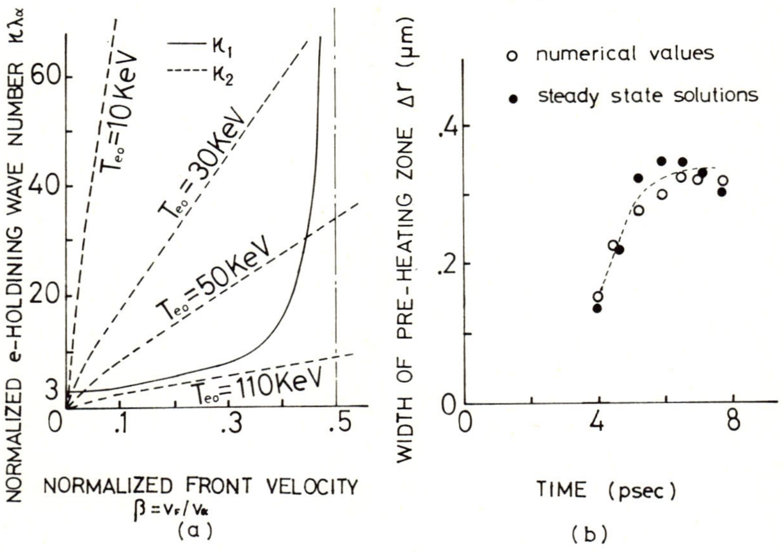

Fig. 10. (a) κ_1 and κ_2 versus $\beta=v_F/v_\alpha$ and T_{e0}. $\overline{T}_{e0}$ is assumed to be $T_{e0}/2$. (b) The width of the pre-heating zone, calculated from eq. (26) and numerical results in the Case A.

$$T_{i1} = \alpha_{i0} E_{\alpha 0}/\kappa_1 \beta, \qquad T_{i2} = 0, \qquad T_{e2} \neq 0.$$

The normalized propagation velocity of the reaction wave, $\beta = v_F/v_\alpha$, is uniquely determined from eqs. (20)∿(25), using the numerical values of T_{e0}, T_{i0} and $E_{\alpha 0}$. The values of β calculated in this way are plotted in Fig. 4. Fairly good coincidence of those values with the other values of β ensures us the validity of the approximation.

The width of the pre-heating zone can be approximately given by

$$\Delta r = \int_{T_{e0}}^{1} \frac{\lambda_\alpha(T_e)}{(dT_e/d\xi)} dT_e \simeq \lambda_\alpha(T_{e0})/\mathrm{Min}(\kappa_1, \kappa_2). \qquad (26)$$

κ_1 and κ_2 are plotted in Fig. 10 (a) as a function of β and T_{e0}. For the temperature, $T_{e0} < 100$ keV, κ_2 is larger than κ_1 as long as β is not too small or not near 1/2. The α-particle conduction is, therefore, the main process in determining the width of the pre-heating zone for $T_{e0} < 100$ keV. Since κ_1 is larger than κ_2 for the higher temperature than 100 keV except for small values of β, the electron thermal conduction will play a dominant role in determing the width of the pre-heating zone for $T_{e0} > 100$ keV. It should be emphasized that the width of the pre-heating zone is smaller than $\lambda_\alpha(T_{e0})/3$ for $T_{e0} < 100$ keV since κ_1 is always larger than 3. The width Δr is plotted in Fig. 10 (b) in comparison with the numerical values.

REFERENCES

1. K. A. Bruckner and S. Jorna, Rev. Mod. Phys. 46, 325 (1974).
2. A. L. Fuller and R. A. Gross, Phys. Fluids 11, 534 (1968).
3. M. S. Chu, Phys. Fluids 15, 413 (1972).
4. N. Asano, K. Nishihara, K. Nozaki and T. Taniuti, J. Phys. Soc. Japan 41, 1774 (1976).
5. K. Nozaki and K. Nishihara, (to be published)
6. G. B. Zimmerman, UCRL Report N. 74811.
7. T. Taniuti, Prog. Theor. Phys. Suppl. No. 55, 1 (1974).
8. E. Ott, W. M. Manherimer, D. L. Book and J. P. Boris, Phys. Fluids 16, 855 (1973).

MODEL FOR IMPLOSION OF STRUCTURED SLAB*

Keishiro Niu and Takashi Yabe

Tokyo Institute of Technology

Meguro-ku, Tokyo, Japan

INTRODUCTION

In the scheme of laser-fusion, it is required to compress the density of Deuterium-Tritium fuel to more than 1000 times of its solid density. One of the possible methods to realize this extremely strong compression is launching successive shock waves from the ablation surface and is controlling the waves to arrive simultaneously at the center. In order to increase the compression ratio, the multi-structured pellet which consists of the D-T microsphere covered with high Z materials is used in the recent experimental works of laser-fusion.[1,2] So far several authors have presented one-dimensional analytical methods for super-compression of the D-T solid slab or sphere.[3-8] Except Kidder's works[4,5], however, their methods cannot be adopted directly to the multi-structured target. Kidder has proposed a method of homogeneous compression, and his method seems to include some obscure physical meaning.

In this paper, we present an analytical model which is applicable to the super-compression of the multi-structured slab. The main purpose of this paper is to clarify; 1) the effects of structured target; 2) the relation between the absorbed laser energy and the mechanical power supply at the critical surface; 3) the validity for neglecting the thermal conduction in the target.

*Presented at the Fourth Workshop on "Laser Interaction and Related Plasma Phenomena" held at Rensselaer Polytechnic Institute, Troy, New York, November 8-12, 1976

In the second section, we derive the condition that shock waves launched from the ablation surface reach simultaneously the center of the multi-structured slab. The deflagration condition at the ablation surface is settled in the third section. The time dependence of the input power of the laser light to obtain an optimal compression of the target is also given in this section. In the fourth section is discussed the relation between the absorbed laser energy in the slab and the mechanical power supply for compression. The effects of thermal conduction are discussed in the fifth section.

MODEL FOR SUPER-COMPRESSION

To simplify the analysis, we limit ourselves to a one-dimensional plane configuration. As shown in Fig.1, the slab consists of a D-T fuel (regionf) and a high Z material (region h).

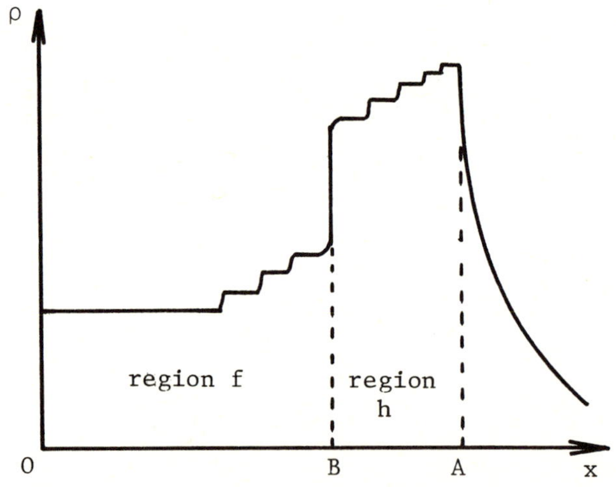

Fig. 1 : Density distribution in the slab.

A large number of successive shock waves are launched from the surface A (the surface of the critical density for the laser light which is thrown from the right hand side in Fig.1), impinge upon the contact surface B and are transmitted through B into region f. These phenomena are schematically shown in Fig.2.

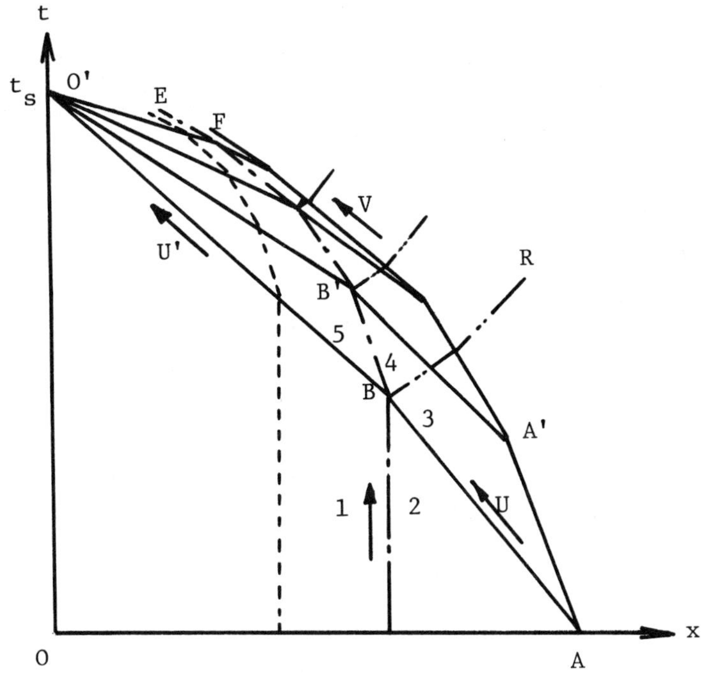

Fig.2 : x-t diagram of the ablation surface (solid line, A-F), the successive shock waves (solid line, A-B, B-O', A'-B' etc.), the rarefaction waves (double chain line, B-R etc.) and the contact surface (chain line, B-B'-E). Dotted lines denote the pathes of particles.

As each shock wave is assumed to be weak, we describe the Mach number M of the shock wave as $M = 1 + \Delta M$, where $0 < \Delta M \ll 1$. Here the Mach number of the shock wave is defined in the reference frame moving with the fluid velocity in front of the shock wave. If we use suffix i to the incident shock wave just in front of the contact surface, and suffix t to its transmitted shock wave from the contact surface, we can derive a relation (see Appendix A)

$$\Delta M_i / \Delta M_t = (\sqrt{\rho_{ho}/\rho_{fo}} + 1)/2 \equiv 1/G, \tag{1}$$

where ρ is the density, the suffixes f and h denote the values in region f and h respectively, the suffix 0 denotes the values before the first shock wave arrives. Equation (1) implies that the transmitted shock wave is weaker than the incident shock wave. In region h, a rarefaction wave is reflected. This wave is represented as a line B-R instead of a fan shape region in Fig.2. We denote the velocity of the ablation surface (A-A'-F) by V, the velocity of the incident shock wave (A-B or A'-B'...) by U, the velocity of the transmitted shock wave (B-O', or B'-O'...) by U' and the velocity of the contact surface (B-B'-E) by W. After the n-th shock wave leaves the surface A-F with the velocity U_n at t_n, it reaches the surface B-E at t'_n and propagates toward O' with the velocity U'_n in the region f. If we assume that the n-th and (n+1)-th shock waves arrive at the point O' at the same time t_s, then the following two equations must be satisfied,

$$U'_n(t_s - t'_n) = U'_{n+1}(t_s - t'_{n+1}) + W_n(t'_{n+1} - t'_n), \tag{2}$$

$$\int_{t_{n+1}}^{t'_{n+1}} U_{n+1}(t)dt - \int_{t_n}^{t'_n} U_n(t)dt$$

$$= W_n(t'_{n+1} - t'_n) - V_n(t_{n+1} - t_n). \tag{3}$$

Since the interaction between the shock wave and the rarefaction wave changes the shock velocity, the shock velocity $U_n(t)$ in region h is time dependent. For the weak shock waves, the left hand side of eq.(3) can be approximately expressed as follows,

$$\int_{t_{n+1}}^{t'_{n+1}} U_{n+1}(t)dt - \int_{t_n}^{t'_n} U_n(t)dt$$

$$= U_n(t'_{n+1} - t'_n - t_{n+1} + t_n) + (U_{n+1}-U_n)(t'_n-t_n)+0(\Delta M_i^2)$$

where U_{n+1} and U_n represent $U_{n+1}(t_{n+1})$ and $U_n(t_n)$ respectively. By use of

$$\tau'_n = t_s-t'_n, \qquad \tau_n = t_s-t_n, \tag{4}$$

eqs.(2)~(3) reduce to

$$\frac{\tau'_{n+1} - \tau'_n}{\tau'_n} = \frac{U'_n - U'_{n+1}}{U'_{n+1} - W_n}, \tag{5}$$

$$\tau'_{n+1}-\tau'_n = \frac{U_n-V_n}{U_n-W_n}(\tau_{n+1}-\tau_n)+\frac{U_{n+1}-U_n}{U_n-W_n}(\tau_n-\tau'_n)+0(\Delta M_i^3). \tag{6}$$

We have the following jump relations for weak shock waves and weak rarefaction waves (see Appendix B),

$$u_n = u_{n-1} + 4c_{n-1}(\Delta M_i + \Delta M_r)/(\gamma+1), \qquad (7\text{-a})$$

$$U_n = u_{n-1} + c_{n-1}(1+\Delta M_i), \qquad (8\text{-a})$$

$$c_n = c_{n-1}[1+2(\gamma-1)(\Delta M_i - \Delta M_r)/(\gamma+1)], \qquad (9\text{-a})$$

$$W_n = W_{n-1} + 4c'_n \Delta M_t/(\gamma+1), \qquad (7\text{-b})$$

$$U'_n = W_{n-1} + c'_{n-1}(1+\Delta M_t), \qquad (8\text{-b})$$

$$c'_n = c'_{n-1}[1+2(\gamma-1)\Delta M_t/(\gamma+1)], \qquad (9\text{-b})$$

where ΔM_r is the decrement in the Mach number of the rarefaction wave (Appendix B), γ the ratio of specific heats, u_n and c_n the fluid and sound velocities in region h behind the n-th shock wave, respectively. The sound velocity in region f behind the n-th shock wave is denoted by c'_n. The Mach number of the incident shock wave is time dependent, because the shock wave interacts with rarefaction waves in region h. Thus we denote the increment in the Mach number of the incident shock wave by $\Delta M_i(t)$. The value ΔM_i represents $\Delta M_i(t)$ just in front of the contact surface. But the difference between ΔM_i and $\Delta M_i(t)$ is $O(\Delta M_i^2)$ as described in Appendix A. The terms of $O(\Delta M_i^2)$ are neglected throughout this paper. By using eqs.(7-a)~(9-b), eqs.(5) and (6) reduce to

$$(\tau'_{n+1} - \tau'_n)/\tau'_n = -2\Delta M_t, \qquad (10)$$

$$(\tau'_{n+1} - \tau'_n) = (1-N_n)(\tau_{n+1} - \tau_n) + \frac{2}{\gamma+1}[4+G(\gamma-3)](\tau_n - \tau'_n)\Delta M_i, \qquad (11)$$

where N_n is the Mach number of the ablation surface at t_n, and is related with the velocity V_n of the ablation surface through

$$V_n = u_n + c_n N_n .$$

Introducing a new variable $\xi(=n\Delta M_i)$ and taking the limit that $n \to \infty$, $\Delta M_i \to 0$ and $n\Delta M_i \to \xi$, we obtain the following differential equations for τ' and τ from eqs.(10) and (11) together with eq.(1),

$$d\tau'/d\xi = -2G\tau' \qquad (12)$$

$$d(\tau'-\tau)/d\xi = -2[4+G(\gamma-3)](\tau'-\tau)/(\gamma+1). \qquad (13)$$

In deriving these equations, we use the fact that $N \ll 1$, whose validity is shown in Appendix D.

The solutions of eqs.(12) and (13) are given by

$$\tau' = \tau'_o \exp(-2G\xi), \tag{14-a}$$

$$\tau = \tau'_o\{1+\tau_o/\tau'_o \exp[-\gamma(1-G)\xi/(\gamma+1)]\}\exp(-2G\xi), \tag{14-b}$$

where τ'_o and τ_o are constant and are expressed as

$$\tau'_o = X_{fo}/c_o\sqrt{\rho_{fo}/\rho_{ho}},\ \tau_o = X_{ho}/c_o,\ (\tau_o+\tau'_o=t_s). \tag{14-c}$$

Here X_{fo} is the initial length of region f, X_{ho} the initial length of region h, c_o and c'_o are the initial sound velocities in region h and f, respectively. Since the pressures on the both sides of the contact surface ballance with each other, we have used a relation $c'_o = \sqrt{\rho_{ho}/\rho_{fo}}c_o$ in eq.(14-c).

INPUT POWER

The power E of the laser light absorbed at the ablation surface can be found as (see Appendix C)

$$E = \frac{\rho_a c_a^3}{2(\gamma^2-1)N}\left[1-\left(\frac{1-\alpha^2}{\alpha^2\gamma+1}\right)^2\right]. \tag{15}$$

A constant α in eq.(15) is defined in Appendix C. The subscript a refers to the values just in front of the ablation surface. The Mach number N of the ablation surface can be expressed as a function of ξ as follows (Appendeix D),

$$N = N_o \exp 2[(\gamma-1)\alpha+2](G-8)\xi/[\alpha(\gamma+1)]\ , \tag{16}$$

where

$$N_o = \alpha\sqrt{\rho_c/\rho_{ho}}(\alpha^2\gamma+1).$$

As shown in Appendix E, $\rho_a c_a$ in eq.(15) is written in the form,

$$\rho_a c_a = \rho_{ho} c_o \exp[2(3\gamma-1)G\xi/(\gamma+1)] \tag{17}$$

Equation (15) combined with eqs.(16) and (17) leads us to

$$E_o = \exp\{[4G(\alpha\gamma-1)+8]\xi/\alpha(\gamma+1)\}, \tag{18}$$

where

$$E_o = \left[1-\left(\frac{1-\alpha^2}{\alpha^2\gamma+1}\right)^2\right]\frac{\rho_{ho}c_o^3}{2(\gamma^2-1)N_o}.$$

Eliminating ξ in eqs.(14) and (18), we obtain a relation between and E as

$$\tau = \frac{X_{fo}}{c_o}\sqrt{\frac{\rho_{fo}}{\rho_{ho}}}\left(\frac{E}{E_o}\right)^{-\ell_1} + \frac{X_{ho}}{c_o}\left(\frac{E}{E_o}\right)^{-\ell_2}, \tag{19}$$

where

$$\ell_1 = \frac{G(\gamma+1)}{2G(\alpha\gamma-1)+4}, \quad \ell_2 = \frac{[G(\gamma+5)-4]\alpha}{2G(\alpha\gamma-1)+4}.$$

Since $\ell_2 > \ell_1$, eq.(19) can be approximated by

$$E = E_o\left(1+\frac{X_{ho}}{X_{fo}}\sqrt{\frac{\rho_{ho}}{\rho_{fo}}}\right)^{-\ell}\left(1-\frac{t}{t_s}\right)^{-\ell}, \tag{20}$$

for large E/E_o. Here

$$\ell = 1/\ell_1$$

and relations

$$\tau = t_s - t, \quad t_s = X_{fo}/c_o\sqrt{\rho_{fo}/\rho_{ho}} + X_{ho}/c_o$$

have been used. The values ℓ are plotted versus ρ_{ho}/ρ_{fo} in Fig.3 for α = 0.5, 1 and 2.

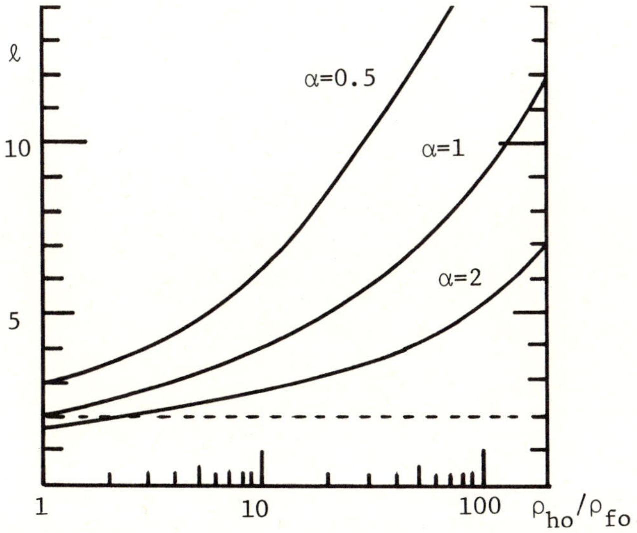

Fig.3 : Exponent ℓ of the tailored pulse [eq.(20)] is plotted versus the density ratio ρ_{ho}/ρ_{fo} for α =0.5, 1 and 2.

MECHANICAL POWER

After the derivation of the relations in Appendix E, the mechanical power E_m for compressing the slab is given by

$$E_m = p_a u_a = E_{mo} \exp\left[\frac{4\gamma}{\gamma+1} G\xi\right]\left\{\exp \frac{2(\gamma-1)}{\gamma+1} G\xi - 1\right\}, \tag{21}$$

where

$$E_{mo} = \frac{2\rho_{ho} c_o^3}{\gamma(\gamma+1)}.$$

Equations (18) and (21) give us

$$\Phi \equiv \frac{E_m}{E} = \frac{4(\gamma-1)N_o}{\gamma\left[1-\left(\frac{1-\alpha^2}{\alpha^2\gamma+1}\right)^2\right]} \exp\left\{\frac{2G[(\gamma-1)\alpha+2]-8}{\alpha(\gamma+1)}\xi\right\}, \tag{22}$$

for $\xi >> 1$. This fraction Φ varies in time. This is, the time dependence of the mechanical power differs from that of the absorbed laser power [eq. (20)]. In the asymptotic case that $\xi >> 1$, we have

$$E_m \propto (1-t/t_s)^{-(3\gamma-1)/(\gamma+1)} = (1-t/t_s)^{-3/2}, \tag{23}$$

for $\gamma = 5/3$.

The self-regulating model for the spherical compression gives us the mechanical power in the form[4,5]

$$E_m \propto (1-t/t_s)^{-2}.$$

This time dependence is not same with that given by eq. (23). However, if we apply the self-regulating model to the slab compression, we can show that the time dependence of the mechanical power is quite same with that given by eq. (23).

(i) Kidder's model[4,5]

To calculate the absorbed laser power E_k in the pellet, Kidder employed a model that the pressure p_a at the ablation surface is proportional to the two-third power of the absorbed laser power E_k. Our analysis derives

$$p_a \propto (1-t/t_s)^{-5/4}. \tag{24}$$

If we apply the Kidder's model to our pressure obtained above, we derive

$$E_k = p_a^{3/2} \propto (1-t/t_s)^{-15/8}, \tag{25}$$

which is nearly equal to eq.(20) provided G = 1, α = 1 and γ = 5/3.

(ii) Brueckner's model[3,7]

Brueckner has assumed that the absorbed laser power E_b depends on p_a and ρ_a in the form

$$E_b \propto p_a^{3/2} \rho_a^{-1/2}. \tag{26}$$

Applying eq.(26) to our result, and substituting ρ_a by $p_a^{1/\gamma}$, we obtain

$$E_b \propto p_a^{(3\gamma-1)/2\gamma} = p_a^{6/5} \propto (1-t/t_s)^{-3/2}, \tag{27}$$

for γ = 5/3. This is in coincidence with eq.(23).

Now we assume that after the time t_g, the laser power is kept at a constant value E_g. Then the total imput energy I and the total mechanical energy I_m are represented as

$$I = \int_0^{t_g} E\,dt + E_g(t_s - t_g), \tag{28}$$

$$I_m = \int_0^{t_g} E_m\,dt + E_{mg}(t_s - t_g). \tag{29}$$

Subscript g is attached to the values at $t = t_g$. By using eqs. (14-b), (18), (21) and

$$\rho_g/\rho_{fo} = \exp[4G\xi_g/(\gamma+1)],$$

we can derive

$$I = \frac{2G(\alpha\gamma-1)+4}{G[\alpha(\gamma-1)-2]+4} E_o \tau'_o \left(\frac{\rho_g}{\rho_{fo}}\right)^{\{G[\alpha(\gamma-1)-2]+4\}/2\alpha G}$$

$$= \frac{E_o \tau'_o (G+3)}{3-G} \left(\frac{\rho_g}{\rho_{fo}}\right)^{2(3-G)/3G}, \tag{30}$$

$$I_m = \frac{3\gamma-1}{2(\gamma-1)} E_{mo} \tau'_o \left(\frac{\rho_o}{\rho_{fo}}\right)^{\gamma-1} = 3E_{mo} \tau'_o \left(\frac{\rho_g}{\rho_{fo}}\right)^{2/3}, \tag{31}$$

in the limit of $\rho_g \gg \rho_{fo}$ and for α= 1, γ = 5/3. Equations (30) and (31) combined with eqs.(16), (18) and (21) yield

$$\Phi_t \equiv \frac{I_m}{I} = \left\{\frac{G[\alpha(\gamma-1)-2]+4}{2G(\alpha\gamma-1)+4}\right\}\left\{\frac{3\gamma-1}{2(\gamma-1)}\right\}\frac{E_{mb}}{E_o}\left(\frac{\rho_g}{\rho_{fo}}\right)^{\{G[\alpha(\gamma-1)+2]-4\}/2\alpha G}$$

$$= \frac{16\sqrt{6}}{5}(3-G)\sqrt{\frac{\rho_c}{\rho_{ho}}}\left(\frac{\rho_g}{\rho_{fo}}\right)^{2(2G-3)/3G}, \qquad (32)$$

for $\alpha = 1$ and $\gamma = 5/3$.

THERMAL CONDUCTION

In the analysis of the preceding sections, we have neglected the effects of thermal conduction, bremsstrahlung and other loss mechanisms. Here we estimate the influence of thermal conduction.

Let us introduce the following nondimensional parameter

$$\Theta = \frac{\dfrac{k\rho}{m_f(\gamma-1)} \left.\dfrac{\partial T}{\partial t}\right|_{ad}}{K_o \dfrac{\partial}{\partial x}\left(T^{5/2}\dfrac{\partial T}{\partial x}\right)}. \qquad (33)$$

Here T is the temperature, m_f the ion mass in region f, k the Boltzmann constant and $K_oT^{5/2}$ the coefficient of thermal conduction defined by

$$K_o = 1.89x(2/\pi)^{3/2}\, k^{7/2}/(m_e^{\ 1/2}e^4 \mathrm{In}\Lambda),$$

where m_e is the electron mass, e the electron charge and InΛ the Coulomb logarithm. In eq.(28), the numerator represents the energy increase due to the adiabatic compression and the denominator represents the energy increase due to the thermal conduction. The temperature T in the denominator is determined approximately by taking only the adiabatic process into account.

Let us first calculate the density and the temperature in region f as functions of x and t. If we suppose that the n-th shock wave reaches x at t, then we have

$$(t_s-t)U'_n = x. \qquad (34)$$

By using the result in the preceding section, we obtain

$$U'_n = \frac{\gamma+1}{\gamma-1}c'_o\exp\left(\frac{2(\gamma-1)}{\gamma+1}G\xi\right) - \frac{2}{\gamma-1}c'_o, \qquad (35)$$

for U_n'. Define ζ by

$$\zeta = t_s - t. \tag{36}$$

Then we have a relation from eqs.(29) and (30) as

$$\left(\frac{x}{\zeta} + \frac{2}{\gamma-1} c_o'\right) \frac{\gamma-1}{(\gamma+1)c_o'} = \exp\left[\frac{2(\gamma-1)}{\gamma+1} G\xi\right]. \tag{37}$$

Since the temperature T and the density ρ in region f just behind the n-th shock wave are written down through ξ, we can derive

$$\frac{T}{T_{fo}} = \frac{p}{p_{fo}} \quad \frac{\rho_{fo}}{\rho} = \exp\left(\frac{4(\gamma-1)}{\gamma+1} G\xi\right) = \left[\frac{\gamma-1}{(\gamma+1)c_o'}\left(\frac{x}{\zeta} + \frac{2}{\gamma-1} c_o'\right)\right]^2, \tag{38}$$

$$\frac{\rho}{\rho_{fo}} = \exp\left[\frac{4}{\gamma+1} G\xi\right] = \left(\frac{\gamma-1}{(\gamma+1)c_o'}\left(\frac{x}{\zeta} + \frac{2}{\gamma-1} c_o'\right)\right)^{\frac{2}{\gamma-1}}. \tag{39}$$

Introducing the non-dimensional variables

$$x^* = \frac{x}{X_{fo}}, \qquad \zeta^* = \frac{\zeta c_o'}{X_{fo}}, \tag{40}$$

and employing eqs.(33) (34), we rewrite eq.(28) as

$$\Theta = \frac{4k\rho_{fo} X_{fo} c_o'}{K_o m_f T_{fo}^{5/2}} \left[\zeta^* - \frac{3\zeta^*}{\frac{x^*}{\zeta^*} + 3}\right]. \tag{41}$$

In view of $x^* > \zeta^*$, eq.(36) has a monomun at $x^* = \zeta^*$, and the minimum value of Θ is given by

$$\Theta_{min} = \frac{k\rho_{fo} X_{fo} c_o'}{K_o m_f T_{fo}^{5/2}} \zeta^*. \tag{42}$$

When $\Theta_{min} < 1$, the phenomena are not approximated by adiabatic ones and the thermal conduction plays a dominant role. In such situations, we cannot expect to compress the fuel any more. Using subscript s for the values when the compression ceases, we have

$$\zeta_s^* \simeq \exp(-2G\xi_s). \tag{43}$$

$$\rho_s/\rho_{fo} = \exp[4G\xi_s/(\gamma+1)]. \tag{44}$$

Putting $\Theta_{min} = 1$ for $\zeta^* = \zeta_s^*$ in eq.(37), and substituting ζ_s^* and ζ_s from eqs.(37) and (38) into eq.(39), we can estimate ρ_s as follows,

$$\rho_s/\rho_{fo} \lesssim k\rho_{fo}X_{fo}c'_o/K_o m_f T_{fo}^{5/2} , \tag{45}$$

DISCUSSION

In the second section, we have used the fact that the shock wave in region f is represented by a straight line in the x-t diagram. On the other hand, Brueckner[3] derived the time dependence of the ablation pressure p_a in the form

$$p_a \propto (1-t/t_s)^{-5},$$

which is based on the fact that the trajectory of fluid particle is straight in the x-t diagram. This Brueckner's result differs from Kidder's and our result [see eq.(24)]. The velocities of the shock waves in region f are faster than that of the particle. Since outrunning of the shock waves accelerated the particle, the assumption of the constant particle velocity seems to be not valid in region f.

REFFERENCES

1. G. Charatis, J. Downward, R. Goforth, B. Guscott, T. Henderson, F. Mayer, S. Segall, L. Siebert, D. Solomon and C. Thomas: in Plasma Physics and Controlled Nuclear Fusion Research (Proc. 5th Int. Conf. Tokyo, 1974) 2, IAEA, Vienna (1975) 317.

2. J.F. Holzrichter, H.G. Ahlstrom, D.R. Speck, E. Storm, J.E. Swain, L.W. Coleman, C.D. Hendricks, H.N. Kornblum, F.D. Seward, V.W. Slivinsky, Y.L. Pan, G.B. Zimmerman and J.H. Nuckolls: Plasma Phys. 18 (1976) 675.

3. K.A. Brueckner and S. Jorna: Rev. Mod. Phys. 46 (1974) 325.

4. R.E. Kidder: Nucl. Fusion 14 (1974) 53.

5. R.E. Kidder: Nucl. Fusion 14 (1974) 797.

6. S. Mikoshiba and B. Ahlborn: Phys. of Fluids 17 (1974) 1198.

7. K.A. Brueckner, P.M. Campbell and R.A. Grandey: Nucl. Fusion 15 (1975) 471.

8. D.E.T.F. Ashby: Nucl. Fusion 16 (1976) 231.

9. R. Courant and K.O. Friedrichs: Supersonic Flow and Shock Waves Interscience (1948).

APPENDIX A

Across a contact surface, there is no pressure jump nor fluid flow. Hence if the each region in Fig.2 is distinguished by the suffix, it follows that

$$p_1 = p_2,\ p_4 = p_5,\ u_1 = u_2,\ u_4 = u_5. \tag{A-1}$$

As the shock waves and the rarefaction waves are weak, we can put $1+\Delta M_t$, $1+\Delta M_i$ and $1-\Delta M_r$ (see Appendix B) for the Mach Number of the transmitted shock wave, incident shock wave and reflected rarefaction wave near the contact surface respectively, where

$$0 < \Delta M_t,\ \ \Delta M_i,\ \ \Delta M_r \ll 1.$$

Then from eq.(A-1), we obtain

$$\Delta M_i = \Delta M_t + \Delta M_r, \tag{A-2}$$

and

$$\Delta M_t\sqrt{\frac{\gamma+1}{2\gamma+4\Delta M_t}} = \Delta M_i\sqrt{\frac{\rho_1}{\rho_2}\,\frac{\gamma+1}{2\gamma+4}} + \Delta M_r\sqrt{\frac{\rho_1 p_3}{\rho_3 p_1}\,\frac{\gamma+1}{2\gamma+4\Delta M_r}}\ . \tag{A-3}$$

Equations (A-2) and (A-3) yield

$$\Delta M_i/\Delta M_t = (\sqrt{\rho_2/\rho_1} + 1)/2,$$

with the aid of the Rankine-Hugoniot relations,

$$p_3 = p_2[1+4\gamma\Delta M_i/(\gamma+1)],\quad \rho_3 = \rho_2[1+4\Delta M_i/(\gamma+1)]. \tag{A-5}$$

The increment $\Delta M_i(t)$ in the Mach number of the incident shock wave is time dependent because of the interaction among the shock wave and the rarefaction waves during the shock propagates in region h. But the interaction gives only a very small contribution to change the Mach number, that is,

$$\Delta M_i = \Delta M_i(t)\{1 + O(\Delta M_i)\}.$$

Thus we can approximate that ΔM_i remains constant in region h. This approximation does not lead the fact that the quantities ρ, p, u and U remain constant in region h. These quantities vary as ΔM_i remains constant. All the quantities in eqs. (A-1)~(A-5) are those in the vicinity of the contact surface.

By using the Rankine-Hugoniot relations

$$\rho_4 = \rho_1 \left(\frac{2-G}{G}\right)^2 \left\{1+ \frac{4}{\gamma(\gamma+1)} \Delta M_i\right\} \left\{1- \frac{4}{\gamma(\gamma+1)} \Delta M_r\right\},$$

$$\rho_5 = \rho_1 \left\{1+ \frac{4}{\gamma(\gamma+1)} \Delta M_t\right\},$$

we obtain

$$\rho_5/\rho_4 = [(2-G)/G]^2+0(\Delta M_i^2) = \rho_2/\rho_1+0(\Delta M_i^2). \tag{A-6}$$

Equation (A-6) means that the ratio of the densities on the both sides of the contact surface remains constant.

APPENDIX B

The velocities u_3 and u_4 on the both sides of the rarefaction wave shown in Fig.2 have the following relation,

$$u_3-u_4 = -2(\sqrt{\gamma p_3/\rho_3}-\sqrt{\gamma p_4/\rho_4})/(\gamma-1). \tag{B-1}$$

Using the isentropic relation across the rarefaction wave, we have

$$\rho_4/\rho_3 = (p_4/p_3)^{1/\gamma}.$$

Defining Δ as

$$p_3 = p_4(1+\Delta), \tag{B-2}$$

then we can reduce eq. (B-1) to

$$u_3-u_4 = -\Delta\sqrt{p_4/\gamma\rho_4}+0(\Delta^2). \tag{B-3}$$

In view of eqs. (B-2) and (B-3), we introduce the decrement ΔM_r in the Mach number of the weak rarefaction wave ($\Delta \ll 1$) as

$$\Delta M_r = (\gamma+1)\Delta/4\gamma. \tag{B-4}$$

Consequently, eqs. (B-2) and (B-3) reduce to respectively

$$p_3 = p_4[1+4\gamma\Delta M_r/(\gamma+1)], \quad u_3-u_4 = -4c_4\Delta M_r/(\gamma+1), \qquad \text{(B-5)}$$

where $c_4 = \sqrt{\gamma p_4/\rho_4}$.

If we substitute the rarefaction wave by a shock wave whose Mach number is $1+\Delta M_s$, the Rankine-Hugoniot relation gives us

$$p_3 = p_4[1-4\gamma\Delta M_s/(\gamma+1)], \quad u_3-u_4 = 4c_4\Delta M_s/(\gamma+1). \qquad \text{(B-6)}$$

Equation (B-5) can be obtained from eq.(B-6) provided that ΔM_s is replaced by $-\Delta M_r$. Thus we define the Mach number of the rarefaction wave by $1-\Delta M_r$.

APPENDIX C

The mass, momentum and energy conservations[6] across the ablation surface shown in Fig.4 and the definition of the sound velocity yield

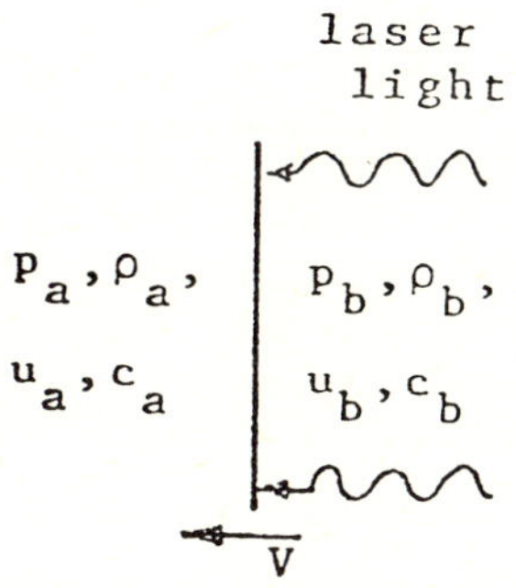

Fig.4 : The ablation surface.

$$\frac{p_b}{p_a} = 1+\frac{\gamma(N^2-1)}{\gamma+1}, \qquad \frac{\rho_a}{\rho_b} = 1-\frac{N^2-1}{(\gamma+1)N^2}\beta,$$

$$\left(\frac{c_b}{c_a}\right)^2 = \frac{p_b\rho_a}{p_a\rho_b}, \qquad \frac{u_b-u_a}{c_a} = \frac{N^2-1}{(\gamma+1)N}\beta, \qquad \text{(C-1)}$$

where

$$\beta = 1-\sqrt{1-w}, \qquad w = \frac{2(\gamma^2-1)NE}{p_a c_a(1-N^2)^2}. \qquad \text{(C-2)}$$

Subscripts a and b refer to the values in front of and behind the ablation surface, respectively.

Equations (C-1) determine the quantities with subscript b, when the quantities with subscript a are given, only in the case that β is definite. We, however, have no method to determine β but the Chapman-Jouguet deflagration condition, which requires that the fluid must go out from the ablation surface with the local sound velocity. We are not sure that this requirement is fulfilled in the laser-driven slab. In this paper we assume that the fluid leaves the ablation surface with the local sound velocity multiplied by ,

$$V = u_b = c_b. \tag{C-3}$$

Since N<<1 (see Appendix D), we have

$$\beta = \alpha^2(\gamma+1)/(\alpha^2\gamma+1), \quad w = 1-[(1-\alpha^2)/(\alpha^2+1)]^2, \tag{C-4}$$

in view of eqs.(C-1), (C-3) and

$$V = u_a + c_a N,$$

which is the definition of N. Especially we obtain the Chapman-Jouguet condition[6,9]

$$\beta = 1, \quad w = 1$$

from eq.(C-4) provided that we put $\alpha = 1$.

Consequently we obtain

$$E = \frac{\rho_a c_a^3}{2(\gamma^2-1)N}\left[1-\left(\frac{1-\alpha^2}{\alpha^2\gamma+1}\right)^2\right], \tag{C-5}$$

from eqs.(C-2) and (C-4) with N<<1.

APPENDIX D

The laser light cannot penetrate into the slab beyond the surface of the critical density. If we neglect the secondary phenomena, the energy deposition occurs only in the under-dense region. This means that initially $(\rho_a/\rho_b) >> 1$. Consequently $N_o << 1$ from eq.(C-1).

Putting $\rho_a = \rho_{ho}$ (cut off density) and $\rho_b = \rho_c$ in eq.(C-1), we may specity the initial Mach number N_o of the ablation surface as

$$N_o = \alpha\sqrt{\rho_c/\rho_{ho}(\alpha^2\gamma+1)} \tag{D-1}$$

Figure 5 shows that the two weak shock waves (whose Mach number are $1+\Delta M_a$ and $1+\Delta M_b$) are generated by the acceleration of the ablation surface from N to N+ΔN.

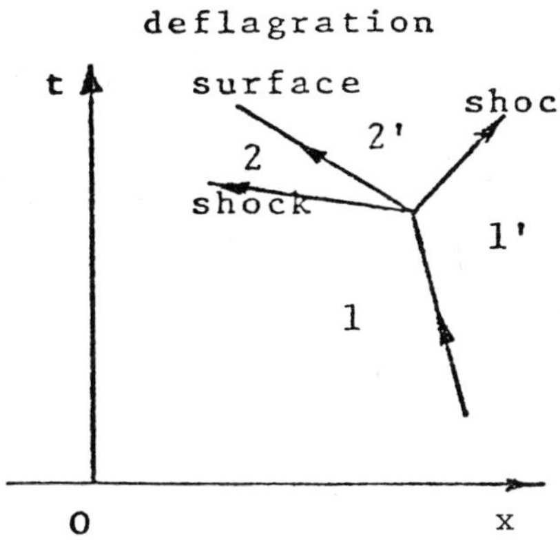

Fig.5 : Interaction among the weak shock waves and the deflagration surface.

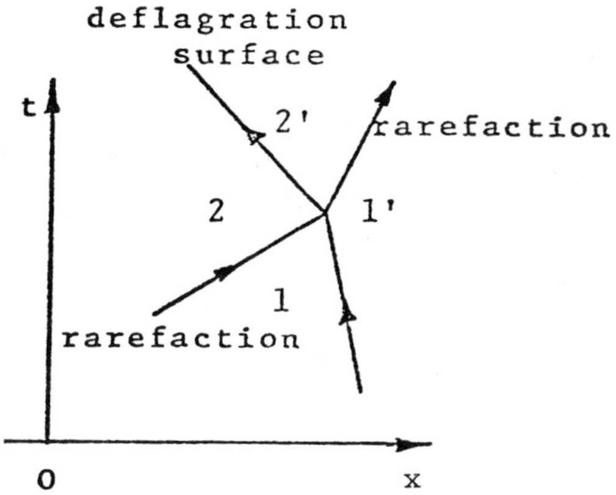

Fig.6 : Interaction among the ablation surface and rarefaction waves.

Figure 6 shows that the ablation surface is accelerated from N to N = ΔN by the interaction with the incident rarefaction wave $(1-\Delta M_a)$ and the transmitted rarefaction wave $(1-\Delta M_b)$.

Splitting the region into 1, 1', 2 and 2' as shown in Figs. 5 and 6, we obtain the identities with respect to u and p as follows,

$$\frac{u_2'}{c_1} = \frac{u_2' - u_2}{c_2}\,\frac{c_2}{c_1} + \frac{u_2 - u_1}{c_1} + \frac{u_1}{c_1}$$

$$= \frac{u_1}{c_1} + \frac{4}{\gamma+1}\Delta M_a - \frac{\beta(1-\Delta N/N)}{(\gamma+1)N}\left(1 \pm \frac{2(\gamma-1)}{\gamma+1}\Delta M_a\right)$$

$$= \frac{u_2' - u_1'}{c_1'}\,\frac{c_1'}{c_1} + \frac{u_1' - u_1}{c_1} + \frac{u_1}{c_1}$$

$$= \frac{u_1}{c_1} - \frac{\beta}{(\gamma+1)N} \mp \frac{4\sqrt{\beta(\gamma+1)-\gamma\beta^2}}{(\gamma+1)^2 N}\Delta M_b, \qquad \text{(D-2)}$$

$$p_2' = p_2\left[1 - \frac{\gamma\beta(2N\Delta N-1)}{\gamma+1}\right] = p_1\left[1 + \frac{\gamma\beta(2N\Delta N-1)}{\gamma+1} \pm \frac{4}{\gamma+1}\left(1-\frac{\gamma\beta}{\gamma+1}\right)\Delta M_a\right]$$

$$= p_1'\left[1 \pm \frac{4\gamma}{\gamma+1}\Delta M_b\right] = p_1\left[1 \pm \frac{4\gamma}{\gamma+1}\Delta M_b\right]\left(\frac{\gamma+1-\gamma\beta}{\gamma+1}\right). \qquad \text{(D-3)}$$

The upper and lower signs of ± in eqs.(D-2) and (D-3) correspond to the processes in Fig.5 and in Fig.6, respectively.

Equations (D-2) and (D-3) yield

$$\frac{\Delta N}{N} = \frac{\pm 2\beta(\gamma-1) - 4\sqrt{\beta(\gamma+1-\gamma\beta)}}{(\gamma+1)\beta}\Delta M_a, \qquad \text{(D-4)}$$

by use of $N<<1$. Combining both the processes, we obtain

$$\frac{\Delta N}{N} = \frac{2\beta(\gamma-1)G - 4\sqrt{\beta(\gamma+1-\gamma\beta)}(2-G)}{(\gamma+1)\beta}\Delta M_i, \qquad \text{(D-5)}$$

where the relations $\Delta M_a = \Delta M_i$ in Fig.5, $\Delta M_a = \Delta M_r$ in Fig.6 and $\Delta M_r = (1-G)\Delta M_i$ are used.

In the limit that $n \to \infty$, $\Delta M_i \to 0$ and $n\Delta M_i \to \xi$, we have

$$N = N_o \exp\left\{\frac{2[(\gamma-1)\alpha+2]G-8}{(\gamma+1)\alpha}\xi\right\}, \qquad \text{(D-6)}$$

from eq.(D-5). Here

$$N_o = \alpha\sqrt{\rho_c/\rho_{ho}(\alpha^2\gamma+1)}$$

APPENDIX E

The successive shock waves launched from the ablation surface compress and heat region h. We have relations

$$c_n = c_0[1+2(\gamma-1)\Delta M_i/(\gamma+1)]^n, \qquad \text{(E-1)}$$

$$\rho_n = \rho_0[1+4\Delta M_i/(\gamma+1)]^n. \qquad \text{(E-2)}$$

In the limit $n \to \infty$, $\Delta M_i \to 0$ and $n\Delta M_i \to \xi$, eqs.(E-1) and (E-2) reduce to

$$c_n = c_0 \exp[2(\gamma-1)\xi/(\gamma+1)], \qquad \text{(E-3)}$$

$$\rho_n = \rho_0 \exp[4\xi/(\gamma+1)]. \qquad \text{(E-4)}$$

On the other hand, the rarefaction waves coming back from the contact surface decrease c_n and ρ_n to c_a and ρ_a respectively, according to

$$c_a = c_n \exp[2(\gamma-1)(G-1)\xi/(\gamma+1)], \qquad (E-5)$$

$$\rho_a = \rho_n \exp[4(G-1)\xi/(\gamma+1)]. \qquad (E-6)$$

These results are based on the fact that the Mach number of the rarefaction waves is given by

$$\Delta M_r = \Delta M_i - \Delta M_t = (1-G)\Delta M_i \qquad (E-7)$$

which is derived from eqs.(A-2) and (A-4). Combining eqs.(E-3), (E-4), (E-5) and (E-6), we obtain

$$\rho_a c_a^3 = \rho_{h0} c_0^3 \exp[2(3\gamma-1)G\xi/(\gamma+1)]. \qquad (E-8)$$

PHYSICAL MODELS AND MATHEMATICAL SIMULATION OF LASER-DRIVEN IMPLOSION AND THEIR RELATIONS WITH EXPERIMENTS*

L. Brun, R. Dautray, F. Delobeau, C. Patou, F. Perrot, J.-M. Reisse, B. Sitt, and J.-P. Watteau

Centre d'Etudes de Limeil, France

ABSTRACT

To define implosion experiments and to interpret their results, we develop physical models and mathematical representation .

We summarize our experimental results of the interaction of a laser beam with a plane solid deuterium target and we discuss them.

An important work has been done on equations of state for different thermodynamical situations and materials. We build self similar models for flows, and study the influence and compatibility of the various terms. We compare our implosion experiment done with the laser C_6 600 J - 1,4ns with those of the numerical simulation.

We analyse the hydrodynamic stability of the different parts of the target during the implosion and develop an entirely Lagrangian stability code coupled to the computation of the unperturbed flow.

* Presented at the Fourth Workshop on "Laser Interaction and Related Plasma Phenomena" held at RPI, Troy, November 8-12,1976.

I - INTRODUCTION

In connection with our experiments, we are developing physical models and numerical simulation, in order to get a better understanding of the phenomena, to prepare and interpret these experiments and to find a good compromise between ideal laser-target caracteristics and present technology limitations.

First of all, we summarize our experimental results of the interaction of a laser beam with a plane solid deuterium target showing that above a given flux, inverse bremsstrahlung absorption cannot any longer explain these results, and that new mechanisms, such as resonant absorption, ponderomotive force, have to be introduced to interpret them.

Then, we present some of the work we are doing in relation with the modelling of the laser implosion. We explain the methods we use to compute the state equations of target materials over the large density and temperature range covered during the implosion. For special boundary conditions and including the thermal conduction term, we establish hydrodynamic similarity laws from which we evaluate the influence of the conduction term on the flow. We compare experimental data of the implosion of a glass microballoon in a nanosecond ablation regime with the results given by an ID code where absorption is treated as inverse bremsstrahlung.

Finally, we examine which types of fluid instabilities can take place in the different target regions during the implosion and we develop a purely Lagrangian analytical model to study them.

II- INTERACTION AT 1,06 AND 10.6 μm

To study absorption mechanisms we have performed simple beam experiments on flat deuterium targets which lead to more simple interpretations than high Z targets. Three wavelengths have been studied in the range of fluxes indicated :

Table 1 : Typical parameters of the single beam experiments

λ	1.06 μm	0.53 μm	10.6 μm
Maximum incident power P_M (W/cm²)	10^{15}	10^{14}	10^{13}
Rise time (ns) $0.1\ P_M \longrightarrow 0.9\ P_M$	0.5 and 1	0.5 and 1	1.5
Pulse width at $P_M/2$ (ns)	1.5 and 3	1.5 and 3	1.5

Experimental results show that for 1.06 μm absorption is mainly due to bremsstrahlung when Pi < 10^{14} W/cm^2. For 10.6 μm those upper fluxes are 10^{12} and even 10^{11} W/cm^2. A comparison is then possible with numerical calculations with only "classical" physical phenomena, in spite of the 2D effects which always appear with single beam experiments.[1]

The following figures indicate for 1.06 and 0.53 μm the results of such a comparison (pulse duration 3 ns ; maximum incoming flux density below 5.10^{13} W/cm^2).

Table 2 : Comparison between experiment and calculation at 1.06 μm and 0.53 μm

Experiments (5.10^{13} W/cm^2)	1.06 μm	0.53 μm
Reflected power	30 - 45 %	5 - 15 %
Number of emitted ion	10^{17}	1.5 - 2.10^{17}
Average energy per ion	1 keV	0.8 keV
Calculations (5.10^{13} W/cm^2)		
Reflected power	50 %	10 %
Average ion energy	1.2 keV	0.9 keV

As we can see, there is a not too bad agreement between the experimental results and the numerical calculation in that range of fluxes.

At higher fluxes, other mechanisms, such as instabilities or resonant absorption take place. Self generated magnetic field and decrease of heat conduction may occur leading to important modifications of the involved processes.

A theoretical explanation of such phenomena can be seen for instance in acceleration of electrons by ponderomotive force due to the gradient of the resonant electrical field near the critical regim.[+] Ions are then dragged by charge separation.

A valuable interpretation would need to take into account :

- the temporal evolution of the resonant electric field profile
- the spatial and temporal variation of the incoming laser light in the focal spot region which allows to know which fraction of this incident laser can be implied in the resonant process.

+ See "Fast ion and neutron emission in CO_2 laser irradiation of solid deuterium targets "J.L. Bocher and alii, presented at this workshop.

III - EQUATIONS OF STATE

The equation of state (EOS) of materials, such as lithium hydride, polyethylene, aluminium, silica,... are needed for simulation of laser implosion experiments. The range of temperatures and densites explored is very large, leading to the investigation of various states of matter. Our approach to this complex problem has been to combine experimental data and theoretical estimates in regions where measurements are not possible. We shall briefly review our present treatments in each temperature-density domain and point out how to improve them in the future.

III.1. Solid state (Region I of figure 1)

For densities ρ greater than the normal density ρ_0 , the zero isotherm is preponderant in the EOS up to rather large temperatures (T). An accurate determination of this isotherm is necessary. The electronic total energy and pressure result from a self-consistent calculation of band structure and electron charge density in the solid. We commonly use two methods :

a) the "Augmented Plane Wave" method (APW) [2,3] ;

b) the "Cellular method", in the spherical version as used by Brooks and Gandel'man [4].

The first method is well-suited for the treatment of compact structures such as metals and some insulators ; the second one, which does not take into account the detailed lattice structure, saves an important amount of computing time for compressed pure materials.

Both methods give results which agree, for very high compressions ρ/ρ_0 , with the Thomas-Fermi (TF) and Thomas-Fermi-Dirac (TFD) models. Table 3 shows in the case of LiH, the electronic pressures as obtained from APW (with a Kohn-Sham exchange potential) and TFD calculations [5].

Table 3 : Electronic pressures (Mbar) in LiH (0°K) resulting from APW and TFD methods.

ρ/ρ_0	0.901	1.063	1.279	1.502	1.815	2.265	3.561	6.241	18.49
P_{APW}	-0.0289	0.0181	0.1076	0.2390	0.4848	0.9726	2.980	11.43	112.6
P_{TFD}	0.0850	0.1425	0.2438	0.3808	0.6276	1.100	3.237	11.30	105.9

Table 4 : Comparison between pressures (Mbar) as given by the APW and PP methods, in aluminium. [6]

ρ/ρ_0	1.166	1.372	1.882
P_{APW}	0.128	0.427	1.705
P_{PP}	0.133	0.442	1.833

None of these methods is well-suited for the case of open structures. So we plan to treat these lattices, when the electronic structure is adapted, by a pseudopotential (PP method) we are now checking in the case of metals. The philosophy behind this method is to replace the true potential by a "small" pseudopotential which gives the same wave-functions outside the ion core, the advantage being in the possibility of using perturbation theory to derive the total energy. The simplest of these pseudopotentials is known as "empty core" pseudopotential [6] . As shown in table 4 for aluminium, it gives an EOS which agrees quite well with the APW EOS.

The lattice vibration contribution to zero temperature isotherm has been calculated, up to now, from a semi empirical Debye model. An estimate of the melting temperature, T_m, is obtained from the Debye temperature using the Lindemann's law. The pseudopotential method is here also of great interest since it allows the calculation of the phonon spectrum of simple metals and consequently the test of the reliability of Debye model. An important effort is done in this direction.

The theoretical energy bands calculations supply us with the density of states parameters which are needed for the estimate of the thermal energy and pressure contributions in the low temperature region : in a metal for instance, the thermal energy increase is proportionnal to $n_f\ T^2$, n_f being the density of states at the Fermi level. The thermal vibration contribution is also treated in the Debye model.

III.2. High temperature fluid phase (Region II of figure 1)

This domain corresponds to temperatures :

- higher than some T_m (melting temperature) when the material is strongly compressed ;
- higher than some 10^4°K when $\rho \leqslant \rho_0$

Table 5 : Energies (in atomic units per mole) from the modified model (MM) and the Thomas-Fermi-Dirac model (TFD), in polyethylene. The sharp increase in the difference $E_{MM}-E_{TFD}$ between 10^6 and 10^7°K is due to the depopulation of the 1s level.

T(°K)	E_{MM}	E_{TFD}	$E_{MM}-E_{TFD}$
10^5	-32.614	-49.054	16.440
2.10^5	-30.622	-47.033	16.441
5.10^5	-22.702	-38.455	15.753
10^6	- 5.602	-20.906	14.504
2.10^6	38.641	18.801	19.840
5.10^6	164.976	142.843	22.133
10^7	358.502	342.364	16.138

The fundamental tool used in this region is the Thomas-Fermi (and Thomas-Fermi-Dirac) theory, modified in order to correct the erroneous features of the TF density of states for negative energies. This density of states is replaced, under the ionization threshold, by a series of lines corresponding to the bound states obtained by integrating the Schrödinger equation [7]. The electron charge density is shared in a bound contribution (described by quantum mechanics) and a free contribution (described in the TF model) and results from a self consistent procedure. This model is intensively used in our EOS calculations. Some difficulties occur when a bound state broadens before its ionization by pressure, but we work to overcome them. The modifications previously described do not strongly affect the pressure, but they have a large influence on the energy, as shown in table 5, for polyethylene at density $\rho = 10\rho_0$ [7].

In this domain, the ions contribute to the EOS as a perfect gas.

III.3. Experimental region (Region III of figure 1)

This is the region surrounding the normal density ρ_0 with temperature up to some 10^3°K. The critical point is enclosed in this region where all the phases coexist. An "a-priori" determination of the EOS would here be very difficult and probably inaccurate. Fortunately, many experimental data are available : cohesive energy, compressibility, specific heats, melting temperature, elastic constants, Hugoniot curves... An empirical approach is used and a so called "Multi-phase equation of state "derived : analytical expressions are established for energy and pressure in each phase, the parameters of which being adjusted from experiment [8].

III.4. Cold gas (Region IV of figure 1)

Less attention has been paid to the EOS in this region up to now. But we plan to take interest in a model for moderately dense fluids at low temperatures. At the present time, we use the Young-Alder formulation[9] which is a generalization of the Van der Waals equation including the hard core repulsive term. A contribution which insures continuity with the "experimental region" is added.

In conclusion, we have attempted to show that our effort concerning EOS covers the full domain of temperatures and densities. Many problems have been treated, some of them using accurate theoretical methods (low temperature solid phase, high temperature fluid), the others qualitatively (melting curve, cold gas), or from experimental results. Several improvements are studied (phonon spectrum ...) and others are planned.

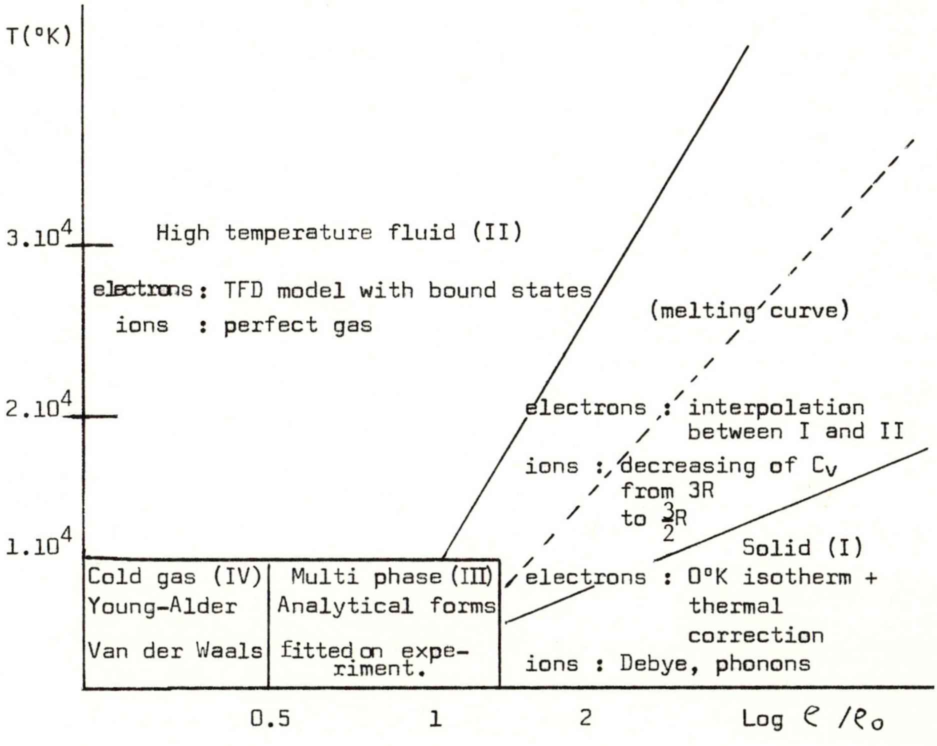

Fig. 1 : Diagram of the various regions where the equations of state are studied, and methods used in each region. Interpolation schemes may be used between zones of validity of different methods.

IV - SIMILARITY LAWS FOR THE PROPAGATION OF A THERMAL WAVE INTO A PLANE TARGET

IV.1. General equations

In Lagrangien coordinates, the 1D hydrodynamic equations with heat transfer are :

$$\left.\begin{aligned} &\frac{\partial R}{\partial t} = u \qquad && \frac{\partial E}{\partial t} = \frac{P}{\rho^2}\frac{\partial \rho}{\partial t} + \frac{\partial}{\partial m}\left(\chi^* \rho \frac{\partial T}{\partial m}\right) \\ &\frac{\partial u}{\partial t} = -\frac{\partial p}{\partial m} && \rho \frac{\partial R}{\partial m} = 1 \end{aligned}\right) \qquad (1)$$

with : R : abscissa, m : mass
u : velocity of matter
$E = K_E T^{\theta_1}$ internal energy per gram
$P = K_p \rho T^{\theta_1}$ pressure
$\chi^* = \chi\, \rho^{n_1} T^{m_1}$ coefficient of thermal conductivity
$\chi^* \frac{\partial T}{\partial R}$ heat conduction flux

We neglect here radiative energy and radiative pressure. We assume that ion temperature equals electron temperature $T_i = T_e = T$. We seek now all the similarity laws for this problem : if T(m, t) ; ρ(m, t) ; u(m, t) ; R(m, t) is a particular solution of the system (1), one can demonstrate that the **general self-similar solution takes the form :**

$\lambda^\theta\ T(\lambda^\alpha m\ ,\ \lambda^\beta t)$	$t^{-\frac{\theta}{\beta}}\ T^*\left(mt^{-\frac{\alpha}{\beta}}\right)$	$m^{-\frac{\theta}{\alpha}}\ T'^*\left(m^{-\frac{\beta}{\alpha}}t\right)$
$\lambda^r\ \rho(\lambda^\alpha m\ ,\ \lambda^\beta t)$	$t^{-\frac{r}{\beta}}\ \rho^*\left(mt^{-\frac{\alpha}{\beta}}\right)$	$m^{-\frac{r}{\alpha}}\ \rho'^*\left(m^{-\frac{\beta}{\alpha}}t\right)$
$\lambda^v\ u(\lambda^\alpha m\ ,\ \lambda^\beta t)$	$t^{-\frac{v}{\beta}}\ u^*\left(mt^{-\frac{\alpha}{\beta}}\right)$	$m^{-\frac{v}{\alpha}}\ u'^*\left(m^{-\frac{\beta}{\alpha}}t\right)$
$\lambda\ R(\lambda^\alpha m\ ,\ \lambda^\beta t)$	$t^{-\frac{1}{\beta}}\ R^*\left(mt^{-\frac{\alpha}{\beta}}\right)$	$m^{-\frac{1}{\alpha}}\ R'^*\left(m^{-\frac{\beta}{\alpha}}t\right)$
	for $\lambda^\beta t = 1$	for $\lambda^\alpha m = 1$

(2)

with the following conditions :

$$\theta = \frac{1 - (n_1 - 1)\, r}{m_1 + 1 - \frac{3}{2}\theta_1} \qquad \alpha = -r - 1$$

$$\beta = -1 + \frac{\theta\ \theta_1}{2} \qquad v = \frac{\theta\ \theta_1}{2} \qquad (3)$$

Boundary conditions :

There are five parameters α, β, θ, v and **r in the four** equations (1), so one parameter, r for example can be choosed as a free parameter.

Initial and boundary conditions are :

- density and temperature of target and outside medium at the origin of time ;
- density and temperature at a fixed point bounded to matter (target - outside medium boundary) at any time.

Then we take :

$$T(m, 0) = K_1 m^{-\frac{\theta}{\alpha}} \qquad T(0, t) = K_3 t^{-\frac{\theta}{\beta}}$$

$$\rho(m, 0) = K_2 m^{-\frac{r}{\alpha}} \qquad \rho(0, t) = K_4 t^{-\frac{r}{\beta}}$$

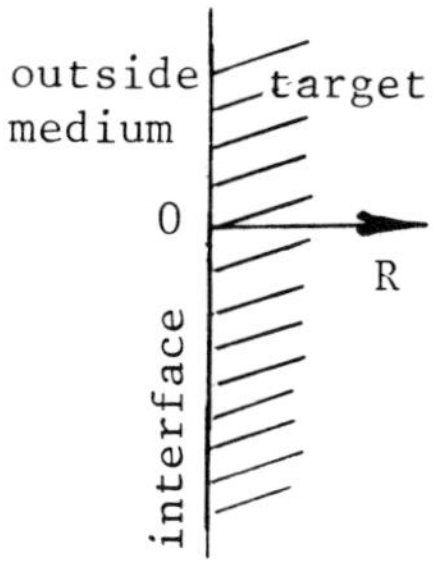

If we choose, for example :

- $K_1 = 0$ initial temperature of the target equals zero ;
- $K_4 = 0$ target in vacuum.

We verify that the boundary temperature laws that lead to self similar solutions are such that $T(0, t) \sim t^{-\theta/\beta}$.

1) Let us try a constant temperature T_o be maintained at the boundary. Using the conditions (3) we can write :

$$\theta = 0 \qquad \beta = -1 \qquad v = 0 \qquad r = \frac{1}{n_1 - 1}$$

this implies, that the target has not a uniform initial density $\rho(m, 0) = K_2 m^{-r/\alpha}$.

In this case, the solution takes the form :

$$T(m, t) = T^*\left(m\, t^{-\frac{n_1}{n_1 - 1}}\right) \qquad T^*(0) = T_o$$

$$\rho(m, t) = \rho^*\left(m\, t^{-\frac{n_1}{n_1 - 1}}\right) \qquad \rho^*(0) = 0$$

$$u(m, t) = u^*\left(m\, t^{-\frac{n_1}{n_1 - 1}}\right) \qquad u^*(0) \to \infty$$

$$R(m, t) = tR^*\left(m\, t^{-\frac{n_1}{n_1 - 1}}\right) \qquad R^*(0) \to \infty$$

For a DT target we have $\theta_1 = 1$, $n_1 = 0$, $m_1 = 5/2$; then $r = -1$ and $\alpha = 0$. Therefore, one cannot have a constant temperature T_o at the boundary with this type of solution.

2) To find a self similar solution for the DT case, let us assume the following boundary conditions :

- uniform initial density $r = 0$;
- zero initial temperature $K_1 = 0$;
- target in vacuum $K_4 = 0$;
- $T(0, t) = K_3 t^{-\theta/\beta}$,

then the solution takes the form :

$$T(m, t) = t^{-\frac{\theta}{\beta}} T^*\left(mt^{-\frac{\alpha}{\beta}}\right) \qquad u(m, t) = t^{-\frac{V}{\beta}} u^*\left(mt^{-\frac{\alpha}{\beta}}\right) \tag{4}$$

$$\rho(m, t) = \rho^*\left(mt^{-\frac{\alpha}{\beta}}\right) \qquad R(m, t) = t^{-\frac{1}{\beta}} R^*\left(mt^{-\frac{\alpha}{\beta}}\right)$$

Using the relations (3) we have :

$$\frac{\theta}{\beta} = -\frac{1}{m_1 + 1 - 2\theta_1} \qquad \frac{V}{\beta} = \frac{\theta_1}{2(-m_1 - 1 + 2\theta_1)}$$

$$\frac{\alpha}{\beta} = \frac{m_1 + 1 - \frac{3}{2}\theta_1}{m_1 + 1 - 2\theta_1} \qquad \frac{1}{\beta} = \frac{m_1 + 1 - \frac{3}{2}\theta_1}{-m_1 - 1 + 2\theta_1}$$

$$\frac{r}{\beta} = 0$$

For DT, we obtain :

$$T(m, t) = t^{2/3} T^*(mt^{-4/3}) \qquad u(m, t) = t^{1/3} u^*(mt^{-4/3})$$

$$\rho(m, t) = \rho^*(mt^{-4/3}) \qquad R(m, t) = t^{4/3} R^*(mt^{-4/3})$$

- the target mass reached by the thermal wave (or the shock front) varies as $t^{4/3}$;
- the pressure in the ablation front varies as $t^{2/3}$;
- the energy flux entering the target varies as t and the energy varies as t^2.

Such a solution can be obtained by irradiating a plane DT target by a laser pulse varying linearly with time. This solution is valid during the beginning of the interaction because we have assumed : $T_i = T_e$ and the critical surface close to the target boundary.

In figures 2 and 3 we have plotted the solution versus the reduced variable $\xi = \frac{m}{M}$(M ~ $t^{4/3}$: target mass reached by the shock) and the time.

IV.2. Variation with coefficient of thermal conductivity

In the system of equations (1) we use $\mu\chi$ instead of χ. If $\left[T(m, t), \rho(m, t), u(m, t), R(m, t)\right]$ is the solution for (1) one can easily demonstrate that the new solution takes the form :

$$T'(m, t) = T\left(\frac{m}{\mu}, \frac{t}{\mu}\right) \qquad u'(m, t) = u\left(\frac{m}{\mu}, \frac{t}{\mu}\right)$$

$$\rho'(m, t) = \rho\left(\frac{m}{\mu}, \frac{t}{\mu}\right) \qquad R'(m, t) = R\left(\frac{m}{\mu}, \frac{t}{\mu}\right)$$

Using the results (4) we obtain :

$$T'(m, t) = t^{-\frac{\theta}{\beta}} \mu^{\frac{\theta}{\beta}} T^{*}\left(mt^{-\frac{\alpha}{\beta}} \mu^{\frac{\alpha}{\beta} - 1}\right)$$

$$\rho'(m, t) = t^{-\frac{r}{\beta}} \mu^{\frac{r}{\beta}} \rho^{*}\left(mt^{-\frac{\alpha}{\beta}} \mu^{\frac{\alpha}{\beta} - 1}\right)$$

$$u'(m, t) = t^{-\frac{V}{\beta}} \mu^{\frac{V}{\beta}} u^{*}\left(mt^{-\frac{\alpha}{\beta}} \mu^{\frac{\alpha}{\beta} - 1}\right)$$

$$R'(m, t) = t^{-\frac{1}{\beta}} \mu^{\frac{1}{\beta}} R^{*}\left(mt^{-\frac{\alpha}{\beta}} \mu^{\frac{\alpha}{\beta} - 1}\right)$$

For DT case and uniform initial density (r = 0), we obtain :

$$T'(m, t) = t^{2/3} \mu^{-2/3} T^{*}(mt^{-4/3} \mu^{1/3})$$

$$\rho'(m, t) = \rho^{*}(mt^{-4/3} \mu^{1/3})$$

$$u'(m, t) = t^{1/3} \mu^{-1/3} u^{*}(mt^{-4/3} \mu^{1/3})$$

$$R'(m, t) = t^{4/3} \mu^{-4/3} R^{*}(mt^{-4/3} \mu^{1/3})$$

- the target mass reached by the thermal wave varies as $t^{4/3} \mu^{-1/3}$;
- the pressure in the ablation front varies as $t^{2/3} \mu^{-2/3}$;
- the energy flux at the interface varies as t/μ, the energy as t^2/μ.

A solution in the (m,t) plane is obtained by direct numerical integration of the partial differential system (1) with the boundary conditions preceding equations (4). We then observe a general trend of the solution towards a self-similar solution of the form given in the second half of page 1068. Figures 2 and 3 depict what appears to be the asymptotic solution for two values of μ: 1 and 0,125.

IV.3. Heat conduction limit value case

$F \sim n_e T_e^{3/2} \sim \rho T^{3/2}$ (assumption of collisionless processes).

The hydrodynamics equations with heat transfer are :

$$\frac{\partial R}{\partial t} = u \qquad \frac{\partial E}{\partial t} = \frac{p}{\rho^2}\frac{\partial \rho}{\partial t} + \frac{\partial}{\partial m}(\chi \rho^{n_1+1} T^{m_1})$$

$$\frac{\partial u}{\partial t} = - \frac{\partial p}{\partial m} \qquad \rho \frac{\partial R}{\partial m} = 1$$

$$F = \chi \rho^{n_1} T^{m_1} \qquad E = K_E T^{\theta_1} \qquad P = K_p \rho T^{\theta_1}$$

As previously, if $|T(m, t), \rho(m, t), u(m, t), R(m, t)|$ is a particular solution, the general similar one takes the form

$$\lambda^{\theta} T(\lambda^{\alpha} m, \lambda^{\beta} t)$$
$$\lambda^{r} \rho(\lambda^{\alpha} m, \lambda^{\beta} t)$$
$$\lambda^{v} u(\lambda^{\alpha} m, \lambda^{\beta} t)$$
$$\lambda R(\lambda^{\alpha} m, \lambda^{\beta} t)$$

with the following relations :

$$\theta = - \frac{n_1 r}{m_1 - \frac{3}{2}\theta_1} \qquad \alpha = - r - 1$$

$$\beta = - 1 + \frac{\theta\theta_1}{2} \qquad v = \frac{\theta\theta_1}{2}$$

We take one of the five coefficient, r , as a free parameter. The initial and boundary conditions are :

$$T(m, 0) = K_1 m^{-\frac{\theta}{\alpha}} \qquad T(0, t) = K_3 t^{-\frac{\theta}{\beta}}$$

$$\rho(m, 0) = K_2 m^{-\frac{r}{\alpha}} \qquad \rho(0, t) = K_4 t^{-\frac{r}{\beta}}$$

For a target with a uniform initial density (r = 0) and with $K_1 = K_4 = 0$, the existence of a self similar solution needs, a constant boundary temperature. This solution takes the form ($\lambda^{\beta} t = 1$) :

$$T(m,\ t) = T^*\left(mt^{-\frac{\alpha}{\beta}}\right) = T^*(mt^{-1}) \qquad T^*(0) = T_o$$

$$\rho(m,\ t) = \rho^*\left(mt^{-\frac{\alpha}{\beta}}\right) = \rho^*(mt^{-1}) \qquad \rho^*(0) = 0 \qquad (5)$$

$$u(m,\ t) = u^*\left(mt^{-\frac{\alpha}{\beta}}\right) = u^*(mt^{-1}) \qquad u^*(0) \to \infty$$

$$R(m,\ t) = t\ R^*\left(mt^{-\frac{\alpha}{\beta}}\right) = t\ R^*(mt^{-1}) \qquad R^*(0) \to \infty$$

Thus :

- The target mass reaches by the thermal wave or the shock front increases linearly with time ;
- the energy received by the target varies in the same way ;
- the pressure in the ablation of the shock front is constant with time.

Let us consider how the solution varies with the boundary temperature T_o. So we write $\lambda'^{1/\theta}$ instead of λ in (2) ; then using (5), we obtain :

$$T'(m,\ t) = T_o T'^*\left(mt^{-1} T_o^{-\frac{\theta_1}{2}}\right) ;\ u'(m,\ t) = T_o^{\frac{\theta_1}{2}} u'^*\left(mt^{-1} T_o^{-\frac{\theta_1}{2}}\right)$$

$$\rho'(m,\ t) = T_o^{-\frac{m_1 - \frac{3\theta_1}{2}}{n - 1}} \rho'^*\left(mt^{-1} T_o^{-\frac{\theta_1}{2}}\right);\ R'(m,\ t) = tT_o^{\frac{\theta_1}{2}} R'^*\left(mt^{-1} T_o^{-\frac{\theta_1}{2}}\right)$$

- The target mass reached by the thermal wave or the shock front varies as

$$t.T_o^{\frac{\theta_1}{2}}$$

- The pressure in the termal or the shock front varies as

$$T_o^{\left(\theta_1 - \frac{m_1 - \frac{3\theta_1}{2}}{n_1 - 1}\right)}$$

In conclusion, we have determined the relations between the characteristic parameters of the equation of state and transport of energy which give a self similar solution for the propagation of a thermal wave into the target. On the other hand, we have studied the variation of the solution when we multiply the thermal conduction coefficient conduction by a constant factor. We have also shown, that the solution strongly varies with the form of the electron transport term (flux limited). The deep origin of these phenomena lies in the invariance properties of these equations of gas dynamics, of the equation of state and of the conduction transport regarding a particular group of Lie. The interest of this class of particular solution is connected with the fact that they are intermediate asymptotic solutions.

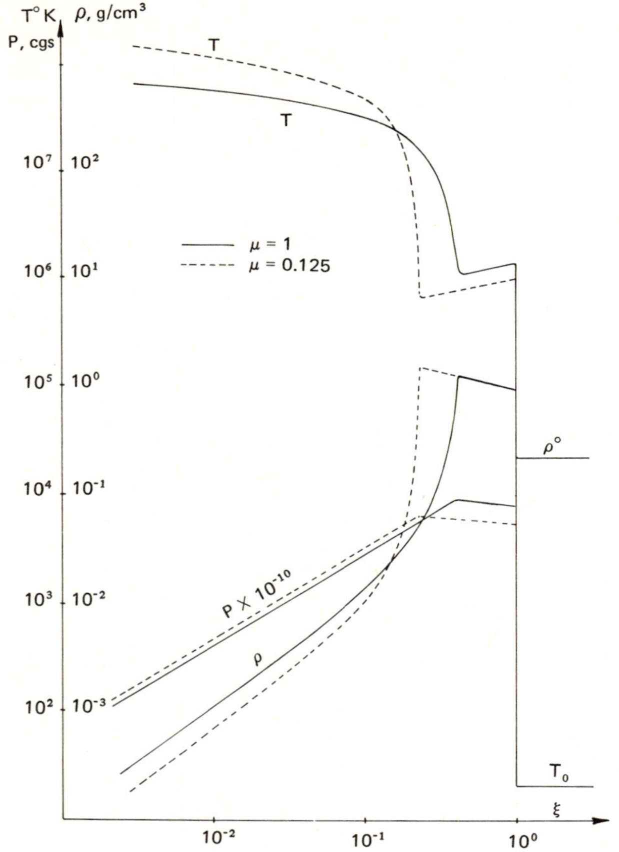

Figure 2 Distribution of temperature T, pressure P, density ρ for the interaction of a thermal wave with a target. (μ: flux limiting parameter)

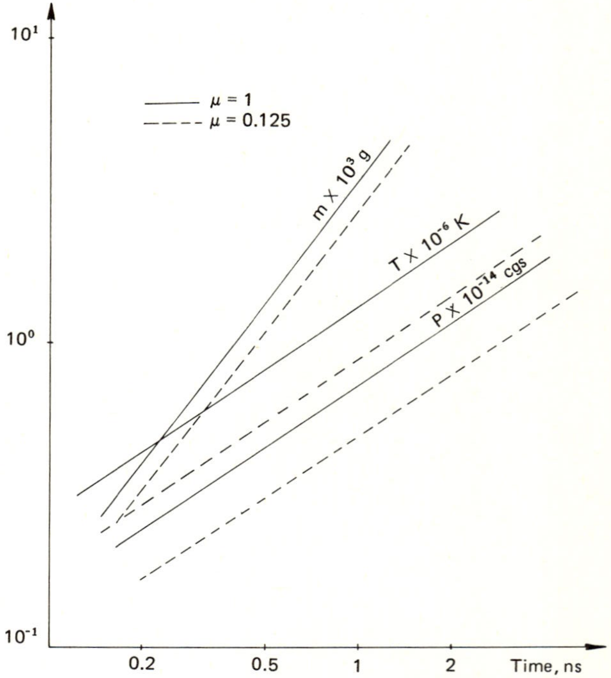

Figure 3 Variation of Pressure, Temperature, mass reached by the shock front for two values of μ.

V - IMPLOSION EXPERIMENTS AND NUMERICAL SIMULATION

As previously mentioned our numerical calculations can especially give account for one dimensionnal experiments at moderate fluxes. This situation is encountered in our four beam experiment IRIS on solid spheres of DLi or on microshell irradiated with our 4 beams C6 laser. Indeed, in order to obtain a good symmetry of the energy deposition we irradiate with an important defocussing [10] ; as a consequence the average incoming flux density is below 10^{13} W/cm^2 at 1.06 µm and bremsstrahlung absorption occurs.

Let us give some example of the comparative results between experiments and calculation restricted to hydrodynamic [11]. Studies of radiation (energy, spectrum) are in progress and will not be reported here.

On figures 4 and 5, we see a schematic drawing of the time evolution of the incident laser-power on the target. The irradiated empty microballoon has an outer diameter of 150 µm and a wall thickness of 1 µm. In order not to destroy the microballoon before the maximum power has been delivered, the maximum absorbed power was choosen as low as 5 GW, that is 7 J absorbed. On the figure below, we can see the time evolution of the radius r_c where the critical density is located. This is compared to the spatio-temporal evolution of the maximum emission of 2 ω_o radiation issueing from the plasma ; experimental and numerical calculations are in pretty good agreement. We can observe that the critical density region begins to move away and then moves toward the center of the target.

On the same figure 5 are drawn the time variation of the ablation front (r_A). Both r_A and r_c are minimum at the same time within experimental errors evaluated to be ± 100 ps.

We also plotted the time variation of the region where the maximum X-ray emission is observed through a 30 mg/cm^2 Beryllium foil. Those results are recorded with an X-ray streak camera [12].

Temporal resolution is about 50 ps or less and spatial resolution 20 µm. Because the incoming flux density is as low as 10^{13} W/cm^2, the X-ray emission is very low and does not allow us to use very small pinholes in front of the camera slit. We observe that the X-ray emitting region is located near the initial surface of the target at the beginning of the interaction ; the distance between r_c and r_A is less than 10 µm. On the contrary, near focussing time, the length of the conduction region increases in agreement with classical heat conduction hypothesis.

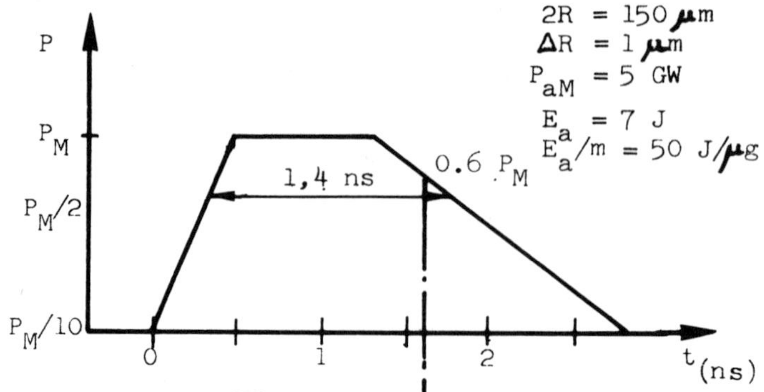

Figure 4

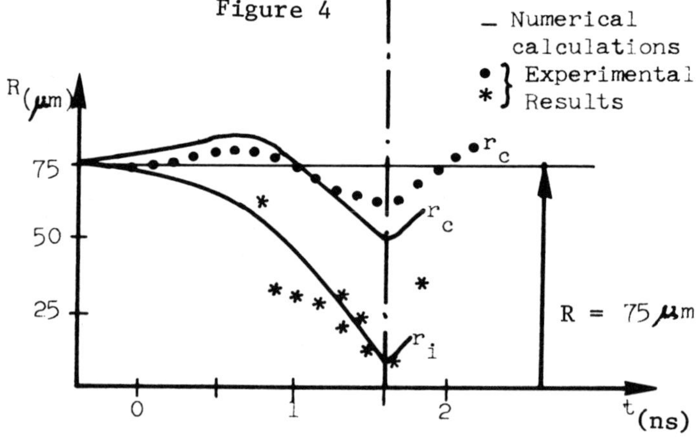

Figure 5 Time evolution:
- of the critical radius and of the ablation front deduded from calculations;
- of the critical radius and of X-ray lumination deduced from the experiments.

VI - IMPLOSION STABILITY

Fluid dynamic instabilities are likely to distort to a considerable extent the symmetry of an otherwise perfectly spherical process, and thus to preclude the attainment of the desired high compression rates. Therefore combined analytical and numerical studies are required to investigate the influence of these phenomena, particularly in complex situations where a number of interrelated physical mechanisms are involved.

VI - 1. DIFFERENT TYPES OF INSTABILITY

The different zones that can be distinguished in a typical implosion structure are all interested by instabilities, together with their mutual boundaries, whether these are high gradient layers or interfaces between different media.

VI.1.1. - Shock front instability. The instability of a shock front has been studied a number of times. Only the stationary plane shock seems to have received a general treatment[13,14]. In spherical geometry a converging self-similar shock in a perfect gas is known to be unstable in shape[15], but this clearly is not sufficient to evaluate the effective influence of the converging shock instability in the laser context, that is, in view of the coupling with the instabilities in the down stream flow structure.

VI.1.2. - Rayleigh-Taylor instability. To be rigorous, let us recall the statement of the standard Rayleigh-Taylor problem :

Consider two incompressible media of different density separated by a plane interface and subject to a normal gravity field in their rest frame : this static configuration is stable if the gravity is directed from the light to the heavy medium, and unstable in the opposite case.

This problem has been reexamined a number of times under less restrictive assumptions : non planar geometry, surface tension, viscosity, compressibility (still quite unsatisfactorily described[16]). Let us simply mention here an interesting extension of the standard problem, which is that of the stratified fluid, where the density jump is replaced by a finite gradient. The conclusions as regards stability, depending upon the relative orientation of the acceleration in the rest frame and the density gradient, are the same as before ; there simply appears a saturation of the growth rate for very short wavelengths[17].

In fact, the laser implosion situation is considerably more complex than any of these simplified models, since it includes all the above mentioned effects ; besides, the acceleration is strongly non stationary and non uniform.

VI.1.3. - Convective instabilities. In the widest sense, convective instabilities are those which take place in a stratified fluid at rest subject to an acceleration field opposite to the density gradient.

The particular case of an incompressible, non-conducting fluid is simply that of the finite density gradient Rayleigh-Taylor problem. With viscosity and heat conduction included, and when the stratification is due to thermal dilatation (or to compressibility) in the presence of a temperature gradient, the situation is that of the classical Bénard problem[18], where thermal conduction and viscosity are known to have a stabilizing effect.

This unstable situation of opposite rest frame acceleration and density gradient obviously corresponds, in the case of non stationary flow, to the case of locally opposite pressure and density gradients. In view of the analogy of this configuration with that of the thermal front region, the instability of the latter has been studied following an approach identical to that of the Bénard problem[19]. But because of the neglect of convection effects due to the propagation of the ablation front, the conclusions of such a study appear questionable.

VI.1.4. - Ablation front instability. Within the class of convective instabilities as meant in the extended sense just defined, the ablation front problem is worth mentioning apart, because it implies the combined influence of all the mechanisms successively introduced above, namely :

Acceleration, stratification, compressibility, thermal conduction, convection.

In view of this situation, it appears hazardous, if of any interest, to investigate separately the influence of each of the classical types of instability, since clearly none of them can be singled out as an independant process. In particular, the analogy between the Rayleigh-Taylor problem and the ablation front problem (or more generally any wave front problem) seems unfounded. Besides, in the laser implosion context, the modelling of such a front as a discontinuity front (i.e. deflagration front) seems questionable, because the acceleration changes sign inside the ablation "layer".

VI - 2. THEORETICAL MODEL

Apart from analytical studies aimed at investigating specific mechanisms in simplified situations, any realistic overall approach must obviously be based on numerical simulation studies.

To treat instability problems, two main types of approach have been followed in the litterature :

- one is the use of two-dimensional numerical codes, which have the great advantage of allowing the computation of non-linear effects [23] ;
- the other is the study of perturbations in their linear phase, where they can be decomposed into normal modes which are generally decoupled and can be studied independently [20,21,22] ; three-dimensional effects are therefore effectively taken into account.

Our model belongs to the second class. Besides the fact that it is lagrangian, the basic idea [24] consists in formulating the perturbations in a purely lagrangian fashion and using the same labelling for the unperturbed and for the perturbed flow. Introducing then for every fluid particle the classical perturbed displacement together with an asymptotic expansion of every flow quantity, one obtains first-order equations of motion which have a very simple form. These turn out to be further simplified if one takes as independant space variables the coordinates of the fluid particle in the zero-order motion. From the numerical point of view, this amounts to computing the first-order motion on a grid following the zero-order motion.

Now, by making use of the first-order momentum equations, one can prove that the transverse part of the perturbed displacement can be decomposed into :

- a solenoïdal component which follows an evolution independant of the rest of the motion, and which can be neglected since it does not generate instabilities (in the absence of viscosity),
- an irrotational component which is coupled exclusively through its potential to the perturbed radial motion and to the thermodynamics.

Then, considering the spherical case and introducing the classical decomposition of the first-order quantities into spherical harmonic (l, m) modes, one finds again the familiar m-degeneracy and l-decoupling. The final form of the equations is of no interest here, and can be found elsewhere [24].

VI - 3. NUMERICAL RESULTS

Starting from this scheme, we have developed a numerical code which uses classical finite difference techniques, and includes a particular treatment of interfaces, and a form of artificial viscosity which allows a correct computation of the perturbations of a shock which converges at the center and then bounces off.

This code can be coupled to the computation of a large class of zero-order spherical flows. In view of the number of physical mechanisms involved, we have found it necessary to clarify the analysis by separating, to a certain extent, the different phenomena.

3.1. - Pressure driven implosion of a perfect gas microballoon. We have been investigating first a purely mechanical case, which is that of a heavy ideal gas thin shell filled with DT imploded by applying some pressure time law (linear ramp and linear plus quadratic ramp) to its outer boundary. Such a simplified scheme is of interest because it can serve as a representative test case for a number of parameter studies.

Without going into details, we will limit ourselves to state some conclusions that can be drawn from a first series of computations, for low and intermediate values ($\lesssim 50$) of the l-number, and for purely geometrical initial defects of the shell [25]:

- the outer and inner shell surfaces and the converging shock are all closely coupled, and then all unstable ; the leading role is clearly played by the outer surface ;
- in particular, the instability of the inner interface in the decelerated phase, before and after being crossed by the outwardly propagating shock, does not change very much the general trend ;
- decreasing the aspect ratio from 41 to 21 introduces no significant difference ;
- increasing the acceleration of the shell by introducing some pulse-shaping increases the perturbation growth rates in the decelerated phase and thus does not seem to improve implosion symmetry.

3.2. - Laser driven implosion, non ideal equation of state. We have also been considering a somewhat less primitive but still not too complicated model :

- aluminium shell with a non ideal, two temperature equation of state [26] ;
- same DT fill as before, with Kerley equation of state [27] ;
- irradiation by laser light at 1.06 μm wavelength with incident power time law following a linear ramp.

The zero-order spherical flow equations include classical absorption by inverse bremsstrahlung, and a two-temperature fluid model with electron and ion thermal conduction, and flux limiting. The first-order model disregards the perturbed heat flux and coupling terms, so that the first-order energy equation for each species is adiabatic. Therefore, the stability problem is, here again, purely mechanical ; in particular, we deprive ourselves from the stabilizing influence of lateral conduction.

The conclusions in this case [25] are not different from those of the preceding one. The interface and the ablation front have practically identical growth rates, which implies that the behaviour of the latter is leading the overall instability process.

While investigating more deeply this coupling process over a wider range of implosion parameters and l values, we now have to go progressively to more general physical models and to investigate the influence of each additional mechanism (including non-classical effects).

REFERENCES

1. A. Salères, Synthesis of experimental results on absorption processes in nanosecond laser interaction at 1.06 μm and 0,13 μm, European Conference on Laser Interaction with Matter, Palaiseau, October, 18-22/1976

2. J.C. Slater, Phys. Rev. 51, 846 (1937)
T.L. Loucks, in "Augmented Plane Wave Method" (Benjamin, New York, 1967)

3. F. Perrot, "Application de la méthode APW self-consistante aux métaux alcalins - Equations d'état de Li, Na et K". Internal report, Commissariat à l'Energie Atomique, France (1970),
F. Perrot, Phys. Lett. 44, 15 (1973),
M. Valadon, "Calcul de l'équation d'état des solides à température nulle par la méthode APW. Application numérique au lithium, à l'aluminium et au cuivre". Internal report, Commissariat à l'Energie Atomique, France (1974)

4. H. Brooks, Suppl. del Nuovo Cimento, 7 (1958)
G.M. Gandel'man, JETP 16, 94 (1969)

5. F. Perrot, Phys. Stat. Sol. (b) 77, 517 (1976)

6. N.W. Ashcroft and D.C. Langreth, Phys. Rev.BV155, 682 (1967)
C. Friedli and N.W. Ashcroft, Phys. Rev. B 12, 5552 (1975)

7. F. Perrot, "Equation d'état du polyéthylène". Internal reports, Commissariat à l'Energie Atomique, France (1974-75)
"Equation d'état de l'aluminium". Internal report, Commissariat à l'Energie Atomique, France (1975)

8. G. Boissière, "Détermination d'une équation d'état du cuivre tenant compte des changements de phase". Internal report, Commissariat à l'Energie Atomique, France (1973)

9. D.A. Young and B.J. Alder, Phys. Rev. A3, 364 (1971)

10. D. Billon, D. Cognard, P.A. Holstein, J. Launspach, C. Patou, D. Schirmann, J.M. Reisse, "Laser driven implosion of microspheres", IXth International Conference of Quantum Electronic, Amsterdam, 14-18 June 1976

11. D. Billon, P.A. Holstein, J. Launspach, C. Patou, J.M. Reisse, D. Schirmann, "Laser driven implosion experiments at Limeil", International Workshop Conference at the Rensselaer Polytechnic Institute, 8-12 November 1976, Troy, N.Y. These Proceedings, pp. 503-534.

12. N. Fleurot, J.P. Gex, M. Lamy, C. Quinnessière, R. Sauneuf, "High time resolution X-ray streak camera with X-ray microscope" XIIth International Congress on High Speed Photography, 1 August 1976, Toronto

13. S.K. Aslanov, Soviet Physics - Doklady, 11, n° 7, 563-565 (1967)

14. G.W. Swan, G.R. Fowles, The Phys. of Fluids, 18, 1, 28-35 (1975)

15. D.S. Butler, A.R.D.E. Report B - 18/56 (1956)

16. L. Brun, CEA unpublished Report DO 121 (1974)

17. S. Chandrasekhar, Hydrodynamic and Hydromagnetic Stability (Clarendon Press, 1961) § 92

18. cf. Ref. 5, Chap. 2

19. K.A. Brueckner, S. Jorna, R. Janda, The Phys. of Fluids, 17, 8 1554-1559 (1974)

20. D.B. Henderson, R.L. Morse, Phys. Rev. Letters, 32, 7, 355-358 (1974) ; D.B. Henderson, R.L. Mc Crory, R.L. Morse, Phys; Rev. Lett., 33, 4, 205-208 (1974) ; G.S. Fraley et al., 5th I.A.E.A. Conf. on Plasma Physics and Controlled Nuclear Fusion Research, Tokyo, paper I.A.E.A. - CN-33/F5-5 (nov. 11-15, 1974)

21. J.N. Shiau, E.B. Goldman, C.I. Weng, Lab. of Laser Energetics, Univ. of Rochester Report n° 14 (1974) ; Physc. Rev. Letters, 32, 7, 352-355 (1974)

22. T. Yabe, K. Niu, I.P.P.J. (June 1975)

23. J.D. Lindl, W.C. Mead, Reports UCRL 75877 and 75879 ; Phys. Rev. Lett., 34, 20, 1273-1275 (1975) ; UCRL 77041 (1975)

24. L. Brun, B. Sitt, CEA unpublished Report DO 133, 1975

25. B. Sitt, N. Wilke, H. Barbry, (Annual Meeting of the A.P.S. Division of Plasma Physics, Nov. 1976) APS Bulletin 21, 9, 1102

26. F. Perrot, CEA unpublished Report DO 137, (1975)

27. G.I. Kerley, L.A. 4776 UC-34 (1972)

CORRECTED PENETRATION LENGTH OF ALPHAS FOR REHEAT CALCULATIONS*

P. S. Ray and H. Hora

Dept. of Theoretical Physics, The University of New South Wales

Kensington, NSW Australia

Abstract

The penetration length R of alphas, protons or other particles from nuclear reactions in high density plasmas is essential for the reaction yields in laser compressed plasmas, high energy particle beam interaction with dense plasmas and other reactions. The often used formula $R \sim T^{3/2}$ gets a remarkable correction for T between 10^1 and 10^3 eV, if the thermalization is described by a Fokker-Planck equation taking into account the elastic as well as inelastic scattering. The results derived from a collective interaction similar to Bagge's generalization of the Bethe-Bloch-formula, differ again from the mentioned former values despite the corrections discussed. The values of the collective model may have the highest probability as they agreed for fast electrons with the only direct measurements available.

I INTRODUCTION

In high temperature plasmas for thermonuclear reactions, the penetration depths of the charged nuclear reaction products are of essential importance for the energy balance of the plasma. This self-generated heating of the plasma is called reheat and plays an important role e.g. for laser compressed plasmas[1] to reach self-ignition[2] or laser induced self-sustained nuclear combustion fronts[3].

The usual method for calculating the penetration depth of protons, deuterons or helium nuclei of MeV initial energy in plasmas uses Coulomb scattering for evaluations in a Boltzmann or in a Fokker-Planck equation[4] where the low angle scattering is to be considered. The numerical evaluation of these mechanisms lead to a penetration depth R in dependence on the temperature T for the MeV

*Presented at the Fourth International Workshop Conference on "Laser Interaction and Related Plasma Phenomena" at Rensselaer Polytechnic Institute, Troy, New York, 8-12 November, 1976.

alphas of a DT reaction in plasma of solid state density[5]

$$R \sim T^{3/2} \tag{1}$$

where R is in cms and T in eV. This result was used for the numerical calculations for laser compression and ignition[2] as the next available relation though the authors involved were aware that the basic theoretical assumptions may be insufficient. The problem of other nuclear reactions taking place with these energetic particles has been expressed in some general considerations by some authors[6].

The problem of reheat is also of essential importance for the design of magnetically confined toroidal configurations, as Tokamaks, where a detailed calculation of the temporal development of the distribution function of the generated alphas from the Fokker-Planck equation illustrate the processes involved[7]. This development is very complex as discussed by some[8] – but it is difficult to draw any definite conclusion from this regarding the penetration depth.

Generally speaking, there are basic problems of the Coulomb scattering involved apart from collective effects of plasmas, which even may be dominant at high densities. But alone the Coulomb processes are connected with the complex question of wide angle and low angle scattering. The wide angle scattering can be assumed as sufficient to calculate collision cross-sections of electrons or other mean free paths at thermal equilibrium[9] in full agreement with the knowledge of resistivity in plasmas (or metals). This agreement holds even for very high frequency resistivity, as e.g. in the calculation of the optical constants based on Spitzer's assumption for the collision[1,10] as can be proved by the agreement with the quantum mechanical calculations of the inverse Bremstrahlung[11]. This agreement even holds for the nonlinear deviations of the optical constants for very high laser intensities [1,12,13,14]. Deviations are within factors of the order ±5, given by the ratio of the Coulomb logarithm of the plasma case to the electrodynamic interaction logarithm of the quantum mechanical case. It has to be noted that for superdense plasma no quantum theory has been presented and the plasma values are based on the collision processes, on the one side and on a phase mixing treatment of the Vlasov equation[15] on the other, both in some agreement.

While for the cases of wide angle scattering, the inelasticity of the Coulomb scattering may not be essential,the slowing down process of energetic heavy particles in a plasma has to include the inelastic contribution for a theoretically satisfactory point of view. The low angle scattering may not be the most effective slowing down process, as can be seen very globally; the penetration depths of particles at temperatures above 10 keV can be 1000 times larger from Eq. (1), than the penetration in cold, unionised materials, as known from the Bethe-Bloch formula[16,17]. This difference did not disturb much, because there were as good as no measurements available to see the differences.

One measurement with relativistic 2 MeV electrons[18], however, showed a discrepancy generally. The currents of 10^5 Amperes of electron beams hitting solid CD_2 - targets transformed these immediately into plasma and it was very remarkable that the penetration depths dropped by a factor of about 20 below the value of unionised cold solid targets. All artificial attempts to explain the measured low penetration of the electrons by the usual models of long lengths with turning of paths could neither arrive at the low values nor could such models[19], be considered as a realistic mechanism.

One can formulate a basically different description of the penetration depth of alphas by including "collective" effects, at least for plasma densities around and above the solid state. The way to this model was given by Bagge by a generalisation of the Bethe-Bloch formula[20] where the averaged binding energy of 14.6 eV of the atoms had to be substituted by the electrostatic energy e^2/λ_D (λ_D Debye length) which we are now able to explain as a plasma binding energy. Bagge calculated by this way drastic changes of the behaviour of cosmic ray electrons in surprising consistency with observations[20]. It was further possible[21] to calculate the measured reduced penetration of the relativistic electrons in high density plasmas directly without any complicated assumptions.

This result encouraged us to study the case of alphas where, however, we did not use the Bethe-Bloch formula but the usual concept of Fermi[22]. We arrived at also in this case penetration depths for high temperature plasma of solid state density values less by a factor 10 to 20 below the range of alphas in cold solids[23,24]. Apart from this treatment, we checked whether the concept of the low angle scattering could be altered by including the properties of inelasticity by including the emitted photons. The suppression of the infra-red catastrophe is possible again by using a "Bagge frequency" in analogy to the mentioned modification of the Bethe-Bloch formula. We also tried a suggestion of Pauli[25], but the penetration depth obtained were all very different. We are now comparing the different models and will draw conclusions for the possible numerical results.

II THE FOKKER-PLANCK FORMALISM

The Fokker-Planck (F.P.) formalism describes statistically the interaction of a system of different species of particles. Let subscripts s, t, . . . denote the physical entities corresponding to the species e.g. $f_s(\vec{r}, \vec{v}, t)$ the distribution function for species s in a collection. Its evolution is governed by the F.P. equation

$$\frac{\partial f_s}{\partial t} + v_i \frac{\partial f_s}{\partial x_i} + \frac{F_i}{m_s} \frac{\partial f_s}{\partial v_i} = \sum_t \sum_{n=1}^{\infty} \frac{(-1)^n}{n!} D^{(n)} (\alpha^{ts}_{(n)} f_s) \qquad (2)$$

Here the sum t extends over all species, F_i the component of the external force ($i = 1, 2, 3$) and we use the notation

$$D^{(n)} \equiv \frac{\partial^n}{\partial v_{i_1} \partial v_{i_2} \ldots \partial v_{i_n}} \quad (\text{each } i_n = 1,2,3)$$

$$\alpha^{ts}_{(n)} = \alpha^{ts} (v_{i_1} v_{i_2} \ldots v_{i_n}) \qquad (3)$$

The α's are the F.P. coefficients. This form of the equation was used by Kramers[26].

We note that there are two different ways of regarding the right side of Eq. (2). In the intuitive picture it represents an approximation to the Boltzmann collision integral. This view is taken by Rosenbluth et. al.[4] who have calculated the F. P. coefficients on the basis of Rutherford scattering. In the other point of view one regards (2) as a description of a stochastic process described by the Chapman-Kolmogorov equation. The right side is then simply a Taylor's expansion. We note that there is no assumption for thermodynamic equilibrium for the validity of (2). We shall denote the heavy particle by suffix H and the electrons by e. Because of the enormous mass difference of the electrons and ions one can neglect the ionic effects for processes of energy exchanges. Let $\psi(\vec{v})$ denote a function of velocity $\vec{v}$ only. Its average $\overline{\psi}$ with respect to the distribution f is then defined as

$$\overline{\psi} = \frac{\int \psi f d\vec{v}}{\int f d\vec{v}} = \frac{1}{n} \int \psi f d\vec{v} \tag{4}$$

$$(d\vec{v} \equiv dv_1 \; dv_2 \; dv_3)$$

where $n = n(\vec{r}, t) = \int f d\vec{v}$ is the local density.

Multiplying (2) by $\psi(\vec{v})$ and by integration one obtains

$$\frac{\partial}{\partial t} (n_H \overline{\psi}^{(H)}) + \frac{\partial}{\partial x_i} (n_H \overline{\psi v_i}^{(H)}) = \frac{1}{m_H} n_H \overline{\frac{\partial (\psi F_i)}{\partial v_i}}^{(H)}$$

$$+ \sum_{m=1}^{\infty} \frac{(-1)^m}{m!} \int \psi \, D^{(m)} (\alpha^{He}_{(m)} f_H) \, d\vec{v} \tag{5}$$

where averaging is taken with respect to the heavy particle distribution function. Here $\alpha^{He}_{(m)}$ denotes the F. P. coefficients of order m describing interaction of electrons and the heavy particle. By partial integration one can write the above as

$$\int \psi \, D^{(m)} (\alpha^{He}_{(m)} f_H) \, d\vec{v} = (-1)^m \int \alpha^{He}_{(m)} D^{(m)} \psi f_H d\vec{v}$$

$$= (-1)^m n_H \overline{\alpha^{He}_{(m)} \; D^{(m)} \psi}^{(H)}$$

In the absence of external force one can then obtain

$$\frac{\partial}{\partial t}\left(n_H \overline{\psi}^{(H)}\right) + \frac{\partial}{\partial x_i}\left(n_H \overline{\psi v_i}^{(H)}\right) = \sum_{m=1}^{\infty} \frac{1}{m!} n_H \overline{\alpha^{He}_{(m)} D^{(m)} \psi}^{(H)} \tag{6}$$

where $\alpha^{He}_{m} D^{(m)} \psi = \alpha^{He}(v_{i_1}\ v_{i_2} \ldots v_{i_m}) \dfrac{\partial^m \psi}{\partial v_{i_1}\, \partial v_{i_2} \ldots \partial v_{i_m}}$

Now setting $\psi(\vec{v}) = \frac{1}{2} m_H \sum_i v_i^2$ one gets from (6) since all derivatives of ψ higher than second order are zero

$$\frac{1}{2}\frac{\partial}{\partial t}\left(n_H\, m_H\, \overline{v^2}^{(H)}\right) + \frac{1}{2}\frac{\partial}{\partial x_i}\left(n_H\, m_H\, \overline{v^2\ v_i}^{(H)}\right)$$

$$= n_H\, m_H\, \left(\overline{a_i\ v_i}^{(H)} + \frac{1}{2}\overline{b_{ij}\ \delta_{ij}}^{(H)}\right) \tag{7}$$

with $a_i = \alpha(v_i)$ and $b_{ij} = \alpha(v_i,v_j)$ are the first two F. P. coefficients. Eq. (7) is an exact relation. To pick a particular particle with coordinate and velocity $(\vec{r}_H, \vec{v}_H)$ we set

$$f_H = \delta(\vec{r} - \vec{r}_H(t))\ \delta(\vec{v} - \vec{v}_H(t))$$

giving $\quad n_H = \delta(\vec{r} - \vec{r}_H(t))$

$$\overline{\psi(\vec{v})}^{(H)} = \psi(\vec{v}_H)$$

One then obtains after integration with respect to $\vec{r}$ and $\vec{v}$

$$\frac{dE_H}{dt} = m_H \left(\sum_i a_i\ v_{H,i} + \frac{1}{2}\sum_i b_{ii}\right) \tag{8}$$

This is the fundamental relation expressing the energy-loss by the F. P. coefficients. Rosenbluth et.al.[4] have calculated the latter when only elastic scattering by Coulomb interaction is taken into account. This result is mostly utilised in plasma physics. One can also use the time correlation technique[2,7] to calculate these. However, both results are the same. These give for the F. P. coefficients the following values

$$a_i = \Gamma \frac{\partial}{\partial v_i} h(\vec{v})$$

$$b_{ij} = \frac{\partial^2}{\partial v_i \, \partial v_j} g(\vec{v}) \qquad (9)$$

where

$$h(\vec{v}) = (1 + \frac{m_H}{m_e}) \int d\,\vec{v}'\, f_e(\vec{v}')\, \frac{1}{|\vec{v} - \vec{v}'|}$$

$$g(\vec{v}) = \int d\,\vec{v}'\, f_e(\vec{v}')\, |\vec{v} - \vec{v}'|$$

$$\Gamma = \frac{4\pi\, Z^2\, e^4}{m_H^2} \ln \Lambda$$

Here Z is the charge of the heavy particle and Λ represents the Coulomb logarithm

$$\Lambda = \frac{3}{2}\, kT\, \frac{\lambda_D}{e^2} = \frac{3}{4} \sqrt{\left(\frac{kT}{e^2}\right)^3 \frac{1}{\pi N_e}}$$

where N_e is the electron density and kT the plasma temperature.

We note that the origin of this term is the divergence at $\theta = 0$, where θ is the scattering angle, of the classical cross-section

$$\left(\frac{d\sigma}{d\Omega}\right)_c = \frac{Z^2\, e^4}{4\, m_r^2\, u^4} \operatorname{cosec}^4 \theta/2 \qquad (10)$$

with m_r denoting the reduced mass and u the relative velocity. To get over this, one is forced to introduce some cut-off $\theta = \theta_{min}$. To estimate θ_{min} one assumes on an intuitive basis that small angle scattering corresponds to large impact parameter and as such one can employ the Debye shielding in plasma. However, this matter has always been of weak foundation in plasma physics. In the following we derive the expression for the range basing on this formalism. For the electron distribution function $f_e(\vec{v})$ we use the Maxwell-Boltzmann distribution

$$f_e(\vec{v}) = N_e \left(\frac{\gamma}{\pi}\right)^{3/2} e^{-\gamma v^2} \qquad \text{with } \gamma = \frac{me}{2kT}$$

One then obtains

$$\frac{1}{2} \sum_i b_{ii} = \frac{1}{2} \Gamma \sum_i \frac{\partial^2 g}{\partial v_i^2}$$

$$= \Gamma\, N_e \left(\frac{\gamma}{\pi}\right)^{3/2} I$$

where
$$I = \int d\vec{v}'\, e^{-\gamma v'^2} \frac{1}{|\vec{v} - \vec{v}'|}$$

$$= \left(\frac{\pi}{\gamma}\right)^{3/2} \frac{1}{v}\, \mathrm{Erf}\,(\gamma^{\frac{1}{2}} v)$$

and $\mathrm{Erf}\,(y) = \frac{2}{\sqrt{\pi}} \int_0^y e^{-t^2}\, dt$ the usual error function.

Similarly
$$\Sigma\, a_i v_i = \Gamma\, \Sigma\, v_i \frac{\partial}{\partial v_i} h(\vec{v})$$

$$= \Gamma\, (1 + \frac{m_H}{m_e})\, N_e \left(\frac{\gamma}{\pi}\right)^{3/2} v \frac{\partial I}{\partial v}$$

Utilising $m_H >> m_e$ one obtains

$$\frac{dE_H}{dt} = \frac{4\pi e^4 Z^2}{m_e} \ln\Lambda\, N_e \left(\frac{2}{\sqrt{\pi}} \sqrt{\frac{m_e}{2kT}}\, e^{-\frac{m_e v^2}{2kT}} - \frac{1}{v} \mathrm{Erf}\left(\sqrt{\frac{m_e}{2kT}}\, v\right)\right) \tag{11}$$

Now the range is defined as the distance travelled until thermal energy $E_{th} = 3/2\ kT$ is reached. Thus if E_o is the inital energy of the heavy particle the range R is given by

$$R = \int_{E_o}^{E_{th}} \frac{dE}{-(\frac{dE}{dx})} = \int_{E_{th}}^{E_o} \frac{v dE}{\frac{dE}{dt}}$$

$$= \frac{m_H}{m_e} \left(\frac{kT}{e^2}\right)^2 \frac{1}{N_e \cdot \ln\Lambda \cdot \pi Z^2} \int_{x_1}^{x_2} \frac{x^3 dx}{f(x)} \tag{12}$$

where $f(x) = \mathrm{Erf}\,(x) - \frac{2}{\sqrt{\pi}}\, x e^{-x^2}$

and

$$x_1 = \sqrt{\frac{3}{2} \cdot \frac{m_e}{m_H}} \qquad x_2 = \sqrt{\frac{E_o}{kT} \cdot \frac{m_e}{m_H}}$$

As kT increases $x_2 \to x_1$ so that the integral in (12) decreases. However, the factor $(kT)^2$ increases much faster giving a range which increases with temperature. The num erical values of this is discussed below.

III PHOTON EMISSION IN SCATTERING

The above result is based solely on the elastic scattering. However, this is not entirely physically correct. As has been very explicitly emphasized by Feynman[28] in the actual process there is always some photon emissions and this should be taken into account. This is represented by the following Feynman diagrams (Fig. 1)

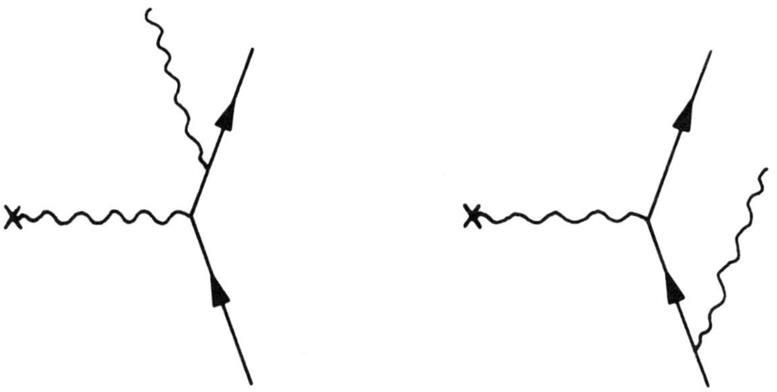

Fig. 1
Feynman diagram showing photon emission in electron-scattering in external field represented by cross. There are two such diagrams contributing to the process.

The cross section for this process was calculated by Schwinger[29] in a relativistic way for the first time. It can be calculated in a simpler way by Feynman rules[30]. In the non-relativistic limit this is given by

$$\left(\frac{d\sigma}{d\Omega}\right)_{ph} = \left(\frac{d\sigma}{d\Omega}\right)_{c} \frac{2\alpha}{\pi} \frac{dk}{k} \frac{4}{3} \beta^2 \sin^2\theta/2 \qquad (13)$$

where α is the fine-structure constant (α = 1/137) and β = u/c. The factor dk/k means that the photon of momentum between k and k + dk is emitted. With respect to the photon frequence ω : dk/k = dω/ω. To find the contributing cross-section one has to integrate over all frequencies emitted say from ω_{min} to ω_{max} giving

$$\left(\frac{d\sigma}{d\Omega}\right)_{ph} = \frac{2\alpha}{3\pi c^2} \frac{Z^2e^4}{m_e^2} \ln\left(\frac{\omega_{max}}{\omega_{min}}\right) \frac{1}{u^2\sin^2\theta/2} \qquad (14)$$

One notes that if $\omega_{min} \to 0$ then dσ/d$\Omega \to \infty$ which is the famous "infra-red" divergence of quantum electrodynamics. Similarly the divergence associated with $\omega_{max} \to \infty$ is the ultra violet catastrophe. One has to use again some cut-off values to get over this difficulty. For ω_{min} we may choose the "Bagge frequency" ω_B defined as $e^2/\lambda_D = h\omega_B$. Bagge[31] first utilised this for the penetration of electrons in cosmic rays as mentioned above. For ω_{max} we utilise the instantaneous energy of the heavy particle: $E = \hbar\omega_{max}$. One thereby obtains

$$\left(\frac{d\sigma}{d\Omega}\right)_{ph} = \frac{2\alpha}{3\pi e^2} \frac{Z^2e^4}{m_e^2} \ln K \frac{1}{u^2 \sin^2\theta/2} \qquad (15)$$

where

$$K = \frac{E}{e^2} \sqrt{\frac{kT}{4\pi N_e e^2}}$$

To incorporate this process one may use the sum of the two differential cross-sections;

$$\frac{d\sigma}{d\Omega} = \left(\frac{d\sigma}{d\Omega}\right)_{ph} + \left(\frac{d\sigma}{d\Omega}\right)_{c} \qquad (16)$$

This will lead to additional terms for the F.P. coefficients a_i's and b_{ij}'s. We work out these in details now. We have

$$a_i = \int d\vec{v}'\, f_e(\vec{v}')\, \{\Delta v_i\}_{He}$$

$$b_{ij} = \int d\vec{v}'\, f_e(\vec{v}')\, \{\Delta v_i\, \Delta v_j\}_{He}$$

Here $\{\Delta\vec{v}\}$ denotes the change in velocity per unit time due to collision. Suppose $\vec{v}$ and $\vec{v}'$ denote the velocity of the heavy particle and the electron before collision.

Then their relative velocity $\vec{u}$ and centre of motion velocity $\vec{V}$ are given by:

$$\vec{u} = \vec{v} - \vec{v}'$$

$$\vec{v} = \vec{V} + \frac{m_e}{M}\vec{u}$$

$$\vec{v}' = \vec{V} - \frac{m_H}{M}\vec{u}$$

$$M = m_e + m_H$$

$$M\vec{V} = m_H\vec{v} + m_e\vec{v}'$$

Now the change in $\vec{u}$ by collision, denoted $\Delta\vec{u}$, can be obtained by considering the figure (Fig. 2)

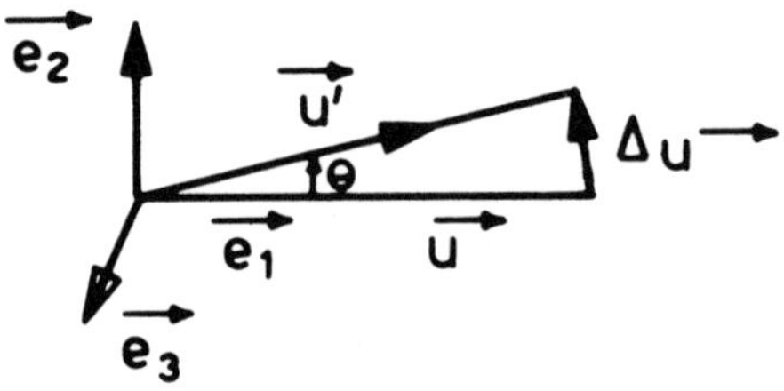

Fig. 2
$\vec{e}_1$, $\vec{e}_2$ and $\vec{e}_3$ local axes system, θ scattering angle, $\Delta\vec{u}$ change in $\vec{u}$ by scattering.

$$|\Delta\vec{u}| = 2u \sin \theta/2$$

$$\Delta\vec{u} = |\Delta\vec{u}|\,[-\vec{e}_1 \sin\theta/2 + (\vec{e}_2 \cos\phi + \vec{e}_3 \sin\phi)\cos\theta/2]$$

Then
$$\Delta\vec{v} = \frac{m_e}{M}\Delta\vec{u} \simeq \frac{m_e}{m_H}\Delta\vec{u}$$

$$\Delta v_i\ \Delta v_j = \left(\frac{m_e}{m_H}\right)^2 \Delta u_i\ \Delta u_j$$

One obtains

$$\{\Delta\vec{v}\} = \frac{m_e}{m_H} \int d\Omega \frac{d\sigma}{d\Omega} u \Delta\vec{u}$$

$$\{\Delta v_i \Delta v_j\} = \left(\frac{m_e}{m_H}\right)^2 \int d\Omega \frac{d\sigma}{d\Omega} u \Delta u_i \Delta u_j$$

Here $d\Omega = \sin\theta \, d\theta d\phi$ and if one substitutes the new cross-section (15) one finds that it is not necessary to introduce a small angle cut-off θ_{min}. Thus the Coulomb logarithm does not appear in this case. However, there is another logarithmic term due to the divergences of quantum electrodynamics. After simplifying the algebra one obtains

$$\{\Delta\vec{u}\} = -\frac{m_e}{m_H} \frac{16\alpha Z^2 e^4}{3m_e^2 c^2} \ln K \, \vec{e}_1$$

$$\vec{a} = \int d\vec{v}' \, f_e(\vec{v}') \, \{\Delta\vec{v}\}$$

$$= -\frac{m_e}{m_H} \frac{16\alpha Z^2 e^4}{3m_e^2 c^2} \ln K \int d\vec{v}' f_e(\vec{v}') \frac{\vec{v}-\vec{v}'}{|\vec{v}-\vec{v}'|}$$

Since $\dfrac{\vec{v}-\vec{v}'}{|\vec{v}-\vec{v}'|} = \dfrac{\partial}{\partial\vec{v}} |\vec{v} - \vec{v}'|$ we can write

$$\vec{a} = -\frac{m_e}{m_H} \frac{16\alpha Z^2 e^4}{3m_e^2 c^2} \ln K \, N_e \left(\frac{\gamma}{\pi}\right)^{3/2} \frac{\partial I}{\partial\vec{v}}$$

where

$$I = \int_{-\infty}^{+\infty} d\vec{v}' \, e^{-\gamma v'^2} |\vec{v} - \vec{v}'|$$

$$= \frac{\pi}{\gamma^{3/2}} \frac{1}{v} \Psi(v)$$

$$\Psi(v) = \frac{v}{\sqrt{\gamma}} (e^{-\gamma v^2} - 1) + \sqrt{\pi} \, \mathrm{Erf}(\sqrt{\gamma} v)\left(v^2 + \frac{1}{2\gamma}\right)$$

Thus

$$\vec{a}\cdot\vec{v} = \frac{m_e}{m_H}\;\frac{16\alpha Z^2 e^4}{3m_e^2 c^2}\;\ln K\;\left(\frac{\gamma}{\pi}\right)^{3/2} N_e\left[I - \frac{\pi}{\gamma^{3/2}}\;\Psi'(v)\right]$$

Similarly $\frac{1}{2}\sum_i b_{ii} = \frac{1}{2}\sum_i \int d\vec{v}'\; f_e(\vec{v}')\;\{\Delta v_i\;\Delta v_i\}$

$$= \left(\frac{m_e}{m_H}\right)^2 \frac{16\alpha Z^2 e^4}{3m_e{}^2 c^2}\;\ln K\; N_e \left(\frac{\gamma}{\pi}\right)^{3/2} I$$

giving $\frac{1}{2}\sum_i b_{ii} \ll \vec{a}\cdot\vec{v}$.

For the rate of energy loss one therefore obtains

$$\frac{dE}{dt} = \frac{4\pi Z^2 e^4}{m_e}\; N_e\; \ln\Lambda\; \frac{1}{v}\left[\frac{2}{\sqrt{\pi}}\sqrt{\frac{m_e}{2kT}}\; v\; \exp\left(-\frac{m_e v^2}{2kT}\right) - \mathrm{Erf}\left(\sqrt{\frac{m_e}{2kT}}\; v\right)\right]$$

$$+ \frac{16\alpha Z^2 e^4}{3\sqrt{\pi} m_e c^2}\; \ln K\; N_e\; \frac{1}{v}\;\frac{1}{\gamma}\; \left[\sqrt{\pi}\;\mathrm{Erf}\left(\sqrt{\frac{m_e}{2kT}}\; v\right) \times \left(-\frac{m_e v^2}{2kT} + \frac{1}{2}\right) - \sqrt{\frac{m_e}{2kT}}\; v\; \exp\left(-\frac{m_e v^2}{2kT}\right)\right]$$

One can then write the following expression for range

$$R = \frac{m_H}{m_e}\;\frac{m_e\; c^2\; kT}{N_e\; Z^2 e^4 \pi \ln\Lambda} \int_{x_1}^{x_2} \frac{dx}{\frac{2m_e c^2}{kT}\left[\frac{\mathrm{Erf}(\sqrt{x})}{x} - \frac{2e^{-x}}{\sqrt{\pi x}}\right] + \frac{16\alpha}{3} \times \frac{\ln K}{\ln\Lambda} \times \left\{\mathrm{Erf}(\sqrt{x})\left(-1 + \frac{1}{2x}\right) - \frac{e^{-x}}{\sqrt{\pi x}}\right\}} \qquad (17)$$

with $\quad x_1 = \frac{3}{2}\;\frac{m_e}{m_H} \quad$ and $\quad x_2 = \frac{m_e}{m_H}\;\frac{E_o}{kT}$

If kT is in electron volts one obtains in cgs units

$$R = \frac{m_H}{m_e} \frac{10^6 \times (6.94\times10^6)^2}{Z^2 \; N_e \; 2\pi \; \ln\Lambda} (kT) \int_{x_1}^{x_2} \frac{dx}{\frac{10^6}{kT}\left[\frac{\mathrm{Erf}(\sqrt{x})}{x} - \frac{2\exp(-x)}{\sqrt{\pi x}}\right] + \frac{16\alpha}{3\pi}\frac{\ln K}{\ln\Lambda}\left\{\mathrm{Erf}(\sqrt{x})\left(-1 + \frac{1}{2x}\right) - \frac{\exp(-x)}{\sqrt{\pi x}}\right\}} \tag{18}$$

where $\ln K = \ln \left(\frac{m_H}{m_e} \frac{x}{\sqrt{4\pi N_e}} \left(\frac{kT}{e^2}\right)^{3/2} \right)$

In quantum electrodynamics one has another conjecture of Pauli[25] which seems to be supported by Landau[31] to use the cut-off

$$\log \frac{\omega_{max}}{\omega_{min}} \sim \frac{1}{2\alpha} = \frac{137}{2}$$

In this case lnK = 68.5. The numerical results are discussed below.

IV COLLECTIVE INTERACTION MODEL

As discussed above the usual F.P. formalism is plagued with the problem of cut-off - be it the small angle scattering or the divergences of quantum electrodynamics. And the range will depend very significantly on this. However, one can consider this problem under the aspect of a "collective" effect as formulated by Gasiorowicz et.al.[32]. One considers here the interaction in basically two parts i.e. one with the immediate neighbourhood where two body type collisions are important and the other with the rest of the plasma. The presence of the energetic particle causes a perturbation of the distribution function of the plasma electrons, some sort of a "polarisation", which then acts back on the particle. Now one can divide the total interaction in two parts - one with the immediate neighbourhood and other with the rest of the plasma. To define the former, one may use the Debye radius λ_D as a cut-off factor within which binary type collisions are important. The component of the force in the direction of motion F_v due to the polarisation effect as calculated by Gasiorowicz et.al.[32] is given by

$$F_v = \frac{2e^2}{m} \log(k_{max}^2 \; \lambda_D^2) \int_{-1}^{+1} d\mu \; \mu \; J(\mu, v) \tag{19}$$

where the function J is defined in their paper. Here k_{max} is a cut-off which is an inverse of a length denoting dynamic correlations of two particles when they are close together. If one uses $k_{max} \sim 1/\lambda_D$ one sees that $F_v \approx 0$. One can use a two-body picture of the energy transferred to electrons within the Debye sphere as given by Fermi . An electron e situated as shown in figure at a distance r with respect to the heavy particle Ze will in general experience a nett force perpendicular to the direction of motion of Ze. If b denotes the impact parameter the average force is

$$\frac{Ze^2}{r^2}\cos\theta = Ze^2\frac{b}{r^3}$$

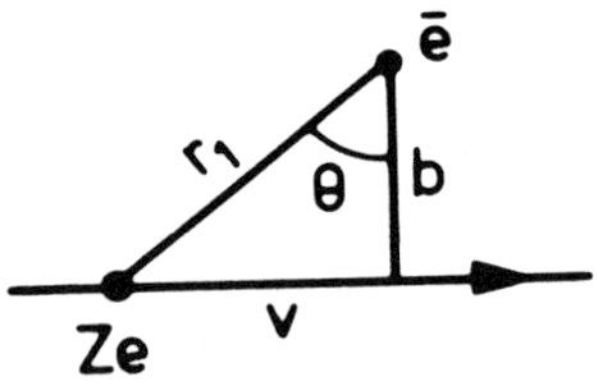

Fig. 3
Ze heavy particle, e electron at an impact parameter b from it.

If it acts for time Δt the gain in momentum by the electron is then

$$\Delta p = Ze^2\frac{b}{r^3}\Delta t = \frac{Ze^2 b}{vr^3}\Delta x$$

where v is the velocity of Ze ($\Delta t/\Delta x = 1/v$) and Δx the distance travelled by it in Δt. The total momentum Δp gained by the electron is then

$$(\Delta p)_{total} = \frac{Ze^2 b}{v}\int_{-\infty}^{+\infty}\frac{dx}{r^3}$$

$$= 2\frac{Ze^2}{bv}$$

If ΔE denote the total energy gain

$$\Delta E = \frac{(\Delta p)^2_{total}}{2\, m_e} = \frac{2Z^2e^4}{b^2v^2m_e}$$

Now the number of electrons situated per path Δx of the heavy particle between impact parameter range b and b + db is $2\pi\, N_e b\, db\, \Delta x$. Thus the total energy loss to the electrons per path Δx is

$$-\Delta E = \int_{b_{min}}^{b_{max}} \frac{4\pi Z^2 e^4}{b^2 v^2 m_e} N_e\ bdb\ \Delta x$$

$$-\frac{dE}{dx} = \frac{4\pi Z^2 e^4}{m_e v^2} \log \frac{b_{max}}{b_{min}}$$

If one uses for minimum and maximum values of the impact parameter the following values

$$b_{min} = \frac{Ze^2}{m_e\ v^2}$$

which is the classical value and $b_{max} = \lambda_D = \sqrt{kt/4\pi N_e e^2}$ the Debye radius as mentioned above one obtains

$$-\frac{dE}{dx} = 2\pi \frac{Z^2 e^4}{m_e\ v^2} N_e \log \left(\frac{kT\ m_e{}^2 v^4}{4\pi N_e\ e^6\ Z^2}\right)$$

which is similar to the Bethe-Bloch formula. Now $v^2 = 2E/m_H$ so that one can write

$$-\frac{dE}{dx} = \pi \frac{m_H\ Z^2\ e^4}{m_e\ E} N_e \log \left(\frac{kT\ m_e{}^2\ E^2}{\pi\ N_e\ Z^2\ m_H^2\ e^6}\right)$$

Using the integral logarithm $Ei(x) = \int_{-\infty}^{x} \frac{e^t}{t} dt$ one can express the range R as

$$R = \int_0^{E_0} \frac{dE}{-\frac{dE}{dx}} = \frac{e^2}{2kT} \frac{m_H}{m_e} Ei\ (\log\ (x\ E_0{}^2)) \qquad (20)$$

where

$$x = \frac{kT\ m_e^{\ 2}}{\pi\ m_H^2\ Z^2\ N_e\ e^6}$$

One criterion to be satisfied is that $b_{max} >> b_{min}$ which is equivalent to

$$kT >> 2.413 \times 10^{-25}\ N_e \tag{21}$$

where kT is in electron volts and N_e is in cm^{-3}.

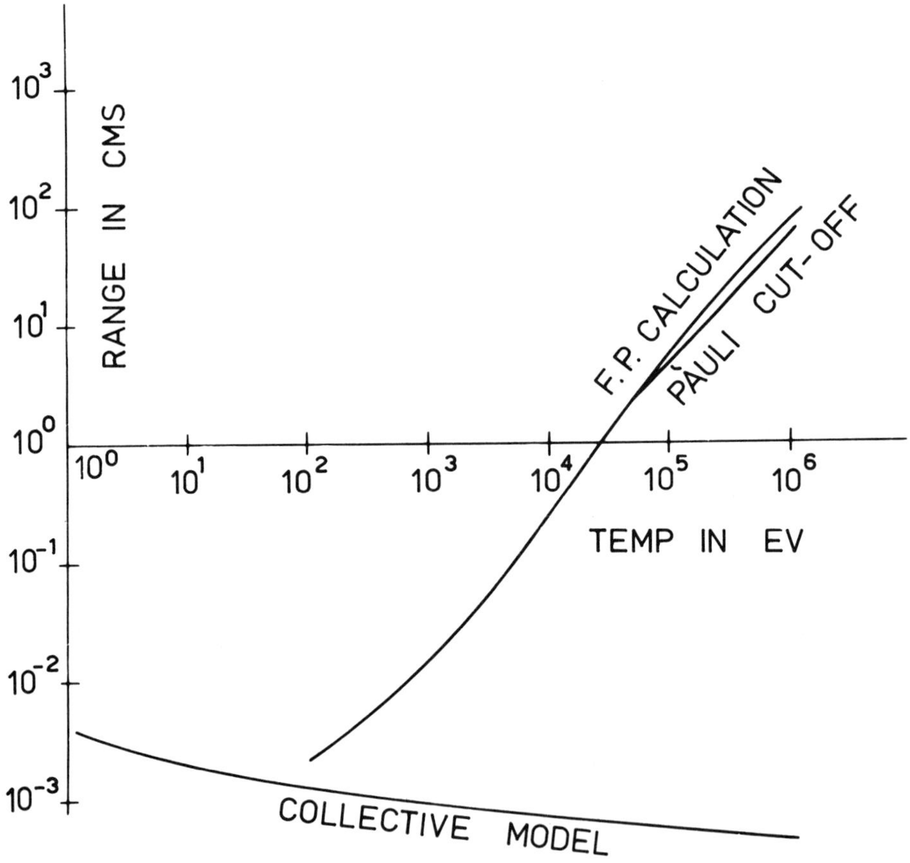

Fig. 4
The range of 2.89 MeV alpha in HB(11) plasma of solid state density.

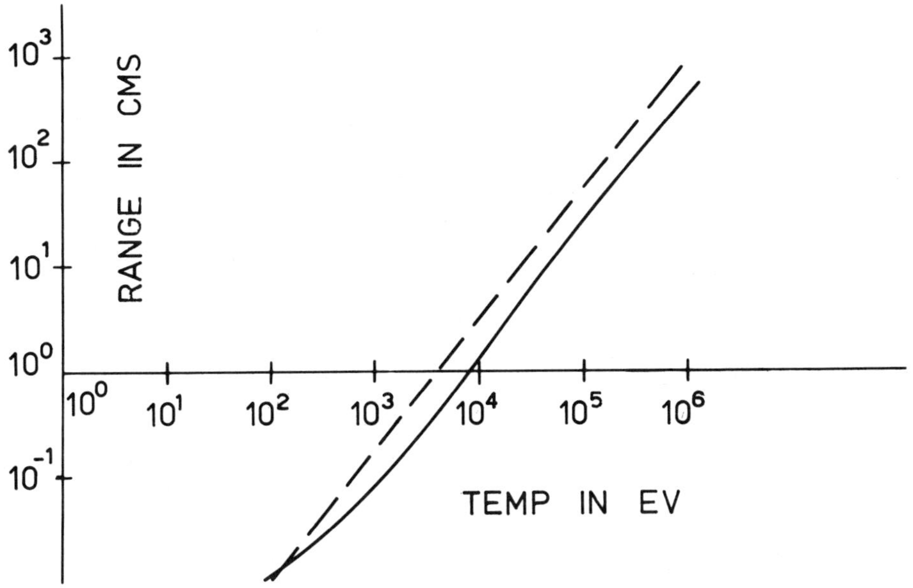

Fig. 5
The range of 3.6 MeV alpha in DT plasma of solid state density based on Fokker-Planck formalism. The dashed line corresponds to Winterberg values.[5]

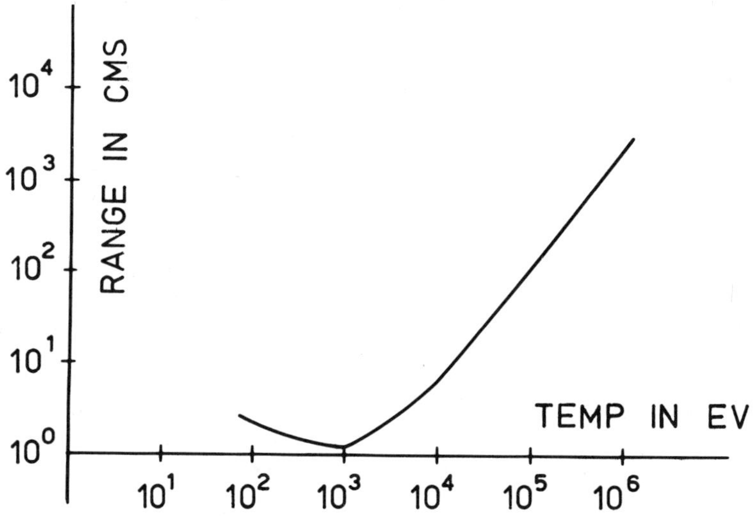

Fig. 6
The range of 14.7 MeV protons in DT-plasma of solid state density calculated on the basis of Fokker-Planck formalism.

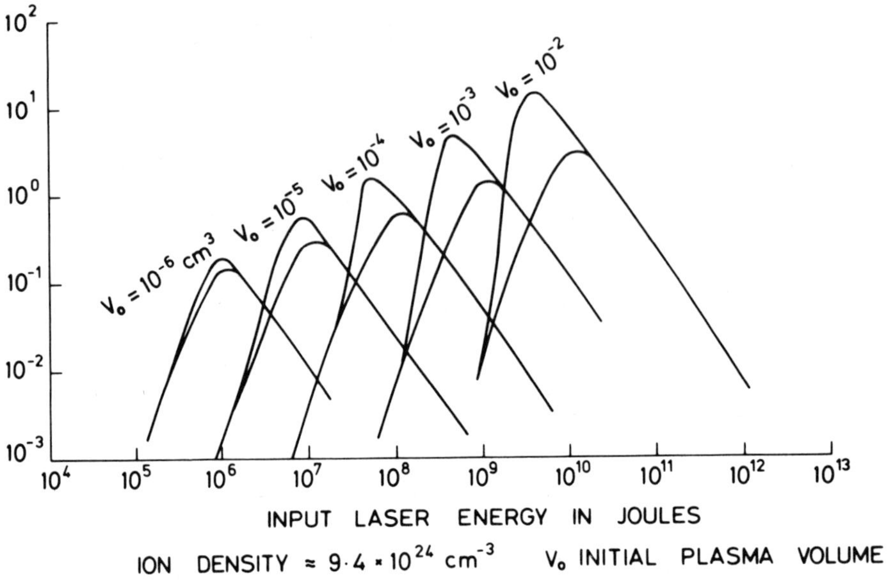

Fig. 7
Fusion yield for $H^1 + B^{11} \rightarrow 3He^4 + 8.682$ MeV reaction. Lower graphs correspond to without alpha reheat, upper ones with alpha reheat.

V. NUMERICAL RESULTS

The usual F.P. calculation together with photon emission correction which is first discussed does not alter the numerical values significantly although this theoretical aspect one should always consider. The Pauli cut-off shows a little lesser range at high temperatures. One remarkable aspect is that the proton range (14.7 MeV) for DT plasma of solid state density first decreases and then increases. Thus the physical character of the range when calculated by the F. P. formalism is susceptible to the parameters as the mass, initial energy and charge of the heavy particle. In the above figures numerical values for certain cases are shown. The effect of the absorpt - ion of alphas produced by fusion reaction

$$H^1 + B^{11} \rightarrow 3\,He^4 + 8.682\text{ MeV}$$

on the fusion yield has also been calculated and shown in Fig. 7.

VI SYNOPSIS

The above ways of calculating give physically different results as can be seen from the fact that the range, in general, increases with the plasma temperature for the F. P. formalism whereas it decreases for the collective type interaction. In the absence of experimental observation it is difficult to form an exact opinion on the correct behaviour on this matter. However, there seems to be good ground to suspect that the collective approach calculation should correspond more to reality. Firstly we emphasize that the divergence of Coulomb cross-section for $\theta = 0$, which has been cut-off by $\theta = \theta_{min}$ and thus forced to be finite, would have implied that the F. P. coefficients are infinite and thus that dE/dx also infinite. This would have meant that this range is zero. Besides the usual scattering cross-section utilizes the classical elastic type collision which is least suited for the purpose of energy exchange. Different cross-sections available from quantum electrodynamics are also objectionable due to the presence of divergances which would again imply zero range. The collective type approach gives the expression for the range close to the Bethe-Bloch formula which is very well in agreement with the nuclear physicists. Here the neglect of interaction beyond the Debye sphere would mean that the range could be even shorter.

One important point to be noted is that for very low temperature regions one cannot apply the usual F. P. calculation since the Coulomb logarithm becomes negative for high density plasma. The restriction (21) is however quite within the range of plasma temperature.

VII CONCLUSION

It is conjectured here that for the calculation of range of energetic charged particles in plasma of high densities one should employ the collective type approach which gives ranges which decrease with the plasma temperature. In the absence of direct experimental evidence, it is difficult to establish this fact. The usual F. P. formalism at least with the known differential scattering cross-section may not be suited for this purpose. Since the energy yield in fusion reactions depends on this alpha particle heating of plasma which in turn depends on the range it might alter the usual values in this respect. Furthermore, certain pictures[2] of nuclear ignition which depend on the range may not correspond to experiments then.

REFERENCES

1. H. Hora, Laser Plasma and Nuclear Energy (Plenum, 1975).

2. K. A. Brueckner and S. Jorna, Rev. Mod. Phys. 46, 325 (1974).

3. J. L. Bobin, Laser Interaction and Related Plasma Phenomena, H. Schwarz and H. Hora, Eds. (Plenum, 1974) Vol. 3B, p.465.

4. M. N. Rosenbluth, et.al., Phys. Rev. 107, 1 (1957).

5. M. S. Chu, D Thesis Columbia University (1971).
F. Winterberg, Desert Research Institute Reprint Series No. 64, March (1969).

6. K. A. Brueckner et.al., J. Plasma Physics, 11, 403 (1974).

7. H. Tsuji et.al., Nuclear Fusion, 16, 287 (1976).

8. D. F. Duchs and D. Pfirsch, Proc. Conf. Plasma Physics and Controlled Nuclear Fusion, Vol. 1, p.669, Tokyo (1974) IAEA.

9. L. Spitzer, Physics of Fully Ionized Gases (Wiley, 1967).

10. H. Hora and H. Wilhelm, Nuclear Fusion, 10, 111 (1970).

11. S. Rand, Phys. Rev., 136B, 231 (1964).

12. T. P. Hughes and M. N. Florence, J. Phys. A (2), 1, 588 (1968).

13. H. Hora, Opto-Electronics, 2, 201 (1970).

14. B. Babuel-Peyrissac, 6th Int. Q. E. Conference (Kyoto, Sept., 1970), Conf. Digest p.14.

15. J. Dawson and C. Oberman, Phys. Fluids, 5, 517 (1962).

16. H. A. Bethe, Ann. d. Phys. 5, 325 (1930).

17. F. Bloch, Ann. d. Phys. 16, 285 (1933).

18. J. R. Kerns, et.al., Bull. Am Phys. Soc. 17, 690 (1972).

19. G. Yanaes, Ettore Majorana Course Erice June (1973).

20. E. Bagge, 13th Int. Cosmic Ray Conf., Denver, Aug. (1973).

21. E. Bagge and H. Hora, Atomkernenergie, 24, 143 (1974).

22. E. Fermi, Nuclear Physics, p.28 (The University of Chicago Press, 1963).

23. P. S. Ray and H. Hora, Nuclear Fusion, 16, 535 (1976).

24. P. S. Ray and H. Hora, Atomkernenergie, 28, 155 (1976).

25. W. Pauli, Helv. Phys. Acta, Suppl IV, p.68 (1956).

26. H. Kramers, Physica, 7, 284 (1940).

27. W. B. Thomson and J. Hubbard, Rev. Mod. Phys. 32, 714 (1960).

28. R. P. Feynman, Quantum Electrodynamics, p.150, (Benjamin, 1962).

29. J. Schwinger, Phys. Rev., 76, 790 (1949).

30. J. D. Bjorken and S. D. Drell, Relativistic Quantum Mechanics, p. 126 (McGraw-Hill, 1964).

31. L. D. Landau, "Niels Bohr and the Development of Physics" (Pergamon-Press, London, 1955).

32. S. Gasiorowicz, et.al., Phys. Rev., 101, 922 (1956).

AUTHOR INDEX

Numbers in parentheses following the text page numbers are reference numbers, and are included to assist in locating a reference at the end of each contribution when the author's name is not cited at the point of reference in the text.

A

C

D

F

G

H

I

J

K

N

S

T

Y

Z

SUBJECT INDEX

A

B

T

Y

Z